Advances in

INORGANIC CHEMISTRY

Volume 37

Advances in INORGANIC CHEMISTRY

EDITOR

A. G. Sykes

Department of Chemistry
The University
Newcastle upon Tyne, England

VOLUME 37

This book is printed on acid-free paper. ∞

Academic Press, Inc.
San Diego, California 92101

United Kingdom Edition published by
ACADEMIC PRESS LIMITED
24-28 Oval Road, London NW1 7DX

Library of Congress Catalog Card Number: 59-7692

ISBN 0-12-023637-0 (alk. paper)

PRINTED IN THE UNITED STATES OF AMERICA
91 92 93 94 9 8 7 6 5 4 3 2 1

CONTENTS

On the Coordination Number of the Metal in Crystalline Halogenocuprates(I) and Halogenoargentates(I)

SUSAN JAGNER AND GÖRAN HELGESSON

I. Introduction 1
II. Structures of Halogenocuprates(I) and Halogenoargentates(I) in the Solid State 2
III. Influence of Cation Properties on Metal(I) Coordination Number . . 32
References 40

Structures of Organonitrogen–Lithium Compounds: Recent Patterns and Perspectives in Organolithium Chemistry

KARINA GREGORY, PAUL VON RAGUÉ SCHLEYER, AND RONALD SNAITH

I. Introduction 48
II. Lithium Imides (Iminolithiums) and Their Complexes: Ring Stacking . . 58
III. "Simple" Lithium Amides (Amidolithiums) and Their Complexes: Ring Laddering 92
IV. Conclusions 131
References 135

Cubane and Incomplete Cubane-Type Molybdenum and Tungsten Oxo/Sulfido Clusters

TAKASHI SHIBAHARA

I. Introduction 143
II. Incomplete Cubane-Type Clusters with $Mo_3O_{4-n}S_n$ Cores 145
III. Cubane-Type Clusters with $Mo_4O_{4-n}S_n$ Cores 156
IV. Incomplete Cubane-Type $W_3O_{4-n}S_n$ Cores 161
V. Cubane-Type Mixed-Metal Clusters with Mo_3MS_4 Cores 163
VI. Note on Preparation of Clusters 165
Abbreviations 170
References 170

Interactions of Platinum Amine Compounds with Sulfur-Containing Biomolecules and DNA Fragments

EDWIN L. M. LEMPERS AND JAN REEDIJK

I. History of *cis*-Pt as an Antitumor Drug 175
II. Aqueous Solution Chemistry of *cis*-Pt 179
III. Antitumor Activity and DNA as the Target 180
IV. Platinum–Sulfur Interactions 189
V. Prospects for Future Studies on Pt Antitumor Compounds . . . 206
Abbreviations 208
References 210

Recent Advances in Osmium Chemistry

PETER A. LAY AND W. DEAN HARMAN

I. Introduction 219
II. Survey of Coordination Complexes 221
III. Electrochemistry of Coordination Complexes 315
IV. Spectroscopic and Magnetic Properties of Coordination Complexes . . 323
V. Reactivity of Coordination Complexes 331
Abbreviations and Trivial Nomenclature 352
References 359

Oxidation of Coordinated Diimine Ligands in Basic Solutions of Tris(diimine)iron(III), -ruthenium(III), and -osmium(III)

O. MØNSTED AND G. NORD

I. Introduction 381
II. Rate Law and Stoichiometry for the Reduction of Dilate Solutions of . . Tris(diimine)iron(III), -ruthenium(III), and osmium(III) in Base . . . 384
III. Stoichiometry and Identification of Oxidized Reaction Products in Concentrated Solutions 385
IV. Intimate Mechanism of Formation of the First Intermediate . . . 387
V. Conclusion 396
References 397

INDEX 399

ADVANCES IN INORGANIC CHEMISTRY, VOL. 37

ON THE COORDINATION NUMBER OF THE METAL IN CRYSTALLINE HALOGENOCUPRATES(I) AND HALOGENOARGENTATES(I)

SUSAN JAGNER and GÖRAN HELGESSON

Department of Inorganic Chemistry, Chalmers University of Technology, S-412 96 Göteborg, Sweden

I. Introduction
II. Structures of Halogenocuprates(I) and Halogenoargentates(I) in the Solid State
  A. Species Containing a Two-Coordinated Metal Center
  B. Species Containing Three-Coordinated Metal Centers
  C. Species Containing Four-Coordinated Metal Centers
  D. Species Containing Five-Coordinated Metal Centers
III. Influence of Cation Properties on Metal(I) Coordination Number
  A. The Effect of Cation Size
  B. The Effects of Charge and Shape
  C. Conclusions
  References

I. Introduction

The aim of this article is to summarize the various structural motifs documented for halogenocuprate(I) and halogenoargentate(I) ions in the solid state, placing special emphasis on variations in the coordination number of the metal. Unlike the situation for many transition metal complexes, anionic configurations in crystalline halogenocuprates(I) and halogenoargentates(I) appear to be strongly dependent on the nature of the cation, thus permitting, in principle, the preparation of a species containing copper(I) or silver(I) with a desired coordination number by pertinent choice of cation.

The interpretation and correlation of spectroscopic properties, in the context of structure and bonding, for compounds of group IB and group IIB metals, have been covered in a comprehensive review (*1*). It is hoped that the present article may serve as a complement by providing somewhat more detail concerning structures of crystalline halogenocuprates(I) and halogenoargentates(I), with special reference to variations in metal coordination numbers. We have restricted the content to anionic species in which halogenide ions are the sole ligands, and no attempt has been made to provide comprehensive coverage concerning preparative methods or physical properties. By far, the most widespread method of preparation is reaction between the metal(I) halide and the halide salt of the appropriate cation dissolved in an organic solvent. Other methods have, however, been employed, in particular for the preparation of iodocuprates(I). For details the reader is referred to the individual papers cited.

In the following presentation, solid-state structures documented hitherto for halogenocuprate(I) and halogenoargentate(I) ions are described in order of increasing coordination number of the metal. Possible correlations between the coordination number of copper(I) or silver(I) in the anion and properties of the cation with which it is coprecipitated, such as size, shape, and exposure of the positive charge, are then discussed.

II. Structures of Halogenocuprates(I) and Halogenoargentates(I) in the Solid State

A. Species Containing a Two-Coordinated Metal Center

Two-coordinated species are of two types: discrete monomers, usually linear, and X—M—X linkages as part of infinite chains.

1. Discrete Monomeric Anions

Although frequent in solution in nonaqueous solvents (*1–12*), discrete monomeric dihalogenocuprate(I) ions and dihalogenoargentate(I) ions appear to occur in the solid state solely in combination with relatively large cations with low effective positive charge.

The first crystal structure determination of a compound containing a monomeric dihalogenocuprate(I) ion was that of $[(C_6H_5)_2PO(CH)_2NH(C_2H_5)_2][CuCl_2]$, reported in 1970 (*13*). Monomeric dibromocuprate(I) was inferred in bis(*N*,*N*-di-*n*-butyldithiocarbamato)gold-(III)dibromocuprate(I) from isomorphism with the corresponding dibromoaurate(I) (*14*). Tetrabutylammonium dibromocuprate(I) and tetraphenylphosphonium dibromocuprate(I) were shown to contain linear, monomeric anions by means of far-infared and Raman spectroscopy

(*2*), the anionic configurations in these compounds being confirmed later by crystal structure determinations (*15, 16*). Despite many attempts to prepare and characterize a monomeric diiodocuprate(I) in the solid state, this species proved elusive until 1985, when its isolation was finally accomplished using the K(18-crown-6)$^+$ and K(dicyclohexano-18-crown-6)$^+$ cations (*17*). Copper(I) lies on a center of symmetry in both compounds, with Cu—I distances in the two compounds of 2.383(1) and 2.394(2) Å, respectively; the iodide ligands are further involved in ionic interactions with potassium, K^+—I distances being 3.598(1) and 3.656(3) Å in the two compounds, respectively (*17*).

The situation regarding documentation of isolated, monomeric dihalogenoargentate(I) ions in the solid state is analogous, with no definitive evidence for the existence of such species until fairly recently. $Cs_2AgAuCl_6$ (*18, 19*) was reported to contain monomeric $[AgCl_2]^-$ (*19*). The structure of the compound can, however, be seen as a somewhat distorted close-packed array of cesium and chloride ions in which alternate octahedral interstices are occupied by silver(I) and gold(III) ions, and a more recent study (*20*) has shown that silver(I), in contrast to gold(I), shows no tendency toward the formation of linear dichloroargentate(I) ions in "$Cs_2AgAuCl_6$." Recently, a discrete, monomeric $[AgCl_2]^-$ anion (Fig. 1) has, however, been isolated with the K(crypt-2,2,2)$^+$ cation (*21*). The anion is well separated from the cation and there are no K^+—Cl contacts <4 Å (*21*). Monomeric dibromoargentate(I) has been determined in $[(Bu_2NCS_2)_2Au][AgBr_2]$ (*22*), there being a cation–anion contact (Ag—S) of 3.16(1) Å in this compound; monomeric $[AgI_2]^-$, however, still remains totally unknown in the solid state. Connectivity relationships for discrete monomeric dihalogenocuprate(I) and dihalogenoargentate(I) anions, characterized hitherto by crystallographic determination, are summarized in Table I (*13, 15–17, 22–46*).

Bonding in linear dihalogenocuprate(I) (*1, 2, 47–49*) and dihalogenoargentate(I) (*47–49*) ions has been discussed in terms of ds hybridization of the metal atom. A theoretical study of the bonding and nuclear quadrupole coupling in $[CuCl_2]^-$ and $[CuBr_2]^-$ has demon-

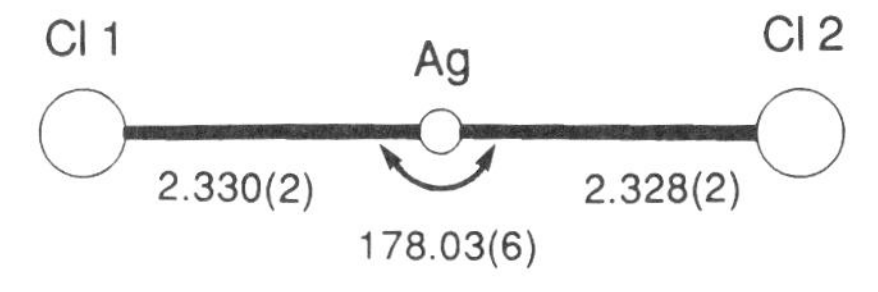

FIG. 1. The anion in $[K(crypt\text{-}2,2,2)][AgCl_2]$.

TABLE I

CONNECTIVITY RELATIONSHIPS IN MONOMERIC DIHALOGENOCUPRATE(I) AND DIHALOGENOARGENTATE(I) IONS

Compound	M — X distance (Å)	X — M — X angle (°)	Comments	Ref.
Dichlorocuprates(I)				
$[(C_6H_5)_2PO(CH)_2NH(C_2H_5)_2][CuCl_2]$	2.086(4), 2.095(4)	175.8(2)	—	*13*
$[N_6P_6(N(CH_3)_2)_{12}CuCl][CuCl_2]$	2.06(1), 2.11(1)[a]	179.5(2)	Cu on 2	*23*
$[Cu(bipy)_2][CuCl_2]_2$	2.091(2)	180	Cu at $\bar{1}$	*24*
$[Cu_4(C_{24}H_{20}NP_2S_2)_3][CuCl_2]$	1.96(1), 2.06(1)	177	—	*25*
$[Cu(C_{18}H_{15}AsO)_4][CuCl_2]_2$	2.079(2)[b]	178.65(8)	—	*26*
$[CuCl(C_{44}H_{60}N_4)][CuCl_2]$	2.066(7), 2.088(8)	179.0(3)	—	*27*
$[CuCl[C_{44}H_{60}N_4)][CuCl_2] \cdot 3C_2H_6O$	2.046(9), 2.10(2)	169.5(1)	Anion disordered	*28*
$[P(C_6H_5)_4][MoOS_3(CuCl)_3][CuCl_2]$	2.091(2), 2.095(2)	178.1(1)	—	*29*
$[Cu(tmeda)_2][CuCl_2]$[c]	2.084(4)	180	Cu at $\bar{1}$ in both independent anions	*30*
	2.088(4)	180		
$[Cu(tmeda)_2][CuCl_2]$	2.095(4)	180	Cu at $\bar{1}$ in both independent anions	*31*
	2.095(4)	180		
$[N(C_4H_9)_4][CuCl_2]$	2.107(1)	180	Cu at $\bar{1}$	*15*
$[As(C_6H_5)_4)][CuCl_2]$	2.069(3), 2.072(3)	176.4(1)	—	*16*
$[P(C_6H_5)_4][CuCl_2]$	2.088(2), 2.090(2)	174.7(1)	—	*16*
$[N(C_3H_7)_4][CuCl_2]$	2.071(2)	178.5(1)	Cu on 2	*32*
$[N(C_6H_5)(CH_3)_3][CuCl_2]$	2.105(2), 2.117(2)	179.63(7)	—	*33*
[BEDT-TTF]$[CuCl_2]$[d]	2.108(1)	180	Cu at $\bar{1}$; Cu partly Cu(I), partly Cu(II)	*34*
$[BEDT\text{-}TTF]_2[CuCl_2]$	2.084(2)	180	Cu at $\bar{1}$	*35*
[(MeCN)Cu(*meso*-L)]$[CuCl_2]$[e]	2.096(6), 2.083(6)	178.0(3)	—	*36*
	2.053(7), 2.077(8)	178.7(3)		
$[\{(mad)_2Cu\}_2Cl][CuCl_2]$[f]	2.09	177.2	—	*37*
$[Cu(2,4,6\text{-}tmpy)_2][CuCl_2]$[g]	2.084(1)	180	Cu at $\bar{1}$	*38*

Dichloroargentates(I)				
[K(crypt-2,2,2)][$AgCl_2$]	2.328(2), 2.330(2)	178.03(6)	—	*21*
Dibromocuprates(I)				
[$N(C_4H_9)_4$][$CuBr_2$]	2.226(1)	180	Cu at $\bar{1}$	*15*
[TSeT][$CuBr_2$][h]	2.267(2), 2.282(3)	153.8(1)	Additional Cu — Br of 2.829(2) Å	*39*
[{$CH_3C(CH_2P(C_6H_5)_2)_3IrP_3$}$_3Cu_5Br_4$][$CuBr_2$]	2.221(10)	178.4(3)	—	*40*
[$Cu(phen)_2$][$CuBr_2$][i]	2.209(2), 2.223(2)	180	Cu and Br (1,2) all lie on 2	*41*
[$P(C_6H_5)_4$][$CuBr_2$]	2.211(2), 2.216(2)	173.62(7)	—	*16*
[$P(C_4H_9)(C_6H_5)_3$][$CuBr_2$]	2.213(2), 2.220(1)	177.67(6)	—	*42*
[$P(C_3H_7)(C_6H_5)_3$][$CuBr_2$]	2.225(1), 2.232(1)	173.18(4)	—	*43*
[$P(C_2H_5)(C_6H_5)_3$][$CuBr_2$]	2.207(2), 2.224(2)	175.0(1)	—	*44*
[$N(C_3H_7)_4$][$CuBr_2$]	2.194(3)	178.4(1)	Cu on 2	*45*
[$P(CH_3)(C_6H_5)_3$][$CuBr_2$][Br]	2.228(1), 2.233(1)	175.29(4)	—	*46*
Dibromoargentates(I)				
[$Au(S_2CN(n\text{-}C_4H_9)_2)_2$][$AgBr_2$]	2.450(4)	179.3(2)	Ag on 2; additional Ag — S of 3.16(1) Å	*22*
Diiodocuprates(I)				
[K(dicyclohexano-18-crown-6)][CuI_2]	2.394(2)	180	Cu at $\bar{1}$	*17*
[K(18-crown-6)][CuI_2]	2.383(1)	180	Cu at $\bar{1}$	*17*

[a] Corrected for libration.
[b] Mean Cu — Cl.
[c] tmeda, Tetramethylethylenediamine.
[d] BEDT-TTF, 3,4;3′,4′bis(ethylenedithio)-2,2′,5,5′-tetrathiafulvalene.
[e] L, 2,6-bis[1-phenyl-1-(pyridin-2-yl)-ethyl]pyridine.
[f] mad, $C_6H_5CH = CH - CH = N - C_6H_4 - p - CH_3$.
[g] 2,4,6-tmpy, 2,4,6-Trimethylpyridine.
[h] TSeT, Tetraselenotetracene.
[i] phen, 1,10-Phenanthroline.

strated involvement of Cu $4p_z$, which undergoes contraction relative to the free-atom orbital on bond formation (*50*).

A mixed bromochlorocuprate(I) monomer has been prepared by dissolving copper(I) chloride in molten tetrabutylammonium bromide, and recrystallizing the product from ethyl acetate. The crystalline phase thus obtained contains the $[CuCl_2]^-$, $[CuBr_2]^-$, and $[CuBrCl]^-$ ions in the statistically most favorable ratio, i.e., 1 : 1 : 2 (*51*).

2. *X—M—X Linkages*

A linear Cl—Cu—Cl linkage as part of an infinite chain of copper(I) chloride tetrahedra occurs in bis(2,2′-bipyridyl)copper(II) bis(dichlorocuprate(I) (*24*). The Cu—Cl distance in the $CuCl_2$ group embedded in the infinite chain, 2.140(2) Å, differs little from that in the isolated $[CuCl_2]^-$ ion in the same compound or from distances determined for other discrete dichlorocuprate(I) anions (cf. Table I). Tris[bis(1,2-diaminoethane)platinum(II)dichlorobis(1,2-diaminoethane)platinum(IV)] tetrakis[tetrachlorocuprate(I)] contains chlorocuprate(I) chains in which linear Cl—Cu—Cl groups, with Cu—Cl = 2.16(1) Å, and Cl—Cu—Cl groups, with Cu—Cl = 3.10(1) Å, alternate (*52*). The latter copper(I) atom is trigonally coordinated by three chloride ligands at 2.291(9) Å, whereas the former has three equivalent chloride neighbors at 3.48(1) Å (*52*). The disordered structure of $(S_4C_6H_4)(Cu_{2/5}Cl_{4/5})$ has been interpreted in terms of three models; the model giving the most satisfactory agreement with the experimental data (R = 0.052) contains a linear $[Cu_2Cl_3]^-$ ion. In the other models (R = 0.054 and R = 0.055), linear Cl—Cu—Cl linkages in which Cu—Cl distances are of the order of 2.4 Å are present (*53*).

In tris(tetraethylammonium) *catena*-μ-chloro-ennea-μ-chloro-heptacuprate(I) (cf. Section II,B,5), six vertex-sharing copper(I) chloride triangles are joined to form an infinite chain via a bent Cl—Cu—Cl linkage in which the Cl—Cu—Cl angle is 153.9(1)° and Cu—Cl = 2.165(2) Å (*54*). The two-coordinated copper(I) atom has two additional intrachain Cu—Cl contacts of 2.896(2) Å, so that it is also possible to regard the coordination polyhedron of this copper(I) atom as being a very distorted tetrahedron (*54*). Two two-coordinated copper(I) centers are present in the $[Cu_6Br_9]^{3-}$ cluster, which is described in more detail in Section II,C,6, the remaining copper(I) atoms being four coordinated (*55*).

There would seem to be no analogous examples of digonal X—Ag—X linkages in halogenoargentates(I).

B. Species Containing Three-Coordinated Metal Centers

The structures of halogenocuprates(I) and halogenoargentates(I), in which the metal is three coordinated, can be described in terms of discrete or of vertex- or edge-sharing metal(I) halide triangles. In all subsequent structural illustrations, halide ligands and metal(I) centers are represented as large circles and small circles, respectively.

1. *Mononuclear Anions*

The simplest species is the discrete $[MX_3]^{2-}$ triangle (Fig. 2), the first example of which to be isolated and characterized crystallographically being $[CuI_3]^{2-}$ in bis(methyltriphenylphosphonium) triiodocuprate(I) (*56*). Since then, a further triiodocuprate(I) ion has been determined in $[Co(Cp)_2]_2[CuI_3]$, Cp = η^5-C_5H_5 (*57*); the bromocuprate(I) analog has been determined in $Cu_4Br_7L_3 \cdot 3H_2O$ [L = tris(1-pyrazolylethylamine)] (*58*) and in the bis(methyltriphenylphosphonium) (*46*) and tetramethylphosphonium salts (*59*), whereas the chlorocuprate(I) analog has been isolated with the tetramethylphosphonium cation (*60*). The triiodocuprate(I) ion does not possess strict threefold symmetry in $[P(CH_3)(C_6H_5)_3]_2[CuI_3]$ and the copper(I) atom lies 0.015(1) Å from the plane through the three iodide ligands (*56*). In $[P(CH_3)_4]_2[CuBr_3]$ (*59*), the tribromocuprate(I) anion has perfect D_{3h} symmetry, whereas the anions in bis(tetramethylphosphonium) trichlorocuprate(I) (*60*) and bis(methyltriphenylphosphonium) tribromocuprate(I) (*46*) both have C_{2v} symmetry, the former with the copper atom and one chloride ligand situated on a crystallographic mirror plane (*60*) and the latter with a twofold axis through the copper atom and one bromide ligand (*46*). $[P(CH_3)(C_6H_5)_3]_2[CuBr_3]$ can be prepared in an additional crystalline form: $[P(CH_3)(C_6H_5)_3]_2[CuBr_2]Br$, containing a monomeric dibromocuprate(I) anion (*46*). A novel environment has been observed for the mononuclear $[CuCl_3]^{2-}$ species in tetrakis(*N*-methylimidazole-*N'*) copper(II) trichlorocuprate(I) hydrate in that, in this compound, it

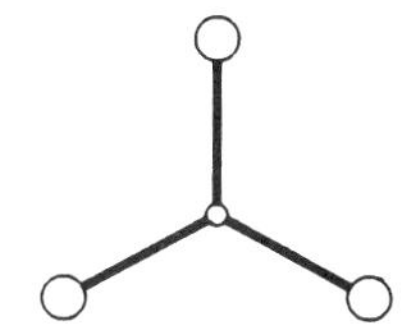

Fig. 2. The discrete $[MX_3]^{2-}$ monomer.

bridges two copper(II) centers to give a Cu(II)—Cu(I)—Cu(II) chain structure (*61*).

For halogenoargentates(I), again, the first mononuclear species to be isolated and fully characterized was the triiodoargentate(I) anion in $[P(CH_3)(C_6H_5)_3]_2[AgI_3]$ (*46*). The $[AgI_3]^{2-}$ anion is not constrained to planarity by crystallographic symmetry, but it is very nearly planar with silver(I) displaced 0.036 Å from the plane through the three iodide ligands (*46*). Recently, mononuclear trichloroargentate(I) and tribromoargentate(I) anions have been isolated with the [Rb(dibenzo-18-crown-6)]$^+$ and [K(dibenzo-18-crown-6)]$^+$ cations (*62*). In all three cases, the anions exhibit perfect D_{3h} symmetry. In Rb(dibenzo-18-crown-6)]$_3$[$AgCl_3$]Cl, the rubidium ions appear to be somewhat too large for the crown cavity and are displaced toward the anion, such that each Rb^+ is involved in two short Rb^+—Cl contacts (Fig. 3). In the potassium analog, however, K^+ is accommodated in the cavity of the crown and the $[AgCl_3]^{2-}$ ion is rotated about the threefold axis, relative to the cations, so that there is only one short K^+—Cl contact per chloride ligand (Fig. 3). As expected, the stronger outer-sphere coordination in the latter complex results in a slight lengthening of the Ag—Cl bond [2.463(2) Å as compared with 2.447(2) Å in the rubidium complex] (*62*). The tribromoargentate(I) isolated with [K(dibenzo-18-crown-6)]$^+$ is completely analogous to [K(dibenzo-18-crown-6)]$_3$[$AgCl_3$]Cl, each bromide ligand being involved in one short K^+—Br contact of 3.199(5) Å (*62*). Comparison with [Rb(dibenzo-18-crown-6)]$_3$[$AgBr_3$]Br (isostructural with [Rb(dibenzo-18-crown-6)]$_3$[$AgCl_3$]Cl) indicates a similar trend toward a longer Ag—Br bond in the potassium compound relative to the rubidium compound. Bond distances and angles determined hitherto for mononuclear $[CuX_3]^{2-}$ and $[AgX_3]^{2-}$ ions are compiled in Table II.

2. *Dinuclear Anions*

By linking two metal(I) halide triangles through a common vertex, planar or folded $[M_2X_5]^{3-}$ anions (Fig. 4) are obtained, only one example of each type of ion being known. The planar ion is a halogenoargentate(I), the μ-chloro-bis[dichloroargentate(I)] ion, determined in $(NH_4)_6[AuCl_4]_3[Ag_2Cl_5]$ (*63*), in which the Cl—Ag—Cl angle involving the terminal chloride ligands is 152.5° and the bridging Ag—Cl distance is 2.69 Å, as opposed to the terminal distance of 2.46 Å (*63*). A description in terms of "chlorine-bridged $AgCl_2^-$" was suggested as being appropriate for this ion (*63*). The folded $[M_2X_5]^{3-}$ species is exemplified by $[Cu_2Br_5]^{3-}$, determined in tris(tetramethylammonium)

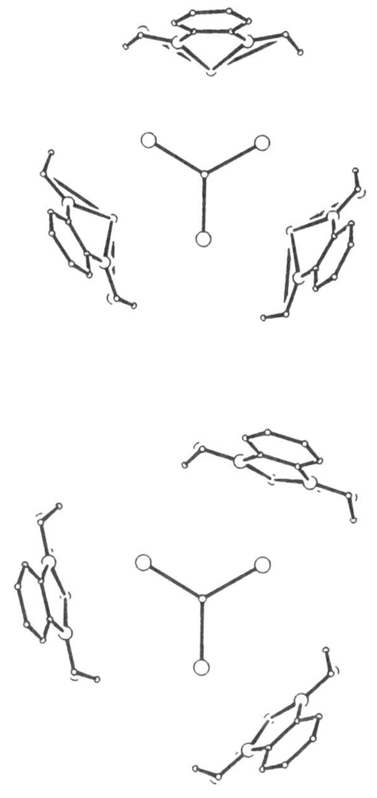

FIG. 3. The packing of $[AgCl_3]^{2-}$ anions and [Rb(dibenzo-18-crown-6)]$^+$ and [K(dibenzo-18-crown-6)]$^+$ cations in [Rb(dibenzo-18-crown-6)]$_3$[$AgCl_3$]Cl (top) and [K(dibenzo-18-crown-6)]$_3$[$AgCl_3$]Cl (bottom), respectively.

μ-bromo-bis[dibromocuprate(I)] (*64*). In this ion, the angle subtended by the copper(I) atoms at the bridging bromide ligand is 72.8(1)° (*64*), i.e., of the same magnitude as in the doubly bridged $[Cu_2Br_4]^{2-}$ ions (see below). The bridging ligand lies on a crystallographic twofold axis and copper(I) is displaced 0.123(2) Å from the trigonal plane toward its symmetry-related equivalent, affording a Cu—Cu separation of 2.837(4) Å; terminal [2.381(3) and 2.397(2) Å] and bridging [2.392(3) Å] Cu—Br distances do not differ (*64*).

The $[M_2X_4]^{2-}$ dimer can be envisaged as two metal(I) halide triangles with a common edge (Fig. 5), this species being unknown until 1981, when the $[Cu_2Br_4]^{2-}$ ion was prepared and characterized in the salt of the tetrathiotetracene cation radical (*65*). The anion in tetraethylam-

TABLE II

BOND DISTANCES AND ANGLES IN MONONUCLEAR MX_3^{2-} IONS

Compound	M — X distance (Å)	X — M — X angle (°)	Comments	Ref.
Trichlorocuprates(I)				
$[P(CH_3)_4]_2[CuCl_3]$	2.215(2)	118.90(5)	C_{2v}; Cu, one Cl in *m*	*60*
	2.232(3)	121.94(10)		
$[Cu(C_4H_6N_2)_4][CuCl_3] \cdot H_2O$	2.224(1)	121.8(1)	Bridges two Cu(II) centers	*61*
	2.246(1)	121.7(1)	Cu(II) — Cl = 2.922(2), 3.007(2) Å	
	2.259(1)	116.4(1)	—	
Trichloroargentates(I)				
$[Rb(dibenzo\text{-}18\text{-}crown\text{-}6)]_3[AgCl_3][Cl]$	2.447(2)	120	D_{3h}; Ag at $\overline{6}$, Cl in *m*; Rb^+ — Cl = 3.272(2), 3.487(2) Å	*62*
$[K(dibenzo\text{-}18\text{-}crown\text{-}6)]_3[AgCl_3][Cl]$	2.463(2)	120	D_{3h}; Ag at $\overline{6}$, Cl in *m* K^+ — Cl = 3.019(3) Å	*62*
Tribromocuprates(I)				
$[CuBr(TPyEA)]_3[CuBr_3]Br \cdot 3H_2O$[a]	2.374(2)	116.1(1)	—	*58*
	2.388(2)	121.0(1)	—	
	2.394(2)	122.9(1)	—	

$[P(CH_3)_4]_2[CuBr_3]$	2.365(3)	120	D_{3h}; Cu site symmetry 32	*59*
$[P(CH_3)(C_6H_5)_3]_2[CuBr_3]$	2.353(3)	118.9(1)	C_{2v}; Cu, one Br on 2	*46*
	2.358(6)	122.2(2)	—	
Tribromoargentates(I)				
[Rb(dibenzo-18-crown-6)]$_3$[$AgBr_3$][Br]	2.550(1)	120	D_{3h}; Ag at $\overline{6}$, Br in m; Rb^+ — Br = 3.420(2), 3.597(2) Å	*62*
[K(dibenzo-18-crown-6)]$_3$[$AgBr_3$][Br]	2.561(2)	120	D_{3h}; Ag at $\overline{6}$, Br in m; K^+ — Br = 3.199(5) Å	*62*
Triiodocuprates(I)				
$[P(CH_3)(C_6H_5)_3]_2[CuI_3]$	2.537(2)	116.55(7)	Cu 0.015(1) Å from ligand plane	*56*
	2.559(2)	120.54(6)	—	
	2.566(2)	122.90(7)	—	
$[Co(C_5H_5)_2]_2[CuI_3]$	2.535(2)	123.53(8)	—	*57*
	2.543(2)	118.74(8)	—	
	2.557(2)	117.61(8)	—	
Triiodoargentates(I)				
$[P(CH_3)(C_6H_5)_3]_2[AgI_3]$	2.742(1)	115.90(3)	Ag 0.036 Å from ligand plane	*46*
	2.746(1)	119.38(4)	—	
	2.755(1)	124.67(3)	—	

[a] TpyEA, Tris (1-pyrazolylethylamine).

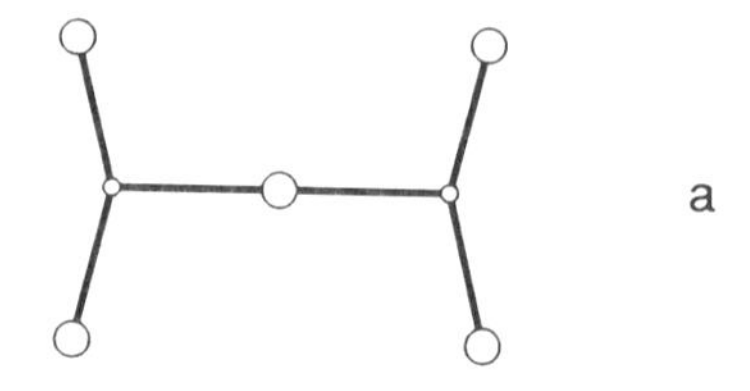

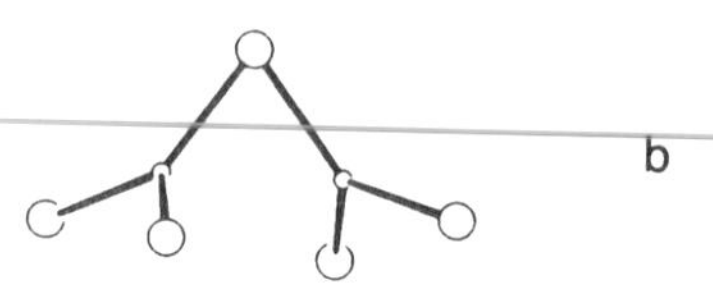

FIG. 4. Planar (a) and folded (b) dinuclear $[M_2X_5]^{3-}$ species.

monium dibromocuprate(I) was shown, by means of far-infrared and low-frequency Raman spectroscopy, to have a polymeric structure that breaks down on dissolution of the solid to yield monomeric $[CuBr_2]^-$ ions (*2*). Subsequent crystal structure determination of the compound showed that it contained a centrosymmetric $[Cu_2Br_4]^{2-}$ dimer (*66*). Iodo- and chlorocuprate(I) counterparts have also been prepared and characterized (cf. Table III). The majority of such ions are approximately planar, but $[Cu_2I_4]^{2-}$ ions folded 147° about the bridging I—I contact (cf Fig. 5) have been found in the tetraphenylarsonium (*67*) and tetraphenylphosphonium (*68*) compounds, the latter cation also crystallizing with a planar, centrosymmetric $[Cu_2I_4]^{2-}$ dimer (*68*). As is seen in Table III, bridging copper(I) halide bonds are invariably longer than terminal Cu—X. There is also considerable variation in the geometry of the four-membered $(Cu—X)_2$ ring, particularly with respect to the magnitude of the Cu—Cu separation, both within a given halogeno-

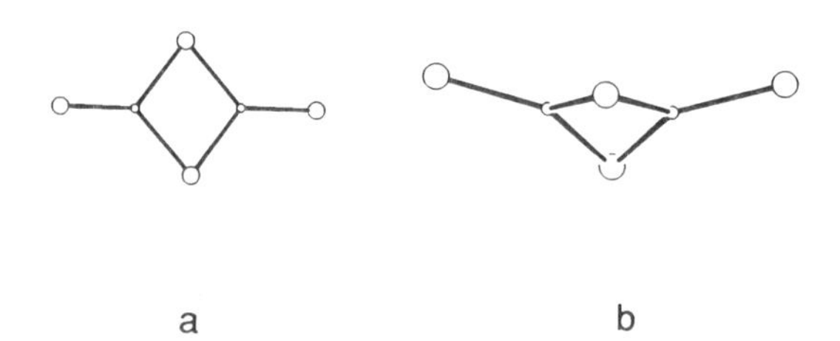

a b

FIG. 5. Planar (a) and folded (b) dinuclear $[M_2X_4]^{2-}$ species.

TABLE III

CONNECTIVITY RELATIONSHIPS REPORTED FOR DI-μ-HALOGENO–DIHALOGENOCUPRATE(I) AND DI-μ-HALOGENO-DIHALOGENOARGENTATE(I) ANIONS $[M_2X_4]^{2-}$ [a]

Compound	$M-X_t$ (Å)	$M-X_b$ (Å)	$M-M$ (Å)	$M-X_b-M$ (°)	X_b-M-X_b (°)	X_b-M-X_t (°)	d (Å)	X_b-X_b (Å)	Ref.
Di-μ-chloro-dichlorodicuprates(I)									
$[VO(SALen)(H_2O)]_2[Cu_2Cl_4]$[b]	2.161(2)	2.224(3) 2.427(3)	—	—	96.8(2)	143.9(2)	—	—	*73*
$[Cu_4(dmtp)_4Cl][Cu_2Cl_4]$[c]	2.116(5)	2.369(4)	2.433	61.8(1)	118.2(1)	120.90(8)	—	—	*74*
$[P(C_2H_5)_4]_2[Cu_2Cl_4]$	2.112(2)	2.135(3) 2.924(3)	3.516(2)	86.58(8)	93.42(8)	165.77(11) 100.68(9)	0.032	3.722(5)	*72*
Di-μ-chloro-dichlorodiargentates(I)									
$[As(C_6H_5)_4]_2[Ag_2Cl_4]$	2.359(2)	2.446(2) 2.809(2)	3.659(2)	87.97(6)	92.03(6)	149.06(6) 118.91(6)	0.004(1)	3.789(3)	*70*
$[P(C_6H_5)_4]_2[Ag_2Cl_4]$	2.358(2)	2.447(1) 2.792(2)	3.657(2)	88.27(5)	91.73(5)	149.53(5) 118.74(5)	0.0014(8)	3.768(3)	*69*
Di-μ-bromo-dibromodicuprates(I)									
$(TTT)_2[Cu_2Br_4]$[d]	2.328(2)	2.472(3) 2.490(2)	2.660(3)	64.7(1)	115.4(1)	125.0(1) 117.9(1)	0.196	—	*65*
$[N(C_2H_5)_4]_2[Cu_2Br_4]$	2.319(2)	2.441(2) 2.454(2)	2.937(3)	73.7(1)	106.3(1)	125.7(1) 127.9(1)	0.06(1)	3.916(3)	*66*
$[N(C_6H_5)(CH_3)_3]_2[Cu_2Br_4]$	2.310(1)	2.417(1) 2.421(1)	2.738(2)	68.95(4)	111.05(4)	124.70(4) 124.12(4)	0.049(1)	3.988(2)	*75*

(*continued*)

TABLE III (*Continued*)

Compound	$M-X_t$ (Å)	$M-X_b$ (Å)	$M-M$ (Å)	$M-X_b-M$ (°)	X_b-M-X_b (°)	X_b-M-X_t (°)	d (Å)	X_b-X_b (Å)	Ref.
$[P(CH_3)(C_6H_5)_3]_2[Cu_2Br_4]$	2.337(2)	2.426(2)	2.697(2)	67.09(5)	112.91(5)	124.60(6)	0.012(1)	4.068(2)	*76*
		2.455(1)				122.47(5)			
$[P(C_2H_5)_4]_2[Cu_2Br_4]$	2.263(4)	2.423(4)	2.870(5)	72.4(1)	107.6(1)	128.2(1)	0.032(3)	3.921(5)	*59*
		2.436(3)				124.1(1)			
$[(MeCN)CuL]_2[Cu_2Br_4]$[e]	2.267(4)	2.435(4)	2.768(5)	69.2(1)	110.8(2)	124.2(2)	—	4.011(3)	*36*
		2.439(4)				125.0(2)			
Di-μ-bromo-dibromodiargentates(I) and heteronuclear (Cu/Ag) counterparts									
$[As(C_6H_5)_4]_2[Ag_2Br_4]$	2.481(1)	2.614(1)	3.549(2)	82.96(3)	97.04(3)	139.16(4)	0.048(1)	4.013(2)	*70*
		2.741(1)				123.69(4)			
$[P(C_6H_5)_4]_2[Ag_2Br_4]$	2.491(1)	2.617(1)	3.578(2)	83.53(4)	96.47(4)	140.38(3)	0.0587(6)	4.006(2)	*69*
		2.752(2)				122.99(4)			
$[P(C_6H_5)_4]_2[AgCuBr_4]$	2.401(2)	2.543(2)	3.449(2)	82.48(5)	97.55(5)	139.06(5)	0.057(1)	3.934(2)	*71*
		2.688(2)				123.25(5)			
Di-μ-iodo-diiododicuprates(I)									
$[As(C_6H_5)_4]_2[Cu_2I_4]$[f]	2.490(3)	2.578(3)	2.663(4)	61.4(1)	114.2(1)	120.9(1)	0.04(1)	4.360(3)	*67*
	2.491(3)	2.584(3)		62.1(1)	114.4(1)	124.6(1)	0.07(1)		
		2.609(3)				120.3(1)			
		2.610(3)				125.3(1)			
$[P(C_6H_5)_4]_2[Cu_2I_4]$ – B[g]	2.471(2)	2.562(2)	2.647(2)	61.55(6)	113.81(6)	121.40(8)	0.04	—	*68*
	2.480(2)	2.588(2)		62.20(6)	113.95(6)	124.71(8)	0.09		
		2.562(2)				121.25(8)			
		2.585(2)				124.42(8)			

$[P(C_6H_5)_4]_2[Cu_2I_4]$ – A	2.497(0)	2.580(0) 2.595(1)	2.957(1)	55.40(3) 54.90(3)	110.29(3)	122.35(4) 127.05(4)	0.08	—	*68*
$[N(C_4H_9)_4]_2[Cu_2I_4]$	2.514(2)	2.566(2) 2.592(2)	2.726(4)	63.8(1)	116.2(1)	117.9(1) 125.8(1)	0.03(3)	4.380(3)	*77*
$[N(C_3H_7)_4]_2[Cu_2I_4]$	2.499(1)	2.571(1) 2.582(1)	2.698(2)	63.14(3)	116.86(3)	120.15(4) 122.98(4)	0.013(1)	4.390(1)	*78*
$[(MeCN)CuL]_2[Cu_2I_4]$[e]	2.52(1)	2.55(1) 2.56(1)	2.78(1)	66.0(3)	114.0(4)	122.1(4) 124.0(3)	—	4.282(5)	*36*
$[(MeCN)CuL]_2[Cu_2I_4] \cdot MeCN$[e]	2.518(1)	2.566(1) 2.579(1)	2.721(2)	63.85(4)	116.15(5)	121.11(4) 122.64(5)	—	4.367(1)	*36*
Di-μ-iodo-diiododiargentates(I)									
$[K(crypt\text{-}2,2,2)]_2[Ag_2I_4]$	2.672(1)	2.789(1) 2.801(1)	3.557(2)	79.02(3)	100.98(3)	127.65(4) 130.75(4)	0.123	4.313(2)	*21*

[a] X_t is a terminal halide ligand and X_b is a bridging halide ligand; d is the displacement (Å) of the metal atom from the plane through the three ligand atoms.
[b] $VO(SALen)^+$, *N,N*′-ethylene-bis(salicylideneiminato)oxovanadium(V).
[c] dmtp, 5,7-Dimethyl[1,2,4]triazolo[1,5-*a*]pyrimidine.
[d] TTT, Tetrathiotetracene.
[e] L, 2,6-Bis[1-phenyl-1-(pyridin-2-yl)ethyl] pyridine.
[f] Anion folded 146.61(4)° about $I_b — I_b$.
[g] Anion folded 146.34° about $I_b — I_b$.

cuprate(I) series and between the chloro-, bromo-, and iodocuprate(I) series, suggesting lack of direct attractive interaction between the copper(I) nuclei.

Dinuclear three-coordinated $[Ag_2X_4]^{2-}$ species (X = Cl, Br) were characterized first in 1988 as the tetraphenylphosphonium salts (*69*); similar species were also isolated subsequently with the tetraphenylarsonium cation (*70*) as well as a heteronuclear $[AgCuBr_4]^{2-}$ ion in $[P(C_6H_5)_4]_2[AgCuBr_4]$ (*71*). Although the $[Ag_2Br_4]^{2-}$ species contain approximately trigonally coordinated metal centers, the $[Ag_2Cl_4]^{2-}$ ions are severely distorted in both di-μ-chloro-dichlorodiargentates(I) such that the ions are perhaps most adequately described as a loose association of two bent $[AgCl_2]^-$ moieties (*69, 70*). A similar description is undoubtedly valid for the $[Cu_2Cl_4]^{2-}$ anion isolated with the tetraethylphosphonium cation (*72*; see Table III). Recently, a dinuclear $[Ag_2I_4]^{2-}$ anion, in which silver(I) exhibits more regular three coordination, has been prepared using $[K(crypt\text{-}2,2,2)]^+$ as cation (*21*). Interatomic distances and angles reported for $[M_2X_4]^{2-}$ species (M = Cu or Ag) are summarized in Table III (*21, 36, 59, 65–78*). In addition to those species for which connectivity relationships are given explicitly in the literature, the crystal structures of $[N(C_2H_5)_4]_2[Cu_2I_4]$ (*79*) and $[Cu(LH)_2]_2[Cu_2I_4]$ (LH = 1,1-di-2-pyridylethanol) (*80*) have also been determined.

3. *Trinuclear Anions*

Discrete trinuclear halogenocuprate(I) or halogenoargentate(I) species containing solely three-coordinated metal centers have not been reported hitherto. An $[Ag_3I_6]^{3-}$ ion in which one of the silver(I) atoms is three coordinated, whereas the remaining two exhibit tetrahedral coordination geometry has, however, recently been isolated (*81*) (cf Section II,C,3).

4. *Tetranuclear Anions*

The existence of tetranuclear halogenoargentate(I) anions composed soley of three-coordinated silver(I) has yet to be documented; discrete tetranuclear halogenoargentate(I) ions that contain four-coordinated or both three- and four-coordinated metal centers are described in Section II,C,4. The tetranuclear halogenocuprate(I) species $[Cu_4X_6]^{2-}$ (Fig. 6) is composed of four copper(I) halide triangles linked through common vertices. The aggregate can also be seen as an octahedron of halide ions containing a tetrahedron of copper(I). The hexa-μ-iodo-

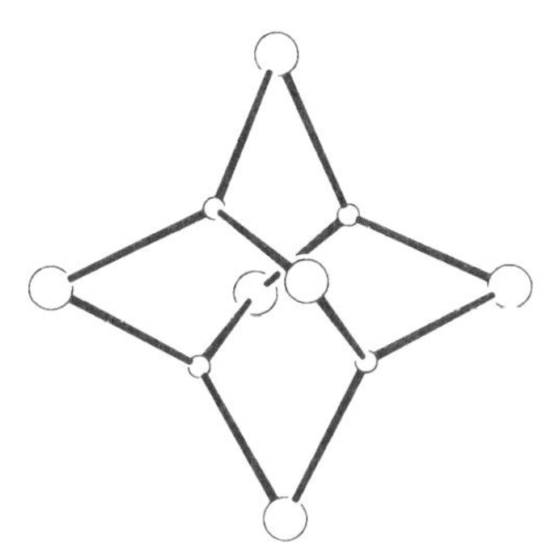

FIG. 6. The tetranuclear $[Cu_4X_6]^{2-}$ cluster; X = Br or I.

tetrahedro-tetracuprate(I) ion was the first of this series to be prepared and characterized, its structure being deduced from infrared and Raman spectra (*82*) and subsequently confirmed by crystal structure analysis (*83*). A further $[Cu_4I_6]^{2-}$ anion has been found in K_7(12-crown-4)$_6[Cu_4I_6][Cu_8I_{13}]$ (*84*). Two bromocuprate(I) analogs are known (*42, 85*) but, as yet, no hexa-μ-chloro-*tetrahedro*-tetracuprate(I) ion has been prepared. The hexa-μ-bromo-*tetrahedro*-tetracuprate(I) cluster can also be seen as being composed of two edge-sharing $[Cu_2Br_5]^{3-}$ ions (cf. Section II,B,2). Distances and angles reported in the literature for $[Cu_4X_6]^{2-}$ clusters are summarized in Table IV. The copper(I) tetrahedra in the anions in $[P(CH_3)(C_6H_5)_3]_2[Cu_4I_6]$ (*83*), $[N(C_3H_7)_4]_2[Cu_4Br_6]$ (*85*), and $[P(C_4H_9)(C_6H_5)_3]_2[Cu_4Br_6]$ (*42*) all exhibit disorder with respect to inversion through the center of the iodide or bromide octahedron.

5. *Polynuclear Anions*

Hitherto, chains in which the dominant coordination number of copper(I) or silver(I) is three have been documented solely for a chlorocuprate(I), i.e., $[Cu_7Cl_{10}]^{3-}$, determined in $[N(C_2H_5)_4]_3[Cu_7Cl_{10}]$ (*54*). The anion is composed of six vertex-sharing copper(I)-chloride triangles, joined by a nonlinear Cl—Cu—Cl linkage (cf. Fig. 7). A rubidium compound with the same stoichiometric Cu : Cl ratio, viz. $Rb_3Cu_7Cl_{10}$, has been shown to exhibit a high Cu^+ conductivity (*86–88*).

In tris[bis(1,2-diaminoethane)platinum(II)dichlorobis(1,2-diaminoethane)platinum(IV)] tetrakis[tetrachlorocuprate(I)] (*52*), the anion is a chain in which two- and three-coordinated copper(I) alternate (see Section II,A,2). Pairs of three- and four-coordinated copper(I) atoms alternate in $[P(CH_3)_4][Cu_2Cl_3]$ (*60*). The $[Cu_2Cl_3]^-$ anion in the tetra-

TABLE IV

DISTANCES AND ANGLES REPORTED FOR $[Cu_4X_6]^{2-}$ AGGREGATES[a]

Compound	Cu — X (Å)	Cu — Cu (Å)	Cu — X — Cu (°)	X — Cu — X (°)	d	Ref.
Hexa-μ-bromo-tetrahedro-tetracuprates(I)						
$[N(C_3H_7)_4]_2[Cu_4Br_6]$	2.373(5)–2.422(5)	2.718(7)–2.750(7)	69.0(2)–70.2(2)	117.7(2)–122.4(2)	0.015(5)–0.029(5)	*85*
$[P(C_4H_9)(C_6H_5)_3]_2[Cu_4Br_6]$	2.358(4)–2.443(4)	2.719(6)–2.755(6)	69.2(2)–69.5(2)	117.9(1)–122.0(2)	0.023(3)–0.030(3)	*42*
Hexa-μ-iodo-tetrahedro-tetracuprates(I)						
$[P(CH_3)(C_6H_5)_3]_2[Cu_4I_6]$	2.539(5)–2.638(5)	2.742(7)–2.757(7)	63.9(2)–65.2(2)	117.9(2)–121.6(2)	0.100–0.240	*83*

[a] For each type of distance or angle the range observed is cited; d is the displacement (Å) of copper(I) from the plane through the three ligand atoms.

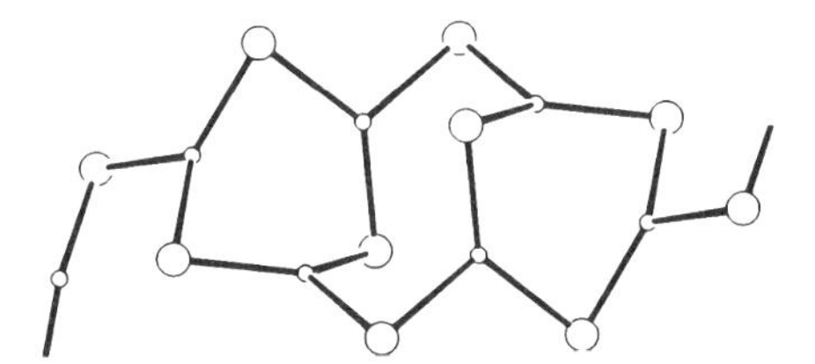

FIG. 7. The $[Cu_7Cl_{10}]^{3-}$ chain polymer.

methylphosphonium salt is essentially a very distorted version of the chain of edge- and face-sharing copper(I) halide tetrahedra (Section II,C,8c), first observed in tetraethylammonium triiododicuprate(I) and in dimethyl-(3-dimethylamino-2-aza-2-propenylidene)ammonium triiododicuprate(I) (*89*). In the $[Cu_2Cl_3]^-$ ion in $[P(CH_3)_4][Cu_2Cl_3]$ (*60*), the closest copper(I) chloride contact associated with the three-coordinated copper(I) atom is >3.8 Å, so that it is not possible to regard this atom as being other than three coordinated. A similar chain composed of alternately three-coordinated and four-coordinated copper(I) has been found in $[S_2C_3(SCH_3)_3][Cu_2I_3]$ (*90*). This is a somewhat less distorted form of the chain composed of edge- and face-sharing copper(I) halide tetrahedra, the shortest Cu—I distance involving the three-coordinated copper(I) atom being 3.452(3) Å.

Several types of $[M_3X_4]^-$ ions in which the metal(I) atoms are four coordinated have been documented (Section II,C,8,d); the anion in [K(dibenzo-24-crown-8)]$[Cu_3I_4]$ (*84*) is, however, a pleated sheat of edge-sharing Cu_2I_2 rhombohedra, in which one of the crystallographically independent copper(I) atoms is three coordinated and the remaining two are four coordinated.

C. SPECIES CONTAINING FOUR-COORDINATED METAL CENTERS

A metal coordination number of four is undoubtedly that most commonly exhibited in halogenocuprates(I) and halogenoargentates(I), vertex, edge, and face sharing of metal(I)–ligand tetrahedra leading to a wealth of structural motifs, especially where polymeric anions are concerned.

1. *Mononuclear Anions*

To our knowledge no mononuclear MX_4^{3-} anions have been proved to exist in the solid state. There would, however, appear to be evidence for the existence of such species in solution (see Ref. *1* and references therein).

2. *Dinuclear Anions*

A dinuclear $[Cu_2I_5]^{3-}$ anion that can be described as a trigonal bipyramid of iodide ions with copper(I) occupying tetrahedral interstices has been determined in $Cs_3[Cu_2I_5]$ (*91, 92*). Cu—I bonds range from 2.575(3) to 3.351(3) Å and 2.561(3) to 2.788(2) Å for the two copper(I) centers, respectively, with a short Cu—Cu contact of 2.538(4) Å (*92*). The anion can, alternatively, be seen as a trigonal-planar CuI_3 unit joined to a CuI_4 tetrahedron through a mutual edge (*92*) and is quite different from the $[Ag_2Cl_5]^{3-}$ (*63*) and $[Cu_2Br_5]^{3-}$ (*64*) anions described in Section II,B,2. The two crystallographically independent cesium ions in $Cs_3[Cu_2I_5]$ are eight and nine coordinated by iodide and there are three Cs^+—Cu contacts, 3.840(6), 3.937(2), and 4.099(5) Å, of the same order of magnitude as the Cs^+—I interactions (*92*).

Dinuclear iodocuprate(I) and iodoargentate(I) $[M_2I_6]^{4-}$ anions, composed of two MI_4 tetrahedra with a common edge (cf. Fig. 8), have been prepared and characterized in bis(dipyridiniomethane) di-μ-iodo-bis[diiodocuprate(I)] (*68*), tetrathallium di-μ-iodo-bis[diiodocuprate (I)] (*93*), and bis(ethylenediamine)diiodoplatinum(IV) di-μ-iodo-bis [diiodoargentate(I)] (*94*). In $[Py_2CH_2]_2[Cu_2I_6]$ (*68*), the anion is centrosymmetric with a long Cu—Cu separation and asymmetry in the Cu—I—Cu bridges [Cu—I = 2.681(1) and 2.850(1) Å]. Two cations are associated with each anion such that there are several I—C or I—N contacts marginally longer than 3.5 Å (*68*). The anion in $Tl_4Cu_2I_6$, on the other hand, exhibits a very short Cu—Cu separation [2.612(3) Å] (*93*); crystallographic symmetry requires that bridging Cu—I distances are exactly equal, nor do these differ from terminal Cu—I distances. Each thallium ion is coordinated by eight iodide ligands, arranged as a capped trigonal prism, a further thallium ion, and a copper(I) center also lying within the coordination sphere affording a short Tl—Cu contact of 3.438(2) Å (*93*).

The $[Ag_2I_6]^{4-}$ ion in $[Pt(C_2H_8N_2)_2I_2]_2[Ag_2I_6]$ (*94*) is well isolated from the cation; equality in bridging Ag—I distances is imposed by symmetry and these bonds are longer than the four terminal Ag—I bonds (also

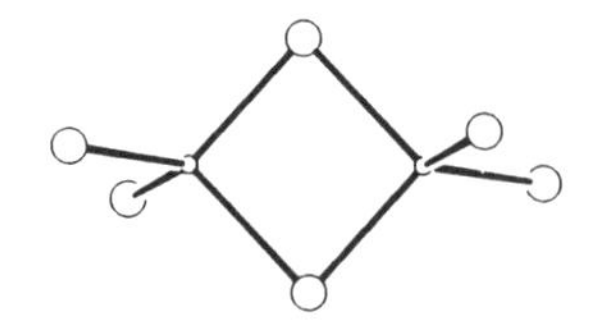

FIG. 8. The dinuclear $[M_2I_6]^{4-}$ ion.

exactly equal owing to crystallographic symmetry), viz. 2.918(1) Å as compared to 2.828(1) Å.

3. *Trinuclear Anions*

The only reported discrete trinuclear halogenocuprate(I) species composed of four-coordinated metal(I) centers is $[Cu_3I_6]^{3-}$, which contains three face-sharing copper(I) iodide tetrahedra (*91, 95*). Terminal Cu—I bonds are 2.505(5) and 2.519(6) Å, whereas bridging Cu—I bonds range from 2.529(6) to 3.205(8) Å; the face sharing of tetrahedra results in very short Cu—Cu contacts [2.518(6) and 2.519(6) Å] involving the middle copper(I) center (*95*).

A somewhat different silver(I) ion has been isolated in $NaAgI_2 \cdot 4DMSO$ (DMSO = dimethyl sulfoxide) (*81*) and is the sole trinuclear halogenoargentate(I) species reported to date. The crystals contain discrete $[Ag_3I_6]^{3-}$ ions that can be described in terms of two face-sharing silver(I) iodide tetrahedra joined by a common edge to an AgI_3 triangle; the Ag—I bonds associated with the three-coordinated silver(I) center are 2.67, 2.76, and 2.80 Å, and for the four-coordinated silver(I) atom Ag—I distances range from 2.66 to 3.08 Å (*81*). The Ag—Ag separation between the face-sharing tetrahedra is 2.96 and 3.49 Å between three- and four-coordinated silver(I) (*81*).

4. *Tetranuclear Anions*

Several tetranuclear ions have been reported. Recently, a $[Cu_4I_7]^{3-}$ cluster has been prepared and characterized in $[N(C_4H_9)_4]_3[Cu_4I_7]$ (*96*). This anion is composed of three edge-sharing Cu—I_4 tetrahedra with a fourth copper(I) in a trigonal-planar site (*96*). $[M_4X_8]^{4-}$ ions containing two four-coordinated and two three-coordinated metal(I) centers (cf. Fig. 9a) exist for copper(I) (*95*), silver(I) (*21, 70*), and as heteronuclear Cu/Ag species (*71*). The first such cluster to be reported was $[Cu_4I_8]^{4-}$ in $[Co(Cp)_2]_4[Cu_4I_8]$ (Cp = η^5-C_5H_5) (*95*). Silver(I) analogs, $[Ag_4I_8]^{4-}$, have been prepared with the tetraphenylarsonium and tetraphenylphosphonium cations (*70*) and a $[Ag_4Br_8]^{4-}$ cluster has been isolated recently with $[K(crypt\text{-}2,2,2)]^+$ as cation (*21*). From solutions of tetraphenylphosphonium iodide, copper(I) iodide, and silver(I) iodide in acetonitrile, it has proved possible to prepare the heteronuclear clusters $[Ag_2Cu_2I_8]^{4-}$ and $[Ag_3CuI_8]^{4-}$ in which copper(I) occupies or partially occupies the three-coordinated metal site (*71*). Selected distances within this type of $[M_4X_8]^{4-}$ cluster are given in Table V.

A completely different isolated tetranuclear $[Ag_4I_8]^{4-}$ ion (cf. Fig. 9b)

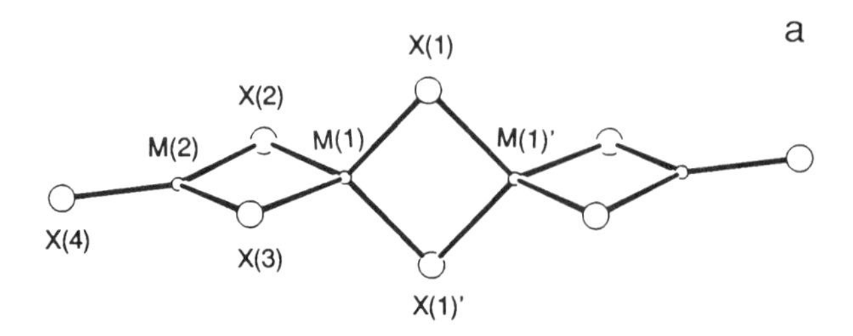

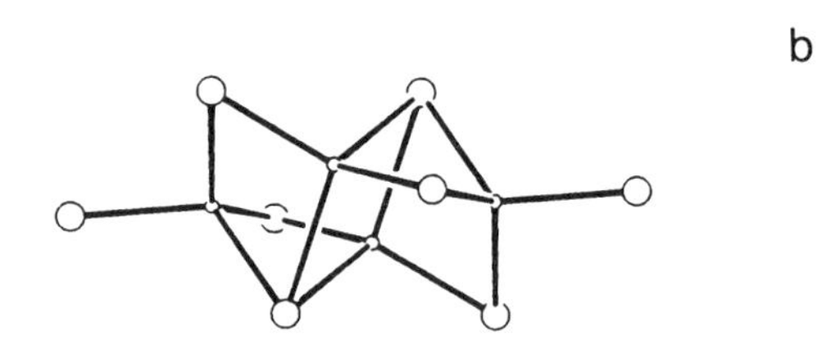

FIG. 9. Tetranuclear $[M_4X_8]^{4-}$ clusters containing (a) three- and four-coordinated metal(I) centers and (b) solely four-coordinated metal centers.

has been determined in $[C_{15}H_{30}N_2]_2[Ag_4I_8]$ (*97*). Here, a centrosymmetric cluster is formed from four edge-sharing silver(I) iodide tetrahedra; the four silver atoms are coplanar, forming a rhomb with Ag—Ag edges of 3.125(4) and 3.283(4) Å and diagonals 2.967(4) and 5.680(4) Å (*97*).

5. *Pentanuclear Anions*

Discrete pentanuclear clusters, $[Cu_5I_7]^{2-}$ (*98*) and $[Cu_5Br_7]^{2-}$ (*99*), have been prepared and characterized in the tetrapropylammonium and methyltributylammonium salts, respectively. A schematic drawing of the $[Cu_5X_7]^{2-}$ anion is shown in Fig. 10. In $[Cu_5I_7]^{2-}$, the seven iodide ions form a pentagonal bipyramid with only slight deviation from C_{5h} symmetry and the anion can be considered to be composed of five face-sharing copper(I) iodide tetrahedra (*98*). The copper(I) atoms exhibit disorder with respect to the pentagonal plane (*98*). In $[N(CH_3)(C_4H_9)_3]_2[Cu_5Br_7]$, however, the ligand pentagonal bipyramid is distorted such that the five equatorial bromide ligands are no longer coplanar (*99*). Metal coordination polyhedra in the latter ion are therefore perhaps more adequately described as two copper(I) atoms with (2 + 2) coordination and three with (3 + 1) coordination rather than as five tetrahedrally coordinated centers (*99*). Cu—Cu contacts are very

TABLE V

SELECTED DISTANCES (Å) IN $[M_4X_8]^{4-}$ CLUSTERS CONTAINING THREE- AND FOUR-COORDINATED METAL CENTERS[a]

Compound	M(1) — X(1) M(1) — X(1)′	M(1) — X(2) M(1) — X(3)	M(2) — X(2) M(2) — X(3)	M(2) — X(4)	M(1) — M(1)′ M(1) — M(2)	*d*	Comment	Ref.
$[Co(Cp)_2]_4[Cu_4I_8]$	2.653(2) 2.690(3)	2.703(3) 2.761(2)	2.580(2) 2.551(2)	2.486(2)	2.690(3) 2.710(3)	—	Cp = $\eta^5 - C_5H_5$; no distinction made between Cu — Cu	*95*
$[K(crypt\text{-}2,2,2)]_4[Ag_4Br_8]$	2.696(2) 2.722(2)	2.750(2) 2.799(2)	2.651(2) 2.631(2)	2.518(2)	3.595(2) 3.562(1)	0.030	—	*21*
$[As(C_6H_5)_4]_4[Ag_4I_8]$	2.847(1) 2.846(1)	2.973(1) 2.897(1)	2.757(1) 2.766(1)	2.721(1)	3.306(2) 3.198(1)	0.026(1)	—	*70*
$[P(C_6H_5)_4]_4[Ag_4I_8]$	2.846(1) 2.842(1)	2.991(2) 2.901(2)	2.765(2) 2.763(2)	2.727(2)	3.256(2) 3.171(2)	0.054(1)	—	*70*
$[P(C_6H_5)_4]_4[Ag_2Cu_2I_8]$	2.824(1) 2.816(1)	2.948(1) 2.864(2)	2.601(2) 2.599(1)	2.577(1)	3.195(2) 3.119(2)	0.050(1)	M(1) = Ag M(2) = Cu	*71*
$[P(C_6H_5)_4]_4[Ag_3CuI_8]$	2.839(1) 2.834(1)	2.968(1) 2.880(1)	2.670(1) 2.678(1)	2.638(1)	3.224(1) 3.147(1)	0.057(1)	M(2) = "AgCu"; Ag, Cu share this site	*71*

[a] The atomic numbering is as in Fig. 9a; *d* is the displacement of the three-coordinated metal from the ligand plane.

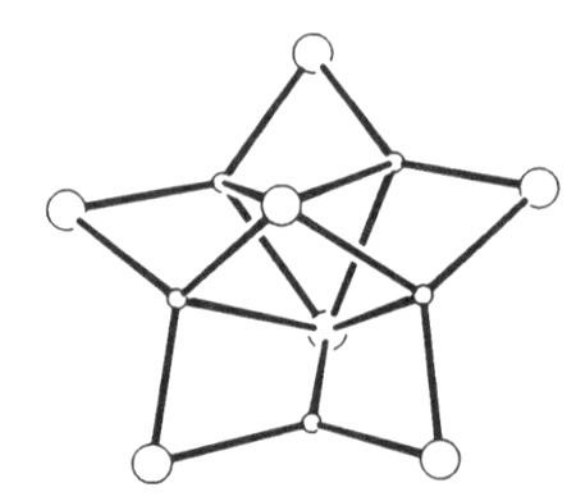

FIG. 10. The pentanuclear $[Cu_5X_7]^{2-}$ cluster; X = Br or I.

similar in both ions, i.e., 2.57–2.62 Å in $[Cu_5I_7]^{2-}$ (*98*) and 2.566(7)–2.637(6) Å in $[Cu_5Br_7]^{2-}$ (*99*).

The chlorocuprate(I) anion in $[Co(NH_3)_6]_4[Cu_5Cl_{16}]Cl$, viz. $[Cu_5Cl_{16}]^{11-}$, has been shown to be composed of five vertex-linked tetrahedra, i.e., a central $CuCl_4$ unit linked via all four vertices to four copper(I) chloride tetrahedra (*100, 101*).

Isolated pentanuclear halogenoargentate(I) species have yet to be documented.

6. *Hexanuclear Anions*

Hexanuclear anions are of two types and again have been reported hitherto solely for halogenocuprate(I) ions. In $[Cu_6I_{11}]^{5-}$ (cf. Fig. 11), determined in $[N(C_2H_5)_4]_6[Cu_6I_{11}]I$ (*79*) and $[Co(Cp)_2]_9[Cu_6I_{11}][(Cu_6I_8)_2]$ (*57*), six copper(I) iodide tetrahedra are connected via five common faces and six common edges to give a trigonal-prismatic metal core with fairly long (≈3 Å) Cu—Cu distances. Cu—I distances vary from 2.567(3) to 2.749(2) Å in $[N(C_2H_5)_4]_6[Cu_6I_{11}]I$ (*79*) and from 2.519(6) to 2.746(6) Å in the $[Cu_6I_{11}]^{5-}$ ion in $[Co(Cp)_2]_9[Cu_6I_{11}][Cu_6I_8)_2]$ (*57*).

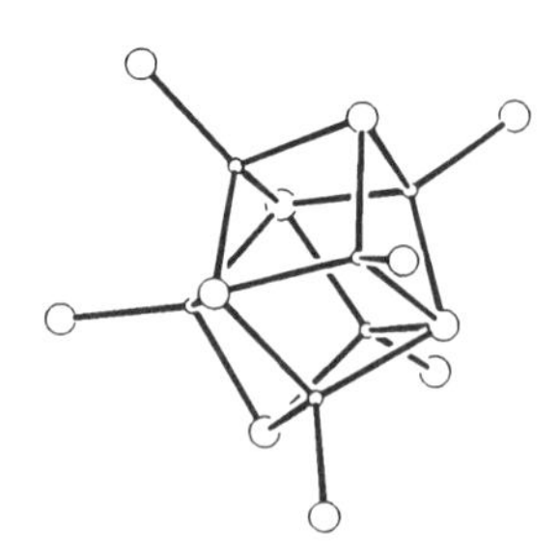

FIG. 11. The hexanuclear $[Cu_6I_{11}]^{5-}$ cluster.

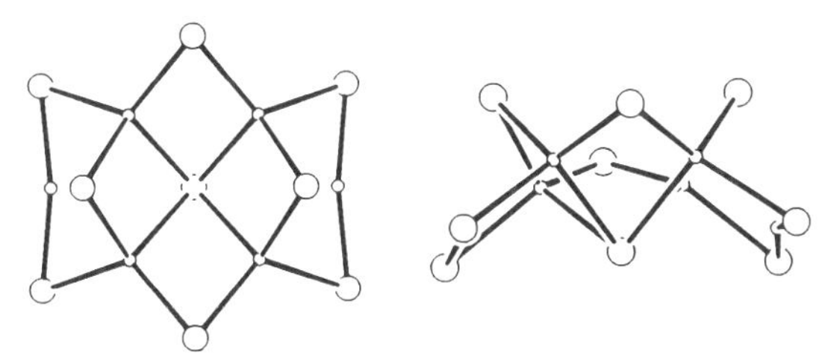

FIG. 12. The $[Cu_6Br_9]^{3-}$ anion viewed along (left) and perpendicular to (right) the twofold axis.

The other hexanuclear cluster is the $[Cu_6Br_9]^{3-}$ ion (Fig. 12), prepared and characterized in $[N(CH_3)(C_2H_5)_3]_3[Cu_6Br_9]$ (*55*). The anion has C_{2v} symmetry and can be visualized as being composed of four edge-sharing copper(I) bromide tetrahedra, two pairs of tetrahedra also being linked by a two-coordinated copper(I) atom situated between vertices. The Cu—Br distance involving this center is 2.285(2) Å, the Br—Cu—Br angle is 162.6(1)°, and the closest contact to a further bromide ligand is 2.866(4) Å, it thus also being possible to regard this copper(I) center as having (2 + 1) coordination (*55*). The Cu—Br distances associated with the four-coordinated copper(I) atom lie in the range 2.386(3)–2.697(3) Å (*55*).

7. *Discrete Clusters with Higher Nucleicity*

Four larger discrete halogenocuprate(I) clusters have been reported: a $[Cu_8I_{13}]^{5-}$ anion, which can be described as a cube of copper atoms bridged on all edges by iodide ligands and centered by a further iodide ligand, has been determined in K_7(12-crown-4)$_6[Cu_4I_6][Cu_8I_{13}]$ (*84*). With the *N,N*-(dimethyl) isopropylidenimmonium cation, a $[Cu_8I_{13}]^{5-}$ cluster with a completely different structure has been isolated (*102*). This anion contains six tetrahedrally coordinated and two trigonal-planar coordinated copper(I) centers (*102*). With dipyridiniomethane, a $[Cu_{19}I_{27}]^{8-}$ anion, which can be described as a centrosymmetric icosahedron composed of face-sharing copper(I) iodide tetrahedra, has been obtained (*103*). The linking of tetrahedra through two and three faces results in very short Cu—Cu contacts, leading to disorder of the copper(I) positions (*103*). Determination of the crystal structure of $(pyH)_2[Cu_3I_5]$ (pyH = pyridinium) revealed the presence of a discrete $[Cu_{36}I_{56}]^{20-}$ cluster, composed of 36 edge-sharing copper(I) iodide tetrahedra, requiring reformulation of the compound as $(pyH)_{24}[Cu_{36}I_{56}]I_4$ (*104*).

8. Polymeric Anions

Extreme diversity is exhibited in the coupling of metal halide tetrahedra to form polymeric halogenocuprate(I) and halogenoargentate(I) ions, there being an abundance of different types of infinite chains, layers, and three-dimensional arrays. Because the object of this article is to focus on variations in metal coordination number, and, in particular, on trends associated with the nature of the cation, structural description will be limited to those types of polymeric anion most frequently encountered hitherto in crystalline halogenocuprates(I) and halogenoargentates(I).

a. Species with Stoichiometry MX_2^-. The most common type of polymeric anion with this stoichiometry is an infinite chain of edge-sharing metal(I) halide tetrahedra (Fig. 13a). This has been found for

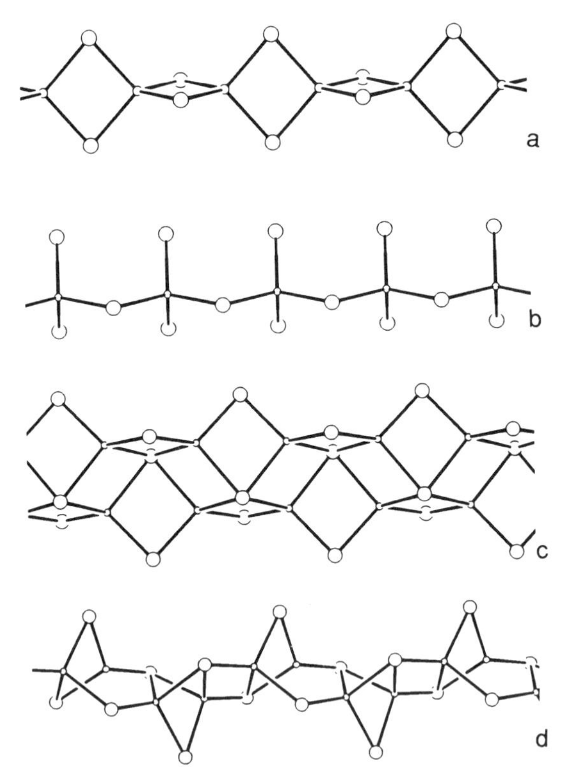

FIG. 13. Infinite-chain polymers: (a) MX_2^-, composed of edge-sharing tetrahedra; (b) MX_3^{2-}, composed of vertex-linked tetrahedra; (c) $M_2X_3^-$, composed of edge-sharing tetrahedra; (d) $M_2X_3^-$, composed of edge- and face-sharing tetrahedra.

chlorocuprates(I) and chloroargentates(I) in (pq)$[CuCl_2]_2$ (pq^{2+} = *N,N'*-dimethyl-4,4'-dipyridylium) (*105*), $[Cu(NH_3)_4][CuCl_2]_2 \cdot H_2O$ (*106*), $[Cu(en)_2][CuCl_2]_2$ (en = ethylenediamine) (*107*), and $[N(CH_3)_4][AgCl_2]$ (*108*); for bromocuprates(I) and bromoargentates(I), in $[Cu(NH_3)_4][CuBr_2]_2$ (*106*), $[Ni(en)_2][AgBr_2]_2$ (*109*), and $[S(CH_3)_3][AgBr_2]$ (*110*); and for iodocuprates(I) and iodoargentates(I), in $[Cu(NH_3)_4][CuI_2]_2$ (*111*), $[Cu(en)_2][CuI_2]_2$ (*112*), $[C_6H_8N][CuI_2]$ (*113*), $[C_6H_{14}N_3][CuI_2]$ (*113*), $[Na(H_2O)_4][CuI_2]$ (*114*), $[Rb_2H_2O][CuI_2]_2$ (*114*), $[(NH_4)_2H_2O][CuI_2]_2$ (*114*), $[Ni(en)_2][AgI_2]_2$ (*109*), $[C_8H_{22}N_2][AgI_2]_2]$ (*115*), $Sr[AgI_2]_2 \cdot 8H_2O$ (*116*), and $[N(CH_3)_4][AgI_2]$ (*117*). Selected interatomic distances reported for these chain anions are given in Table VI. The mixed-valence

TABLE VI

DISTANCES AND ANGLES REPORTED FOR SINGLE CHAINS OF STOICHIOMETRY $[MX_2]^{-a}$

Compound	M—X (Å)	M—M (Å)	X—M—X (°)	Ref.
Chlorocuprates(I)				
(pq)$[CuCl_2]_2$[b]	2.34–2.38	2.95	101–113	*105*
$[Cu(NH_3)_4][CuCl_2]_2 \cdot H_2O$	2.359(6)	2.734(4)	109.2(1)– 109.6(1)	*106*
$[Cu(en)_2][CuCl_2]_2$	2.297(3)–2.531(2)	3.119(2)	—	*107*
Chloroargentates(I)				
$[N(CH_3)_4][AgCl_2]$	2.599(1)–2.618(1)	3.137(1)–3.433(1)	98.08(4)–113.31(2)	*108*
Bromocuprates(I)				
$[Cu(NH_3)_4][CuBr_2]_2$	2.503(4)	2.857(4)	109.0(2)–110.4(2)	*106*
Bromoargentates(I)				
$[Ni(en)_2][AgBr_2]_2$	2.689(2)–2.742(2)	2.99–3.59	97.71(8)–113.30(6)	*109*
$[S(CH_3)_3][AgBr_2]$	2.661(4)–2.789(3)	—	102.4(1)–123.0(1)	*110*
Iodocuprates(I)				
$[Cu(NH_3)_4][CuI_2]_2$	2.575(15)–2.959(15)	3.506(2)	98.5(5)–118.9(5)	*111*
$[Cu(en)_2][CuI_2]_2$	2.646–2.706	2.987–3.581	97.2–111.6	*112*
$[C_6H_8N][CuI_2]$	2.61(1)–2.74(1)	3.01(3)–3.54(3)	98.7(6)–116.2(3)	*113*
$[C_6H_{14}N_3][CuI_2]$	2.680(1)–2.685(1)	3.362(1)	102.12(2)–117.34(3)	*113*
Iodoargentates(I)				
$[Ni(en)_2][AgI_2]_2$	2.845(2)–2.895(1)	3.09–3.79	98.29(5)–114.46(5)	*109*
$[C_8H_{22}N_2][AgI_2]_2$	2.856(3)–2.857(3)	3.81	96.3(1)–112.8(1)	*115*
$Sr[AgI_2]_2 \cdot 8H_2O$	2.879(2)	2.78(1)	—	*116*
$[N(CH_3)_4][AgI_2]$	2.756(3)–2.806(2)	—	97.5(1)–112.5(1)	*117*

[a] In those cases for which more than one value has been reported a range is given.
[b] pq^{2+}, *N,N'*-dimethyl-4,4'-dipyridylium.

Cu(I)–Cu(II) chain in $[N(C_2H_5)_4][Cu_2Cl_4]$ (*118*) is also of similar type, with Cu(I) and Cu(II) localized at alternate tetrahedral sites.

Recently, evidence has been provided for the existence of polymeric $[AgI_2]^-$ ions in fairly concentrated solutions of sodium diiodoargentate(I) in acetonitrile (*119*). In bis(2,2-bipyridyl)copper(II) bis[dichlorocuprate(I)] (*24*), the polymeric anion is composed of pairs of edge-sharing tetrahedra linked via linear Cl—Cu—Cl units (see Section II,A,2).

b. Species with Stoichiometry MX_3^{2-}. Such anions are typically infinite chains of vertex-linked tetrahedra (Fig. 13b) and are most often found in conjunction with alkali metal ions. Many of these compounds were the first halogenocuprates(I) and halogenoargentates(I) characterized by means of crystal structure determination. The crystal structures of $K_2[CuCl_3]$, $Cs_2[AgCl_3]$, and $Cs_2[AgI_3]$ were reported in 1949 (*120*). Other compounds that have similar structures are $K_2[AgI_3]$ (*121–123*), $Rb_2[AgI_3]$ (*121, 124*), $(NH_4)_2[CuCl_3]$ (*125*), and $(NH_4)_2[CuBr_3]$ (*125*), and the relationships between these and other structures derived from complex halogenides of type R_2MX_3 have been reviewed (*126*). More recently, the structure of $(NH_4)_2[AgI_3]\cdot H_2O$ has been shown (*127*) to be isomorphous with that of $K_2[AgI_3]$.

c. Species with Stoichiometry $M_2X_3^-$. Here the most common anionic configuration for both halogenocuprates(I) and halogenoargentates(I) is a band, or double chain, of edge-sharing metal(I) halide tetrahedra (Fig. 13c). Like the MX_3^{2-} chain polymer, this anion is most often found in compounds with relatively small cations. For chlorocuprates(I), anions of this type have been documented in $Cs[Cu_2Cl_3]$ (*128, 129*) and $[N(CH_3)_4][Cu_2Cl_3]$ (*130*); for bromocuprates(I), in $[C_6H_5N_2][Cu_2Br_3]$ (*131*), $[(CH_3)_2N_2CHN_2(CH_3)_2][Cu_2Br_3]$ (*132*), and $Cs[Cu_2Br_3]$ (*129*), and for iodocuprates(I), in $Cs[Cu_2I_3]$ (*133*), $Rb[Cu_2I_3]$ (*92*), $[C_6H_8N][Cu_2I_3]$ (*134*), $[S(CH_3)_3][Cu_2I_3]$ (*135*), $[(CH_3)_2NH_2][Cu_2I_3]$ (*102*), $[(CH_3)_3NH][Cu_2I_3]$ (*102*), and $[(CH_3)_2CHNH_3][Cu_2I_3]$ (*102*). In addition, the mixed halide species $[Cu_2Cl_2I]^-$ and $[Cu_2ClI_2]^-$ have also been prepared and characterized as the cesium salts (*136*).

Analogous halogenoargentate(I) $[M_2X_3]^-$ ions have been determined in $[N(C_2H_5)_4][Ag_2Cl_3]$ (*137*), $[N(CH_3)_4][Ag_2Br_3]$ (*138*), $[N(C_2H_5)_4][Ag_2Br_3]$ (*137*), $Cs[Ag_2I_3]$ (*128*), and $[N(CH_3)_4][Ag_2I_3]$ (*127, 139*). A common feature of these anions, with some exceptions, e.g., $[N(C_2H_5)_4][Ag_2Cl_3]$ (*137*), is that M—M separations perpendicular to the length of the chain are slightly shorter than the corresponding distances along the chain. Selected interatomic distances reported for

$[M_2X_3]^-$ double chains of edge-sharing metal(I) halide tetrahedra are given in Table VII.

Another type of $[M_2X_3]^-$ chain, which was unknown until 1981 but which has since been documented to occur in several halogenocuprates(I), is depicted in Fig. 13d. To our knowledge this single chain of edge- and face-sharing metal(I) halide tetrahedra has not been reported for a halogenoargentate(I). The anion was first characterized as the triiododicuprate(I) in the tetraethylammonium and dimethyl-(3-dimethylamino-2-aza-2-propenylidene)ammonium salts (*89*). Similar $[Cu_2I_3]^-$ anions were subsequently determined in the 2,4,6-triphenylthiopyrylium (*140*), 3,4,5-tris(methylthio)-1,2-dithiolium (*90*), tetramethylammonium (*141*), K(15-crown-5)$^+$ (*84*), and *N*-isopropylidenisopropylimmonium (*102*) compounds. A very distorted variant has been found in $[P(CH_3)_4][Cu_2Cl_3]$ (*60*), in which pairs of three- and four-coordinated copper(I) atoms alternate. The $[Cu_2I_3]^-$ anion has been documented in an unusual layer structure, composed of edge- and vertex-linked tetrahedra, in $[(CH_3)_3CNH_3][Cu_2I_3]$ (*102*).

d. Species with Stoichiometry $M_3X_4^-$. There appear to be at least seven types of polymeric $[M_3X_4]^-$ anion, and relationships between four of these have been discussed in a recent publication (*142*). A chain anion composed of edge- and face-sharing copper(I) iodide tetrahedra has been isolated in $[N(C_3H_7)_4][Cu_3I_4]$ (*134*); of face-sharing tetrahedra, in $[P(C_6H_5)_4][Cu_3I_4]$ (*142*); and of edge-sharing metal(I) halide tetrahedra, in $[P(CH_3)(C_6H_5)_3][Cu_3I_4]$ (*134*), $[As(C_6H_5)_4][Ag_3I_4]$ (*70*), and $[P(C_6H_5)_4][Ag_3I_4]$ (*70*). The anion in $[P(C_6H_5)_4][Ag_3I_4]$ (*70*) is depicted in Fig. 14a. Several heteronuclear, mixed-halide species (*142*) assume a fourth $[M_3X_4]^-$ structural type (Fig. 14b), a triple chain of edge-linked tetrahedra, in which the peripheral halogens bridge two metal atoms and the inner halogens bridge four, i.e., analogous to the $[M_2X_3]^-$ double chain (Fig. 13c). The compounds are $[(DMF)_2H][M_3I_{4-n}X_n]$, where M = Cu/Ag, X = I/Br/Cl, and n = 0, 1, and 2 (*142*). A fifth type of chain, composed of pairs of edge-sharing metal(I) halide tetrahedra linked to similar units through the four free vertices, has been found in $[N(C_4H_9)_4][Ag_3I_4]$ (*143*), whereas in $[(CH_3)_3NCH_2CH_2N(CH_3)_3][Ag_3I_4)_2]$ (*144*), edge-linked tetrahedra form a layer. A chain containing one three-coordinated and two four-coordinated copper(I) centers has been determined in $[K(dibenzo\text{-}24\text{-}crown\text{-}8)][Cu_3I_4]$ (*84*) (cf. Section II,B,5).

e. Species with Stoichiometry $M_4X_5^-$. The majority of these species, such as $Rb[Ag_4I_5]$ (*145*) and its analogs, are important solid electro-

TABLE VII

DISTANCES AND ANGLES REPORTED FOR DOUBLE CHAINS OF STOICHIOMETRY $[M_2X_3]^-$ [a]

Compound	M—X (Å)	M—M (Å)	X—M—X (°)	Ref.
Chlorocuprates(I)				
$Cs[Cu_2Cl_3]$	2.283–2.490	2.800–3.113	102.7–119.2	*129*
$[N(CH_3)_4][Cu_2Cl_3]$	2.254(2)–2.574(2)	2.869(2)–3.197(2)	96.28(7)–125.15(9)	*130*
Chloroargentates(I)				
$[N(C_2H_5)_4][Ag_2Cl_3]$	2.514(2)–2.778(2)	3.348(2)–3.486(2)	94.45(7)–123.09(8)	*137*
Bromocuprates(I)				
$[C_6H_5N_2][Cu_2Br_3]$	2.45–2.57	2.86–3.09	—	*131*
$[(CH_3)_2N_2CHN_2(CH_3)_2][Cu_2Br_3]$	2.427(2)–2.609(2)	2.904(2)–3.224(3)	103.55(6)–119.40(7)	*132*
$Cs[Cu_2Br_3]$	2.426–2.587	2.909–3.094	106.5–117.0	*129*
Bromoargentates(I)				
$[N(CH_3)_4][Ag_2Br_3]$	2.612(2)–2.833(2)	3.078(3)–3.526(3)	99.05(6)–122.35(9)	*138*
$[N(C_2H_5)_4][Ag_2Br_3]$	2.630(2)–2.844(2)	3.250(2)–3.547(2)	97.62(5)–120.96(6)	*137*
Iodocuprates(I)				
$Cs[Cu_2I_3]$	2.599(2)–2.694(2)	3.036(3)–4.361(3)	—	*133*
$Rb[Cu_2I_3]$	2.592(5)–2.716(4)	2.864(2)–3.427(6)	101.8(2)–115.8(2)	*92*
$[C_6H_8N][Cu_2I_3]$	2.615(3)–2.714(3)	2.947(4)–3.144(2)	106.6(1)–114.1(1)	*134*
$[S(CH_3)_3][Cu_2I_3]$	2.601(4)–2.753(4)	2.941(6)–3.432(7)	101.0(1)–118.1(1)	*135*
$[(CH_3)_2NH_2][Cu_2I_3]$	2.554(7)–2.811(7)	2.942(6)–3.260(7)	101.4(2)–120.4(2)	*102*
$[(CH_3)_2CHNH_3][Cu_2I_3]$	2.599(2)–2.768(2)	2.882(1)–3.310(2)	103.10(7)–116.72(7)	*102*
$[(CH_3]_3NH][Cu_2I_3]$	2.594(7)–2.745(7)	2.894(8)–3.470(11)	100.5(3)–117.0(2)	*102*
Iodoargentates(I)				
$Cs[Ag_2I_3]$	2.79–2.90	—	—	*128*
$[N(CH_3)_4][Ag_2I_3]$	2.780(2)–2.966(2)	—	99.84(5)–119.86(7)	*127*

[a] For each type of distance or angle the range observed is cited.

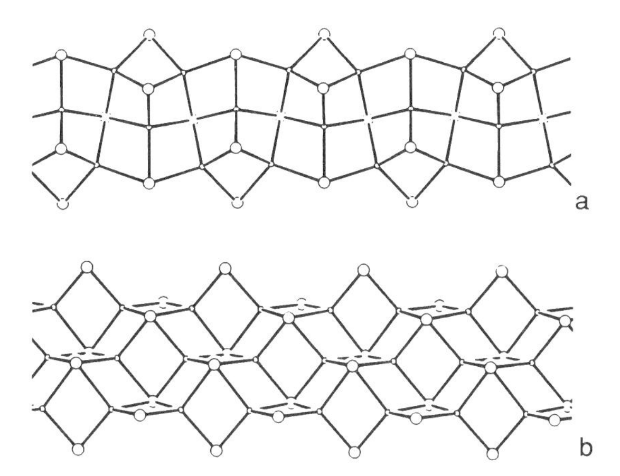

FIG. 14. The triple $[M_3X_4]^-$ chains of edge-linked tetrahedra as in (a) $[P(C_6H_5)_4][Ag_3I_4]$ and (b) $[(DMF)_2H][M_3I_{4-n}X_n]$.

lytes. In the unit cell of $Rb[Ag_4I_5]$ (*145*), Ag^+ diffusion paths are provided by 56 face-linked iodide tetrahedra that form a three-dimensional network. A large number of such phases, including $CsCu_4Cl_3I_2$ (*146*), $RbCu_4Cl_3(I_{2-x}Cl_x)$ (*87*), and $(Cs_{1-y}Rb_y)Cu_4Cl_3(I_{2-x}Cl_x)$ (*147*), have been prepared and characterized, showing a variety of three-dimensional patterns. The solid electrolytes $[N(CH_3)_4]_2[Ag_{13}I_{15}]$ (*148*), $[N(CH_3)_2(C_2H_5)_2]_2[Ag_{13}I_{15}]$ (*149*), and $[C_5H_5NH]_2[Cu_5Br_7]$ (*150*) can also be considered to be of this type. Recently, a one-dimensional solid electrolyte $DMM[Ag_4I_5]$ (DMM = *N,N'*-dimethylmorpholinium), in which Ag^+ ions are constrained to move only in one direction via an infinite chain of face-sharing icosahedra of iodide ions, has been reported (*151*). In $[Te(C_2H_5)_3][Ag_4I_5]$, the anion is an infinite layer of linked trigonal bipyramids formed from pairs of face-sharing tetrahedra (*152*). It is, however, often somewhat difficult to distinguish the presence of halogenocuprate(I) or halogenoargentate(I) anions, as such, in these solid electrolytes, which are primarily matrices for mobile Cu^+ or Ag^+ ions, and detailed treatment of the rich structural chemistry exhibited by this type of compound is therefore considered to be beyond the scope of the present article.

f. Miscellaneous Polymeric Anions. Unusual polymeric anions reported include a $[Ag_5Br_8]^{3-}$ band that is related to the $[M_3X_4]^-$ triple chain (cf. Fig. 14b) by loss of every second spinal metal atom (*153*), an $[Ag_5I_6]^-$ chain that can be seen as a decagonal tube (*154*), and an infinite $[Cu_3Cl_5]^{2-}$ cylinder (*155*). In $[Pt(dapn)_2][Pt(dapn)_2Br_2]$

$[(Cu_3Br_5)_2]$ (dapn = 1,2-diaminopropane) (*156*), a single-strand $[CuBr_2]^-$ polymer is intertwined with a linear Br—Cu—Br chain, so that one copper(I) atom is tetrahedrally coordinated and the other has distorted trigonal-bipyramidal coordination geometry (cf. Section II,D). A polymeric $[Cu_6I_8]^{2-}$ layer is present in $[Co(Cp)_2)]_9[Cu_6I_{11}][(Cu_6I_8)_2]$ (*57*) and a $[Ag_7I_{11}]^{4-}$ layer is present in $[(CH_3)_3CNH_3]_4[Ag_7I_{11}]$ (*102*).

D. Species Containing Five-Coordinated Metal Centers

Halogenocuprates(I) or halogenoargentates(I) in which the metal is five coordinated are rare. One of the copper(I) centers in tris[bis(1,2-diaminoethane)platinum(II)dichlorobis(1,2-diaminoethane)platinum (IV)] tetrakis[tetrachlorocuprate(I)] (*52*) [cf. Section II,A,2], i.e., that trigonally coordinated by three chloride ligands at 2.291(9) Å and with apical contacts where Cu—Cl = 3.10(1) Å, could be regarded as having trigonal-bipyramidal coordination. Similar considerations apply to one of the crystallographically independent copper(I) atoms in $[Pt(dapn)_2][Pt(dapn)_2Br_2][Cu_3Br_5)_2]$ (dapn = 1,2-diaminopropane) (*156*) (Section II,C,7,f). Silver(I) has distorted trigonal-bipyramidal coordination geometry in $Cs[AgCl_2]$ with equatorial Ag—Cl distances of 2.49 and 2.67 (×2) Å and apical Ag—Cl bonds of 2.87 Å, the Cl—Ag—Cl apical angle being 164° (*157*). Layer formation is effected by linking through four vertices of the polyhedra (*157*). Comparison of X-ray powder patterns indicates that $Cs[AgBr_2]$ has an analogous structure (*158*).

III. Influence of Cation Properties on Metal(I) Coordination Number

It has been recognized that cationic properties, such as size, shape, and distribution of the positive charge, are of importance for the anionic configurations assumed in iodocuprates(I) (*89*) and iodoargentates(I) (*159*). Several independent systematic studies have been carried out in order to investigate possible correlations between cationic properties, in particular, cation size, and the structure of the halogenocuprate(I) or halogenoargentate(I) anion coprecipitated. Thus many of the anions whose structures have been described in the preceding section have been prepared and characterized in connection with such investigations. In this section an attempt will be made to summarize those correlations that have been previously reported.

A. The Effect of Cation Size

Investigations concerning the role of cation size have mainly been carried out using bulky unipositive cations (R^+) with a well-screened positive charge. The most usual method of preparation is reaction between the appropriate metal(I) halide and RX in an organic solvent. In the majority of cases a single product is obtained independent of the solvent used and of the total molar ratios M : X in the system. In some systems, however, solvent effects have been shown to be important, leading to different products. This seems to be particularly true for iodocuprates(I) (see, e.g., Ref. *57* and references therein) and for iodoargentates(I) (*70*). A cation that has proved exceptionally versatile in the preparation of halogenocuprates(I) and halogenoargentates(I) is methyltriphenylphosphonium, which crystallizes with discrete anions such as $[Cu_4I_6]^{2-}$ (*82, 83*), $[CuI_3]^{2-}$ (*56*), $[CuBr_3]^{2-}$ (*46, 56*), $[AgI_3]^{2-}$ (*46, 56*), $[CuBr_2]^-$ (*46*), and $[Cu_2Br_4]^{2-}$ (*76*), as well as with a $[Cu_3I_4]^-$ triple chain (*134*). In addition, for a few tetraalkylammonium and related chlorocuprates(I) and bromocuprates(I), for which total molar ratios Cu : X of 1 : 2 have been employed during preparation and for which the primary product has a stoichiometry different from this value, a second phase has been obtained at the end of the crystallization, invariably containing an anion with a lower copper(I) coordination number than that in the major product (*42, 45, 60*).

Comparison of halogenocuprate(I) anions obtained with symmetrically substituted unipositive quaternary ammonium, phosphonium, and arsonium cations indicates a distinct correlation between cation size and the coordination number of copper(I) in the resulting anion (*32*), further substantiated for chlorocuprates(I) (*60*) and bromocuprates(I) (*59*). The trend is such that copper(I) assumes digonal, trigonal, and finally tetrahedral coordination geometry with decreasing size of the cation. Calculation of the concentration of halide ligand in the crystalline phase, which can be considered to be an indicator of cation size, shows that this parameter also increases regularly with copper(I) coordination number (*32, 59, 60*). This is illustrated in Fig. 15. In addition to the values of the ligand concentration given in the references cited, a chloride concentration of 10.6 mol dm^{-3} for $[P(C_2H_5)_4]_2$-$[Cu_2Cl_4]$ (*72*) and iodide concentrations of 5.7 mol dm^{-3} for $[P(C_6H_5)_4]_2$-$[Cu_2I_4]$ (planar and folded anions) (*68*) and of 6.2 mol dm^{-3} for $[N(C_4H_9)_4]_3[Cu_4I_7]$ (*96, 103*) have been calculated. In $[N(C_2H_5)_4]_6$-$[Cu_6I_{11}]I$ (*79*), the total concentration of iodide is estimated to be 10.0 mol dm^{-3}. For those anions in which copper(I) exhibits more than

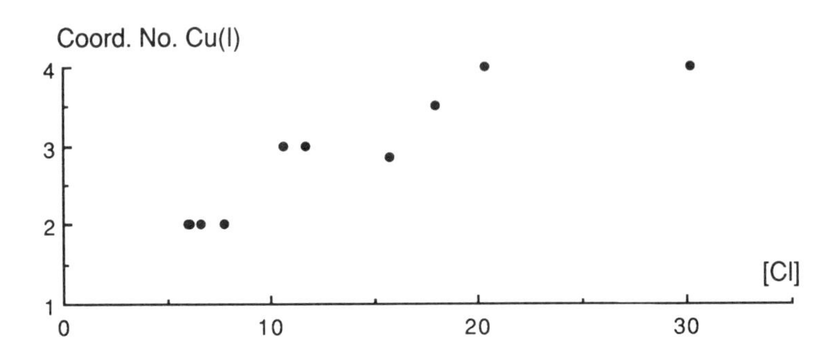

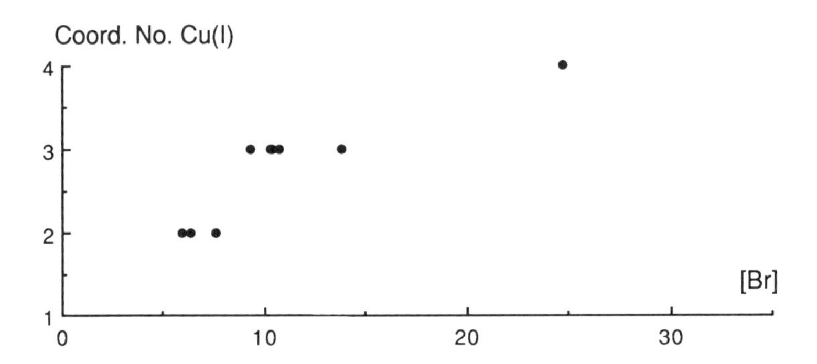

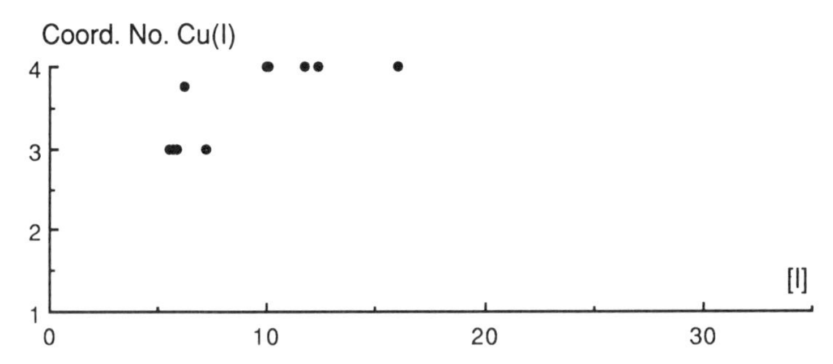

FIG. 15. The copper(I) coordination number as a function of the concentration of the halogenide ligand (mol dm^{-3}) in halogenocuprates(I) crystallizing with symmetrically substituted quaternary ammonium, phosphonium, and arsonium cations.

one coordination number, a mean coordination number has been employed. As is apparent from the diversity in stoichiometry of the various halogenocuprate(I) species obtained, there is no correlation between the coordination of copper(I) and the concentration of this metal ion in the crystalline phase (cf. Ref. *32*).

Examination of the three graphs in Fig. 15 shows that the coordination number of copper(I) increases with increasing concentration of ligand in the solid (decreasing cation size) in all three halogenocuprate(I) series, larger cations being required to promote lower coordination numbers the heavier the halogenide ligand. It has not yet proved possible to prepare an iodocuprate(I) in which copper(I) is two coordinated using cations of this type. As mentioned in Section II,A,1, monomeric $[CuI_2]^-$ has, however, been isolated with the K(18-crown-6)$^+$ and K(dicyclohexano-18-crown-6)$^+$ cations (*17*). These crown ether-based cations differ in shape from the effectively spherical quaternary ammonium, phosphonium, and arsonium cations under discussion, and, consequently, also with respect to the exposure of the positive charge. Both, however, give rise to relatively low solid-state ligand concentrations, viz. 6.2 and 4.8 mol dm^{-3}, respectively.

That there exists a correlation between the copper(I) coordination number in the relevant anion and the concentration of halide ligand in the solid has been interpreted as suggesting that dilution of the ligand ions by the cations is a determinative factor for the attainment of a particular copper(I) coordination number and thus for the resulting configuration of the anion (*32*). Large cations hinder the accumulation of a high ligand concentration and thus suppress catenation, leading to the formation of small, discrete anions in which the metal exhibits a low coordination number. Small cations, on the other hand, are less effective in the suppression of catenation and tend to promote the formation of polynuclear species in which copper(I) attains a high coordination number.

Indirect evidence for the importance of cation–halogenide ligand packing for the determination of a particular coordination number has been provided by showing that the halogenocuprate(I) species in solution at the onset of crystallization may differ from that present in the crystals (*45*). Thus, infrared spectroscopic investigation of ethanolic solutions from which the $[Cu_4Br_6]^{2-}$ cluster, composed of three-coordinated copper(I) (cf. Section II,B,3), crystallizes ($[Br^-]_{max.} \approx 0.16$ *M*) demonstrate that the centrosymmetric $[CuBr_2]^-$ monomer, $\nu_3 = 323$ cm^{-1}, is the dominant and probably the sole bromocuprate(I) species present in solution (*45*). More concentrated solutions ($[Br^-] \approx 0.30$ *M*) of solids containing bromocuprate(I) anions in which copper(I) is three coordinated, obtained by dissolving $[N(C_3H_7)_4]_2[Cu_4Br_6]$ (*85*), $[P(CH_3)_4]_2[CuBr_3]$ (*59*), and $[N(CH_3)_4]_3[Cu_2Br_5]$ (*64*) in nitromethane, also contained $[CuBr_2]^-$ as the sole observable species (*45*). Identical results had been obtained previously for the dissolution of the tetraethylammonium dibromocuprate(I) in nitromethane (*2*), the solid compound subsequently being shown to contain a dinuclear anion in which copper(I) is trigonally coordinated (*66*). Although the existence of three-coordinated halogenocuprate(I) species such as $[CuCl_3]^{2-}$ (*160–162*) and $[Cu_2I_4]^{2-}$ (*163*) in solution has been demonstrated by means of techniques other than vibrational spectroscopy, media permitting far higher concentrations of halogenocuprates(I) have been employed. Indeed the total halogenide concentrations in some cases (*163*) approach the lower limits of those estimated for the relevant halogenocuprate(I) species, here $[Cu_2I_4]^{2-}$, in the solid state. Similarly, monomeric $[AgI_3]^{2-}$ ions (*81*) and various oligomeric (*81*) or polymeric (*119*) iodooargentate(I) species have been shown to exist in concentrated solutions, and also in fused $(K,Na)NO_3$ (*164*).

It has been suggested that cation–halogenide ligand packing may be envisaged as the primary process occurring during crystal nucleation, copper(I) attaining the appropriate coordination number by diffusion into available interstices, subsequent rearrangement then resulting in the specific anion (*45*). Such a "mechanism" would imply a rapid ligand exchange rate for copper(I) in solution, the dihalogenocuprate(I) anion providing a source of the "naked" metal ion. Although it has not yet proved possible to substantiate the latter hypothesis, the extreme reversibility of copper(I) complexes is well documented in electrochemical applications. Moreover, the high polarizability associated with Cu^+ (*165*) is not inconsistent with easy adaptability of copper(I) to the wide variety of coordination situations that might be envisaged to ensue from cation–halogenide ligand packing.

Silver(I) is another ion commonly associated with electrochemical reversibility and is also readily polarizable (*165*). One might therefore expect an analogous correlation between the coordination number of silver(I) in crystalline halogenoargentates(I) and the size of the cation. The variation in silver(I) coordination number as a function of the concentration of ligand in crystalline halogenoargentates(I) containing symmetrically substituted quaternary ammonium, phosphonium, and arsonium cations is illustrated in Fig. 16, the ligand concentrations having been taken from Ref. *70*. The graphs demonstrate the dominance of coordination number four for silver(I), coordination number three being found only with the larger cations. Silver(I) would, however, appear to show a tendency toward the trends observed for

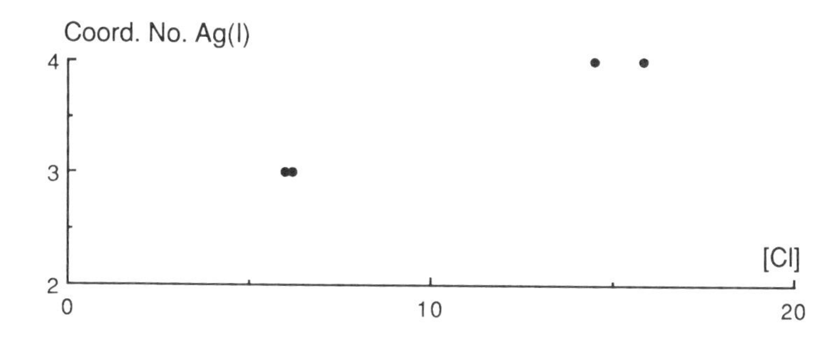

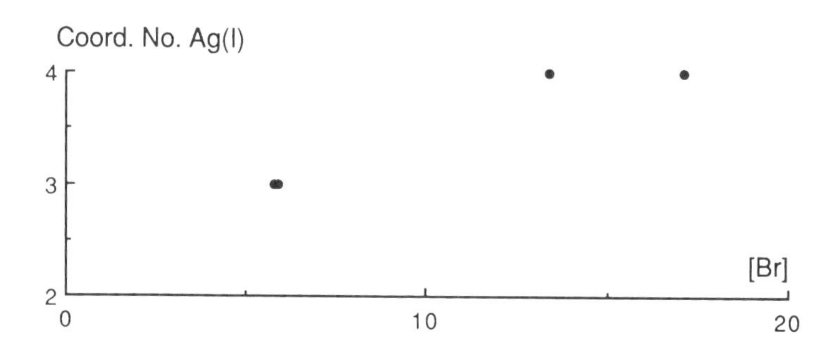

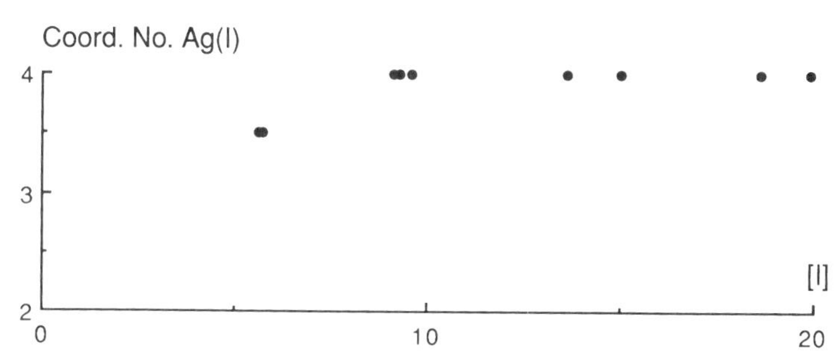

FIG. 16. The silver(I) coordination number as a function of the concentration of the halogenide ligand (mol dm^{-3}) in halogenoargentates(I) crystallizing with symmetrically substituted quaternary ammonium, phosphonium, and arsonium cations.

copper(I) in halogenocuprates(I) with this type of cation, even if there is less variation in the silver(I) coordination number. By employing a larger cation of spherical type, viz. $K(crypt\text{-}2,2,2)^+$, it has proved possible to isolate a linear, monomeric dichloroargentate(I) anion in the solid state (*21*); the chloride concentration in this compound is

5.1 mol dm^{-3}. This, in itself, suggests that cation–halogenide packing is also of importance for the determination of the metal coordination number in crystalline halogenoargentates(I).

The successful preparation in $[P(C_6H_5)_4]_4[Ag_2Cu_2I_8]$ and $[P(C_6H_5)_4]_4[Ag_3CuI_8]$ (*71*) of heteronuclear $[Ag_2Cu_2I_8]^{4-}$ and $[Ag_3CuI_8]^{4-}$ clusters, in which copper(I) is three coordinated, whereas silver(I) is four coordinated or both three and four coordinated, respectively, from the metal(I) iodides and tetraphenylphosphonium iodide in acetonitrile solution, may also be seen as lending support to the hypothesis that the primary process at the solution–crystal interface is cation–halide packing, the "naked" metal ion diffusing into the appropriate interstices.

B. The Effects of Charge and Shape

In general, it is true to say that relatively small cations with exposed positive charge, such as the alkali metal ions, tend to crystallize with polymeric halogenocuprates(I) or halogenoargentates(I) in which the metal exhibits tetrahedral coordination geometry, the most common anion being the dinegative MX_3^{2-} chain (Section II,C,8,b). Complexed cations with higher positive charge, e.g., $[Cu(en)_2]^{2+}$ or $[Ni(en)_2]^{2+}$, are also associated with polymeric anions in which the metal is four coordinated, but in this case the MX_2^- chain of edge-sharing tetrahedra appears to be prevalent (Section II,C,8,a). For a series of iodocuprates (I) containing polymeric $[Cu_3I_4]^-$, $[Cu_2I_3]^-$, and $[CuI_2]^-$ anions composed of shared copper(I) iodide tetrahedra (Section II,C,8), Hartl and Mahdjour-Hassan-Abadi have demonstrated that with decreased polarizing ability of the cation, anions with progressively lower negative charge distribution, i.e., decreased I^- : CuI ratio, are obtained (*134*). Thus it would seem that for a given metal(I) coordination number in a polymeric species, the general trend is that the higher the effective charge of the cation, the higher the net charge of the anion, whereas within a group of anions bearing the same negative charge, the charge distribution and thus the I^- : MI ratio tend to increase with increased polarizing power of the cation.

Whereas the metal(I) coordination number would appear to be largely determined by cation size, distribution of positive charge is undoubtedly determinative for the detailed structure of the anion. This has been further demonstrated by Hartl and co-workers in studies involving pairs of isomeric cations, e.g., $[N(CH_3)_4]^+$ and $[CH_3)_3CNH_3]^+$ and $[(CH_3)_3NH]^+$ and $[(CH_3)_2CNH_3]^+$ (*102*). All four cations crystallize with iodocuprate(I) anions of stoichiometry $[Cu_2I_3]^-$ composed of copper(I) tetrahedra. Whereas $[N(CH_3)_4]^+$ gives rise to a single chain of edge- and face-sharing tetrahedra (*141*), a layered anion is formed with the isomeric cation $[(CH_3)_3CNH_3]^+$ (*102*); both $[(CH_3)_3NH]^+$ and $[(CH_3)_2CNH_3]^+$ crystallize with double chains of edge-sharing tetrahedra (cf. Section II,C,8,c), differing slightly from one another with respect to distortion (*102*).

Investigation of a series of bromocuprates(I) crystallizing with unsymmetrically substituted quaternary ammonium and phosphonium cations, with well-screened positive charge (see Ref. *55* and references therein), has shown that, whereas the copper(I) coordination number increases with increasing concentration of bromide ligand in the crystalline phase, correlation between the copper(I) coordination number and decreasing overall volume of the cation, as estimated from volume increments for the relevant substituents (*166*), is somewhat less pronounced (*55*), perhaps reflecting the role of shape in cation–halogenide packing.

C. Conclusions

A feature of halogenocuprates(I) and, to a somewhat lesser degree, of halogenoargentates(I) that is consistent with their apparent adaptability to cation dictates is the observed variation in metal–ligand distances for a given anionic species. This is particularly marked for anions in which the metal exhibits coordination numbers of three or four (cf. Tables II–VII). The observed spread in bond distances suggests that halogenocuprates(I), in particular, are rather polar assemblies of metal(I) cations and individual halide ions held together by electrostatic forces. Although there is less variation in the connectivity relationships shown by a given discrete, digonal MX_2^- ion in different environments, nuclear quadrupole resonance (NQR) spectra for monomeric $[CuCl_2]^-$ and $[CuBr_2]^-$ indicate an ionic character of 82 and 80% in the Cu—Cl and Cu—Br bonds, respectively (*2*), which can be compared with an ionic character of 68% for the Au—Cl bond in $[AuCl_2]^-$ (*167*).

That there is a relationship between the coordination number of the metal in crystalline halogenocuprates(I) and halogenoargentates(I) and the properties of the cation with which it is coprecipitated would now seem to be well established. The tuning of anionic configurations to cation properties reflects the versatility in coordination requirements not only of the metal but also of the ligands. Cation size would appear to be of prime importance for the determination of a particular metal(I) coordination number, the tendency to attainment of a higher coordination number and to catenation increasing with decreasing cation size and also from Cl through Br to I. Larger cations are required to promote the formation of monomeric or dimeric halogenoargentates(I) with a low silver(I) coordination number than is the case for halogenocuprates(I).

Rationalization of the detailed geometry of the particular halogenocuprate(I) or halogenoargentate(I) ion in terms of the influence of cation shape or positive charge distribution is, however, much more complex. Whereas it has proved possible to employ cations predictively for anion design in series of compounds containing comparable cations, e.g., symmetrically substituted tetraalkylammonium and phosphonium analogs (cf. *60, 71, 72*), it seems doubtful that cations of disparate shape can be used as other than predictors of metal(I) coordination number.

Acknowledgments

We thank Solveig Olson for preparing the figures and for assistance in surveying the literature. Financial support from the Swedish Natural Science Research Council (NFR) is gratefully acknowledged.

References

1. Bowmaker, G. A., *Adv. Spectrosc.* **14,** 1 (1987).
2. Bowmaker, G. A., Brockliss, L. D., and Whiting, R., *Aust. J. Chem.* **26,** 29 (1973).
3. Creighton, J. A., and Lippincott, E. R., *J. Chem. Soc.* p. 5134 (1963).
4. Specker, H., and Pappert, W., *Z. Anorg. Allg. Chem.* **341,** 287 (1965).
5. Ahrland, S., Bläuenstein, P., Tagesson, B., and Tuhtar, D., *Acta Chem. Scand., Ser. A* **A34,** 265 (1980).
6. Ahrland, S., Nilsson, K., and Tagesson, B., *Acta Chem. Scand., Ser. A* **A37,** 193 (1983).
7. Waters, D. N., and Basak, B., *J. Chem. Soc. A* p. 2733 (1971).
8. Ahrland, S., Ishiguro, S., and Persson, I., *Acta Chem. Scand., Ser. A.* **A40,** 418 (1986).
9. Berglund, U., and Sillén, L. G., *Acta Chem. Scand.* **2,** 116 (1948).
10. Althin, B., Wåhlin, E., and Sillén, L. G., *Acta Chem. Scand.* **3,** 321 (1949).
11. Ålin, B., Evers, L., and Sillén, L. G., *Acta Chem. Scand.* **6,** 759 (1952).
12. Dallinga, G., and Mackor, E. L., *Recl. Trav. Chim. Pays-Bas* **75,** 796 (1956).
13. Newton, M. G., Caughman, H. G., and Taylor, R. C., *J. Chem. Soc., Chem. Commun.* p. 1227 (1970); *J. Chem. Soc., Dalton Trans.* p. 258 (1974).
14. Beurskens, P. T., Cras, J. A., Hummelink, T. W., and van der Linden, J. G. M., *Recl. Trav. Chim. Pays-Bas* **89,** 984 (1970).
15. Asplund, M., Jagner, S., and Nilsson, M., *Acta Chem. Scand., Ser. A* **A37,** 57 (1983).
16. Andersson, S., and Jagner, S., *Acta Chem. Scand., Ser. A.* **A39,** 297 (1985).
17. Rath, N. P., and Holt, E. M., *J. Chem. Soc., Chem. Commun.* p. 311 (1986).

18. Ferrari, A., *Gazz. Chim. Ital.* **67,** 94 (1937).
19. Elliot, N., and Pauling, L., *J. Am. Chem. Soc.* **60,** 1846 (1938).
20. Berthold, H. J., and Ludwig, W., *Z. Naturforsch., B: Anorg. Chem., Org. Chem.* **35B,** 970 (1980).
21. Helgesson, G., and Jagner, S., *Inorg. Chem.* **30,** in press (1991).
22. Cras, J. A., Noordik, J. H., Beurskens, P. T., and Verhoeven, A. M., *J. Cryst. Mol. Struct.* **1,** 155 (1971).
23. Marsh, W. C., and Trotter, J., *J. Chem. Soc. A* p. 1482 (1971).
24. Kaiser, J., Brauer, G., Schröder, F. A., Taylor, I. F., and Rasmussen, S. E., *J. Chem. Soc., Dalton Trans.* p. 1490 (1974).
25. Siiman, O., Huber, C. P., and Post, M. L., *Inorg. Chim. Acta* **25,** L11 (1977). Huber, C. P., Post, M. L., and Siiman, O., *Acta Crystallogr., Sect. B* **B34,** 2629 (1978).
26. Francisco, R. H. P., de Almeida Santos, R. H., Lechat, J. R., and Massabni, A. C., *Acta Crystallogr., Sect. B* **B37,** 232 (1981).
27. Sakurai, T., Kobayashi, K., Masuda, H., Tsuboyama, S., and Tsuboyama, K., *Acta Crystallogr., Sect. C* **C39,** 334 (1983).
28. Tsuboyama, S., Kobayashi, K., Sakurai, T., and Tsuboyama, K., *Acta Crystallogr., Sect. C* **C40,** 1178 (1984).
29. Clegg, W., Garner, C. D., Nicholson, J. R., and Raithby, P. R., *Acta Crystallogr., Sect. C.* **C39,** 1007 (1983).
30. Engelhardt, L. M., Papasergio, R. I., and White, A. H., *Aust. J. Chem.* **37,** 2207 (1984).
31. Garbauskas, M. F., Haitko, D. A., and Kasper, J. S., *J. Crystallogr. Spectrosc. Res.* **16,** 729 (1986).
32. Andersson, S., and Jagner, S., *Acta Chem. Scand., Ser. A* **A40,** 52 (1986).
33. Andersson, S., and Jagner, S., *Acta Chem. Scand., Ser A* **A39,** 799 (1985).
34. Kawamoto, K., Tanaka, J., and Tanaka, M., *Acta Crystallogr., Sect. C* **C43,** 205 (1987).
35. Geiser, U., Wang, H. H., Hammond, C. E., Firestone, M. A., Beno, M. A., Carlson, K. D., Nuñez, L., and Williams, J. M., *Acta Crystallogr., Sect. C* **C43,** 656 (1987).
36. Canty, A. J., Engelhardt, L. M., Healy, P. C., Kildea, J. D., Minchin, N. J., and White, A. H., *Aust. J. Chem.* **40,** 1881 (1987).
37. Stamp, L., and tom Dieck, H., *Inorg. Chim. Acta* **147,** 199 (1988).
38. Healy, P. C., Kildea, J. D., Skelton, B. W., and White, A. H., *Aust. J. Chem.* **42,** 115 (1989).
39. Shibaeva, R. P., Kaminskii, V. F., Yagubskii, E. B., and Kushch, L., *Kristallografiya* **28,** 92 (1983).
40. Cecconi, F., Ghilardi, C. A., Midollini, S., and Orlandini, A., *Angew. Chem.* **95,** 554 (1983); *Angew. Chem., Suppl.* p. 718 (1983).
41. Healy, P. C., Engelhardt, L. M., Patrick, V. A., and White, A. H., *J. Chem. Soc., Dalton Trans* p. 2541 (1985).
42. Andersson, S., and Jagner, S., *Acta Chem. Scand., Ser. A* **A40,** 210 (1986).
43. Andersson, S., and Jagner, S., *Acta Chem. Scand., Ser. A.* **A39,** 577 (1985).
44. Andersson, S., and Jagner, S., *Acta Chem. Scand., Ser. A* **A39,** 515 (1985).
45. Andersson, S., Håkansson, M., and Jagner, S., *J. Crystallogr. Spectrosc. Res.* **19,** 147 (1989).
46. Bowmaker, G. A., Camus, A., Skelton, B. W., and White, A. H., *J. Chem. Soc., Dalton Trans.* p. 727 (1990).
47. Orgel, L. E., *J. Chem. Soc.* p. 4186 (1958).
48. Dunitz, J. D., and Orgel, L. E., *Adv. Inorg. Chem. Radiochem.* **2,** 1 (1960).

49. Jørgensen, C. K., and Pouradier, J., *J. Chim. Phys. Chim. Biol.* **67,** 124 (1970).
50. Bowmaker, G. A., Boyd, P. D. W., and Sorrenson, R. J., *J. Chem. Soc., Faraday Trans. 2* **81,** 1627 (1985).
51. Asplund, M., Jagner, S., and Nilsson, M., *Acta Chem. Scand., Ser. A.* **A38,** 57 (1984).
52. Endres, H., Keller, H. J., Martin, R., and Traeger, U., *Acta Crystallogr., Sect. B* **B35,** 2880 (1979).
53. Sanz, F., *An. Fis.* **72,** 43 (1976).
54. Asplund, M., and Jagner, S., *Acta Chem. Scand., Ser. A.* **A38,** 807 (1984).
55. Andersson, S., and Jagner, S., *Acta Chem. Scand.* **43,** 39 (1989).
56. Bowmaker, G. A., Clark, G. R., Rogers, D. A., Camus, A., and Marsich, N., *J. Chem. Soc., Dalton Trans.* p. 37 (1984).
57. Hartl, H., and Brüdgam, I., *Z. Naturforsch., B: Chem. Sci.* **44,** 936 (1989).
58. Bencini, A., and Mani, F., *Inorg. Chim. Acta* **87,** L9 (1984).
59. Andersson, S., and Jagner, S., *Acta Chem. Scand., Ser. A.* **A41,** 230 (1987).
60. Andersson, S., and Jagner, S., *Acta Chem. Scand., Ser. A* **A42,** 691 (1988).
61. Clegg, W., Nicholson, J. R., Collison, D., and Garner, C. D., *Acta Crystallogr., Sect. C* **C44,** 453 (1988).
62. Helgesson, G., and Jagner, S., in preparation.
63. Bowles, J. C., and Hall, D., *J. Chem. Soc., Chem. Commun.* p. 1523 (1971); *Acta Crystallogr., Sect. B* **B31,** 2149 (1975).
64. Asplund, M., and Jagner, S., *Acta Chem. Scand., Ser. A.* **A39,** 47 (1985).
65. Shibaeva, R. P., and Kaminskii, V. F., *Kristallografiya* **26,** 332 (1981).
66. Asplund, M., and Jagner, S., *Acta Chem. Scand., Ser. A* **A38,** 135 (1984).
67. Asplund, M., and Jagner, S., *Acta Chem. Scand., Ser. A.* **A38,** 297 (1984).
68. Hartl, H., Brüdgam, I., and Mahdjour-Hassan-Abadi, F., *Z. Naturforsch., B: Anorg. Chem., Org. Chem.* **40B,** 1032 (1985).
69. Helgesson, G., and Jagner, S., *J. Chem. Soc., Dalton Trans.* p. 2117 (1988).
70. Helgesson, G., and Jagner, S., *J. Chem. Soc., Dalton Trans.* p. 2413 (1990).
71. Andersson, S., Helgesson, G., Jagner, S., and Olson, S., *Acta Chem. Scand.* **43,** 946 (1989).
72. Andersson, S., Håkansson, M., and Jagner, S., in preparation.
73. Banci, L., Bencini, A., Dei, A., and Gatteschi, D., *Inorg. Chim. Acta* **84,** L11 (1984).
74. Haasnoot, J. G., Favre, T. L. F., Hinrichs, W., and Reedijk, J., *Angew. Chem., Int. Ed. Engl.* **27,** 856 (1988).
75. Andersson, S., and Jagner, S., *Acta Chem. Scand., Ser. A* **A39,** 423 (1985).
76. Andersson, S., and Jagner, S., *Acta Crystallogr., Sect. C* **C43,** 1089 (1987).
77. Asplund, M., Jagner, S., and Nilsson, M., *Acta Chem. Scand., Ser. A.* **A36,** 751 (1982).
78. Asplund, M., and Jagner, S., *Acta Chem. Scand., Ser. A* **A38,** 411 (1984).
79. Mahdjour-Hassan-Abadi, F., Hartl, H., and Fuchs, J., *Angew. Chem.* **96,** 497 (1984).
80. Basu, A., Bhaduri, S., Sapre, N. Y., and Jones, P. G., *J. Chem. Soc., Chem. Commun.* p. 1724 (1987).
81. Gaizer, F., and Johansson, G., *Acta Chem. Scand., Ser. A.* **A42,** 259 (1988).
82. Bowmaker, G. A., Brockliss, L. D., Earp, C. D., and Whiting, R., *Aust. J. Chem.* **26,** 2593 (1973).
83. Bowmaker, G. A., Clark, G. R., and Yuen, D. K. P., *J. Chem. Soc., Dalton. Trans.* p. 2329 (1976).
84. Rath, N. P., and Holt, E. M., *J. Chem. Soc., Chem. Commun.* p. 665 (1985).
85. Asplund, M., and Jagner, S., *Acta Chem. Scand., Ser. A* **A38,** 725 (1984).

86. Takahashi, T., and Yamamoto, O., *Abstr. 5th Symp. Solid State Ionics,* Nagoya, Japan, *1977,* p. 199 (1977).
87. Nag, K., and Geller, S., *J. Electrochem. Soc.* **128,** 2670 (1981).
88. Kanno, R., Takeda, Y., Masuyama, Y., Yamamoto, O., and Takahashi, T., *Solid State Ionics* **11,** 221 (1983).
89. Hartl, H., and Mahdjour-Hassan-Abadi, F., *Angew. Chem.* **93,** 804 (1981).
90. Asplund, M., and Jagner, S., *Acta Chem. Scand., Ser. A* **A38,** 129 (1984).
91. Hartl, H., *Z. Kristallogr* **178,** 83 (1987).
92. Bigalke, K. P., Hans, A., and Hartl, H., *Z. Anorg. Allg. Chem.* **563,** 96 (1988).
93. Hoyer, M., and Hartl, H., *Z. Anorg. Allg. Chem.* **587,** 23 (1990).
94. Keller, H. J., Keppler, B., and Pritzkow, H., *Acta Crystallogr., Sect. B* **B38,** 1603 (1982).
95. Hartl, H., *Angew. Chem.* **99,** 925 (1987); *Angew. Chem., Int. Ed. Engl.* **26,** 927 (1987).
96. Hartl, H., Brüdgam, I., and Mahdjour, F., *Int. Conf. Coord. Chem., 28th,* Gera, *1990,* p. 5–34 (1990).
97. Estienne, J., *Acta Crystallogr., Sect. C* **C42,** 1512 (1986).
98. Hartl, H., and Mahdjour-Hassan-Abadi, F., *Angew. Chem.* **96,** 359 (1984).
99. Andersson, S., and Jagner, S., *J. Crystallogr. Spectrosc. Res.* **18,** 591 (1988).
100. Murray-Rust, P., Day, P., and Prout, C. K., *J. Chem. Soc., Chem. Commun.* p. 277 (1966).
101. Murray-Rust, P., *Acta Crystallogr., Sect. B* **B29,** 2559 (1973).
102. Herrschaft, G., Ph.D. Thesis, Freie Universität Berlin (1990).
103. Mahdjour-Hassan-Abadi, F., Ph.D. Thesis, Freie Universität Berlin (1985).
104. Hartl, H., and Fuchs, J., *Angew. Chem.* **98,** 550 (1986).
105. Prout, C. K., and Murray-Rust, P., *J. Chem. Soc. A* p. 1520 (1969).
106. Baglio, J. A., and Vaughan, P. A., *J. Inorg. Nucl. Chem.* **32,** 803 (1970).
107. Simonsen, O., and Toftlund, H., *Acta Crystallogr., Sect. C* **C43,** 831 (1987).
108. Helgesson, G., Josefsson, M., and Jagner, S., *Acta Crystallogr., Sect. C* **C44,** 1729 (1988).
109. Stomberg, R., *Acta Chem. Scand.* **22,** 2022 (1968); **23,** 3498 (1969).
110. Shirshova, L. V., Lavrent'ev, I. P., and Ponomarev, V. I., *Koord. Khim.* **15,** 1048 (1989).
111. Baglio, J. A., Weakliem, H. A., Demelio, F., and Vaughan, P. A., *J. Inorg. Nucl. Chem.* **32,** 795 (1970).
112. Freckmann, B., and Tebbe, K.-F., *Z. Naturforsch., Anorg. Chem. Org. Chem.* **35B,** 1319 (1980).
113. Hartl, H., Brüdgam, I., and Mahdjour-Hassan-Abadi, F., *Z. Naturforsch., B: Anorg. Chem., Org. Chem.* **38B,** 57 (1983).
114. Hoyer, M., and Hartl, H., *Proc. Int. Conf. Coord. Chem., 28th* Gera, *1990* p. 6–36 (1990).
115. Thackeray, M. M., and Coetzer, J., *Acta Crystallogr., Sect. B* **B31,** 2341 (1975).
116. Geller, S., and Dudley, T. O., *J. Solid State Chem.* **26,** 321 (1978).
117. Peters, K., von Schnering, H. G., Ott, W., and Seidenspinner, H.-M., *Acta Crystallogr., Sect. C* **C40,** 789 (1984).
118. Willett, R. D., *Inorg. Chem.* **26,** 3423 (1987).
119. Nilsson, K., and Persson, I., *Acta Chem. Scand., Ser. A.* **A41,** 139 (1987).
120. Brink, C., and MacGillavry, C. H., *Acta Crystallogr.* **2,** 158 (1949).
121. Brink, C., and Stenfert Kroese, H. A., *Acta Crystallogr* **5,** 433 (1952).

122. Thackeray, M. M., and Coetzer, J., *Acta Crystallogr., Sect. B* **B31,** 2339 (1975).
123. Shoemaker, C. B., *Acta Crystallogr., Sect. B* **B32,** 1619 (1976).
124. Brown, I. D., Howard-Lock, H. E., and Natarajan, M., *Can. J. Chem.* **55,** 1511 (1977).
125. Brink, C., and van Arkel, A. E., *Acta Crystallogr.* **5,** 506 (1952).
126. Shoemaker, C. B., *Z. Kristallogr.* **137,** 225 (1973).
127. Kildea, J. D., Skelton, B. W., and White, A. H., *Aust. J. Chem.* **39,** 171 (1986).
128. Brink, C., Binnendijk, N. F., and van de Linde, J., *Acta Crystallogr.* **7,** 176 (1954).
129. Meyer, G., *Z. Anorg. Allg. Chem.* **515,** 127 (1984).
130. Andersson, S., and Jagner, S., *Acta Chem. Scand., Ser. A* **A40,** 177 (1986).
131. Rømming, C., and Wærstad, K., *J. Chem. Soc., Chem. Commun.* p. 299 (1965).
132. Boehm, J. R., Balch, A. L., Bizot, K. F., and Enemark, J. H., *J. Am. Chem. Soc.* **97,** 501 (1975).
133. Jouini, N., Guen, L., and Tournoux, M., *Rev. Chim. Miner.* **17,** 486 (1980).
134. Hartl, H., and Mahdjour-Hassan-Abadi, F., *Z. Naturforsch., B: Anorg. Chem., Org. Chem.* **39B,** 149 (1984).
135. Asplund, M., Jagner, S., and Nilsson, M., *Acta Chem. Scand., Ser. A* **A39,** 447 (1985).
136. Geller, S., and Gaines, J. M., *J. Solid State Chem.* **59,** 116 (1985).
137. Helgesson, G., and Jagner, S., *Acta Crystallogr., Sect. C* **C44,** 2059 (1988).
138. Jagner, S., Olson, S., and Stomberg, R., *Acta Chem. Scand., Ser. A* **A40,** 230 (1986).
139. Meyer, H-J., *Acta Crystallogr.* **16,** 788 (1963).
140. Batsanov, A. S., Struchkov, Y. T., Ukhin, L. Y., and Dolgopolova, N. A., *Inorg. Chim. Acta* **63,** 17 (1982).
141. Andersson, S., and Jagner, S., *Acta Chem. Scand., Ser. A* **A39,** 181 (1985).
142. Frydrych, R., Muschter, T., Brüdgam, I., and Hartl, H., *Z. Naturforsch., B: Chem. Sci.* **45,** 679 (1990).
143. Gilmore, C. J., Tucker, P. A., and Woodward, P., *J. Chem. Soc. A* p. 1337 (1971).
144. Coetzer, J., and Thackeray, M. M., *Acta Crystallogr., Sect. B* **B31,** 2113 (1975).
145. Geller, S., *Science* **157,** 310 (1967).
146. Geller, S., Ray, A. K., Fardi, H. Z., and Nag, K., *Phys. Rev. B: Condens. Matter* [3] **25,** 2968 (1982).
147. Geller, S., Ray, A. K., and Nag, K., *J. Solid State Chem.* **48,** 176 (1983).
148. Geller, S., and Lind, M. D., *J. Chem. Phys.* **52,** 5854 (1970).
149. Gaines, J. M., and Geller, S., *J. Phys. Chem. Solids* **48,** 1159 (1987).
150. Chan, L. Y. Y., Geller, S., and Skarstad, P. M., *J. Solid State Chem.* **25,** 85 (1978).
151. Xie, S., and Geller, S., *J. Solid State Chem.* **68,** 73 (1987).
152. Chadha, R. K., *Inorg. Chem.* **27,** 1507 (1988).
153. Kildea, J. D., and White, A. H., *Inorg. Chem.* **23,** 3825 (1984).
154. Peters, K., Ott, W., and von Schnering, H. G., *Angew. Chem.* **94,** 720 (1982) *Angew. Chem., Suppl.* p. 1479 (1982).
155. Baker, R. J., Nyburg, S. C., and Szymański, J. T., *Inorg. Chem.* **10,** 138 (1971).
156. Keller, H. J., Martin, R., and Traeger, U., *Z. Naturforsch., B: Anorg. Chem., Org. Chem.* **33B,** 1263 (1978).
157. Gaebell, H.-C., Meyer, G., and Hoppe, R., *Z. Anorg. Allg. Chem.* **497,** 199 (1983).
158. Gaebell, H.-C., and Meyer, G., *Z. Anorg. Allg. Chem.* **513,** 15 (1984).
159. Bustos, L. A., and Contreras, J. G., *J. Inorg. Nucl. Chem.* **42,** 1293 (1980).
160. Malik, W. U., and Tyagi, J. S., *Indian J. Chem., Sect. A* **20A,** 1208 (1981).
161. Davis, C. R., and Stevenson, K. L., *Inorg. Chem.* **21,** 2514 (1982).
162. Stevenson, K. L., Braun, J. L., Davis, D. D., Kurtz, K. S., and Sparks, R. I., *Inorg. Chem.* **27,** 3472 (1988).

163. Persson, I. and Sandström, M., *In* "Proceedings of the 13th Nordiske Strukturkemikermøde," p. 37. H. C. Ørsted Institutet, University of Copenhagen, Copenhagen, 1990.
164. Holmberg, B., *Acta Chem. Scand.* **27,** 3657 (1973).
165. Ammlung, R. L., Scaringe, R. P., Ibers, J. A., Shriver, D. F., and Whitmore, D. H., *J. Solid State Chem.* **29,** 401 (1979).
166. Immirzi, A., and Perini, B., *Acta Crystallogr., Sect. A* **A33,** 216 (1977).
167. Bowmaker, G. A., and Whiting, R., *Aust. J. Chem.* **29,** 1407 (1976).

STRUCTURES OF ORGANONITROGEN–LITHIUM COMPOUNDS: RECENT PATTERNS AND PERSPECTIVES IN ORGANOLITHIUM CHEMISTRY

KARINA GREGORY,* PAUL von RAGUÉ SCHLEYER,* and RONALD SNAITH†

* Institut für Organische Chemie der Universität Erlangen-Nürnberg, W-8520 Erlangen, Germany
† Department of Chemistry, University Chemical Laboratory, University of Cambridge, Cambridge CB2 1EW, England

I. Introduction
  A. Uses of N—Li Compounds; Scope of the Review
  B. Structure and Bonding
II. Lithium Imides (Iminolithiums) and Their Complexes; Ring Stacking
  A. Uncomplexed Lithium Imides $(RR'C{=}NLi)_n$: Solid-State Structures
  B. Iminolithium Complexes: Solid-State Structures
  C. Solution Structures of Lithium Imides and Their Complexes
  D. Calculations on Iminolithiums and on Lithium Species with Related Structures
  E. Ring Stacking: General Applications to the Structures of Organolithium Compounds
III. "Simple" Lithium Amides (Amidolithiums) and Their Complexes; Ring Laddering
  A. Introduction
  B. Uncomplexed Lithium Amides
  C. Complexed Lithium Amides $[(RR'NLi{\cdot}xL)_n]$
IV. Conclusions
  A. Definitions and Nomenclature
  B. Bonding
  C. Implications for Organic Reaction Mechanisms
  D. Structures of Uncomplexed Organolithiums
  E. Structures of Complexed Organolithiums
  References

I. Introduction

The first X-ray crystal structure of an organolithium compound [$(EtLi)_4$] was published (*1*) just over a quarter of a century ago. These intriguing compounds have received increasing attention ever since (*2*, *3*). The definition of "organolithium" has broadened. The term now means not only compounds with C—Li bonds [e.g., alkyl- and aryllithiums, RLi; and alkynyllithiums (lithium acetylides), RC≡CLi], but also lithium derivatives of organic molecules in general, e.g., N—Li species such as amidolithiums (R_2NLi, lithium amides) and O—Li species such as lithium enolates [RC (=CH_2)OLi]. Indeed, the two most recently published reviews have been largely concerned with lithium enolates (*4*) and with lithiated sulfones (*4a*). A very thorough review now in press (*5*) describes the syntheses and structures of carbanions of alkali (and alkaline earth) metal cations, and includes related species with organic anions such as amide, enolate, and alkoxide. This present review concentrates on compounds with N—Li bonds. These are chiefly lithium imides [iminolithiums $(RR'C{=}NLi)_n$] and their complexes with added Lewis bases (L), and lithium amides [amidolithiums $(RR'NLi)_n$] and their complexes. For the lithium amide species in particular, we discuss only those whose R,R′ groups do not contain additional functionalities, i.e., the R,R′ groups remain largely uninvolved with lithium centers. These species are termed "simple" lithium amides (see Section III,A). Lithium amide structures exhibiting R,R′-group involvement will be covered in a (pending) follow-up review.

A. Uses of N—Li Compounds; Scope of the Review

Organonitrogen–lithium compounds, and particularly lithium amides (R_2NLi), are widely used both in organic and in organometallic syntheses. For the former, these strong bases are employed as proton abstractors (*5–8*), to generate new organolithiums. These can then be derivatized with so-called electrophiles, e.g., alkyl and acyl halides ($E^+ = R^+$ and $R{-}C^+{=}O$) and trimethylsilylchloride ($E^+ = Me_3Si^+$) [Eq. (1)].

$$\rangle CH + R_2NLi \xrightarrow{-R_2NH} \rangle CLi \xrightarrow{E^+} \rangle CE \qquad (1)$$

Alternatively, the newly generated organolithium can be first added to an unsaturated organic electrophile (notably, to a carbonyl compound) prior to "quenching" [Eq. (2)].

$$\geq CLi + R_2C{=}O \longrightarrow \overset{|}{\underset{R_2C-OLi}{\geq C-}} \xrightarrow{E^+} \overset{|}{\underset{R_2C-OE}{\geq C-}} \qquad (2)$$

These R_2NLi bases, especially those with bulky R groups, have the advantage over C—Li reagents (Bu^nLi, MeLi, etc.) of having lower nucleophilic character. Proton abstraction is favored over addition to unsaturated functional groups in the organic precursor. Furthermore, such proton abstractions are often not only regiospecific but are also enantiospecific. Hence, there is considerable interest in the use of these bases [including chiral lithium amides (*9*)] in asymmetric syntheses (*10*).

Several R_2NLi bases are now available commercially, e.g., bis(trimethylsilyl)amidolithium, *N*-lithiohexamethyldisilazide [$(Me_3Si)_2NLi$], diisopropylamidolithium [Pr^i_2NLi (LDA)], and (2,2,6,6-tetramethylpiperidinato)lithium, [$Me_2\overline{C(CH_2)_3CMe_2N}Li$], in the form of solutions of specified molarity in polar solvents such as Et_2O or tetrahydrofuran (THF). These reagents actually are quite specific complexes, $(R_2NLi{\cdot}xL)_n$, with x defining the degree of solvation and with n defining the degree of association. The preparations and use of reagents of this type have been described in a short article (*11*) and in two recent texts (*12, 13*). These also include details of solvent purification, reagent solubilities, and safety precautions. Procedures for the syntheses of specific lithium reagents also can be found in compilations such as King and Eisch (*14*), whereas Shriver and Drezdzon (*15*) cover all the practical aspects of manipulating oxygen- and moisture-sensitive materials (i.e., all organic lithium compounds).

Lithium amides and lithium imides (iminolithiums) ($R_2C{=}NLi$) have also been used to prepare amides and imides of other metals and metalloid elements, M (*16*), e.g., by reaction with precursors having M—X bonds [e.g., X = Hal, OR, or H; Eq. (3)].

$$M{-}X + \begin{matrix} R_2NLi \\ R_2C{=}NLi \end{matrix} \longrightarrow \begin{matrix} M{-}NR_2 \\ M{-}N{=}CR_2 \end{matrix} + LiX \qquad (3)$$

Despite their extensive synthetic use, it is only quite recently that the NLi species have been isolated and examined. The first single-crystal structure of a lithium amide was the $[(Me_3Si)_2NLi]_3$ trimer. This structure was fully described in 1978 (*17*) [although a powder diffraction study in 1969 also indicated a trimer to be present (*18*)]. Some synthetic advantages of isolating crystalline lithium amides prior to use were first described in 1983 (*19*). Isolation can ensure reagent purity and allows nonpolar solvents to be used as the reaction medium. For example, $[(Me_3Si)_2NLi\cdot OEt_2]_n$, a dimer ($n = 2$) in the solid state, can be prepared by lithiation of $(Me_3Si)_2NH$ in Et_2O. The crystalline solid isolated from such a medium can then be weighed and dissolved in a hydrocarbon solvent for reaction with an organic substrate. As has been pointed out more recently (*20, 21*), prior isolation of not only lithium amides but also of *all* organic lithium reagents generated in solution allows their proper *identification*. Such species are generally solvated, e.g., *not* $(R_2NLi)_n$, but $(R_2NLi\cdot xL)_n$, where L is the polar solvent employed during the lithiation of the amine. Isolation often also allows structural characterization of the reagent, in the solid state by X-ray diffraction and in nonpolar solvents (e.g., benzene or toluene) by variable-concentration colligative measurements (cryoscopy, vapor-pressure barometry, etc.) as well as by multinuclear (chiefly, ^6Li, ^7Li, ^{13}C, and ^{15}N) nuclear magnetic resonance (NMR) spectroscopy. The extent of ligand complexation (the value of x) and the concentration- and temperature-dependent degrees of association (the values of n) in solutions of the reagent can thus be ascertained. These are essential data for the full understanding of the reactions of these reagents.

Although synthetic uses have dominated the interest in NLi compounds, our primary concern in this review is the *structures*. The compounds considered usually originate *via* lithiations of organic precursors with N—H bonds (chiefly imines and amines) or *via* additions of organolithiums to CN double or triple bonds, e.g., imines and nitriles. Two other classes of species, with so-called dative NLi linkages, are not treated here and are only mentioned for completeness. First, there are many examples of lithiated organic compounds with central bonds from lithium to anions such as R^- (alkyl, aryl), $RC{\equiv}C^-$ (alkynyl), and RO^- (alkoxy, enolato), whose lithium centers also interact with polar N compounds such as bidentate TMEDA, TMPDA, TMHDA [$Me_2N(CH_2)_nNMe_2$, with $n = 2$, 3, and 6, respectively], and tridentate PMDETA [$Me_2N(CH_2CH_2NMe_2)_2$] ligands. In general, these result in linked cubanelike tetramer, ring dimer, or monomer structures. Schematic representations of these structures, with selected examples, are given in Fig. 1 (*22–27*). Many other examples can be found in recent reviews of the structures of organolithium compounds (*2, 4, 5*). Second, there are a growing number of examples of organic lithium compounds whose anions (R^-, $RC{\equiv}C^-$, RO^-) also contain one or more NMe_2 groups, i.e., in effect, *internal* coordination sites. These also often give rise to tetramers (these are truly isolated, because all Me_2N—Li coordinations take place intramolecularly) and to ring dimers, as illustrated in Fig. 2 (*28–33*).

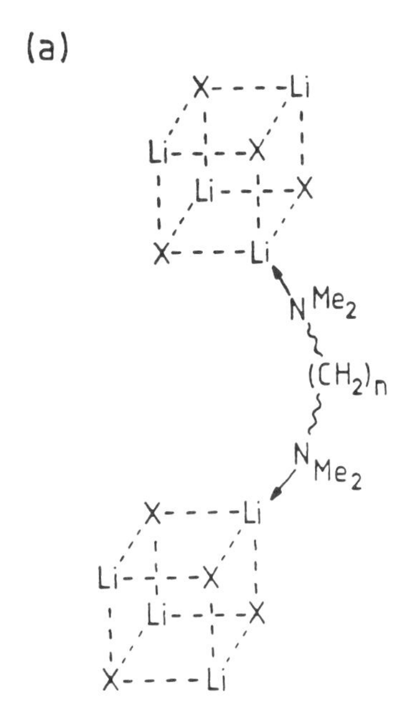

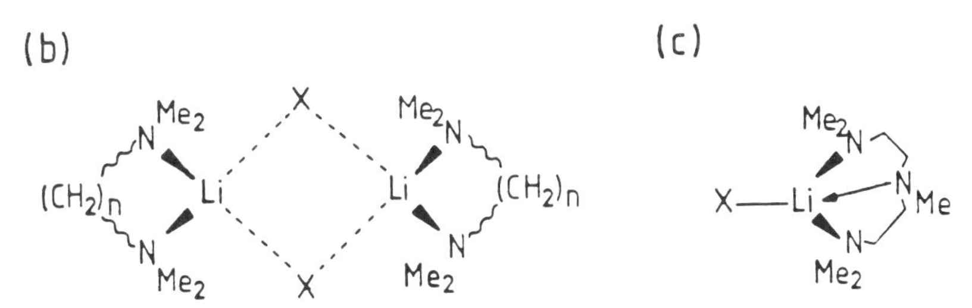

FIG. 1. Structural types found for organolithiums containing neutral N donors: (a) linked tetrameric cubanes, e.g., X = alkyl, $n = 2$ [$(MeLi)_4 \cdot 2TMEDA$ (*22*)]; X = alkynyl, $n = 6$ [$(PhC{\equiv}CLi)_4 \cdot 2TMHDA$ (*23*)]; (b) ring dimers, e.g., X = aryl, $n = 2$ [$(PhLi \cdot TMEDA)_2$ (*24*)]; X = alkynyl, $n = 3$ [$(Ph{\equiv}CLi \cdot TMPDA)_2$ (*25*)]; X = enolato, $n = 2$ {$[Bu^tOC({=}CMe_2)OLi \cdot TMEDA]_2$ (*26*)}; (c) monomers, e.g., X = aryl [$PhLi \cdot PMDETA$ (*27*)].

Other examples of intramolecularly coordinated (by O as well as by N groups) organolithium compounds can be found in Setzer and Schleyer (*2*) and Seebach (*4*). Two recent reviews are also pertinent. Klumpp (*34*) deals with O- and N-assisted lithiation and carbolithiation of non-aromatic compounds, and Snieckus (*34a*) deals with directed (by amide and carbamate groups) ortho metalation in polysubstituted aromatics.

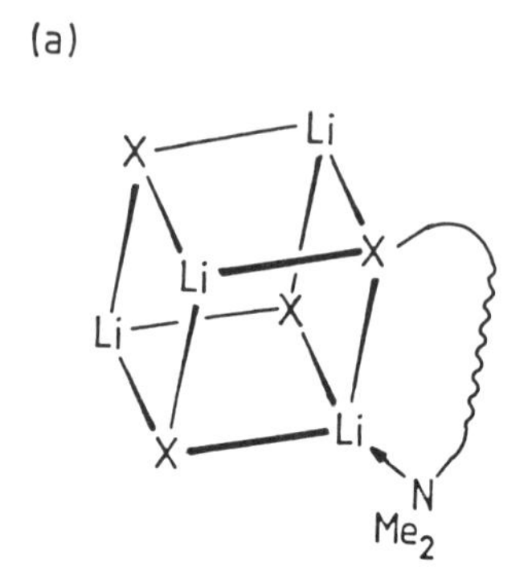

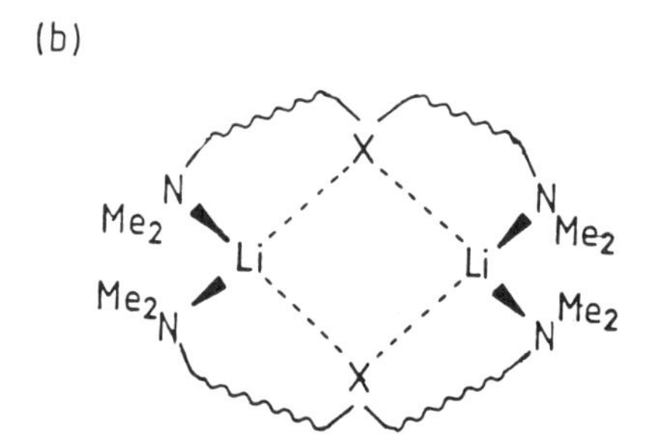

FIG. 2. Structural types found for organolithiums containing internal—NMe_2 donor centers: (a) discrete tetrameric cubanes, e.g., an alkyllithium [$(Me_2NCH_2CH_2CH_2Li)_4$ (*28, 29*)], an aryllithium {[$(2\text{-}Me_2NCH_2)C_6H_4Li]_4$ (*30*)}, and an enolatolithium {[$(2\text{-}Me_2NCH_2)C_6H_4{\cdot}C({=}CH_2)OLi]_4$ (*31*)}; (b) ring dimers, e.g., an alkyllithium {[$(Me_2NCH_2)_2CH{\cdot}CH(Me)Li]_2$ (*32*)} and an aryllithium {[$(2,3,5,6\text{-}Me_2NCH_2)_4C_6HLi]_2$ (*33*)}.

Selected NMR studies of aryllithium compounds with *ortho*-NMe_2 and -CH_2NMe_2 groups have also been described (*35*).

We will also include the recently determined structures of imides and amides of other alkali metals (*viz.* Na, K, Rb, and Cs). Only a few structures of derivatives of the heavier alkali metals are known (*3*), but this area can be expected to develop rapidly.

B. STRUCTURE AND BONDING

Many textbooks still state that organic lithium compounds have appreciable covalent character. This misconception arises from physical properties such as relatively low melting points and solubility in hydrocarbons or other nonpolar solvents. It is true that these properties are usually associated with covalent, discrete molecular species rather than with typical ionic, three-dimensional latticed "salts" such as $(LiCl)_\infty$. However, such comparisons are misleading. Most of the physical properties of organolithium compounds (the marked exception being conductance) arise due to the overall size and shape of the units making up these materials, and the nature of the peripheries of these units. The lithium bonding within the clusters and rings does not influence these properties. It is now generally agreed that C—Li bonds are essentially ionic (*2, 36, 37*). On electronegativity grounds alone, N—Li bonds must also be similar. For example, natural population analysis of H_3CLi and H_2NLi indicates the C—Li bond to have 89%, and the N—Li bond 90%, ionic character (*38*). Results of MO calculations on N—Li species, presented later in this review, agree in revealing large positive charges on lithium. However, these electrostatic predilections do not necessarily result in polymeric three-dimensional latticed structures. These characterize many common inorganic lithium salts, e.g., $(LiCl)_\infty$, because the anions have spherically symmetrical negative charge distributions and only are constrained by radius ratio criteria (i.e., cation–anion attractions are optimized vis-à-vis cation–cation and anion–anion repulsions). However, there is a crucial difference for organolithium salts. The anions (with C^-, N^-, O^-, etc. centers) are generally large groups whose bulk limits the directionality of the electrostatic interactions. These limitations are increased if the organic lithium compounds are complexed by polar solvents or ligands. These add yet further steric constraints and often decrease the degree of aggregation from that of the pure organolithium.

For the above reasons, organolithium compounds and complexes have been termed "supramolecules" ["complex molecules held together by noncovalent bonds" (*4*)]. What we have here are ionically bonded (at least as far as the central metal–organic anion linkages are concerned) yet often discrete molecular species. Most of their physical properties reflect their limited aggregation and their organic peripheries. These points are stressed in Fig. 3, a schematic presentation of the major structural types of uncomplexed organic lithium compounds.

Several previous reviews have also considered some of these structural generalities (*2, 3, 39–41*). The structural options include polymers. Lithiation of organics (e.g., an "acidic" hydrocarbon, an imine, and a methyl ketone) in nonpolar media (benzene, toluene, and hexanes) often leads to precipitation of amorphous, insoluble products. These can be thought of as arising from continuous vertical association of $(LiX)_n$ rings ("stacking," Fig. 3a) or from continuous lateral association ("laddering" or "fencing," Fig. 3b; $n = 2$ is shown for reasons

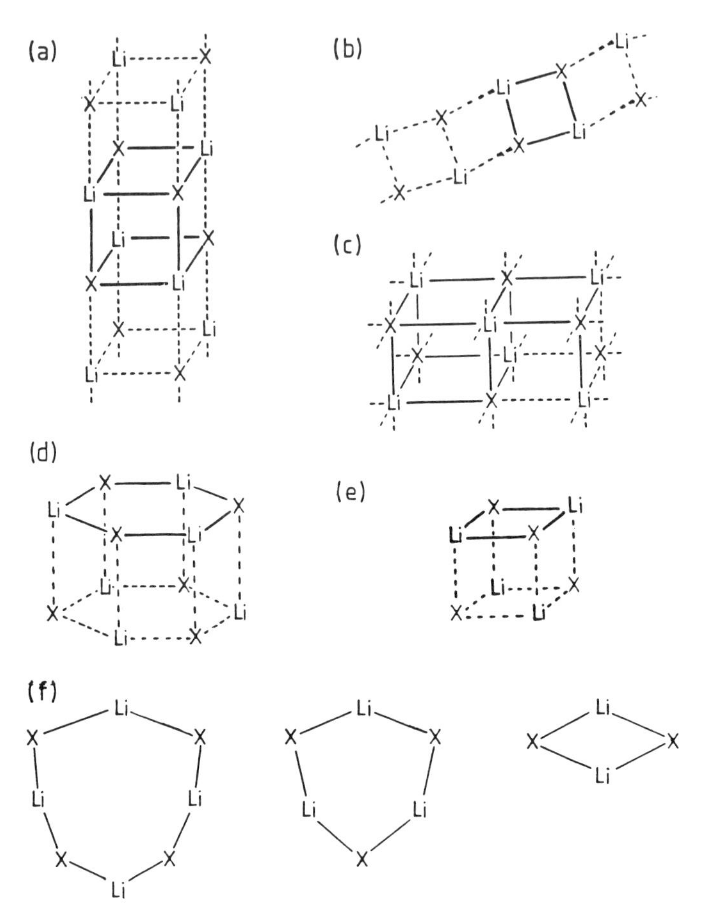

FIG. 3. Structural options for uncomplexed organolithiums: (a) infinite stack of $(LiX)_2$ rings, (b) infinite ladder of $(LiX)_2$ rings, (d) hexamer, a limited stack of two $(LiX)_3$ rings, (e) tetramer, a limited stack of two $(LiX)_2$ rings, and (f) isolated $(LiX)_4$, $(LiX)_3$, and $(LiX)_2$ rings. For contrast, (c) illustrates a segment of an infinite three-dimensional lattice, X being an inorganic "atomic" anion such as Cl^-.

explained later). The reasons for the preference for stacked or laddered structures depend upon the precise stereochemical requirements and orientations of the organic groups (R) in the anion. This will be discussed later, but the crucial point here is the restricted three-dimensional nature of these polymers. In contrast, a typical inorganic lithium salt having a point-charge "atomic" anion (e.g., Cl^-, Fig. 3c) forms a three-dimensional lattice that can be viewed as a stereochemically allowed combination of $(LiX)_2$ ring stacking and laddering. Stereochemical features can also rationalize the other structural types represented in Fig. 3. Particularly large, sterically demanding R groups within X^- can prevent continuous association, and so result in oligomeric (usually crystalline and organically soluble) materials. For example, stacking can then be limited to just two rings (trimeric ones giving a hexamer, Fig. 3d; dimeric ones giving a tetrahedral tetramer, Fig. 3e). In the limit, stacking and laddering can be prevented altogether, so giving the isolated rings themselves [$(LiX)_n$, with $n = 2, 3$, or 4, Fig. 3f]. The effects of complexation have been illustrated in Figs. 1 and 2. Whether complexation is achieved by addition of neutral Lewis bases or by the presence of base functions within the organic anion, in general it can be said that extensive stacking or laddering is then prevented. This is due largely to the increased steric constraints within the system. The result is that the complexed structures found have limited aggregation. Two-ring stacks of usually dimeric rings, and not of trimeric ones (Figs. 1a and 2a), dimeric rings themselves (significantly, not usually trimeric or tetrameric ones; Figs. 1b and 2b), and, in the limit, monomers (Fig. 1c) are found.

The still quite widespread representations of organolithium reagents as isolated carbanions (R^-) or as organoanions (RO^-, R_2N^-, etc.) are inappropriate if one wishes to understand the nature of these reagents. "Carbanions" must be monomers, but monomeric organolithiums are quite rare. Even for the few known monomers, it is still quite unusual to find totally ion-separated species, even in solution (solvent-separated ion pairs, SSIP). Linked to these points, a general feeling is that many of these structures are in fact in part dictated by the electronic requirements of the metal. However, such dictates are tempered by the steric demands of the organic ligands, both the anionic ones and any Lewis bases present. If this is so, one consequence is that many organic reactions commonly represented as proceeding *via* carbanionic/nucleophilic attack of R^- (of the RLi reagent) on an electrophile or electrophilic center (e.g., the carbon center of a carbonyl

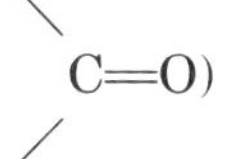

may in fact proceed in the reverse manner, i.e., nucleophilic attack (better termed coordination) of the organic on the electrophilic (Li^+) center of the organolithium. Detailed discussion of this area is beyond the scope of this review, but note that *ab initio* molecular orbital (MO) calculations on the mechanisms and energetics of additions of MeLi and of $(MeLi)_2$ to formaldehyde show that end-on $H_2C{=}O{-}(LiMe)_{1,2}$ complexes form first, prior to $H_2C{-}CH_3$ bond formation *via* a cyclic transition state (*42*). We stress a major point. None of the structures discussed in this review make much sense if the lithium is excluded. The reactions of these species depend intimately on the metal cation.

Two other bonding features of organolithiums in general require a preliminary note. In addition to their central bonds (Li—C, Li—N, Li—O, etc.), numerous such species appear to have additional and stabilizing Li···H—C interactions. These usually occur when the organic groups restrict association, so leaving lithium centers with low coordination numbers (in general, two or three; e.g., structural types illustrated in Fig. 3d, e, and f). In this sense, they can be considered as compensatory interactions. They were first manifested in short Li···H—C distances found in several crystal structures, some of which are represented in Fig. 4. Indeed, such distances were remarked upon (*43*) regarding the structure of $(MeLi)_4$ [(*44*); Fig. 4a] long before the importance of transition metal and hydrocarbon "agostic" interactions was recognized (*45*).

In methyllithium, the individual tetramers associate further through these intermolecular Li—H—C contacts. Short Li to hydrocarbon group distances are also apparent in the linked structures of several lithium'ates, e.g., $(LiBMe_4)_\infty$ [(*46*); Fig. 4b], and, intramolecularly, in the hexamers $(cyclohexylLi)_6$[(*47*); Fig. 4c] and $(Me_3SiCH_2Li)_6$ (*48*). Full structural details are given in Barr *et al.* (*49*), concerning semiempirical MO calculations on these and related species. Such interactions have also been pinpointed by minimum neglect of differential overlap (MNDO) calculations and by ^6Li–^1H two-dimensional HOESY NMR experiments (*50–51a*). For example, these methods revealed a close Li···$H(C_8)$ contact in a complexed mixture of MeLi and 1-lithionaphthalene (but not in the latter alone), and indeed second lithiation does occur specifically at this 8-(peri) position (*51*). Interactions such as these can also be important in N—Li structures, as will be discussed later. The second bonding feature concerns the importance or otherwise of Li···Li interactions. In many organolithium crystal structures, relatively short Li—Li distances are observed, usually much shorter than the distance found in lithium metal (3.04 Å) and frequently shorter than that found in the Li_2 molecule (2.74 Å). For example, in $(MeLi)_4$ this distance is 2.68 Å (*44*), whereas in $(LiX)_2$ dimers in general ($X = R^-$, $RC{\equiv}C^-$, R_2N^-, RO^-, etc.), which characteristically have rather acute angles (~75°) at the bridging X groups, the Li—Li cross-ring distance can be as small as 2.30 Å (*2*). The key question is, do such short contacts reflect metal–metal bonding? Most MO calculations indicate not; for example, recent optimizations and topological analyses of 23 RLi structures found no Li—Li bond paths (*52*). NMR experiments can give only somewhat inconclusive results. Thus, using the COSY

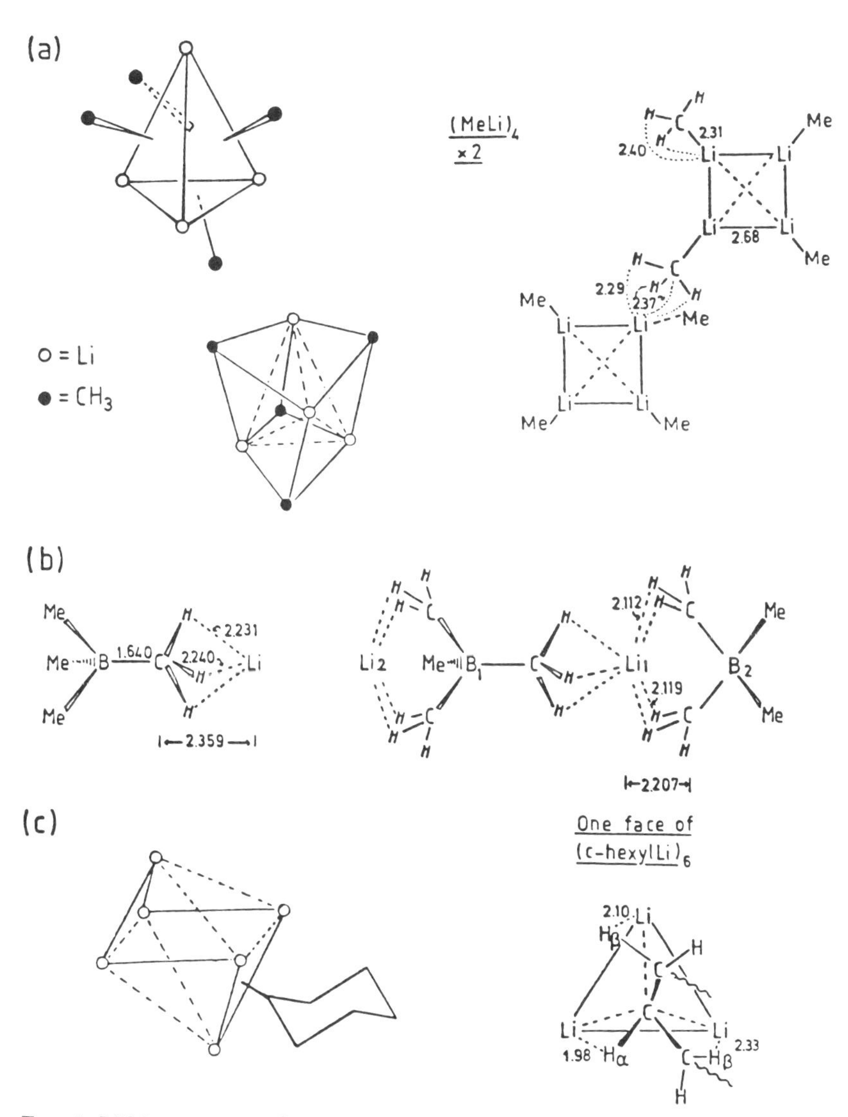

FIG. 4. Lithium compounds exhibiting short Li···H—C contacts in the solid state: (a) representations of $(MeLi)_4$, and of the intercubane contacts found, (b) Li—H_3C and Li—H_2C contacts found in $(LiBMe_4)_\infty$ and (c) part of the structure of $(cyclohexyllithium)_6$, and Li—αHC, —βHC contacts found over its Li_3 faces.

technique, the first homonuclear ^{6}Li,^{6}Li coupling (~0.1 Hz) was recently detected in tetrahydrofuran (THF) solutions of 3,4-dilithio-2,5-dimethyl-2,4-hexadiene (a tetramer in the solid, with four inner and four outer Li atoms). However, it is not known if such coupling is direct or if it occurs *via* the lithiated carbon atoms (*50, 53*). For several reasons, it is likely that short Li—Li contacts in organolithiums have little or no bonding significance. First, there is the problem that it is not possible to find a proper reference distance for a "single" lithium–lithium bond within such species. Cross-references over very disparate bonding situations, ionic (organolithiums), to metallic (Li metal), to covalent (the Li_2 molecule), are hardly valid. Second, in organolithiums, which are essentially ionically bonded species, the Li centers bear quite substantial $\delta+$ charges. By inference, the essential interaction between such centers is a repulsive/antibonding one. Of course, Li^+ centers are found in lithium metal, but there they have ($1e^-$) electron density between them, so mediating against repulsions. In organolithiums this $1e^-$ per metal atom is largely transferred to, and used in bonding to, the organic groups. A final point is that, even if a fair proportion of the $1e^-$ available to each Li atom was concentrated between the metal centers, the thermodynamic value of a resulting fractional Li—Li bond would be small in comparison with the value of the $Li^{\delta+}$–organic anion$^{\delta-}$ bonding; witness the weak single metal–metal bonds found in lithium itself.

The above discussion of the types of organic lithium compounds, their synthetic uses, and the structural and bonding features they possess sets the scene for the bulk of the review. Here we turn our attention specifically to the structures of NLi compounds.

II. Lithium Imides (Iminolithiums) and Their Complexes; Ring Stacking

A. Uncomplexed Lithium Imides $(RR'C{=}NLi)_n$: Solid-State Structures

Lithium imides [$(RR'C{=}NLi)_n$] have until recently been used without prior isolation. Such use has been as reagents in the syntheses of iminoderivatives of many other metals and metalloids (M) *via* reactions with M—X bonds [X = halide, in particular; see Eq. (3)]; e.g., of main group metals and metalloids Be (*54*), B (*55, 56*), Al (*57, 58*), Si (*59*), and P (*60*), and of transition metals Mo, W (*61*), and Fe (*62*). Addition of organolithiums to nitriles [Eq. (4)] and the lithiation of imines, commonly by Bu^nLi [Eq. (5)], are the most common preparative routes to lithium imides.

$$RLi + R'C{\equiv}N \longrightarrow 1/n(RR'C{=}NLi)_n \qquad (4)$$

$$Bu^nLi + RR'C{=}NH \longrightarrow 1/n(RR'C{=}NLi)_n + Bu^nH \qquad (5)$$

The lithium imides, produced, e.g., by PhLi additions to PhC≡N, *p*-$ClC_6H_4C{\equiv}N$ (*63*), and $Me_2NC{\equiv}N$ (*64*), have been reacted with organosilylchlorides [R''_3SiCl, to give $Ph(R)C{=}NSiR''_3$ derivatives] or have been hydrolyzed [to give, e.g., the imine $Ph(Me_2N)C{=}NH$]. The imine precursors for Eq. (5) have generally been prepared by addition of Grignard reagents to nitriles, followed by hydrolysis [Eq. (6)].

$$RMgHal + R'C{\equiv}N \longrightarrow RR'C{=}NMgHal \xrightarrow{H_2O} RR'C{=}NH \qquad (6)$$

Examples include imines with R = R′ = Bu^t, Ph, and *p*-MeC_6H_4 (*65*). For an imine in which R ≠ R′, e.g., $Ph(Bu^t)C{=}NH$, successful syntheses have employed either PhC≡N + Bu^tMgBr (*65*) or $Bu^tC{\equiv}N$ + PhMgBr (*61*) additions, for example.

The first isolation and characterizations (largely by elemental analyses and infrared (IR) spectroscopy) of lithium imides were described in 1968 (*66*). $(Ph_2C{=}NLi)_n$, a yellow amorphous solid, was prepared from reactions in ether of PhLi + PhC≡N or of MeLi + $Ph_2C{=}NH$. $[Ph(Me)C{=}NLi)]_n$, also a yellow powder was obtained by reacting MeLi and PhC≡N in ether. $[(Me_2N)_2C{=}NLi]_n$, a crystalline hydrocarbon-soluble material, was produced by MeLi lithiation of the imine $(Me_2N)_2C{=}NH$. The syntheses of crystalline $(Bu^t_2C{=}NLi)_n$ (*55, 67*) from Bu^tLi + $Bu^tC{\equiv}N$ in hexane, and of (incompletely characterized) $[Ph(Me_2N)C{=}NLi]_n$ and $\{Ph[(Me_3Si)_2N]C{=}NLi\}_n$ from additions to PhC≡N of the lithium amides Me_2NLi and $(Me_3Si)_2NLi$, respectively (*68*), were described shortly afterward.

In the last decade, the syntheses, characterization, and solid-state structures of four crystalline iminolithiums have been published. All of them are hexamers: $(Bu^t_2C{=}NLi)_6$ (**1**) (*69–72*), $[(Me_2N)_2C{=}NLi]_6$ (**2**) (*70–72*), $[Bu^t(Ph)C{=}NLi]_6$ (**3**) (*71, 72*), and $[Me_2N(Ph)C{=}NLi]_6$ (**4**) (*71, 72*). Both of the methods shown in Eqs. (4) and (5) gave over 80% yields (*72*). The structures of (**1**)–(**4**) are all extremely similar. One typical view, emphasising the six lithium atoms, is shown in Fig. 5 for hexamer (**4**). The metal core can be viewed as a folded chair-shaped Li_6 ring, or as a pseudooctahedron with the apical lithium atoms extended slightly. The dihedral angles between the chair "seat" and "back" in the Li_6 cores of (**2**)–(**4**) average 80 ± 2° [(**1**) is excluded, because its structure was rather imprecisely determined (*69*)]. This contrasts with the 130.7° value for the chair form of cyclohexane and 54.7° for a pure octahedron. In the related hexameric structures of $(cyclohexylLi)_6$ (*47*) and of $(trimethylsilylLi)_6$ [$(Me_3SiLi)_6$] (*73*), these mean angles are 72.9° and

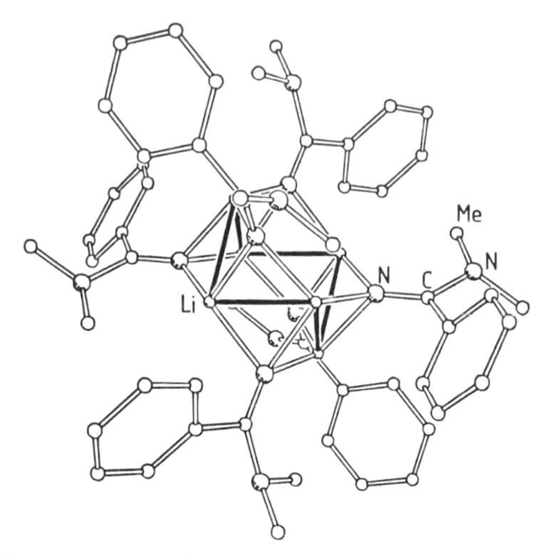

FIG. 5. Molecular structure of the iminolithium hexamer $[Ph(Me_2N)C{=}NLi]_6$ (**4**), highlighting the Li_6 core.

70.5°, respectively, so that the $(iminoLi)_6$ chairs are rather more upright. Each of the six smaller Li_3 triangular faces is bridged by a μ_3-imide ligand; the two larger faces are vacant. Each Li atom then has two short contacts to other Li atoms [mean Li—Li distance 2.48 ± 0.03 Å for (**2**)–(**4**)], two longer Li contacts that form part of each open face (3.21 ± 0.05 Å), and one cross-ring distance (4.05 ± 0.05 Å). As noted in Section I,B, even the shortest contacts do not imply Li—Li bonding.

These hexameric $(RR'C{=}NLi)_6$ molecules are highly ionic. The best quantitative evidence for this comes from *ab initio* MO calculations (6-31G level) on $H_2C{=}NLi$, $(H_2C{=}NLi)_2$, and $(H_2C{=}NLi)_3$ *(72, 74)*. [Calculational results are discussed in full later (see Section II,D)]. A Mulliken population analysis (which tends to underestimate the ionicity) for all three species gave charges on Li of +0.62 to +0.65, and on N of −0.74 to −0.81*e*. When the dimer calculation was repeated with only Li^+ centers present (i.e., the 2s and 2p orbitals were omitted in the lithium basis set) the optimized geometry was very similar (angles within 1°, bond lengths within 0.02 Å) to that using the full basis set. The calculated dimerization energies also were quite close, 73.3 kcal mol^{-1} for the partial basis set and 66.0 kcal mol^{-1} for the full basis set. Furthermore, the lengths of the C=N bonds in the RR′C=N ligands of (**2**)–(**4**) (averaging 1.244, 1.255, and 1.261 Å, respectively) are like those of double bonds in imines. The ν(C=N) stretching frequencies for these compounds also are similar to those of the precursor imines [e.g.,

1615 cm^{-1} for (**4**) and 1590 cm^{-1} for $Me_2N(Ph)C{=}NH$]. Clearly, the imide ligands do not use the π electron density within their C=N linkages. But the largely electrostatic interactions between the Li^+ cations and the imide ligands, $RR'C{=}N^-$, do not explain why hexamers of a certain type form. This problem extends beyond iminolithiums; such hexamers are found for the uncomplexed $(c\text{-}C_6H_{11}Li)_6$ (*47*) and $(Me_3SiLi)_6$ (*73*), as well as for (tetramethylcyclopropylmethylLi)$_6$ $[(Me_2\overline{CMe_2C}CHCH_2Li)_6]$ (*75*) and several lithium enolates (see Sections II,E and I,B and Fig. 3e).

Reasons for the formation of hexameric organolithium "clusters" have emerged from an analysis of the geometries of the $(iminoLi)_6$ species (**2**)–(**4**) (*71, 72*). When these hexamers [e.g., (**4**), Fig. 5] are viewed through unbridged Li_3 faces, the perspective [as shown in Fig. 6, also for (**4**)] is quite striking; the hexamer appears to be constructed from two vertically associated ("stacked") puckered trimeric rings $[(RR'C{=}NLi)_3]$. The metal atoms of one ring nearly eclipse the nitrogen atoms of the other. This has some physical meaning. Figure 7a shows the orientation of an imino ligand over one Li_3 face of (**4**) (the view direction being approximately along the C=N bond). The basal Li atoms in this Li_3 face belong to the same trimeric ring, and the apical Li atom is in the other ring. This orientation is typical for all the Li_3 faces of (**4**), and indeed for all such faces of hexamers (**2**) and (**3**). This

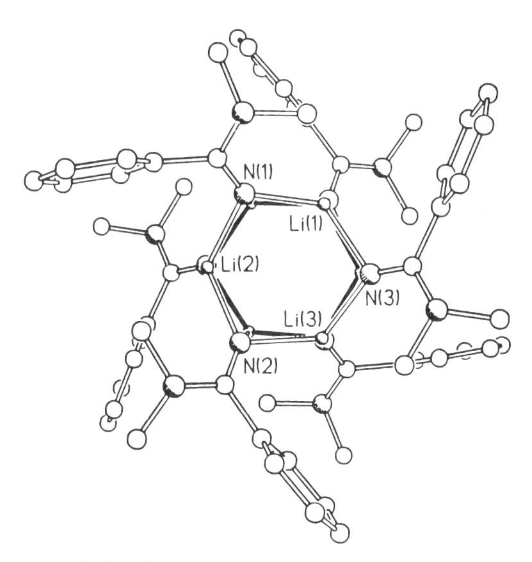

FIG. 6. View of $[Ph(Me_2N)C{=}NLi]_6$ through the unbridged Li_3 faces.

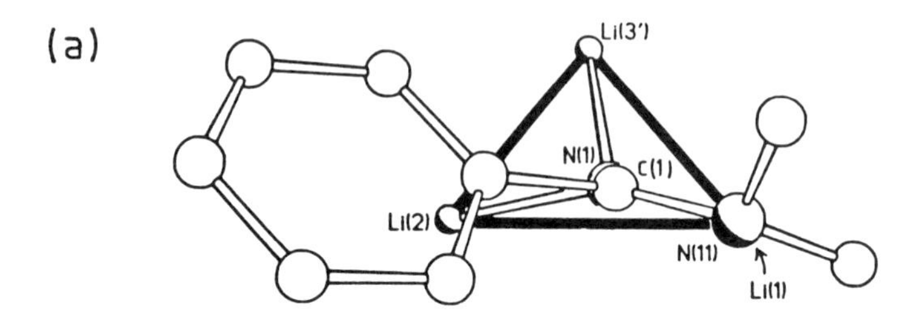

* view direction along C(1)=N(1) bond is inclined at 12.8° to the normal of this Li_3 plane

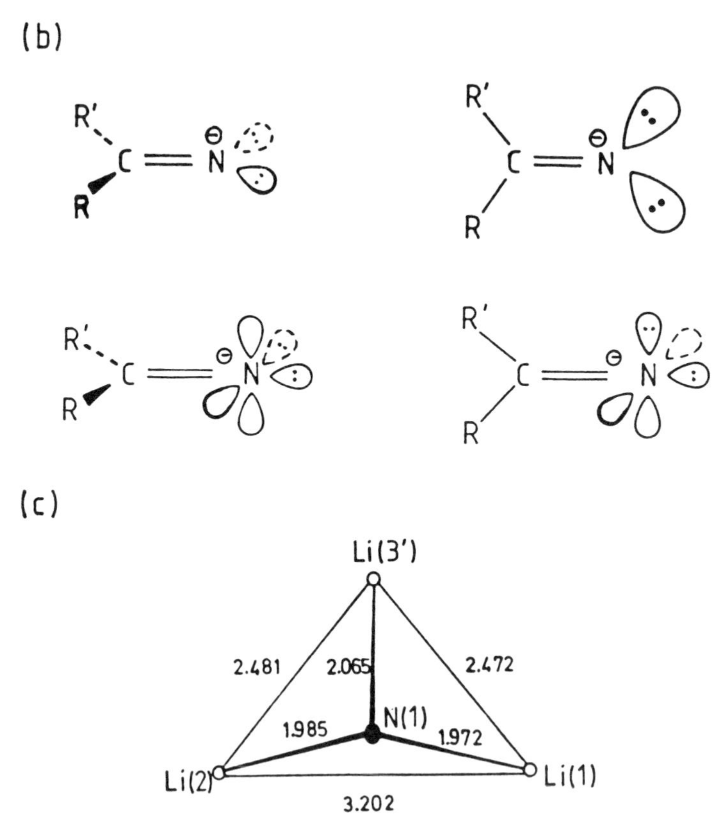

FIG. 7. (a) Orientation of an imino ligand over one Li_3 face of $[Ph(Me_2N)C{=}NLi]_6$ (**4**); (b) representations depicting the electron density at the imino nitrogens; though there is cylindrical character in this distribution, the in-plane trigonal density is highest; (c) N—Li bond lengths in the Li_3 face shown in (a).

amounts to 24 faces in all, because (**3**) has two independent molecules in its unit cell. Why does the imino ligand tilt with regard to its skeletal plane [e.g., as defined in Fig. 7a, α-C of phenyl·C(1)=N(1)·N(11) of NMe_2]? One possible answer comes from consideration of Fig. 7b. Due to the sp character of the imino N and C atoms, the orientation of the ligands signals the electrostatic directionality (i.e., the dispositions of the lone-pair electron densities on the imino N center). The Li atom which lies very close to the ligand skeletal plane [Li(1) in Fig. 7a] has one lobe on N very much directed toward it. The electrostatic attraction should be strong, and this N-Li link should be relatively short. The other lobe on N bisects the remaining Li centers, but the second basal one [Li(2), which is in the same trimeric ring as Li(1); see Fig. 6] is favored over the apical one [Li(3′), which is in the ring below]. Hence N—Li bonds of unequal lengths are expected, longer and longer still than the more direct N(1)—Li(1) bond. These expectations are confirmed for the face of (**4**) being especially considered (Fig. 7a). As shown in Fig. 7c, quite distinct relatively short [N(1)—Li(1)], medium-length [(N(1)—Li(2)], and long [N(1)—Li(3′)] N—Li bonds are indeed found. A similar relationship between the imino ligand orientations and the N—Li bond lengths is found, without exception, for all the (imino)Li_3 faces of hexamers (**2**)–(**4**).

The N—Li bond lengths given in Table I show that each hexamer has three quite distinct sets of distances, one of each set being found in each of its bridged Li_3 faces: a short bond and a medium-length bond (involving basal Li atoms in terms of Li_3 faces; *cf.* Fig. 7a, involving Li atoms within a given trimeric ring; *cf.* Fig. 6), and a longer bond (to an apical Li atom; an intertrimer bond). Figure 8 emphasises these N—Li bond lengths for (**4**); the perspective and atom labeling are the same as in Fig. 6. During a traverse of each trimer, all the RR′C=N ligands are twisted the same way. This gives alternating short and medium-length bonds. On comparing the trimers, it is seen that every short bond in a given trimer lies above or below a medium-length bond in the other trimer. The result is idealized local D_3, rather than D_6, symmetry. This arrangement is obviously adopted so as to minimize R,R′ repulsions both within and between the trimers.

The above detailed analysis shows that iminolithium hexamers can be regarded as being composed of two stacked trimeric rings. Figure 9 helps explain how the constituent rings associate. Figure 9a shows part of an uncomplexed trimeric ring; the key feature is the general flatness of the system. The $(NLi)_3$ ring is fully planar, and the imino C and (at least) the primary (α) atoms of the groups R,R′ lie also in this plane. [Reasons for this preference for pseudoplanar systems, and for trimeric

TABLE I

N–Li DISTANCES IN THE IMINOLITHIUM HEXAMERS (**2**), (**3**), AND (**4**)[a]

Hexamer	Range of N—Li distances Å	Range of N—Li distances in each trimer[b]		Range of N—Li distances between trimers[b]
		Short bonds	Medium-length bonds	Long bonds
(**2**)[$(Me_2N)_2C{=}NLi]_6$	1.97–2.04	1.97–1.99 (1.98)	2.00–2.01 (2.00)	2.01–2.04 (2.02)
(**3**) $[Bu^t(Ph)C{=}NLi]_6$[c]	1.98–2.05	1.98–1.99 (1.99)	2.01–2.02 (2.02)	2.04–2.05 (2.05)
(**4**)$[Me_2N(Ph)C{=}NLi]_6$	1.96–2.08	1.96–1.98 (1.97)	1.99–2.02 (2.01)	2.03–2.08 (2.06)

[a] From Shearer *et al.* (*69*), Clegg *et al.* (*70*), Barr *et al.* (*71*), and Armstrong *et al.* (*72*).
[b] Mean values in parentheses. Compare Figs. 6 and 8, specifically for hexamer (**4**).
[c] Values given are for one of the two independent molecules found in the unit cell of hexamer (**3**).

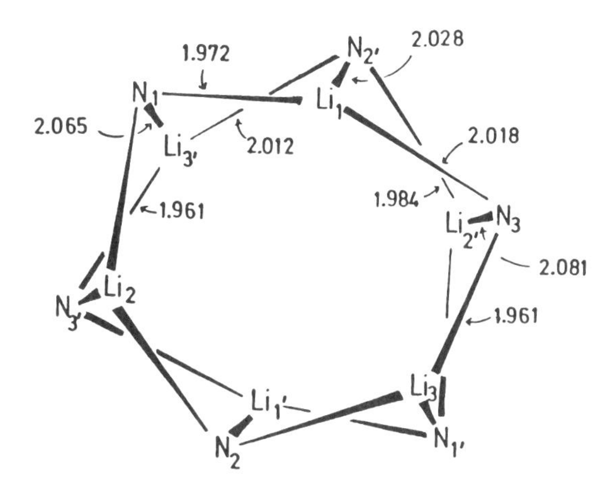

outer N_6 ring N··N 3.178–3.187 Å av. Li$\hat{N}$Li in trimers 107.5°

inner Li_6 ring Li··Li 2.472–2.487 Å av. N$\hat{Li}$N " 124.4°

av. Σ ring angles 695.7°

FIG. 8. N—Li bond lengths in $[Ph(Me_2N)C{=}NLi]_6$ (**4**), viewed as two stacked trimeric rings.

over dimeric rings, will be substantiated in Section II,B, where the results of MO calculations on $(H_2C{=}NLi)_2$ and $(H_2C{=}NLi)_3$ are presented.] The Li centers in such rings are merely two coordinate. The N atom of each imide anion bears a high charge density. Hence, each ring is set up perfectly to associate further (by stacking). When two or more such rings come together, the coordination number of their Li cations will increase (Fig. 9b). The imino ligands will tilt (and the rings pucker) so that the N-centered regions of electron density are directed toward three Li centers, two in the original ring and the other in the second (*cf.* Fig. 7a and b). As noted above [see Table I and, specifically for hexamer (**4**), Fig. 8], the result is alternating short and medium-length bonds within each constituent ring, and longer N—Li distances between rings.

No iminolithium trimers are known experimentally (the R,R′ substituents would need to be very bulky to prevent stacking). However, two amidolithium trimeric rings have been structurally characterized in the solid state. The lengths of their N—Li bonds give some indication of the strength of the bonding within the hexamers: the mean N—Li bond lengths in $[(PhCH_2)_2NLi]_3$ (*20, 76*) and in $[(Me_3Si)_2NLi]_3$ (*17*) are

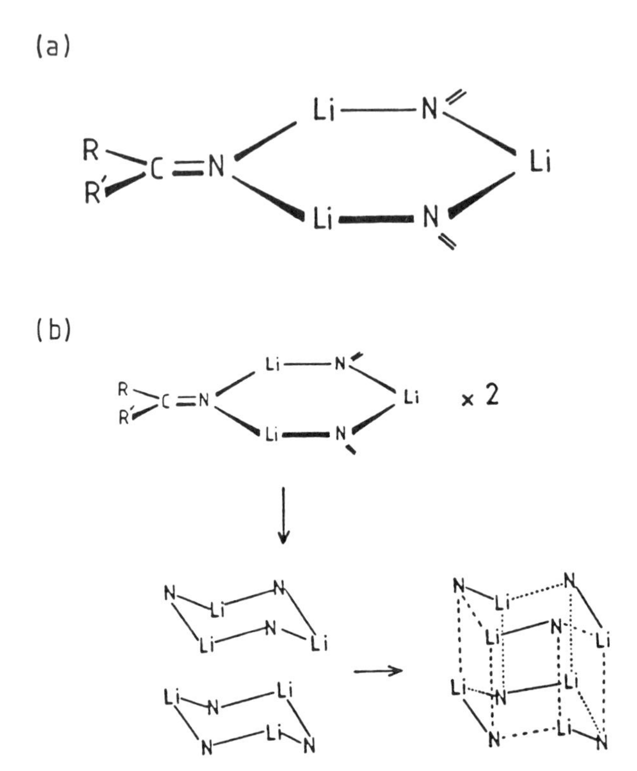

FIG. 9. (a) An isolated, flat $(iminoLi)_3$ trimeric ring (one imide ligand shown); (b) a representation of how two such rings associate, giving a hexamer.

1.95 and 2.01 Å, respectively. Hence the shorter, well-directed N—Li bonds in the constituent trimers of (**2**)–(**4**) (*cf.* Table I) are barely affected by further association: their average length is 1.98 Å. The same is true for the intermediate bonds in each trimer [average length 2.01 Å in (**2**)–(**4**)]. Furthermore, even the bonds between the rings remain quite strong (average, 2.04 Å). This reflects the limited degree of twisting of the RR′C=N ligands [and hence the limited puckering of the $(NLi)_3$ rings] needed to facilitate association. The sum of the angles within the $(NLi)_3$ rings in hexamers (**2**)–(**4**) average 697.5°, 699.0°, and 695.7°, respectively (720° defines total planarity). Nonetheless, even this restricted ligand tilting and ring puckering projects R,R′ groups outward from the central $(NLi)_6$ cluster [note the diagonal view of the structure of (**4**), as shown in Fig. 5]. This is sufficient to preclude the stacking of more than two trimeric rings (at least in the solid state).

This ring-stacking concept, used to explain $(iminolithium)_6$ hexamers, can be applied widely, for example, to higher aggregates in solution (Section II,C) and to the widespread occurrence of clustered species (hexamers and tetramers) and rings for the majority of organolithium compounds (Section II,E). Two further experimental observations can be rationalized. First, none of the crystalline hexamers (**1**)–(**4**) are di*aryl*iminolithiums (R,R′ are both Bu^t, both Me_2N, or combinations of these with a single Ph group); in fact, when R,R′ are *both* aryl groups [e.g., Ph (*63, 65, 66*) and *p*-MeC_6H_4 (*65*)], amorphous, seemingly polymeric materials result. These $(RR'C{=}NLi)_n$ ring systems (trimeric or otherwise) are flat. This allows their extensive or continuous association (as illustrated in Fig. 10). Second, the crystal structure of the first uncomplexed mixed-metal imide has recently become available (*77*). This also is a hexamer, $[Bu^t(Ph)C{=}N]_6Li_4Na_2$ (**5**). However, as shown in Fig. 11, it consists of a *triple*-layered stack of *dimeric* $(N{\cdot}metal)_2$ rings, the two outer rings containing essentially Li and the inner one containing Na. It may be that smaller, homocyclic $(NLi)_2$ and $(NNa)_2$ rings provide a better fit for stacking than would mixed (N_3Li_2Na) cycles. However, the considerable substitutional disorder of Li and Na precludes detailed geometric analysis.

B. Iminolithium Complexes: Solid-State Structures

As the lithium centers in the iminolithium hexamers (**1**)–(**4**) are only three coordinate, they might be expected to add monodentate complexing agents (e.g., Et_2O, THF, or pyridine) and, with bi- and tridentate donors (e.g., TMEDA or PMDETA), to give trimers, dimers, or even monomers. In fact, the hexamers are recovered intact after treatment with these various bases (*78*). This possibly reflects a combination of their seemingly quite strong N—Li bonds (including those between the trimeric rings) and the relative steric unavailability of the Li centers in

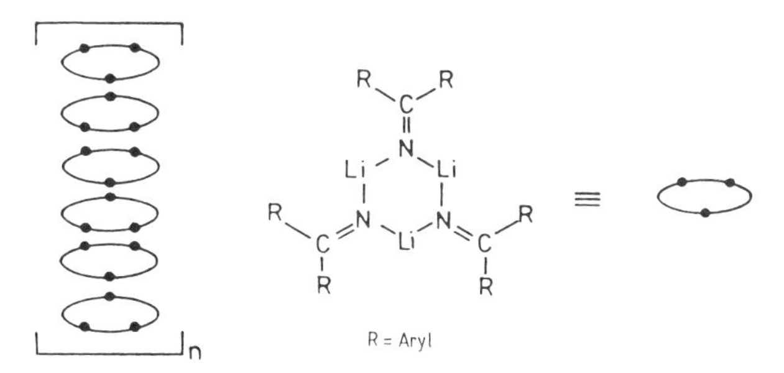

Fig. 10. Association of diaryliminolithium rings into continuous, polymeric stacks.

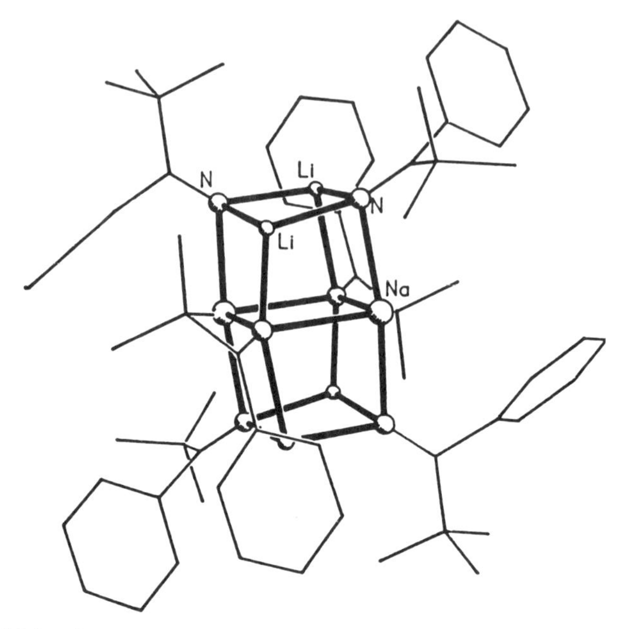

FIG. 11. Molecular structure of the mixed-metal imide $[Bu^t(Ph)C{=}N]_6Li_4Na_2$ (**5**).

these hexamers [e.g., for (**4**), as shown in Fig. 8, the inner Li_6 ring (mean Li—Li distance, 2.48 Å) is surrounded by a larger N_6 ring (mean N—N distance, 3.18 Å)]. Only hexamer (**1**), $(Bu^t_2\,C{=}NLi)_6$, which on steric grounds might be expected to give the most labile stack, has so far given a complex, and then only with the strong donor HMPA. This complex, the solid-state dimer $(Bu^t_2C{=}NLi{\cdot}HMPA)_2$ (**6**), is shown in Fig. 12 (*78, 79*). Its structure raises two questions. Why is there a switch from a cyclic trimer [as found in (**1**) and in all other hexamers] to a dimer on complexation? Why does this dimer not stack to give a tetrameric pseudocubane structure (as found for many kinds of organolithiums complexed by effectively monodentate donors; *cf.* Figs. 1 and 2, and see also Section II,E)?

A qualitative answer to the first question is provided by consideration of Fig. 13. This shows that $(RR'C{=}NLi)_2$ dimers (Fig. 13a) provide much more room (a larger coordination arc) for further complexation than do $(RR'C{=}NLi)_3$ trimers (Fig. 13b). Moreover, the former require minimal reorganization in terms of bond angle modification at their Li centers. The angles given at Li, ~105° for both a bare dimer and a complexed one and ~145° for a bare trimer, come from several sources. *Ab initio* optimized structures of $(H_2C{=}NLi)_2$ and

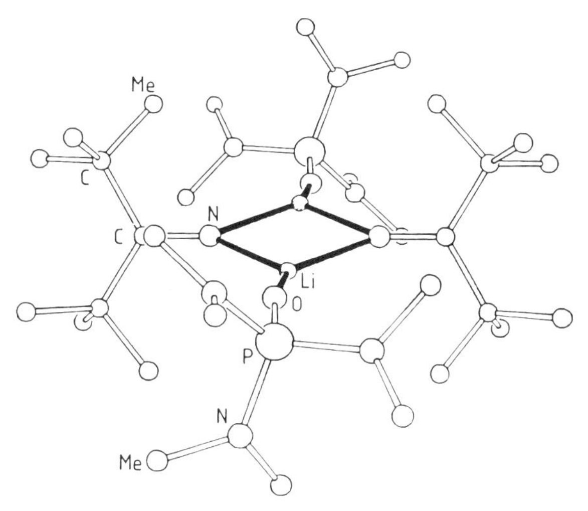

FIG. 12. Molecular structure of the complexed iminolithium ring dimer ($Bu^t{}_2C$ $=NLi \cdot HMPA)_2$ (**6**).

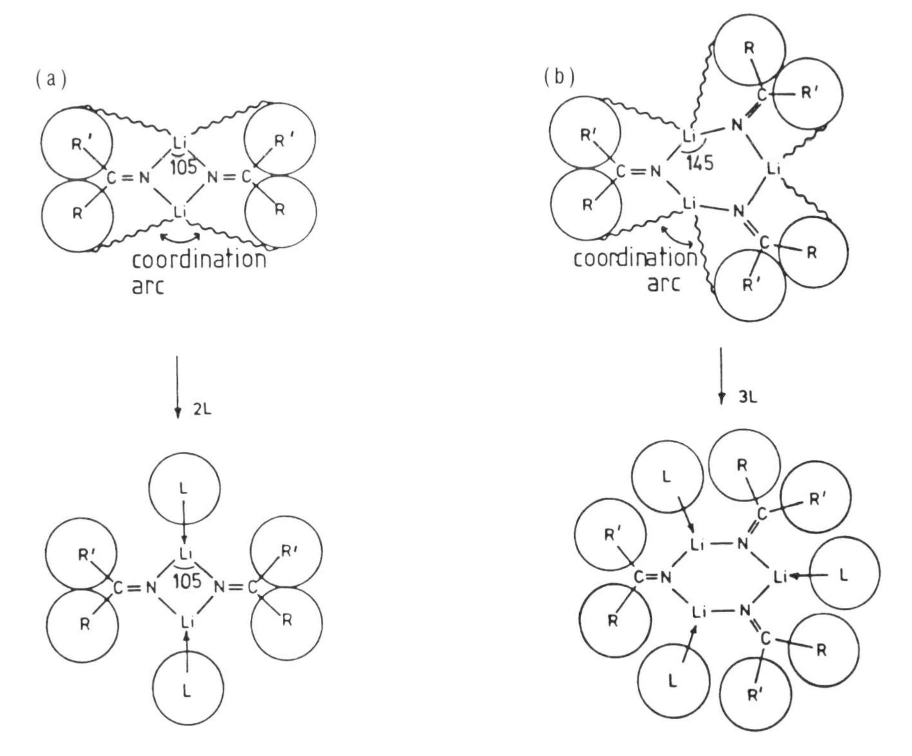

FIG. 13. The steric and geometric consequences of complexing (a) $(iminoLi)_2$ and (b) $(iminoLi)_3$ rings.

$(H_2C{=}NLi)_3$ (*72, 74, 78;* see also Section II,D) have Li angles of 104.4° and 141.2°, respectively. Although uncomplexed iminolithium dimers and trimers are not known experimentally (stacks are favored), a limited number of amidolithiums [$(RR'NLi)_n$, n = 2 and 3] have been reported. Thus [$(Me_3Si)_2NLi]_n$ is a dimer (n = 2) in the gas phase with a ring Li angle of ~100° (*80*), and a trimer (n = 3) in the solid (Li angle 148°) (*17*). In the trimer [$(PhCH_2)_2NLi]_3$, the angle at Li averages 144° (*20, 76*). Numerous solid-state structures of complexed amidolithium ring dimers also are available. A typical example, [$(Me_3Si)_2NLi \cdot OEt_2]_2$ has angles at Li of 105° (*19, 81*). In the complexes [$(PhCH_2)_2NLi \cdot L]_2$, L = Et_2O or HMPA, these angles average 103° (*20, 76*). In (**6**), the ring angle at Li is 104.6°.

The failure of (**6**) to associate further is due to the severe twisting (58.6°) of the $C_2C{=}N$ moiety of the $Bu^t{}_2C{=}N$ ligands with respect to the $(NLi)_2$ ring plane. The resulting projection of the Bu^t groups above and below the ring plane precludes stacking. Such twisting is needed to avoid a steric clash between the Me groups of the HMPA ligands and those of the Bu^t groups. Yet, somewhat perplexingly, this sterically required realignment of the imino ligands away from their usually preferred orientations seems to have little effect on resulting N—Li bond strengths; indeed these N—Li bonds are the shortest (1.92 and 1.95 Å) known in the solid state. This is due to the largely electrostatic nature of the interaction and is discussed further in the computational section (Section II,D).

The structure of the α-cyanobenzyllithium complex [Ph(H)C⋯C≡NLi·TMEDA$]_2$ (*82*) is related to (**6**). This also is a $(NLi)_2$ ring dimer (there are no Li—C bonding contacts). The near-planar anions are at angles of 95.7° and 103.7° to the least-squares plane of the slightly puckered $(NLi)_2$ ring. Hence, as in (**6**), the C substituents (H, Ph) project above and below this ring. The same is true of the chelating TMEDA ligands, so that stacking is doubly prevented. Judging from the bond lengths (C—C, 1.38 Å; C=N, 1.15 Å) within the near-linear anions, $Ph(H)C^-$—C≡N is a better formulation than Ph(H)C=C=N^-. In any event, the N—Li bond lengths (average 2.04 Å) are much longer than in (**6**). A $(NLi)_2$ ring is also found in the structure of the THF complex of 1-cyano-2,2-dimethylcyclopropyllithium. [$Me_2CCH_2C(CN)Li \cdot THF]_\infty$ (*4a, 83*). However, the CN groups are now attached to tetrahedral αC^- (of *c*-propyl) centers, and these centers form short (2.14 Å) contacts to Li^+ cations. In this way, each $(NLi)_2$ ring is also part of an (C—C—N—Li·C—C—N—Li) eight-membered ring. These repeating four- and eight-membered rings constitute the polymer. The crystal structure of a complex *di*lithiated trimethylsilylacetonitrile has also been reported (*4a, 84*). This species crystallizes from ether–hexane solutions as a hexane solvate of $[Me_3Si{\cdot}C(CN)Li_2]_{12}{\cdot}(Et_2O)_6$. Twelve dianions, 24 Li^+ cations, and six Et_2O molecules form, not surprisingly, "an aggregate of unusual size." This contains N—Li, αC—Li, *and* cyano—C—Li bonds.

The general intransigence of the crystalline iminolithium hexamers toward further interaction with Lewis bases is not shown by the amorphous diaryliminolithiums. This may reflect their extensively stacked nature (Section II,A; Fig. 10), which, while raising the lithium coordination number to four in all but the outer rings of the polymer, will presumably weaken individual N—Li bonds. These materials dissolve quite readily in several polar solvents, e.g., THF (*66*), pyridine (*66, 78, 85*), and HMPA (*86*). Crystalline complexes can be recovered from these solutions. Two of these, both derivatives of $(Ph_2C{=}NLi)_n$, have been characterized structurally in the solid state. The tetrameric cubane $(Ph_2C{=}NLi{\cdot}pyridine)_4$ (**7**) is depicted in Fig. 14 (*78, 85*).

For reasons just discussed, complexation has led to the basic ring now being a dimeric $(NLi)_2$ one. However, unlike (**6**), these rings can stack, because both the neutral ligands and the imide substituents are flat. This steric compatibility is illustrated in Fig. 15a, a view of (**7**) looking through the face containing atoms Li 1 and Li 2 [*cf.* the through-face

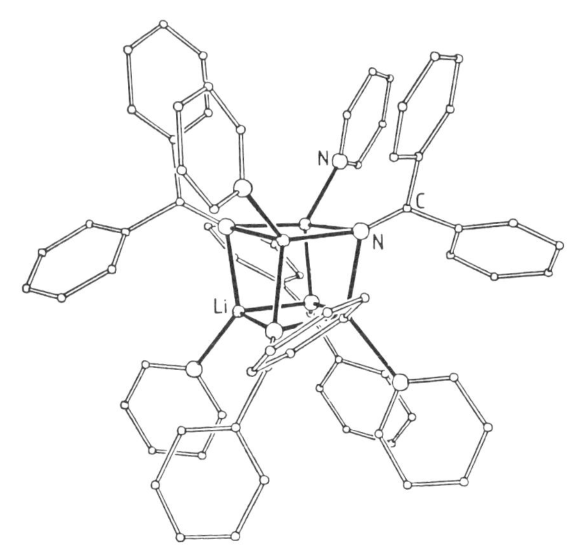

FIG. 14. Molecular structure of the complexed iminolithium tetramer, $(Ph_2C{=}NLi{\cdot}pyridine)_4$ (**7**).

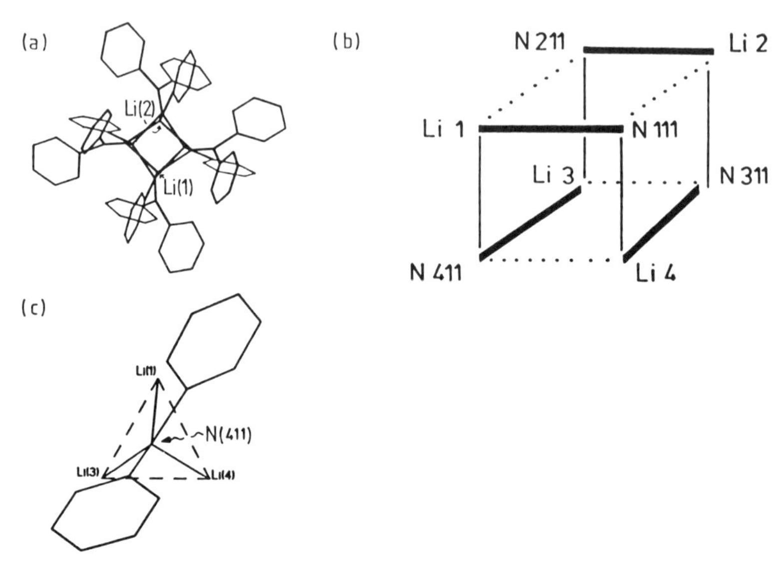

FIG. 15. (a) A view of the ($Ph_2C{=}NLi\cdot pyridine)_4$ cubane (**7**), through the face containing Li(1) and Li(2); (b) N—Li bond lengths in ($Ph_2C{=}NLi\cdot pyridine)_4$: heavy lines, two center (2.02–2.03 Å, av. 2.03 Å); light lines, between dimers (2.07–2.10 Å, av. 2.08 Å); dotted lines, within dimers (2.15–2.17 Å, av. 2.16 Å); (c) orientation of an imino ligand over one Li_3 face of ($Ph_2C{=}NLi\cdot pyridine)_4$.

view for hexamer (**4**) shown in Fig. 6]. The clue to this further association is provided by the (imide)N—Li bond lengths in (**7**). As shown in Fig. 15b, these fall into three distinct sets, *viz.* relatively short (range 2.02–2.03 Å), medium length (2.07–2.10 Å), and long (2.15–2.17 Å). A consistent pattern is seen when one considers the cube as being formed from a top dimer (containing Li 1 and Li 2) stacking upon a lower dimer (containing Li 3 and Li 4) [*cf.* the view of (**7**) in Fig. 15a]. Bonds within a constituent dimer are then alternating short and long ones, with medium-length bonds occurring between dimers. This pattern arises due to the orientation of the $Ph_2C{=}N$ ligands over the Li_3 faces of the tetramer. A typical orientation is shown in Fig. 15c, for imino N 411 bonding to Li3 and Li 4 (within the lower dimer) and to Li 1 (in the top dimer). As illustrated earlier for iminolithium hexamers (Section II,A, especially Fig. 7), the orientation of the $RR'C{=}N^-$ ligand defines the direction of its electrostatic interactions with Li^+ cations. Hence, the bond from N 411 to Li 3 is expected to be relatively short, that to Li 1 rather longer, and that to Li 4 longer still. This is what is observed (Fig. 15b). The changed pattern in going from the hexamers (short and medium-length bonds within rings, long ones between) to this tetramer

(**7**) (short and long bonds within rings, medium ones between) is due to the much greater tilt of RR′C=N ligands over the Li_3 faces in (**7**) (*cf.* Fig. 15c with Fig. 17b). This is caused by the presence of a pyridine molecule on each Li center in (**7**).

The other structurally characterized crystalline complex of (Ph_2C=NLi)$_n$, a yellow powder, originated from treatment with HMPA in diethyl ether–toluene. Formulated empirically as (Ph_2C=N-Li$\cdot\frac{5}{6}$HMPA), the complex exists in the solid state as the ion pair [Li-$(HMPA)_4]^+\cdot[(Ph_2C{=}N)_6Li_5\cdot HMPA]^-$ (**8**) (*86*). In the anion of (**8**) (Fig. 16), four uncomplexed Li^+ cations form a tetrahedron (Li—Li mean edge distance, 3.02 Å). The fifth Li^+, which is complexed by an HMPA molecule, caps one face of this tetrahedron, forming much shorter Li—Li contacts (mean distance, 2.60 Å). In the resulting distorted trigonal-bipyramidal anion, three of the imide ligands μ_3 bridge Li_3 faces (those which incorporate the $Li^+\cdot$HMPA unit) and the other three μ_2 bridge the remaining Li—Li edges. ’Ate complexes (or “triple ions”) were first proposed in the 1950s (*87*), but have proved to be rather elusive species (*4*). Apart from (**8**), a crystal structure is available for $[Li(THF)_4]^+\cdot\{[(Me_3Si)_3C]_2Li\}^-$ (*88*), and the structure of (Ph_4Li-$Na_3\cdot$3TMEDA) has recently been interpreted in terms of an ’ate, $[Ph_4Li]\cdot[Na(TMEDA)]_3$ (*89*). Similar triple ions in solution have been

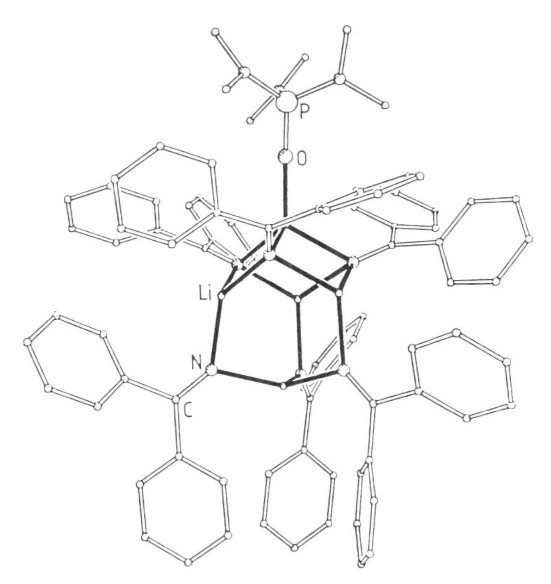

FIG. 16. Structure of the anion of $[Li(HMPA)_4]^+\cdot[(Ph_2C{=}N)_6Li_5\cdot HMPA]^-$ (**8**).

detected by NMR spectroscopy, e.g., a combination of ^6Li, ^{13}C, and ^{15}N NMR spectra have shown that lithium *N*-isopropylanilide in ether solutions containing HMPA exists partly as $[Li(HMPA)_4]^+ \cdot [(PhPr^iN)_2\text{-}Li \cdot HMPA]^-$ *(90)*. It is not known why (**8**) forms an 'ate structure. However, HMPA and macrocyclic ligands (cryptands, etc.) seem to favor 'ate formation *(4)*. It may be pertinent that polar molecules such as these dissolve lithium metal to give $Li(donor)_x^+$ complexed cations and solvated electrons *(91, 92)*.

The crystal structure of a mixed-metal imide $[(Me_2N)_2C{=}N]_4Li\text{-}Na_3 \cdot 3HMPA$ (**9**) *(93)* merits special discussion. Here, complexation leads to formation of a tetrameric pseudocubane structure (Fig. 17) [*cf.* (**7**), Fig. 14]. However, this complex has several unusual features. Although (**9**) was synthesized by reacting two equivalents of $(Me_2N)_2C{=}NH$ with one equivalent each of Bu^nLi and Bu^nNa in an excess (five equivalents) of HMPA, this $LiNa_3$ stoichiometry results, and only the Na^+ ions bear HMPA ligands. The reasons for adoption of an $LiNa_3$ stoichoiometry are not known. However, the large $LiNa_3$ core allows cubane formation even in the presence of bulky HMPA ligands [*cf.* dimeric (**6**)]. Further, it is significant that HMPA ligands are found only on the Na^+ centers. Because of the relatively long Na—O distances

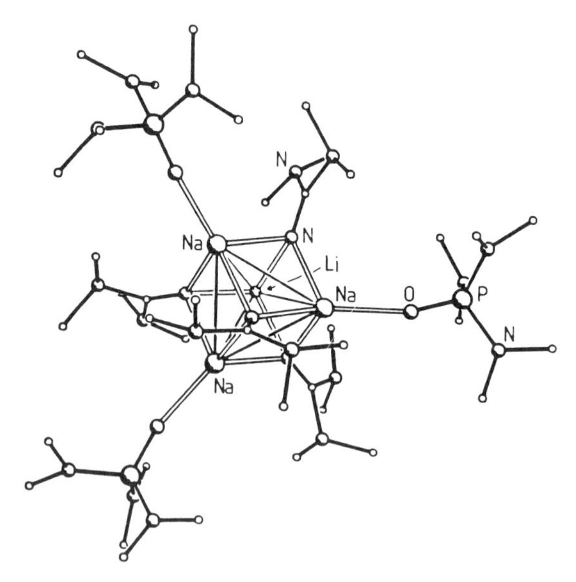

FIG. 17. Molecular structure of the complexed mixed-metal imide $[(Me_2N)_2C{=}N]_4Li\text{-}Na_3 \cdot 3HMPA$ (**9**).

[mean, 2.25 Å; *cf.* in (**6**), Li—O, 1.86 Å] there is no steric crowding between the Me_2N groups of the HMPA ligands and those of the imide ligands (see Fig. 17). The lack of a ligand on the lithium may be partially compensated by "agostic" Li···H—C interactions (see Section I,B and Fig. 4). Each $Me_2N)_2C{=}N$ ligand bridging the $LiNa_2$ triangular faces of (**9**) has one Me group leaning toward the Li atom, giving Li—C distances of 2.9–3.1 Å and inferred Li—H distances of ~2.5 Å. Unfortunately, the M—N bond lengths observed in this structure do not allow an analysis similar to that performed for the pure N—Li clusters (**1**)–(**4**) and (**7**). The N—Na distances only span the 2.40- to 2.46-Å range and they are distributed randomly within the cluster.

C. Solution Structures of Lithium Imides and Their Complexes

The nature of lithium imides and their complexes in solution has been deduced from cryoscopic measurements in benzene [the relative molecular mass (rmm) values give the degree of association (n)] and from high-field ^{7}Li(139.96 MHz) NMR spectroscopic data (*79, 94, 95*). ^{7}Li in conjunction with ^{13}C NMR spectroscopy has been used to study the solution equilibria of C—Li bonded organolithiums, e.g. (*s*- and *t*-butyllithium)$_n$ (*95a*) and (*n*-propyllithium)$_n$ (*96*). However, ^{6}Li-enriched samples often are superior in revealing the nature of the species present in solution *via* observations of ^{6}Li–^{13}C couplings and coupling constants (*97, 98*). ^{6}Li–^{1}H HOESY NMR experiments have also pinpointed the close proximities of Li and H nuclei in several organolithiums (e.g., *51, 51a*). Applications of NMR spectroscopy to the study of organolithiums, with particular emphasis on the recent development of two-dimensional techniques, have been reviewed (*50*).

The solid-state hexamers (**2**)–(**4**) at first appeared to dissolve intact in benzene (*94*). Cryoscopic rmm measurements over a range of concentrations (0.03–0.09 *M*, molarity expressed relative to the empirical formula mass) implied n values of 5.9–6.1. Furthermore, their room-temperature ^{7}Li NMR spectra in d_8-toluene each consisted of broad singlets within the narrow chemical shift (δ) range of +0.6 to −0.2 ppm (relative to external phenyllithium in the same solvent). However, variations in temperature and concentration affected the ^{7}Li NMR spectra of (**2**) and, in particular, of (**4**) (*95*). Figure 18a shows these spectra for three d_8-toluene solutions of (**4**) at ~−100°C. The most concentrated solution has a dominant signal at δ ~+0.7, though five or six other signals (indicated by asterisks) are apparent. On dilution,

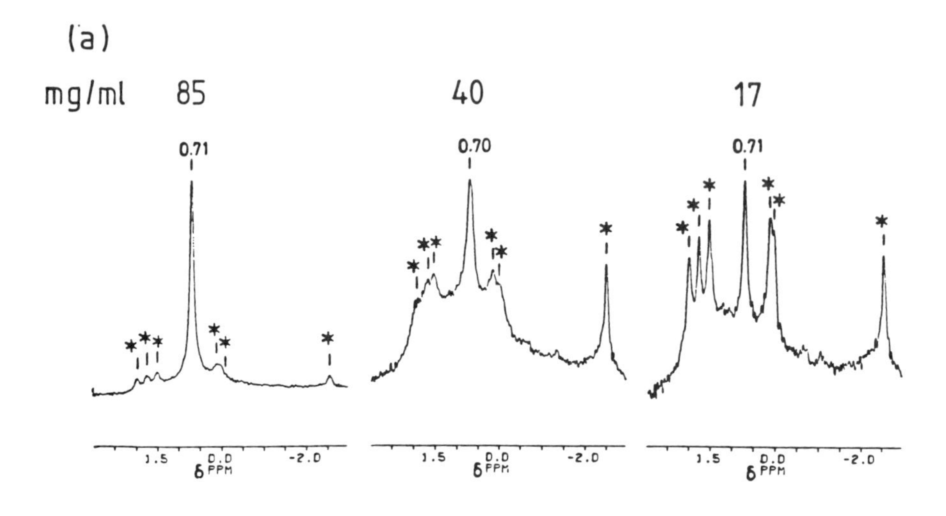

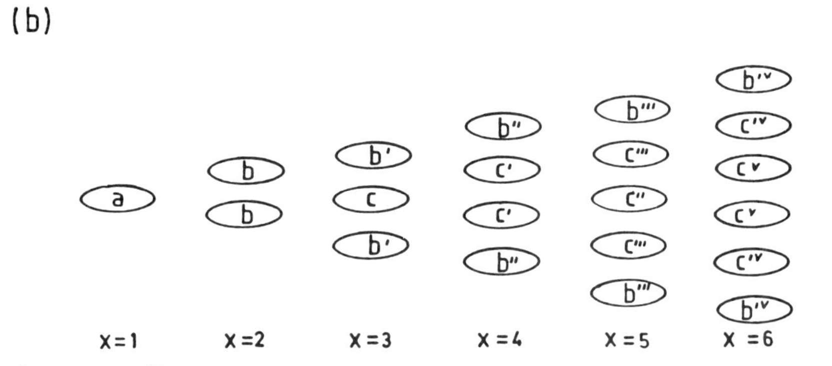

FIG. 18. (a) ^{7}Li NMR spectra (139.96 MHz, $-100°C$) of three solutions of [Ph-$(Me_2N)C{=}NLi]_6$ (**4**), in d_8-toluene; (b) representation of Li environments in $(RR'C{=}NLi)_3$ stacks of various extents.

these smaller signals increase in intensity relative to that at δ 0.7. Raising the temperature leads to coalescence; at $-60°C$, only one resonance is observed. These results may be explained using ring-stacking ideas outlined earlier (Section II,A) and described in full below (Section II,E). Thus, Fig. 18b depicts an iminolithium trimer $(RR'C{=}NLi)_3$ as an ellipsoid and also shows how the Li environments of type *a* (two coordinate) will vary as trimers $(RR'C{=}NLi)_{3x}$ stacked to various extents are formed. For example, a hexamer ($x = 2$) of S_6 symmetry should give only one ^{7}Li signal (type *b*, three-coordinate Li), whereas a nonamer ($x = 3$) will contain six "outer" Li atoms (with environment b', similar to b) and three equivalent "inner" Li atoms (type c, four coordinate). On this basis, the dominant singlet at δ 0.7 ppm in the ^{7}Li NMR spectrum of the most concentrated solution of (**4**) was assigned to the intact hexamer. The very distinctly separated singlet at $\delta \sim -2.5$ ppm, which increases in relative intensity markedly on dilution, can reasonably be assigned to the trimer (of unique ^{7}Li environment a). The remaining five signals on either side of δ 0.7 ppm must then be due to higher aggregates ($x > 2$). Their formation appears to involve prior dissociation of the hexamer to the trimer, because the growth of these five ^{7}Li signals is accompanied by an increase in the relative intensity of the trimer resonance and a decrease in that of the hexamer signal. Though the precise nature of these higher stacks has not been established, their five ^{7}Li signals fall into two sets. The relative intensities within each set are concentration independent, i.e., two signals at δ +0.12 and 0.04 ppm of ratio 1 : 0.5 and three at δ +1.99, 1.77, and 1.53 ppm of ratio 1 : 1 : 1. These values could correspond to the nonamer $[Me_2N(Ph)C{=}NLi]_9$ (Fig. 18b; $x = 3$, environments $b' \times 2$ and c) and the "octadecamer" ($x = 6$, environments $b^{IV} \times 2$, $c^{IV} \times 2$, and $c^{V} \times 2$]. A nonamer (as well as a hexamer and an octamer) was indicated to be present (by ^{6}Li and ^{13}C NMR data) in hydrocarbon solutions of n-propyllithium (*96*).

Much more conclusive results were obtained for solutions of $(Bu^t{}_2C{=}NLi{\cdot}HMPA)_2$ (**6**) (*79, 94*), which is a dimer in the solid state. Cryoscopic measurements in benzene gave n values of 1.33 (0.033 M) and of 1.12 (0.017 M); this implied the existence of a monomer $\rightleftharpoons$ dimer equilibrium with approximate monomer : dimer ratios of 2 : 1 and 6 : 1 for the more concentrated and more dilute solutions, respectively. This inference was supported by the ^{1}H and ^{7}Li room-temperature NMR spectra of (**6**) in benzene-d_6; typical spectra are shown in Fig. 19. Unusually for ^{7}Li NMR spectra recorded at ambient temperature, two signals are observed (Fig. 19a and c). The peak at $\delta \sim +0.4$ ppm is proportionately larger in the dilute (Fig. 19c, 0.05 M) than in the concentrated solution (Fig. 19a, 0.25 M). This suggests that this δ 0.4-ppm signal is due to the monomer and that the $\delta \sim 0.15$-ppm peak corresponds to a dimer. The ^{1}H NMR spectra provide support (Fig. 19b and d). Apart from the doublet due to HMPA, the two signals at $\delta \sim 1.4$ and 1.2 ppm can be assigned to the Bu^t groups in the dimeric and monomeric species, respectively. Further, both types of spectra agree closely on the relative molar ratios of dimer : monomer in each solution; ~1.45 : 1 in the more concentrated solution (hence a relative integration of ~2.9 : 1, Fig. 19a and b), giving $n \sim 1.6$, and ~0.7 : 1 in the more

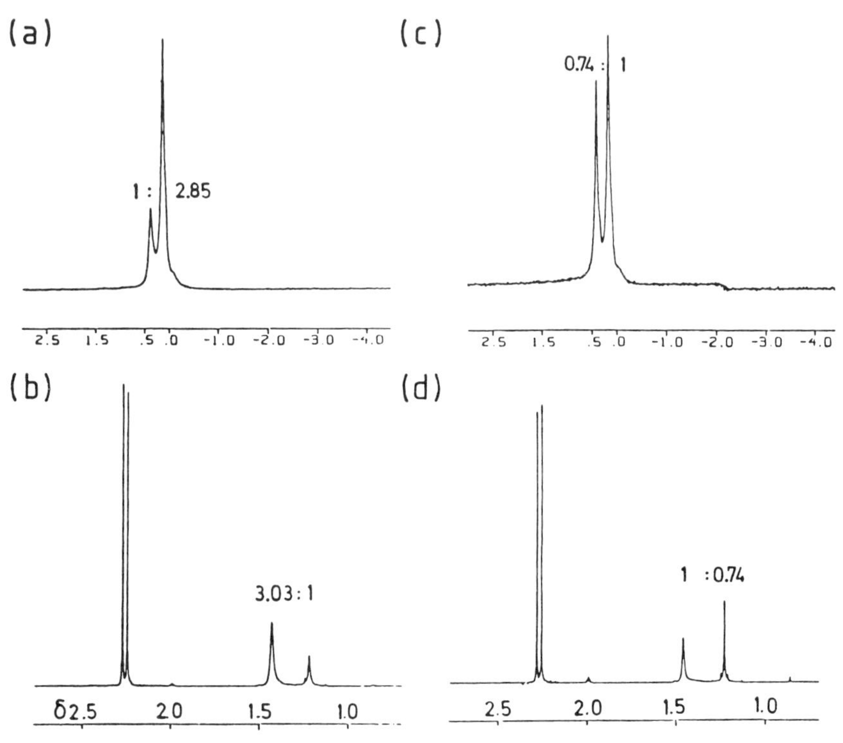

FIG. 19. NMR spectra (25°C, d_8-toluene) of solutions of $(Bu^t{}_2C{=}NLi \cdot HMPA)_2$ (**6**): (a) ^7Li, 0.25 *M*, (b) ^1H, 0.25 *M*, (c) ^7Li, 0.05 *M*, and (d) ^1H, 0.05 *M*.

dilute one, so $n \sim 1.4$. The dimer must be similar to the solid-state structure, whereas the monomer probably has only a single mole of HMPA, $Bu^t{}_2C{=}NLi \cdot HMPA$. If more HMPA were present, ^7Li NMR signals at δ 0.62–0.64 ppm found for solutions of the hexamer $(Bu^t{}_2C{=}NLi)_6$ (**1**) would be apparent. Remarkably, $Bu^t{}_2C{=}N$ group and ^7Li atom exchange between the monomer and the dimer of $Bu^t{}_2C{=}NLi \cdot$ HMPA is too slow to be observed on the NMR time scale. This implies stabilization of the monomer, despite its two-coordinate Li. Presumably Li···H—C or Li···NMe_2 (of HMPA) interactions are responsible.

Unlike (**6**), the related dimeric "keteniminate" $[Ph(H)C{\cdots}C{\doteq}NLi \cdot TMEDA]_2$ may dissolve intact (*82*). Earlier rmm measurements on Ph(H)C·CN·Li in THF (at 18.5°C), 1,2-dimethoxyethane, or dimethyl sulfoxide indicated dimers to be present (*99*). However, in THF at −108°C, a monomer is found (*82, 100*). MNDO calculations on model species indicate that the dimer $[Ph(H)C \cdot CN \cdot Li \cdot 2NH_3]_2$ (and $2NH_3$) is only ~5 kcal mol^{-1} more stable than two Ph(H)C·CN·Li·3NH_3 monomers. Hence, monomers could be favored, entropically, at low temperatures.

A few solution studies have been carried out on the iminolithium tetramer, ($Ph_2C{=}NLi$·pyridine)$_4$ (**7**) (*94*). Molecular mass measurements in benzene gave *n* values ranging from ~2.2 (0.07 *M*) to ~1.4 (0.03 *M*), implying extensive dissociation of the tetramer. However, the species in benzene must still be ligated by pyridine because ($Ph_2C{=}N$-Li)$_n$ did not precipitate. The ^{7}Li NMR spectrum of (**7**) in toluene-d_8 at 25°C shows a sharp singlet at δ ~+1.7 ppm and a broader peak at δ ~+1.5 ppm. Because the relative intensity of the latter increases on dilution, from ~1 : 0.25 at 0.40 *M* to ~ 1 : 1.4 at 0.05 *M*, it presumably is due to a species with lower association. The correspondence of the *n* values obtained cryoscopically and those gained by integration of NMR signals (for solutions of identical concentrations) can only be achieved if it is assumed that (**7**) engages in a tetramer ⇌ monomer equilibrium.

D. Calculations on Iminolithiums and on Lithium Species with Related Structures

MNDO and *ab initio* molecular orbital calculations on organolithium compounds have proved of enormous value in understanding the structures of these species (*2, 36, 37, 101, 102*). They have furnished thermochemical data (rarely available experimentally) such as association, complexation and solvation (*103*), and bond (*103a*) energies. Furthermore, they have both rationalized known solid-state structures and, in many cases and with remarkable accuracy, predicted such structures prior to their solutions by X-ray crystallography. The semiempirical MNDO method (*104–106*), despite having a tendency to overestimate C—Li bonding (*106, 107*), has been particularly successful in giving reliable structural information on lithiated organics, not least because it can deal with relatively large molecules.

Regarding uncomplexed iminolithiums, the structures of ($H_2C{=}NLi$)$_n$, n = 1, 2, and 3, have been investigated by both MNDO (*108*) and *ab initio* [3-21G basis set (*108*) and 6-31G basis set (*72, 74*)] methods. Figure 20 shows the optimized structures (6-31G) for the two associated species. For the dimer, all the calculations show that the planar D_{2h} form (Fig. 20a) is preferred to a form with the $H_2C{=}N$– ligand plane perpendicular to the $(NLi)_2$ ring plane (Figure 20b). The energy difference calculated by MNDO is quite small, 6.8 kcal mol^{-1} (*108*), though at the 6-31G level (*72, 74*) it is somewhat larger, 17.0 kcal mol^{-1}. As described earlier, the only known example of a solid-state

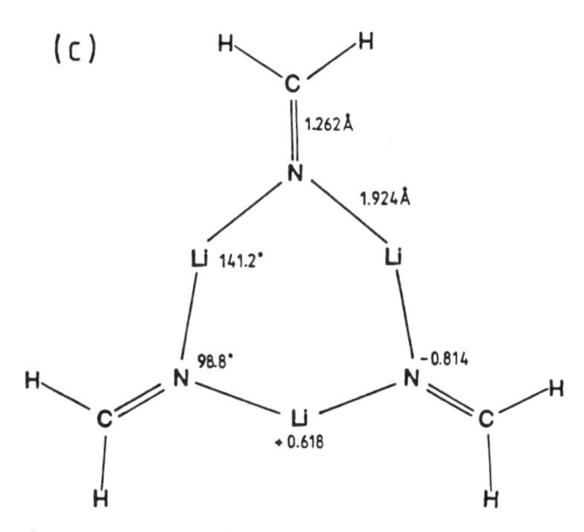

FIG. 20. Optimized structures (6-31G basis set level) of (a) $(H_2C{=}NLi)_2$ dimer, planar, (b) $(H_2C{=}NLi)_2$ dimer, perpendicular, and (c) $(H_2C{=}NLi)_3$ trimer, planar.

iminolithium dimer is $(Bu^t_2C{=}NLi{\cdot}HMPA)_2$ (**6**) (*78, 79*), which, for steric reasons, adopts a compromise geometry between planar and perpendicular (see Section II,B, especially Fig. 12). The rather low MNDO energy difference between the two forms may explain the retention of strong N—Li bonding in this structure. Thus, it has been argued (*108*) that in the $H_2C{=}N^-$ anion the two lone pairs on N and the π electrons of the C═N bond can combine to give a quasicylindrical arrangement of electronic charge at N (see Fig. 7). In that case, the Li^+ placements in the dimer are not very critical. What is clear-cut is that dimerization of $H_2C{=}NLi$ is very exothermic. The MNDO dimerization energy (-51.7 kcal mol^{-1}) agrees quite well with the 3-21G/3-21G value (-71.4 kcal mol^{-1}) taking into account the basis set superposition error (*109, 110*) in the quite low-level *ab initio* calculation. At the 6-31G level, this value is -66.0 kcal mol^{-1} (*72, 74*).

For the trimer $(H_2C{=}NLi)_3$ (Fig. 20c), at the 6-31G level the planar form is very strongly preferred (by 91.0 kcal mol^{-1}) to the perpendicular one. This is largely because the latter's six-membered ring is distorted much more from a regular hexagon (*74*). A planar trimeric ring is also preferred to a planar dimer (trimerization energy −119.2 or −39.7 kcal mol^{-1} per $H_2C{=}NLi$ monomer unit, *cf.* for a dimer, −33.0 kcal mol^{-1} per monomer unit). These two preferences, for a trimer over a dimer and for planar forms, lead to the stacking of two or more (trimeric) rings and hence the formation of hexameric or polymeric species (see Section II,B, especially Figs. 6, 9, and 10).

MNDO calculations have also been performed on a model tetrameric lithium imide, $(H_2C{=}NLi)_4$ (*108*). On increasing the association state (n) to four, a planar ring geometry is no longer found. Thus, a tetrahedral cubanelike structure (tetramerization energy, −144.4; −36.1 kcal mol^{-1} per monomer unit) is favored over an eight-membered planar $(NLi)_4$ ring structure. This preference is also found experimentally in the solid-state structures of $(Ph_2C{=}NLi{\cdot}pyridine)_4$ (**7**) and $[(Me_2N)_2C{=}N]_4{\cdot}LiNa_3{\cdot}3HMPA$ (**9**) (see Section II,B, especially Figs. 14 and 17). These dual preferences, for formation of planar dimeric rings and then for their further association [*cf.* for $(H_2C{=}NLi)_2$, the dimerization energy per monomer is only −25.9 kcal mol^{-1}] lead to clustered species. The related tetramers $(LiF)_4$, $(LiOH)_4$ (*111*), and $(MeLi)_4$ (*102*) also favor the tetrahedral arrangement over the planar ring one. These conclusions have been verified experimentally in the case of $(MeLi)_4$, whose solid-state structure is indeed a pseudocubane (*44*). They reflect essentially the dominance of electrostatic interactions in these systems: a cubanelike arrangement gives each Li^+ cation a coordination number of three rather than just two as in a ring structure. In fact, purely electrostatic approaches, using point plus and minus charges in the form of two interpenetrating tetrahedra, can describe these tetrameric clusters quite well (*112–113a; 102*).

Iminolithium hexamers have not so far been examined by MO calculations. Thus there are no *direct* theoretical results pertaining to the structures of the experimentally found hexamers (**1**)–(**4**) [D_{3d}-type clusters viewed as two stacked trimeric rings (*69–72*)] and the mixed-metal hexamer (**5**) [a stack of three dimeric rings (*77*)]. However, MNDO and *ab initio* computations have been carried out on hexamers with smaller anions, e.g. $(LiH)_6$ (*114*), $(LiF)_6$ and $(LiOH)_6$ (*115*), and $(MeLi)_6$ (*102, 116*). In all cases, distorted octahedral (D_{3d}) structures are found to be more stable than hexagonal planar ring (D_{6h}) structures. For example, in the case of $(LiF)_6$, the former structure is favored by 32.1 kcal mol^{-1} (at the HF/6-31G+sp+d level). The merits of further association are also borne out by the ever-increasing binding energies per monomer unit. For the $(LiF)_n$ systems, these are 32.5 (dimer), 42.0 (trimer), 47.0 (cubic tetramer), and 52.1 (octahedral hexamer) kcal mol^{-1} (*115*). Further, force-field models (*117–121*), which consider only hard-sphere volume exclusion and point-charge electrostatic interactions, have proved capable of predicting the relative dimensions of $(RLi)_6$ hexamers with reasonable accuracy (*113a*).

Calculations on lithium amides, using $(H_2NLi)_n$ models (n = 1–4; 6), are discussed in Section III,B,2.

E. Ring Stacking: General Applications to the Structures of Organolithium Compounds

We have shown in Section II,A how uncomplexed iminolithium hexamers (**1**)–(**4**) can be regarded as double stacks of trimeric rings, and that flat diaryliminolithium ring systems can stack extensively to give polymers (see especially Figs. 5–10). For iminolithium complexes, the constituent ring size is four-membered (Section II,B; see Fig. 13). These rings also can stack, affording tetrameric cubic arrangements such as (**7**) (Figs. 14 and 15) and (**9**) (Fig. 17). Ring stacking takes place because the *exo*-RR′C═N ligands are essentially coplanar with the $(NLi)_{2,3}$ rings, particularly to and including the primary (α) atoms of the R,R′ groups. However, this stereochemistry is by no means unique to imide anions. It occurs widely, as shown in Fig. 21a, for a range of organoli-

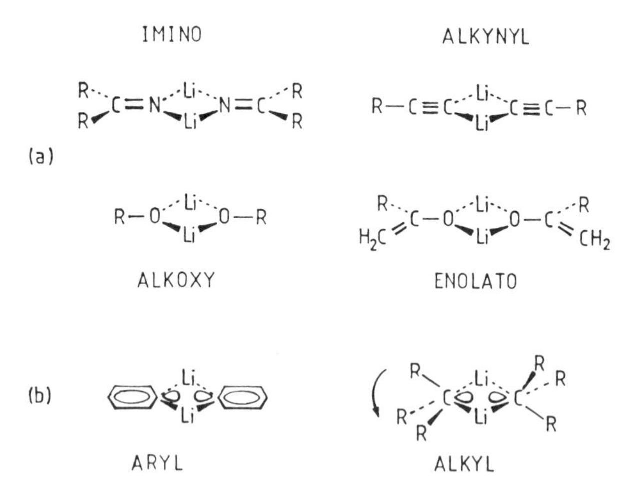

Fig. 21. Organolithium ring systems capable of stacking: (a) imino, alkynyl, alkoxy/aryloxy, and enolato systems; (b) aryl and alkyl systems.

thium ring systems involving alkynyl ($RC{\equiv}C^-$), alkoxy/aryloxy (RO^-), and enolato [$RC({=}CH_2)O^-$] anions. Although dimeric rings are shown in Fig. 21a, the same stereochemistry would obviously be found for trimeric or any other size rings. The ring size will depend largely on the bulk of the anion, i.e., the size of the R groups. For alkyl and aryl anions (Fig. 21b), the bonding situation is slightly different. In both, the α-C^- centers have lobes with pseudocylindrical symmetry projecting into the $(CLi)_2$ or $(CLi)_3$ ring. Hence the precise orientation of the ligand as a whole will have little effect on the strength of ring bonding. Thus, aryl ligands can be coplanar with the ring, whereas the substituents of alkyl ones can take up positions that best allow interlocking and stacking.

The stereochemical feature noted above leads to widespread ring stacking in organolithium chemistry. Such structures do not result as a matter of course from a preference of lithium centers for four coordination (though this is often stated to be so). There are many organolithium compounds whose metal centers are only two or three coordinate, and a fair number whose metal centers exceed four coordination (even excluding consideration of Li—Li contacts) (*2, 3*). Rather, ring-stacking propensities are merely a manifestation of electrostatic principles: a Li^+ cation within any structure will interact, inevitably, with as many anionic centers or polar ligands as possible. Just how many will be present (see Section I) depends on the *steric* features of the anions and ligands. Monoatomic anions, such as Hal^-, allow full three-dimensional association (stacking *and* laddering; Fig. 3c). As just discussed, many organic anions allow only vertical association. This can be extensive when R groups are small or flat (Fig. 3a); bulkier groups limit association to double stacks of trimeric and dimeric rings [hence clustered hexamers (Fig. 3d) and tetramers (Fig. 3e)]. Complexation of such systems, whether by addition of polar ligands ("external" complexation; Section I, Fig. 1) or with internal sites in the organic anion (Fig. 2), usually results in ring dimers. For steric reasons, monodentate ligands allow two such rings to stack to give cubic tetramers. Bidentate ligands, in contrast, result in isolated dimers, or monomers when the R groups in the organic anions are large. Tridentate complexants often favor monomer formation.

Crystal structures illustrating these ring-stacking ideas are given in Table II (aggregated alkyl- and aryllithium species), Table III (alkynyllithium species), Table IV (oxycarbanion–lithium species), and Table V (lithium halides and lithium halide–organolithium mixed species). Figures 1–4 (Section I) have already shown representations of many of the structural types from these tables. For other types, repre-

TABLE II

CRYSTAL STRUCTURES OF ASSOCIATED ALKYL-AND ARYLLITHIUM COMPOUNDS AND COMPLEXES AND RELATED SPECIES

Empirical formula[a]	Formula number[b]	Association state (n)	Figure	Ref.
MeLi	(**10**)	4	3e, 4a	*44*
EtLi	(**11**)	4	3e	*1*
Me_3SiCH_2Li	(**12**)	6	3d	*48*
c-$C_6H_{11}Li$	(**13**)	6	3d, 4b	*47*
$Me_2C{\cdot}Me_2C{\cdot}CH{\cdot}CH_2Li$	(**14**)	6	3d	*75*
Me_3SiLi	(**15**)	6	3d	*73*
$MeLi{\cdot}\frac{1}{2}TMEDA$	(**16**)	4	1a	*22*
$PhLi{\cdot}OEt_2$	(**17**)	4	22a	*122*
PhLi·TMEDA	(**18**)	2	1b	*24*
$Me_2N(CH_2)_2CH_2Li$	(**19**)	4	2a	*28, 29*
2-$Me_2NCH_2{\cdot}C_6H_4Li$	(**20**)	4	2a	*30*
$MeOCH_2CH_2{\cdot}(Me)CHLi$	(**21**)	4	2a	*123*
$(Me_2NCH_2)_2CH{\cdot}(Me)CHLi$	(**22**)	2	2b	*32*
$(2,3,5,6,$-$Me_2NCH_2)_4C_6HLi$	(**23**)	2	2b	*33*
$Me_2N(CH_2)_3NMe{\cdot}C(CH_2)_4{\cdot}({=}C)Li$	(**24**)	2	22b	*124*
8-$Me_2NC_{10}H_6Li{\cdot}OEt_2$	(**25**)	2	22c	*125*
$(2,6$-$MeO)_2C_6H_3Li$	(**26**)	4	22d	126, 127
$(2,6$-$Me_2N)_2C_6H_3Li$	(**27**)	3	22e	*133*

[a] With respect to the organolithium component.
[b] See text.

sentations or specific structures are given in Figs. 22–25. Brief comments on the compounds and complexes listed in these tables are now given.

Uncomplexed alkyllithiums [(**10**)–(**14**)] are tetramers or hexamers (Table II). Small alkyl groups (Me, Et) allow dimeric $(CLi)_2$ rings to form. Larger ones, and the large $Si^{\delta-}$ centers of Me_3Si groups in (**15**), dictate trimeric rings. Two such rings can stack, leading to cubic tetramers or pseudooctahedral hexamers. Because each Li atom then becomes only three coordinate, Li···H—C interactions often are prompted (*43–50*). The tetramers (**10**) and (**11**) associate further (hence their insolubility in hydrocarbon solvents) by means of intercube Li···H—C interactions; these were illustrated for (**10**) in Fig. 4a. The hexamers (**12**) and (**13**) have been noted to adopt intramolecular interactions, as detailed for (**13**) in Fig. 4b. Significantly, these examples [(**10**)–(**15**)] do not include an aryllithium. No uncomplexed *oligomeric* aryllithium species are known, presumably because their $(arylLi)_{2,3}$ ring systems are flat enough to allow more extensive association. For example, $(PhLi)_n$, prepared in a noncomplexing solvent such as toluene, is an amorphous and insoluble material.

"External" complexation by addition of Lewis bases gives three kinds of associated structures. The first is a tetramer [(**16**)] whose individual cubes are linked by a usually bidentate ligand, TMEDA, acting only as a monodentate one; in effect, the $Li \cdots H_3C$ interactions seen in (**10**) have been replaced by these Me_2N coordinations. With a monodentate complexant as such, an isolated cubic tetramer (**17**) results *(122)* (Figure 22a); the Et_2O molecules preclude further linkages. Bidentate ligation, e.g., by TMEDA, results usually in dimeric complexes, e.g. (**18**). Ring stacking is prevented by the ligand's Me_2N groups projecting above and below the $(CLi)_2$ ring plane, giving a pseudotetrahedral Li environment. Complexations by functional groups within the anion result in similar structures to those just noted. A single functionality per anion leads to tetramers (**19**), (**20**), and (**21**) *(123)*, while two such functionalities per anion afford dimers (**22**), (**23**), and (**24**); in (**24**) *(124)*, both complexing sites of a given anion attach to the same lithium cation (Fig. 22b). The dimer (**25**) *(125)* provides an example of "mixed" coordination, involving both internal (Me_2N) and external (Et_2O) complexants (Fig. 22c). Significantly, all these complexed aggregates are dimers or stacks of two dimers; trimers and their stacks no longer are found. The final two species in Table II [(**26**) and (**27**)] are aryllithium complexes, each with two functional groups per aryl anion, and they provide an interesting contrast. The structure of (**26**) *(126, 127)* substantiated the theoretical prediction of a planar, rather than a tetrahedral, arrangement when two Li atoms are attached to the same C atom *(128–132)*. Thus the MeO groups (two per Li) lie in the $(CLi)_2$ ring plane (Fig. 22d). Therefore, association to a tetramer is possible sterically, and each Li becomes five coordinate. For (**27**) *(133)*, the larger $(Me_2N)_2C_6H_3^-$ anions cause a trimeric ring to form, and the Me_2N groups project above and below the $(CLi)_3$ ring plane (Fig. 22e), preventing stacking. Complex (**27**) is the only known example of a *complexed* trimer in lithium chemistry.

Alkynyllithium compounds form flat ring systems (Fig. 21) whose vertical association is uninhibited (Fig. 3a): lithiations of alkynes (acetylenes) in nonpolar media produce amorphous materials. When polar ligands (Lewis bases) are added, crystalline complexes can be isolated, their structures depending on the effective denticity of the base and on its molar proportions relative to lithium (Table III). A complexant such as TMHDA, $Me_2N(CH_2)_6NMe_2$, with a long methylene chain, fails to behave in a bidentate manner toward a given Li center; hence complex (**28**) results [*cf.* (**16**)], with the TMHDA linking

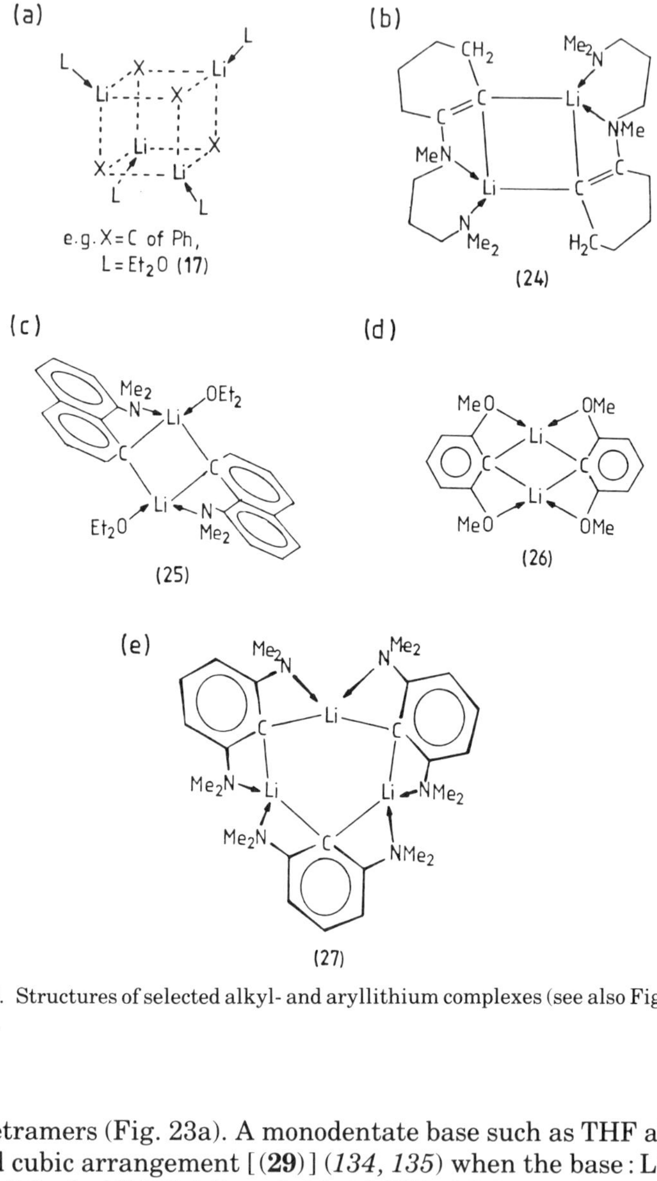

FIG. 22. Structures of selected alkyl- and aryllithium complexes (see also Figs. 1–4 and Table II).

cubic tetramers (Fig. 23a). A monodentate base such as THF allows an isolated cubic arrangement [(**29**)] (*134, 135*) when the base : Li ratio is 1 : 1. In (**30**), the TMPDA ligand, $Me_2N(CH_2)_3NMe_2$, with a short methylene chain, can chelate, and a dimer is formed. The totally new structural option is provided by (**31**) (*134, 135*). This dodecamer exhibits sixfold stacking of $(CLi)_2$ rings, terminated by complexants (Fig. 23b). A structure like this can be regarded as an intercepted polymeric stack

TABLE III

CRYSTAL STRUCTURES OF ALKYNYLLITHIUM COMPLEXES

Empirical formula [a]	Formula number[b]	Association state (n)	Figure	Ref.
PhC≡CLi·½TMHDA	(**28**)	4	1a, 23a	*23*
Bu^{t}C≡CLi·THF	(**29**)	4	22a	*134, 135*
PhC≡CLi·TMPDA	(**30**)	2	1b	*25*
Bu^{t}C≡CLi·⅓THF	(**31**)	12	23b	*134, 135*

[a] With respect to the organolithium component.
[b] See text.

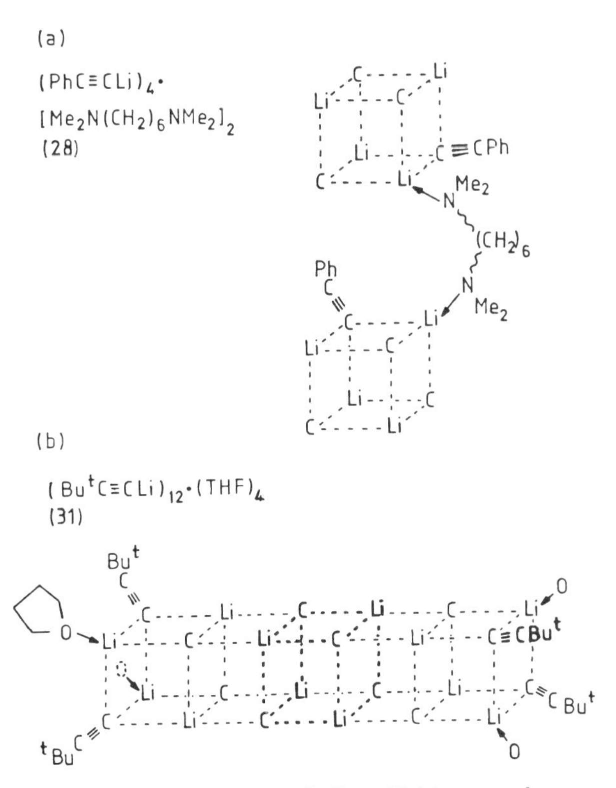

FIG. 23. Structures of selected alkynyllithium complexes.

(Fig. 3a). As in the 1:1 THF:Li analog [(**29**)], each Li center in (**31**) reaches four coordination, though obviously the internal ones do so in a different way. Again significantly, all of complexes (**28**)–(**31**) are dimeric rings or multiples thereof; no trimer rings are present.

The crystal structures of numerous organolithium compounds and complexes with O—Li bonds are now available (*2–5*). Table IV lists a number of these species, as well as two derivatives of heavier alkali metals. As with the C—Li derivatives just discussed (Tables II and III), clustered $(ROLi)_n$ tetramers and hexamers, as well as ring dimers, are prevalent. Note that $(OLi)_{2,3}$ ring systems also are pseudoplanar (Fig. 21a). However, extensive stacking leading to polymers will only occur if the substituents on O are small and if polar ligands are absent. Otherwise, limited (double) stacks or unstacked rings form.

Two lithium enolate hexamers are known [*cf.* (**12**)–(**15**) in Table II]. In (**32**) (*136, 137*), the large Bu^t substituents result in the preference for

TABLE IV

CRYSTAL STRUCTURES OF OXYCARBANION–ALKALI METAL COMPOUNDS AND COMPLEXES WITH O—M BONDS

Empirical formula[a]	Formula number[b]	Association state (n)	Figure	Ref.
$Bu^tC(=CH_2)OLi$	(**32**)	6	3d, 24a	*136, 137*
$Et_2N{\cdot}CH=C(OEt)OLi$	(**33**)	6	24b	*138*
$Bu^tC(=CH_2)OLi{\cdot}THF$	(**34**)	4	22a	*139*
$H_2\overline{C(CH_2)_2CH=C}OLi{\cdot}THF$	(**35**)	4	22a	*139*
$2\text{-}Me_2NCH_2{\cdot}C_6H_4{\cdot}C(=CH_2)OLi$	(**36**)	4	*cf.* 2a; 24c	*31*
$Bu^tC(=O){\cdot}CH_2{\cdot}CH(Bu^t)OLi$	(**37**)	4	*cf.* 2a	*140*
$Bu^t_2Si(NH_2)OLi$	(**38**)	4	*cf.* 2a	*141*
$MeO{\cdot}C(=CHBu^t)OLi{\cdot}THF$	(**39**)	4	22a	*26*
$Bu^tC(=CH_2)OLi{\cdot}TMEDA$	(**40**)	2	1b	*142*
$Bu^tO{\cdot}C(=CMe_2)OLi{\cdot}TMEDA$	(**41**)	2	1b	*26*
$Bu^tO{\cdot}C(=CHMe)OLi{\cdot}TMEDA$	(**42**)	2	1b	*26*
$MeCH=C(NMe_2)OLi{\cdot}TMEDA$	(**43**)	2	1b	*142*
$H\overline{C=CH{\cdot}(CH=CH)_2C}=C(NMe_2)OLi{\cdot}2THF$	(**44**)	2	*cf.* 1b	*143*
$Bu^t_3COLi{\cdot}THF$	(**45**)	2	24d	*144*
$2,6\text{-}Bu^t_2\text{-}4\text{-}MeC_6H_2OLi{\cdot}OEt_2$	(**46**)	2	24d	*145*
$Bu^tC(=CH_2){\cdot}CH=C(Bu^t)OLi{\cdot}DMPU$[c]	(**47**)	2	24d	*146*
$Bu^tC(=CH_2)ONa{\cdot}O=C(Bu^t)Me$	(**48**)	4	22a	*137*
$Bu^tC(=CH_2)OK{\cdot}THF$	(**49**)	6	24e	*137*

[a] With respect to the organometallic component.
[b] See text.
[c] DMPU, $Me\overline{N(CH_2)_3NMe{\cdot}C}=O$.

an $(OLi)_3$ ring. Moreover, only two of these rings can stack (Fig. 24a). Each $=CH_2$ unit appears to π bond to Li; this also deters further stacking. In (**33**), the enolate of ethyl-*N*,*N*-diethylglycinate (*138*), a similar double stack of $(OLi)_3$ trimers is present. The Et_2N groups coordinate to the Li^+ cations in the rings (Fig. 24b). In contrast to other complexed lithium stacks [but note (**49**) below], the constituent rings in (**33**) are trimeric rather than dimeric. This presumably reflects the size and the extent of the enolato substituents. With smaller substituents, and with monocomplexation by "external" polar molecules, cubic tetramers form [(**34**) and (**35**)] (*139*). The presence of a polar group in the oxyanion ligand leads to the same result. Note the tetrameric structures of (**36**) (Fig. 24c), (**37**) (*140*), and (**38**) (*141*), containing Me_2N—Li,

$$>C{=}O{-}Li,$$

and H_2N—Li coordinations, respectively. In (**39**), a tetramer also results because only the "external" THF ligands, and not the MeO groups, complex to lithium. The dimers (**40**)–(**44**) result from there being two polar groups (as well as the enolato anion) attached to each Li. In (**40**) (*142*) there are no internal complexing sites, and the two polar groups

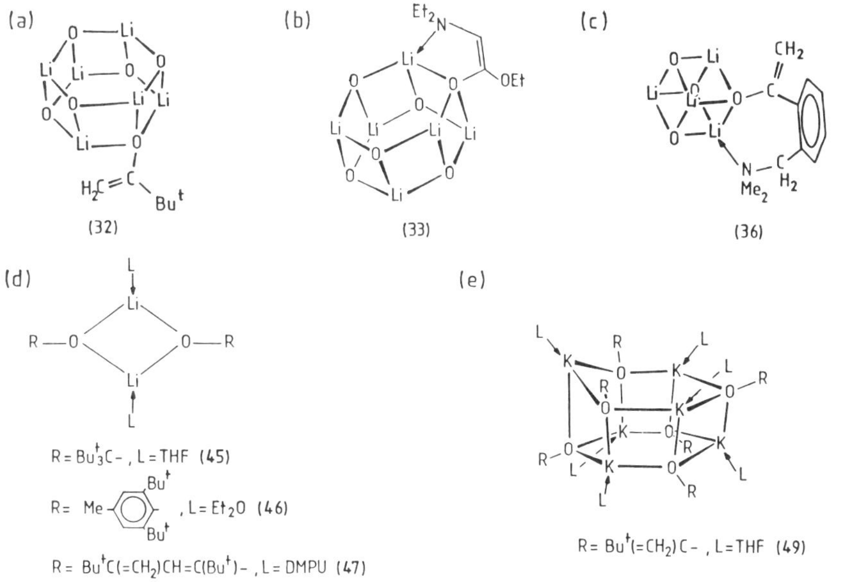

(a) (32) (b) (33) (c) (36)

(d) $R = Bu^t_3C-$, L = THF (45); R = 2,6-Bu^t_2-4-Me-C_6H_2, L = Et_2O (46); $R = Bu^tC(=CH_2)CH=C(Bu^t)-$, L = DMPU (47)

(e) $R = Bu^t(=CH_2)C-$, L = THF (49)

FIG. 24. Structures of selected oxycarbanion–alkali metal compounds and complexes with O—M bonds.

are those of TMEDA (i.e., two Me_2N groups). The same is true for (**41**), (**42**), (**43**) (*142*), and (**44**) (*143*) because their internal and potentially coordinating sites (Bu^tO and Me_2N, respectively) are not capable, for steric reasons, of attaching to lithium. Complexes (**45**) (*144*), (**46**) (*145*), and (**47**) (*146*) all are dimers (Fig. 24d) despite having only one added monodentate ligand per Li center, and no additional internal ligand sites. The particularly bulky R groups attached to the O^- centers extend above and below each $(OLi)_2$ ring. This precludes stacking [*cf.* (**34**) and (**35**)]. The final two examples in Table IV illustrate what happens due to the switching from lithium to heavier alkali metals (*137*). The behavior of the sodium enolate (**48**) mirrors that of similar lithium derivatives. Complexation of each Na^+ cation by a single monodentate base leads to dimeric $(ONa)_2$ rings, two of which stack. The larger potassium cations favor formation of trimeric $(OK)_3$ rings; double stacking then gives hexamer (**49**) (Fig. 24e). It is interesting that the larger cations dictate the ring size here, but not the extent of stacking.

Finally we turn to lithium halides and their coaggregates with organolithiums (Table V). By themselves, lithium halides can form dimeric rings that are perfectly set up to both stack and to ladder. This is so for both electrostatic and steric reasons: the Hal^- anions have high electron densities and there are no substituents *exo* the ring. Three-dimensional infinite ionic lattices (Fig. 3c) result. Addition of polar molecules might break down these systems, at least partially. Note the major complexed structural types described earlier for N—Li, C—Li, and O—Li bonded derivatives: ring dimers and double stacks (cubic tetramers) of such dimers. For example, $(LiBr)_\infty$ provides the dimer (**50**) (*146*) (Fig. 25a) with acetone. The rings can neither stack, because the acetone molecules (two per lithium center) project above and below the $(LiBr)_2$ ring plane, nor can they ladder, because these complexants use

TABLE V

CRYSTAL STRUCTURES OF LITHIUM HALIDE AND MIXED LITHIUM HALIDE–ORGANOLITHIUM COMPLEXES

Structural formula[a]	Formula number[a]	Figure	Ref.
$[LiBr\cdot(O{=}CMe_2)_2]_2$	(**50**)	25a; *cf.* 2b	*146*
$(LiCl\cdot HMPA)_4$	(**51**)	25b; *cf.* 22a	*85, 147, 148*
$(PhLi\cdot OEt_2)_3\cdot LiBr$	(**52**)	25c; *cf.* 22a	*122*
$(\overline{CH_2CH_2CH}Li\cdot OEt_2)_2\cdot(LiBr\cdot OEt_2)_2$	(**53**)	*cf.* 25b, c; 22a	*149*

[a] See text.

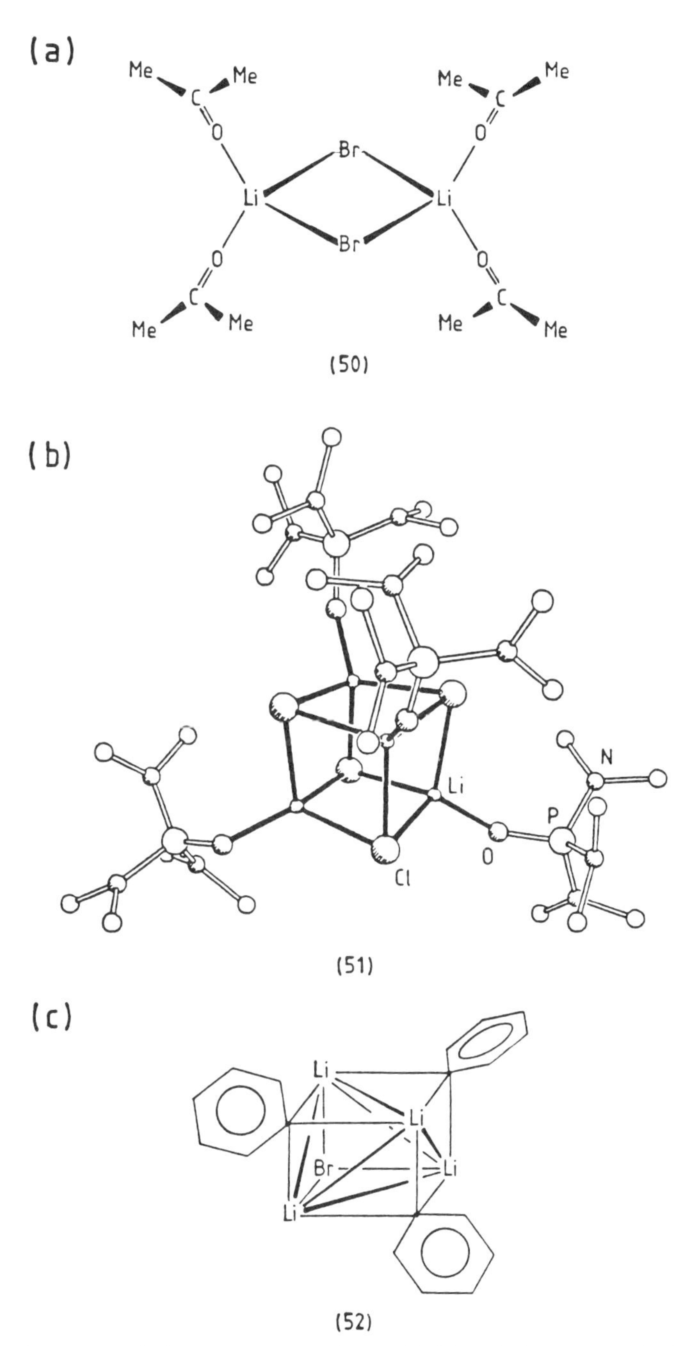

FIG. 25. Structures of selected lithium halide and lithium halide–organolithium complexes.

up lateral space as well. The presence of just one monodentate base per Li center will allow two such rings to stack, as in (**51**) (Fig. 25b). Many anhydrous salts [$(LiHal)_\infty$] cannot be dissolved directly in the polar solvents, but small $(LiHal)_n$ units can be captured using *in situ* methods, e.g., by reactions, in the presence of complexants, of organolithium reagents in hydrocarbon solvents with organic or metal halides (*85*), or of solid ammonium halides suspended in toluene with Bu^nLi, with solid LiH, or with metallic Li (*147, 148*). Two tetrameric mixed aggregates also have been structurally characterized, (**52**) (*122*) and (**53**) (*149*). These species are relevant to the well-known and distinct differences in reactivity between "pure" organolithium (RLi) reagents and those containing lithium halides (by virtue of their preparation from RHal plus two equivalents of Li metal). Deliberate addition of lithium halides to solutions of lithium amides (see Section III) and of lithium enolates is also known to lead to the formation of mixed aggregates (*4*). In (**52**), which can be regarded as a stack of $(CLi)_2$ and Li(C,Br)Li rings (Fig. 25c), the unique Li^+ cation attached to three Ph^- anions is not fully ligated. The other lithium centers, each bonded to two Ph^- and a Br^- anion, have more open coordination sites. Complex (**53**) has a structure similar to (**52**), although every Li^+ is bonded to at least one Br^-.

III. "Simple" Lithium Amides (Amidolithiums) and Their Complexes; Ring Laddering

A. Introduction

The products obtained by lithiating amines RR′NH, usually by employing commercially available organolithium reagents, e.g., MeLi and Bu^nLi, are commonly referred to as lithium amides. Unfortunately, the "amide" designation refers to two distinctly different classes of molecules: (1) those containing RR′N·C(=O)·R″ groups and (2) salts of ammonia or of amines, $RR'N^-M^+$. A "lithiated amide" would be, e.g., RCONHLi, whereas a "lithium amide" would be RR′NLi. This rather schizophrenic nomenclature is widely used, and has been employed in this present review. The "simple" in the title of this section should also be explained. We restrict ourselves here to describing the products of lithiating monoamines RR′NH whose R,R′ groups contain no additional functionalities that involve themselves with N-attached Li centers in the resulting structures. Such functionalities, inherent in RR′NH precursors, can be of several types, e.g. (1) further N centers when R or R′ is pyridyl (C_5H_4N), when an —NMe_2 side-arm is present, or when the precursor is a di- or triamine, (2) O centers provided by —OMe side-arms or by O atoms of aza crowns, (3) S centers, as in the lithiation of mercaptoamines, and (4) potentially coordinating π systems, when heterocyclic amines (e.g., pyrrole, carbazole, and indole) are lithiated. Studies on such functionalized lithium amides are in general quite recent. To date, they have involved mainly the elucidation of solid-state structures, with rather less emphasis on calculational and solution investigations. We will discuss such species in a follow-up review.

Section II of this review concentrated on iminolithium species and showed that their ring-stacking propensities are widespread in organolithium chemistry (see especially Section II,E). These structural preferences occur because substituents attached to most $(NLi)_n$, $(CLi)_n$, or $(OLi)_n$ rings ($n = 2, 3$) usually also lie in the same plane. This allows vertical association ("stacking"). Rings also are basic building blocks for lithium amides, but there is a crucial stereochemical difference. Because of the near-tetrahedral geometries at the amide N centers in these $(RR'NLi)_n$ systems, the R and R′ groups project above and below

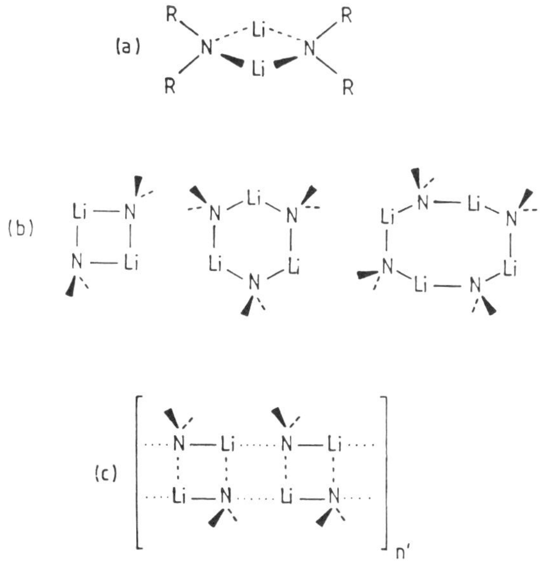

FIG. 26. Structural types for uncomplexed lithium amides $(RR'NLi)_n$: (a) dimeric ring ($n = 2$), showing the projection of R,R′ groups above and below the ring plane; (b) rings with $n = 2$, 3, and 4; (c) further association of rings into ladders.

the $(NLi)_n$ ring plane (Figure 26a), shown for $n = 2$). Therefore stacking is prevented. Several $(RR'NLi)_n$ rings ($n = 2, 3, 4$) have been observed (Fig. 26b). However, further association of such rings may occur *laterally,* by so-called ring laddering (*20, 21, 150*) (Fig. 26c).

We describe first the structures found for uncomplexed lithium amides (i.e., those not containing added neutral Lewis bases such as THF, HMPA, or TMEDA) by X-ray crystallography, by molecular orbital calculations, and by solution methods (Section III,B). In Section III,C, the structures of lithium amide complexes are discussed similarly.

B. Uncomplexed Lithium Amides

1. *Solid-State Structures*

As described above, ring oligomers or ladders are structural options for uncomplexed lithium amides $(RR'NLi)_n$. Isolated rings occur when the R and R′ groups are bulky and occupy much of the lateral space around the ring. Such sterically hindered lithium amides are widely used in organic syntheses as strong bases (see Section I,A). Significantly, they can be both prepared and used in nonpolar solvents. Notable examples include those with R = R′ = Pr^i, Me_3Si, *c*-hexyl; R=Ph, R′=Bu^t,*t*-amyl, *l*-adamantyl; R=Bu^t, R′=*t*-octyl; R=Pr^i, R′=*c*-hexyl; and RR′N=$Me_2\overline{C(CH_2)_3CMe_2N}$ (*4, 6–8, 11, 13, 151–155*). Although most of these compounds can be expected to form rings, only five solid-state structures are available (Table VI). The ring size (Fig. 26b) depends on the size of the R and R′ groups. The large tetramethylpiperidinato anion leads to the formation of $[Me_2\overline{C(CH_2)_3CMe_2N}Li]_4$ (**54**) (*19*), the only simple ring tetramer known in organolithium chemistry. The N—Li—N arrangements in the ring are almost linear. When R,R′ groups are less sterically demanding, trimeric rings are formed: $[(PhCH_2)_2NLi]_3$ (**55**) (*20, 76*), $[(Me_3Si)_2NLi]_3$ (**56**) (*17, 18*), and $[(Me_3Ge)_2NLi]_3$ (**57**) (*156*). The two-coordinate nature of the Li^+ cations in (**55**) (Fig. 27a) allows (or prompts) additional Li···H—C interactions (*43–49*). The $PhCH_2$ groups pivot toward the Li^+ cations (Fig. 27b) and quite short Li—H_2C, Li—α—C, and Li—*o*—CH contacts result (shortest distance to C, 2.60 Å; to H, 2.32 Å). The $[(c\text{-hexyl})_2NLi]_n$ amide has been isolated as a colorless crystalline material (melting point 209–211°C), but twinning has so far prevented solution of the structure (*157*). An unsolvated $(AsLi)_3$ ring trimer $\{[(Me_3Si)_2CH]_2AsLi\}_3$ has also been reported (*158*). However, in contrast to the $(NLi)_3$ species (**55**)–(**57**), the ring is not planar and deviates from threefold symmetry. One of the

TABLE VI

STRUCTURES AND KEY PARAMETERS FOUND FOR UNCOMPLEXED LITHIUM AMIDES

Structural formula[a]	Formula number[b]	Range of N–Li distances (Å)[c]	Range of NLiN angles (°)[c]	Figure	Ref.
$[Me_2C(CH_2)_3CMe_2NLi]_4$	(**54**)	(2.00)	(169)	26b	*19*
$[(PhCH_2)_2NLi]_3$	(**55**)	1.91–2.04 (1.95)	141–147 (144)	26b; 27	*20,76*
$[(Me_3Si)_2NLi]_3$	(**56**)	1.98–2.02 (2.00)	144–151 (147)	26b	*17,18*
$[(Me_3Si)_2NLi]_2$[d]		(1.99)	(100)	26b	*80*
$[(Me_3Ge]_2NLi]_3$	(**57**)	1.88–1.99 (1.92)	142–157 (149)	26b	*156*
$[H_2C(CH_2)_5NLi]_6$	(**58**)	1.99–2.00 (2.00)	—	28	*164*
		2.06–2.12 (2.09)	—		

[a] By X-ray crystallography unless otherwise stated.
[b] See text.
[c] Mean value in parentheses.
[d] By electron diffraction in the gas phase.

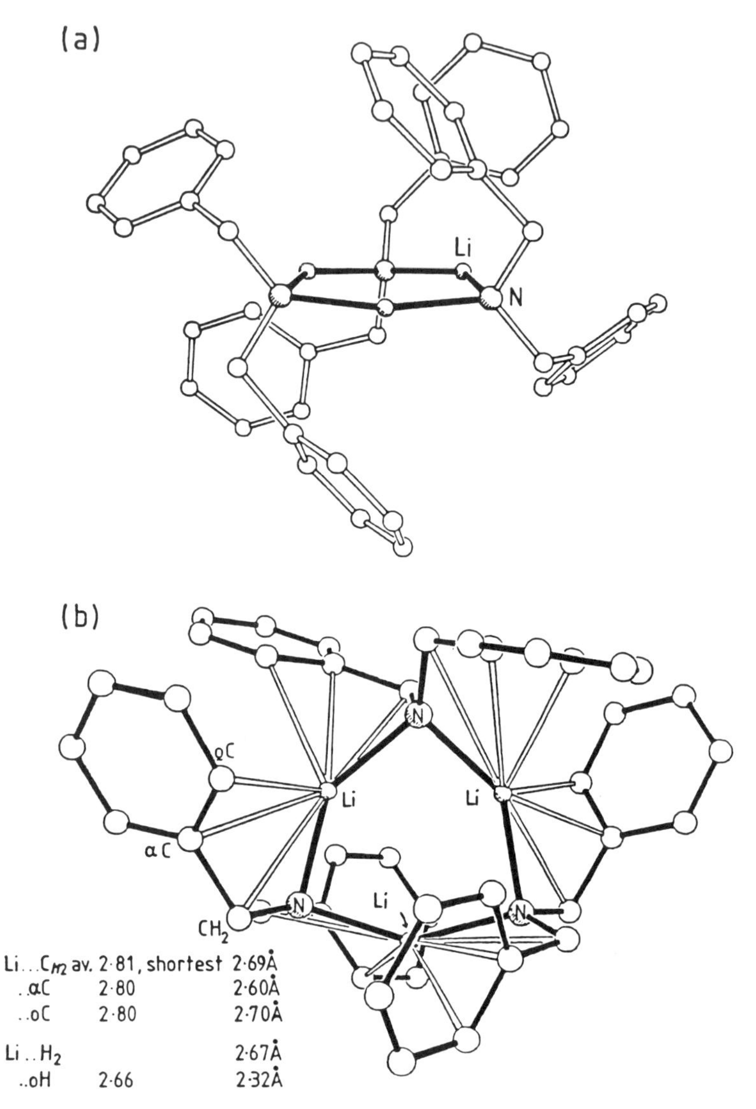

FIG. 27. (a) Molecular structure of $[(PhCH_2)_2NLi]_3$ (**55**); (b) Li···H_2C, ···αC, and ···*o*-CH interactions in (**55**).

Li—As—Li angles is significantly smaller [98(1)°] than the other two [103(1)/104(1)°].

Isolated uncomplexed lithium amide dimers are not known in the solid state. However, electron diffraction shows (**56**) to be dimeric in the gas phase (*80*). The major change in structural parameters with ring size occurs for the NLiN angles. Comparisons can be made with the phosphidolithium compound $\{[(Me_3Si)_2CH]_2PLi\}_2$, a dimer even in the solid state (*159*). Presumably the larger P atom (hence longer P—Li bonds versus N—Li bonds) allows adequate separation of the R and R′ groups and so removes any steric necessity to adopt a larger ring.

Heavier alkali metal analogs of (**55**) and (**56**) are noteworthy. Dibenzylamidosodium $[(PhCH_2)_2NNa]_n$ has been prepared as red crystals (*160*). A structure is not yet available, but its low melting point (98–101°C) and reasonable solubility in hydrocarbons suggest an oligomeric ring structure. The sodium and potassium analogs of (**56**) also are known, $[(Me_3Si)_2NNa]_\infty$ (*161*) and $[(Me_3Si)_2NK \cdot toluene]_2$ (*162*), respectively. The former is a polymer with N—Na—N—Na zig-zag chains (angles at Na, ~150°). For the latter, the larger size of K versus Li allows dimer formation despite the decreased NKN ring angle (94°).

Lithium amides with relatively small or flat R and R′ groups will give ladder structures (Figure 26c). Further association occurs laterally by joining N—Li ring edges [*cf.* $(NLi)_{2,3}$ ring faces in the case of stacks]. The internal Li centers become three coordinate. In many cases the lithiation of amines in nonpolar solvents results in the formation of amorphous materials, insoluble in such solvents and having high melting points, e.g., R=Ph, R′=Ph, naphthyl, $PhCH_2$, and Me, all with melting points >250°C (*20, 163*) and RR′N=$H_2\overline{C(CH_2)_3N}$ (pyrrolidide) (*21, 150*). All of these lithium amides presumably form very long ladders. Only one oligomeric lithium amide ladder is known so far. Lithiation of hexamethyleneimine in hexane/toluene produces the crystalline and hydrocarbon-soluble compound $[H_2\overline{C(CH_2)_5N}Li)]_n$, (**58**) (*164*), a hexamer (n = 6) in the solid state (Fig. 28a). This is the first structurally characterized uncomplexed lithium amide with $n > 4$ and the first structure that is not a simple planar $(NLi)_n$ ring. Other views of (**58**) make clear that it could be regarded as a stack of two six-membered $(NLi)_3$ rings (Figure 28b) or as a cyclized ladder of six N–Li rungs/three $(NLi)_2$ rings (Fig. 28c).

In this regard, one can predict the N—Li bond length pattern expected for a cyclized ladder (Fig. 29). The shortest N—Li bond length is expected in uncomplexed RR′NLi monomers, e.g., in the gas phase or in calculational studies. Longer, but still short, N—Li bonds are expected in isolated rings because each $RR'N^-$ ligand presents two lobes, each containing a pair of electrons, for bonding to two Li^+ centers [*cf.* the N—Li bond lengths for (**54**)–(**57**), Table VI]. However, when such rings associate, the electron density on the amide ligands must interact with three Li^+ centers, and this requires an electronic reorientation. Hence, one expects alternating shorter and longer N—Li bonds for the ladder edges, and longer bonds also for the N—Li ladder rungs.

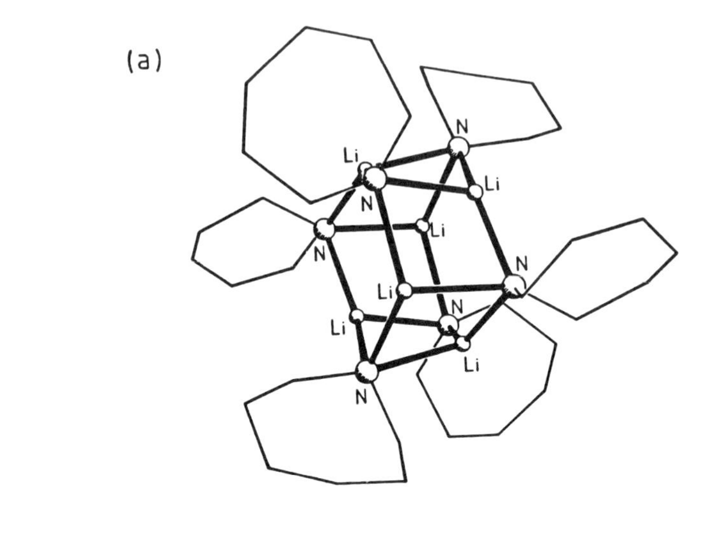

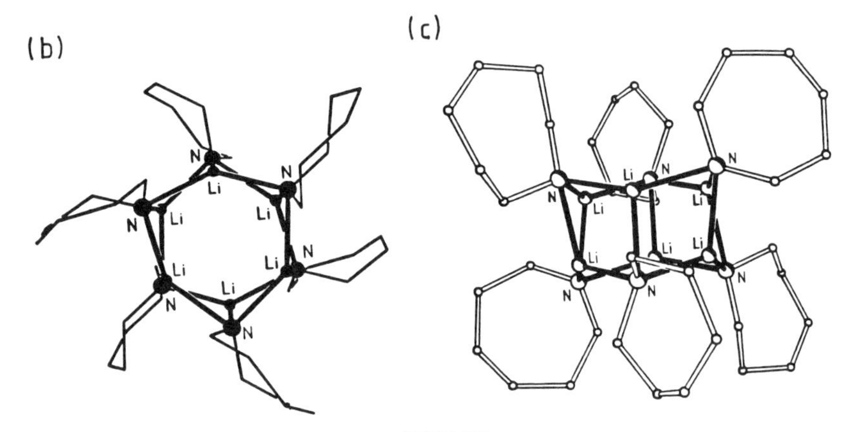

FIG. 28. (a) Molecular structure of $[H_2\overline{C(CH_2)_5N}Li]_6$ (**58**); (b) view of (**58**) as two stacked trimeric rings; (c) view of (**58**) as a cyclized ladder of three dimeric rings/six N—Li rungs.

This is observed for (**58**), whose N—Li bond lengths fall into two very distinct sets (Table VI). In fact, the general pattern is one of alternating shorter (1.99–2.00 Å, mean 2.00 Å) and longer (2.07–2.12 Å, mean 2.10 Å) bonds within the six-membered rings (Fig. 28b), and medium distances (2.06–2.09 Å, mean 2.07 Å) between the rings. This is different from the pattern found for iminolithium hexamers (**1**)–(**4**), whose structures were interpreted in terms of two stacked trimeric rings. There, alternating shorter and medium-length bonds are found within each

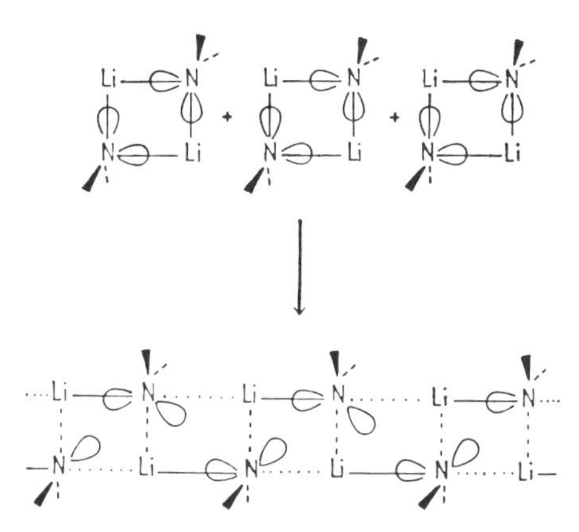

FIG. 29. Representation of the association of $(NLi)_2$ rings into a ladder structure, and of the bond length pattern expected.

ring and longer bonds are found between the rings (Section II,A; Table I). Hence (**58**) is better regarded as a cyclized ladder (Fig. 28c). We will return to this point when the MO optimized structures of $(H_2NLi)_6$ are discussed (Section III,B,2).

2. Calculational Studies

Ab initio and MNDO calculations have been performed on uncomplexed $(H_2NLi)_n$ models. Structures with $n = 1$ (*103*), with $n = 2$ and 3 (*108, 165–168*), with $n = 4$ (*21, 108, 111, 150, 168*), and with $n = 6$ (*115, 169*) have been optimized at various theoretical levels. For the oligomers with $n = 2$ and 3, the geometries are in excellent agreement with the experimental X-ray and electron diffraction data (Section III,B,1; Table VI). All calculations show the $(NLi)_n$ rings with H atoms perpendicular to the ring plane (Fig. 30a and c) to be preferred to the forms with all atoms in the same plane (Fig. 30b and d).

The relative energies of the two forms of each ring are given in Table VII. The electronic reason for these preferences is clear. The amido N^- electron density is tetrahedral-like and interactions with two lithium centers in the same plane as H_2N^- are much less effective (see Section III,A and Fig. 26a). Contast this with the directionality of the imide N^- (sp^2) electron density, which dictates the formation of ring systems with substituent atoms in the same plane. For example, planar $(H_2C{=}NLi)_2$ is preferred to the perpendicular form by 17.0 kcal mol^{-1} [6-31G (*72, 74*)] or by 6.8 kcal mol^{-1} [MNDO (*108*)] (Section II,D; Fig. 20).

The association energies for the favored perpendicular forms of

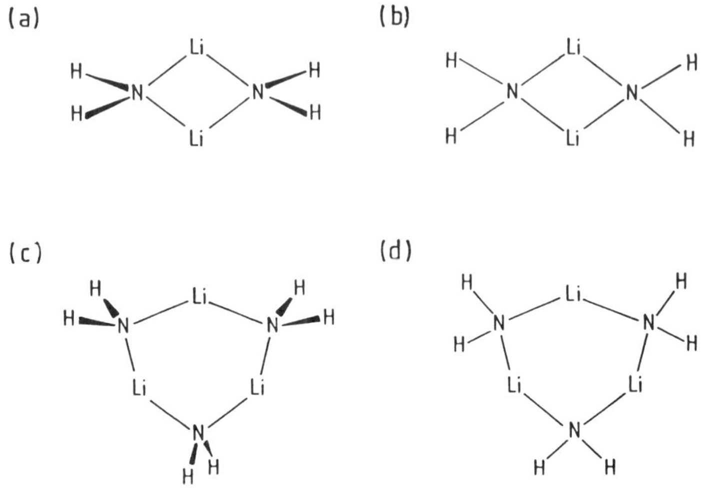

FIG. 30. Perpendicular and planar forms of $(H_2NLi)_2$ and $(H_2NLi)_3$ rings.

$(H_2NLi)_2$ and $(H_2NLi)_3$ (Table VIII) show that a trimer is preferred over a dimer. This is most simply ascribed to the more favorable Li—N—Li angles in the larger ring (Table IX) and to the reduction in repulsive interactions (see below). Though these calculations neglect the bulk of the substituent groups and crystal packing effects, no crystalline $(RR'NLi)_2$ dimers are known experimentally (Table VI), only trimers (**55**), (**56**), and (**57**) and a tetramer (**54**). The calculated N—Li bond lengths (Table IX) are similar for the dimer and the trimer. However, the expansion in ring size is manifested particularly by the large increase in the angle at lithium (approaching 40°), as the Li—N—Li

TABLE VII

RELATIVE ENERGIES OF PERPENDICULAR AND PLANAR FORMS OF $(H_2NLi)_2$ AND $(H_2NLi)_3$

	Perpendicular D_{nh} (kcal mol^{-1})	Planar D_{nh} (kcal mol^{-1})	Theoretical level	Figure	Ref.
$(H_2NLi)_2$	0.0	21.9	6-31 + G*/6-31G*//6-31G*[a]	30a and b	*165*
	0.0	22.0	MNDO[b]		*108,165*
$(H_2NLi)_3$	0.0	42.6	3-21 + G/3-21G//3-21G[a]	30c and d	*165*
	0.0	45.4	MNDO		*165*

[a] The diffuse orbitals were omitted from Li in these calculations.
[b] Minimum neglect of differential overlap.

TABLE VIII

ASSOCIATION ENERGIES OF PERPENDICULAR FORMS OF $(H_2NLi)_2$ AND $(H_2NLi)_3$

	Association energy (kcal mol^{-1} per monomer unit)	Theoretical level	Ref.
$(H_2NLi)_2$	−32.6	6-31 + G*/6-31G*//6-31G*[a]	*165*
	−31.4	MNDO[b]	*108, 165*
	−35.9	MP2/6-31G*//3-21G	*166a*
	−36.8	6-31G//6-31G	*167*
$(H_2NLi)_3$	−41.3	3-21 + G/3-21G//3-21G[a]	*165*
	−38.0	MNDO	*165*
	−44.6	6-31G//6-31G	*167*

[a] Diffuse functions on N, but not on Li.
[b] MNDO, Minimum neglect of differential overlap.

angle, is less variable. 6-31G* calculations on $H_2NLi_2^+$ show that the geometry with the Li—N—Li angle being ~130° is the lowest point in energy. The energy rises similarly on deviation above or below this ideal Li—N—Li angle (*168*). Also, as the Li—Li and N—N distances are increased on ring expansion, the reduction in these repulsive interactions favors the trimer. These $(H_2NLi)_n$ species are obviously highly ionic. A natural population analysis (*38*) indicates lithium charges of +0.9 for monomeric, planar H_2NLi (*170*). The ionic nature of N—Li bonds, and X—Li bonds in general, is also exemplified by the calculated dimerization energies of lithium compounds with first-row substituents. The association energies correlate well with the electronegativity of the substituent, e.g., they become more negative with increasing electronegativity (*166*).

For the tetramer $(H_2NLi)_4$, three structures have been examined.

TABLE IX

SELECTED BOND PARAMETERS FOR PERPENDICULAR FORMS OF $(H_2NLi)_2$ AND $(H_2NLi)_3$

	Distance (Å)			Angle (°)		Theoretical level	Ref.
	N—Li	Li—Li	N—N	Li$\hat{N}$Li	N$\hat{L}$iN		
$(H_2NLi)_2$	1.943	2.314	3.122	73.1	106.9	6-31G*	*165*
	1.931	—	—	72.3	107.7	6-31G	*167*
$(H_2NLi)_3$	1.906	2.781	3.649	93.7	146.3	3-21G	*165*
	1.95	2.94	3.69	98	142	6-31G	*165*
	1.928	—	—	94.5	145.5	6-31G	*167*

These are depicted in Fig. 31a showing a planar eight-membered ring, with NH_2 groups perpendicular to the ring plane, in Fig. 31b showing a ladder formed by lateral association of two dimeric rings, and in Fig. 31c showing a tetrahedral form, which can be viewed as two stacked dimeric rings. All studies find the large ring structure to be more stable than the stack. The energy difference depends on the calculational method used (Table X). Low-level (3-21G) *ab initio* calculations on $(H_2NLi)_n$ species result in energies that are far too negative due to the basis set superposition error (BSSE) (*171, 172*). It is now well established that diffuse as well as polarization functions are necessary to eliminate BSSE and hence to provide a more nearly complete description of first-row molecules with lone pairs (*173–176*). One study (*21*) found the ladder structure to be a local minimum, intermediate in

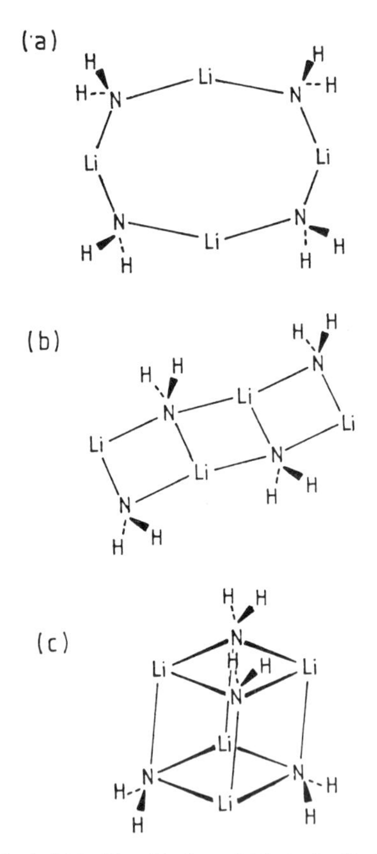

FIG. 31. (a) Ring (D_{4h}), (b) ladder (C_{2h}), and (c) stack (D_{2d}) forms of $(H_2NLi)_4$.

TABLE X

RELATIVE ENERGIES AND ASSOCIATION ENERGIES FOR OPTIMIZED STRUCTURES OF $(H_2NLi)_4$[a]

Relative energies			Association energy for ring form (kcal mol^{-1})	Theoretical level	Ref.
Ring (D_{4h})	Ladder (C_{2h})	Stack (D_{2d})			
0.0	—	8.9	−41.8	HF/6-31G + sp + d	*111*
0.0	—	11.5	−52.9	3-21G	*111*
0.0	—	28.4	−40.0	MNDO[b]	*108*
0.0	7.4	11.4	—	6-31G	*21,150*

[a] See Fig. 31.
[b] Minimum neglect of differential overlap.

energy between the ring and the stack. The association energies found for the tetrameric ring are large, comparable to those noted earlier for a trimer (Table VIII).

The above results are significant for several reasons. Calculations on most other lithium-containing tetramers favor stacked, tetrahedral forms, e.g. $(LiF)_4$, $(LiOH)_4$ (*111*), and $(MeLi)_4$ (*101, 102*). Why then is a ring preferred to a stack for $(H_2NLi)_4$? One possible answer emphasizes the importance of lone-pair orientation effects (*111*). The two lone-pair lobes on H_2N^- are unable to interact with the three Li^+ centers on a tetrahedral face as effectively as with the two centers of a dimer. Compare this with the situation for F^- and HO^-, which have four or three lone pairs, respectively. Moreover, H_3C^- (with only one lone-pair lobe) also has effective conical (threefold) symmetry. However, $RR'C{=}N^-$ (with two lobes, the same as H_2N^-) manages to bond to three Li^+ ions, and forms stacks. This is due to the contribution from the C=N π bond, which again results in effective conical symmetry. Another argument can be based on stereochemistry. Consider the approach and subsequent stacking of two $(RLi)_2$ rings, with R=F, OH, or NH_2 (Fig. 32). Only in the case of R=NH_2 (Fig. 32c) are there substituents *between* the two $(RLi)_2$ rings. The NH_2 groups cannot twist much in order to allow closer approach of the two rings. If they did, the N—Li bonding *within* each ring would be weakened (Fig. 30a and b). These points are quantified when one examines bond lengths within the optimized ring and stack structures of $(H_2NLi)_4$ (Table XI). The N—Li bonds are considerably shorter in the ring, and the interdimer N—Li contacts in the stack are particularly long. However, though there are only eight N—Li electrostatic interactions in the ring, stack formation generates 12 such contacts, albeit rather longer ones. The summation of

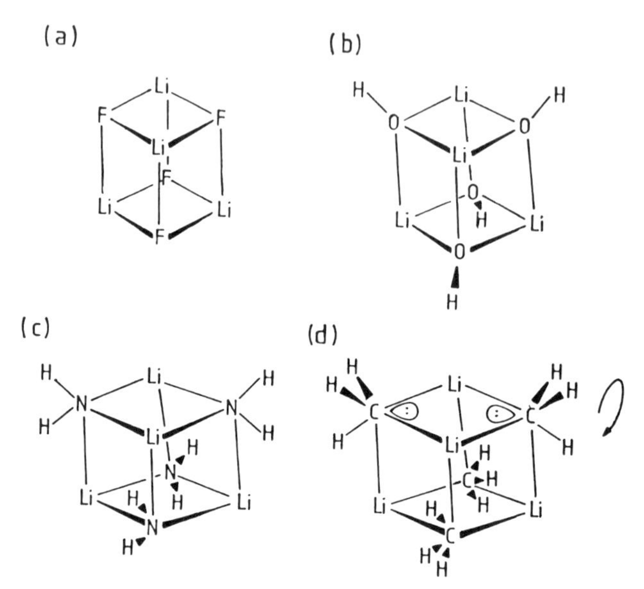

FIG. 32. Representations of the stacking of two $(RLi)_2$ rings: (a) R=F, (b) R=OH, (c) R=H_2N, and (d) R=CH_3.

TABLE XI

SELECTED BOND PARAMETERS FOR OPTIMIZED STRUCTURES OF $(H_2NLi)_4$

	Figure	Distance (Å)		NL̂iN angle (°)	Theoretical level	Ref.
		N—Li	Li···Li			
$(H_2NLi)_4$	31a	1.906	3.003[a]	166.0	3-21G	*111*
ring		1.927	3.058[a]	165.0	6-31G	*21, 150*
$(H_2NLi)_4$	31b	1.882–2.115	2.431/2.444/	—	6-31G	*21,150*
ladder		(mean 2.008	3.780[b]			
		for 10 bonds)	(mean 2.855)			
$(H_2NLi)_4$	31c	1.962/2.027[c]	2.239/2.426[d]	—	3-21G	*111*
stack		2.013/2.042[c]	2.343/2.528[d]	—	6-31G	*21,150*
		(mean 2.023	(mean 2.405)			
		for 12 bonds)				

[a] Two closest contacts per Li.
[b] Across an outer ring/across the inner ring/going along the ladder, respectively.
[c] Within dimeric rings/between them, respectively.
[d] Between Li atoms in different rings/within each ring, respectively.

N—Li bond energies could perhaps be similar for both forms. But repulsions also are important. Eight-membered ring formation allows $Li^{\delta+}$ (and $N^{\delta-}$) centers to be widely spaced. In contrast, Li—Li distances in the stack are much shorter, both within each ring (now only four-membered) and, especially, between rings (Table XI).

These theoretical results help explain why the only known lithium amide tetramer is a ring [(**54**), Table VI], and also why pseudocubane tetramers, so prevalent elsewhere in organolithium chemistry [Section II,B; (**7**) and (**9**); Section II,E; Tables II–V], are unknown for lithium amides. If two $(H_2NLi)_2$ rings prefer not to stack (as shown by the MO calculations), it is unlikely that two "real" $(RR'NLi)_2$ rings, i.e., with more bulky R and R′ substituents, will do so, especially if the lithiums are coordinated further.

The ladder structure of $(H_2NLi)_4$ (Fig. 31b) is intermediate in energy between an eight-membered ring and a stacked arrangement (Table X). Though RR′NLi aggregation avoids three-dimensional stacking, lateral (ladder) association (see Section III,C) may be a viable alternative to the formation of open rings. The close relationship between the two forms is obvious: if the two central N—Li rungs of the ladder are broken, an eight-membered ring will form (Fig. 31a). The bond distances for the ladder (Table XI) are intermediate. The N—Li bonds in the ladder are, on average, longer than those in the ring, but shorter than those in the stack. Antibonding contacts are more important (shorter Li—Li and N—N distances) than those in the ring, but are less important than those in the stack.

Related MO calculations on $(LiH)_4$ tetramers are relevant (*114*). As in $(H_2NLi)_4$, an eight-membered ring was found to be the lowest energy form. Higher in energy were a stack of two dimers (a tetrahedral structure, termed a "ring dimer," relative energy +3.7 kcal mol^{-1}) and then a ladder structure (termed a "fence," +11.0 kcal mol^{-1}). The reversal of stack–ladder preference versus $(H_2NLi)_4$ is understandable. There are no substituents in $(LiH)_2$ rings [*cf.* Fig. 32a, for $(LiF)_2$ rings], so their vertical approach is not discouraged.

Ab initio calculations have also been carried out on hexamers $(RLi)_6$ with R═H (*114*), Me (*102, 116*), F, OH, and NH_2 (*115, 169*). For these highly associated species, rings become disfavored. For example, $(LiH)_6$ prefers a stack of two trimers (relative energy 0.0 kcal mol^{-1}) versus the preference of a ring for $(LiH)_4$ (*114*). Next in energy comes a stack of three dimers (termed a "fence dimer," relative energy +3.5 kcal mol^{-1}), followed by a twelve-membered ring (+7.7 kcal mol^{-1}), and then a ladder (+19.5 kcal mol^{-1}). Specifically for $(H_2NLi)_6$ (*115*), it is not surprising that a stack of two puckered trimers (Fig. 33a) is favored by

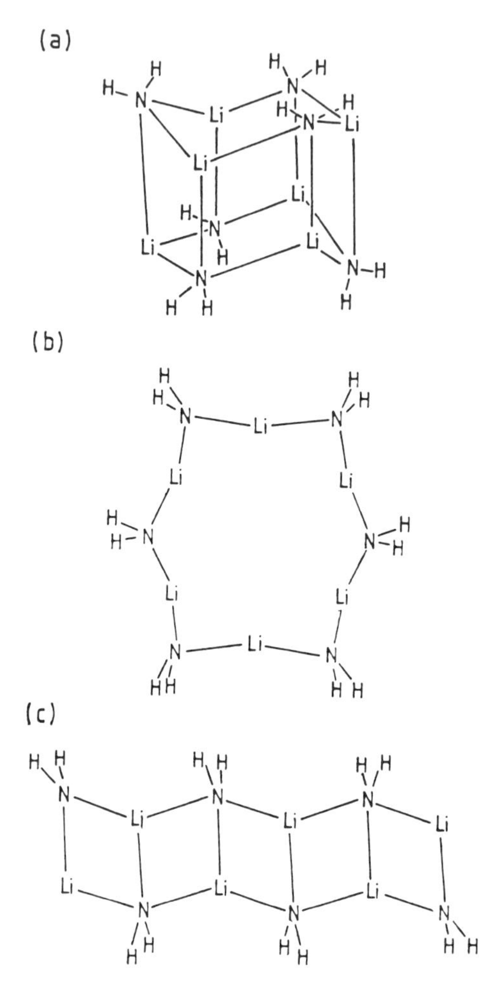

FIG. 33. (a) Stack, (b) ring, and (c) ladder forms of $(H_2NLi)_6$.

20.2 kcal mol^{-1} (6-31G basis set augmented by a set of diffuse sp functions and a set of six d-type functions on N) over a large ring structure (Fig. 33b). The "star" shape shown in Fig. 33b (inverted lithiums) results from the preference of the Li—N—Li angle for lower values. Comparisons were drawn between the stacked $(H_2NLi)_6$ structure and those of the iminolithium hexamers $(RR'C{=}NLi)_6$ (**1**)–(**4**) (see Section II,A). In $(H_2NLi)_6$, just *two* types of N—Li bonds are found, of lengths 1.989 Å within constituent rings and 2.055 Å between the rings. The fact that *three* types of N—Li bond lengths are found in the experimental iminolithium hexamers (see especially Fig. 8 and Table I) was attributed to crystal packing effects (*115*). This was refuted (*177*). The stereochemical propensities of RR′N^- and RR′C=N^- ligands are different. The former favors ring formation or, if this is unattainable, laddering; the latter ligand favors stacking. The $(H_2NLi)_6$ ladder (Fig. 33c) is intermediate in energy between the stack and the ring, i.e., the stack is still preferred by 15 kcal mol^{-1} (*169*). The apparent preference of $(H_2NLi)_6$ for a stack over a ladder may be reversed in experimental systems $(RR'NLi)_6$, where R ≠ H. As has been pointed out (*164*), the structure of the only known amidolithium hexamer, $[H_2\overline{C(CH_2)_5N}Li]_6$ (**58**), which displays three sets of N—Li bond lengths, is better viewed as a cyclized ladder than as a stack (see Section III,B,1, Table VI, and Figs. 28 and 29).

3. *Structures in Solution*

Very few studies on *uncomplexed* lithium amides have been carried out in solution, e.g., in hydrocarbon solvents. Only five oligomeric, ligand-free compounds of this type have been characterized structurally [(**54**)–(**58**); Table VI; Section III,B,1]. Studies carried out on solutions in polar solvents would, in effect, involve complexes. For example, the unsolvated crystalline trimers (**55**) and (**56**) both form dimeric etherate complexes (*19, 20, 76, 81*). Solution studies of complexed lithium amides will be described in Section III,C,3.

The solid-state trimer $[(PhCH_2)_2NLi]_3$ (**55**) dissolves in arene solvents to give pink solutions (*20, 76*). Cryoscopic measurements of benzene solutions gave association state values (n) of 2.87–2.66 (0.033–0.025 M concentrations, respectively). ^{7}Li NMR spectra of relatively concentrated solutions at 20°C only show a single resonance (e.g., at −0.66 ppm for a 0.17 M solution). However, two singlets were observed in more dilute solutions (e.g., δ ≠ − 0.71 and δ −2.91 ppm, ratio ~10 : 1, for a 0.06 M solution). Because n is <3, and two well-separated ^{7}Li NMR resonances are observable, an equilibrium between the trimer and a dimer or a monomer must be present. The electronic spectra of (**55**) gave more information. In benzene solution, absorption occurs in the visible region at λ_{max} 525–530 nm. The absorptivity of this band increased on dilution, e.g., for 5.3×10^{-3} and 5.3×10^{-4} M solutions, a = 100 and 300 M^{-1} cm^{-1}, respectively. Hence, the pink color of solutions of (**55**) seems due to the lower association species (dimer or monomer) present. In fact, configuration interaction MO calculations (*178*) predicted that only a monomer, $(PhCH_2)_2NLi$, should give a visible region HOMO–LUMO transition, at 545 nm. The calculated λ_{max} values are 255 and 320 nm for the trimer and dimer, respectively. As already noted, in the crystal structure of (**55**) the benzyl groups bend in toward the two-coordinate Li centers of the trimeric ring (see Fig. 27b). It was concluded that in solution the trimer is in equilibrium with a monomer, and that the latter (containing one-coordinate Li) has much increased Li···benzyl interactions. These interactions allow a charge transfer mechanism to operate, producing the solution color observed.

The only other relevant study concerns the uncomplexed hexamer, $[H_2\overline{C(CH_2)_5N}Li]_6$ (**58**). This dissolves in arene solvents apparently to give even higher aggregates (*164*). In benzene, cryoscopic measurements gave association state (n) values of 5.4 ± 0.2 to 10.1 ± 0.5 (0.06–0.14 M solutions, respectively). The ^{7}Li NMR spectra of d_8-toluene solutions of (**58**) at −90°C are concentration dependent. Relatively dilute solutions give two sharp singlets [0.19 M, δ −0.50, −1.63 ppm (ratio 1:1); 0.25 M, δ −0.49, −1.61 ppm (ratio 2:1)]. At higher concentrations, only the higher field resonance occurs (e.g., at 0.61 M, δ −0.48 ppm). Clearly, at least two species are present in solution: presumably one is the hexamer (^{7}Li signal at ~δ −1.6 ppm) and the second is an aggregate with $n > 6$ (^{7}Li signal at ~δ −0.5 ppm).

C. Complexed Lithium Amides $[(RR'NLi \cdot xL)_n]$

1. *Solid-State Structures*

We showed earlier that the basic structural units of uncomplexed lithium amides are rings, $(RR'NLi)_n$, with $n = 2, 3, 4$ (Section III, A; Fig. 26). Only rarely, when R,R′ groups are particularly large, can such rings be isolated [Table VI; (**54**)–(**57**)]. More usually, these small rings associate further, laterally, joining N—Li ring edges to give ladders. Although one exception is known [(**58**); Fig. 28], such laddering appears generally to be extensive and gives amorphous, hydrocarbon-insoluble materials. Therefore most lithium amide reagents are generated in polar solvents (L), e.g., Et_2O, THF, HMPA, DMPU, and TMEDA (*6, 7, 11–13, 179*). This results in complexes of type $(RR'NLi \cdot xL)_n$. The polar Lewis base molecules (L) coordinate to the Li centers and limit the degree of association sterically; in particular, extensive laddering is prevented and crystalline, hydrocarbon-soluble materials can be isolated. For example, whereas lithiation of Ph_2NH in hexane or toluene results in the precipitation of a white powder [$(Ph_2NLi)_\infty$, melting point 326°C], the same reaction in Et_2O or Et_2O/toluene media affords colorless crystals of $(Ph_2NLi \cdot OEt_2)_2$ (melting point 50°C). These are very soluble (~2 g ml^{-1}) in toluene or benzene (*157*). Furthermore, the lithium amide reagents made in the presence of these complexants (commonly termed "additives") are much more reactive in subsequent metalations. It has long been assumed that this enhancement is kinetic in origin, reflecting the low association states of the complexes produced (*179–182*).

The major structural types found for lithium amide complexes in the solid state are illustrated in Fig. 34. These comprise ladders of limited extent when the L : Li ratio is less than 1 : 1 (Fig. 34a), dimeric $(NLi)_2$ rings, when this ratio is 1 : 1 and, usually, when the complexants are monodentate (Fig. 34b), and monomers, both contact-ion pairs (CIPs) and solvent-separated ion pairs (SSIPs) (Fig. 34c). Monomers occur always when there are two or more monodentate complexants per Li. This also is usual with bidentate ligands, and is always found when the ligands have higher denticity.

Lithiation of pyrrolidine, $H_2\overline{C(CH_2)_3N}H$, in hexane produces a white insoluble powder, $[H_2\overline{C(CH_2)_3N}Li]_n$ (*21, 150*). This is assumed to have an extensive ladder structure (Section III,A; Fig. 26c); compare the cyclized-ladder structure of crystalline $[H_2\overline{C(CH_2)_5N}Li]_6$ (*164*). However, if lithiation is carried out in the presence of the ligands PMDETA or TMEDA, crystalline complexes are isolated: $\{[H_2\overline{C(CH_2)_3\text{-}N}Li]_3\cdot PMDETA\}_2$ (**59**) (*21, 150*) and $\{[H_2\overline{C(CH_2)_3N}Li]_2\cdot TMEDA\}_2$ (**60**)

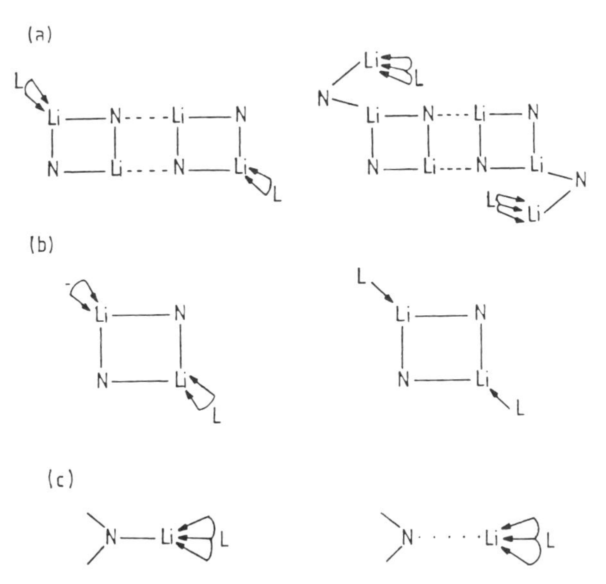

FIG. 34. Structural types for complexed lithium amides $(RR'NLi\cdot xL)_n$, L = a Lewis base: (a) ladders with n = 2 and 3; (b) dimeric rings; (c) monomers (CIPs and SSIPs).

(*150*). The molecular structures are shown in Fig. 35 (see also Fig. 34a). Both are limited-length ladders and only their terminal Li centers are complexed. These provide indirect evidence that association of $(RR'NLi)_n$ rings to give polymers occurs by continuous laddering. Indeed, structures (**59**) and (**60**) may be formed *via* interception of longer ladders [note the structure of (**31**), an intercepted stack, in Fig. 23b]. The structures of both (**59**) and (**60**) contain central ladders made up of two attached $(NLi)_2$ rings or, alternatively, four N—Li rungs. The internal ring of each ladder, N(1)Li(1)·N(1′)Li(1′) in each case, is planar. The N atoms of the outer rungs, N(2) and N(2′), lie near this inner ring plane, but the terminal Li atoms, Li(2) and Li(2′), deviate considerably from this common plane [by ±0.71 Å for (**59**), and by ±1.01 Å for

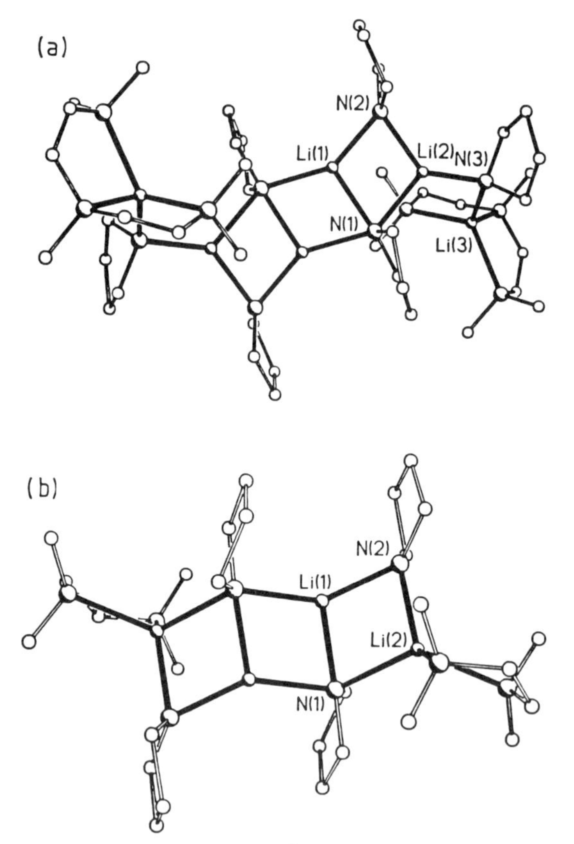

FIG. 35. Molecular structures of (a) $\{[H_2C(CH_2)_3NLi]_3\cdot PMDETA\}_2$ (**59**) and (b) $\{[H_2C(CH_2)_3NLi]_2\cdot TMEDA\}_2$ (**60**).

(**60**)]. In (**60**), these end-Li atoms are each complexed by a TMEDA molecule. For (**59**), there are two N—Li(PMDETA) units attached to these end-Li atoms. Thus, interception (by ligation) of the continuous ladder envisaged for $[H_2\overline{C(CH_2)_3N}Li]_\infty$ leads to its partial disconnection.

The N—Li bond lengths and ring internal angles for (**59**) and (**60**) are given in Fig. 36. These will be analyzed in detail below along with the results of MO calculations on $(H_2NLi)_4 \cdot (H_2O)_n$ complexes ($n = 2$ and 4) (Section III,C,2). We note here that the expected N—Li bond length pattern is found (see Section III,B,1; Fig. 29) when two or more $(NLi)_2$ rings join to give a ladder. Alternating relatively short and long bonds occur along the uncomplexed ladder edges [1.95 and 2.03 Å for (**59**), 1.96 and 2.03 Å for (**60**)], and relatively long bonds are also found for the inner rungs [2.05 Å for (**59**), 2.04 Å for (**60**)].

Lateral association is not restricted to lithium amides. Lithium phosphide rings $(RR'PLi)_n$ will have a stereochemistry similar to $(RR'NLi)_n$ rings. The R,R′ groups perpendicular to the $(PLi)_n$ ring plane will preclude stacking, but facilitate laddering. The presence of a deficiency of Lewis base (less than one per Li) already precludes the formation of

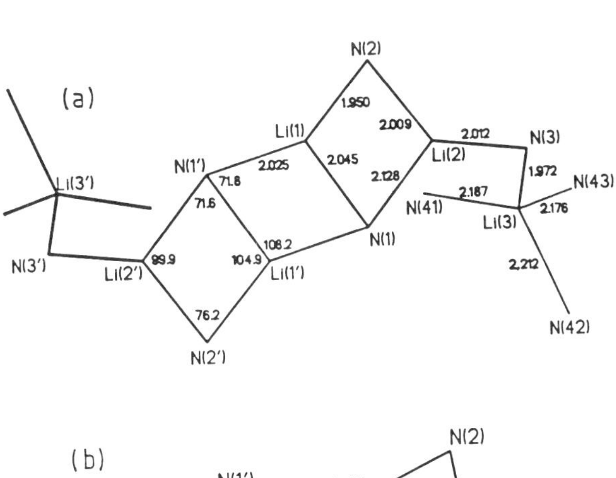

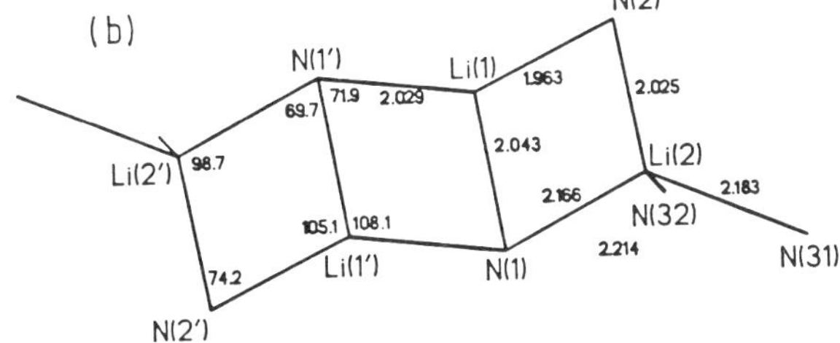

FIG. 36. Key bond lengths and angles in the structures of (a) (**59**) and (b) (**60**).

continuous ladders. Hence, the crystalline complexes $[(Me_3Si)_2PLi]_4{\cdot}2THF$ (*183*) and $(Bu^t_2PLi)_4{\cdot}2THF$ (*184*) have been isolated and structurally characterized. Each has a limited ladder structure with four PLi rungs. Because the complexants (one per terminal Li atom) are monodentate, the P_4Li_4 frameworks are essentially planar.

When the Lewis base (L):Li ratio is 1:1 or greater, ladder-type structures no longer persist (see Section III,C,2). Dimeric rings or monomers (C.I.P.s or S.S.I.P.s) are formed instead (Fig. 34b and c). Structural details for the known solid-state dimers (**61**)–(**68**) are shown in Table XII (*185–188*). Excluding (**68**) for the time being, all contain one monodentate oxyligand (Et_2O, HMPA, or THF) per Li center. The $(NLi)_2$ rings are planar and rhomboidal, with angles at Li of 102–106°. The relatively short Li—Li cross-ring contacts (~2.4–2.5 Å) are due to the acute angles (74–78°) at the μ_2-amido N^- centers, but have little or no metal–metal bonding significance.

The Li centers in these rings are only three coordinate. What limits further association? Two views of $[(PhCH_2)_2NLi{\cdot}OEt_2]_2$ (**61**) provide an answer. Figure 37a shows very clearly the perpendicular projection of R,R′ groups (here, $PhCH_2$) relative to the $(NLi)_2$ ring plane. Such rings cannot stack. Figure 37b, a side-on view of the $(NLi)_2$ ring, demonstrates equally clearly how a single monodentate ligand per Li atom occupies the lateral space around such a ring. Laddering is precluded as well. This contrasts with most other $(RLi{\cdot}L)_2$ systems, e.g., with R = imino, alkyl, aryl, alkynyl, and enolato; L is a monodentate

TABLE XII

DIMERIC LITHIUM AMIDE COMPLEXES: KEY STRUCTURAL PARAMETERS (X-RAY DIFFRACTION DATA)

Complex	Formula number[a]	Distance (Å) N—Li	Li—O	Li—Li	N—Li—N ring angle (°)	Ref.
$[(PhCH_2)_2NLi{\cdot}OEt_2]_2$	(**61**)	1.98–1.99	2.01	2.45	104.0	*20, 76*
$[(PhCH_2)_2NLi{\cdot}HMPA]_2$	(**62**)	2.00–2.01	1.85	2.51	102.7	*20, 76*
$[(Me_3Si)_2NLi{\cdot}OEt_2]_2$	(**63**)	2.06	1.94	—	104.9	*19, 81*
$[(Me_3Si)_2NLi{\cdot}THF]_2$	(**64**)	2.03	1.88	2.43	106.3	*185*
$[2,4,6\text{-}Bu^t_3C_6H_2{\cdot}N(H)Li{\cdot}OEt_2]_2$	(**65**)	1.99–2.04	1.91	—	102.2	*186*
$[Mes_2B{\cdot}N(H)Li{\cdot}OEt_2]_2$	(**66**)	2.01–2.02	1.96	—	105.1	*187*
$[1,8\text{-}C_{10}H_6NH{\cdot}SiMe_2{\cdot}NLi{\cdot}OEt_2]_2$	(**67**)	2.00–2.03	1.94	2.43	106	*188*
$[Ph{\cdot}N(Me)Li{\cdot}TMEDA]_2$	(**68**)	2.08	—[b]	2.74	97.6	*163*

[a] See text.
[b] Li—N (of TMEDA) distances, 2.30 Å.

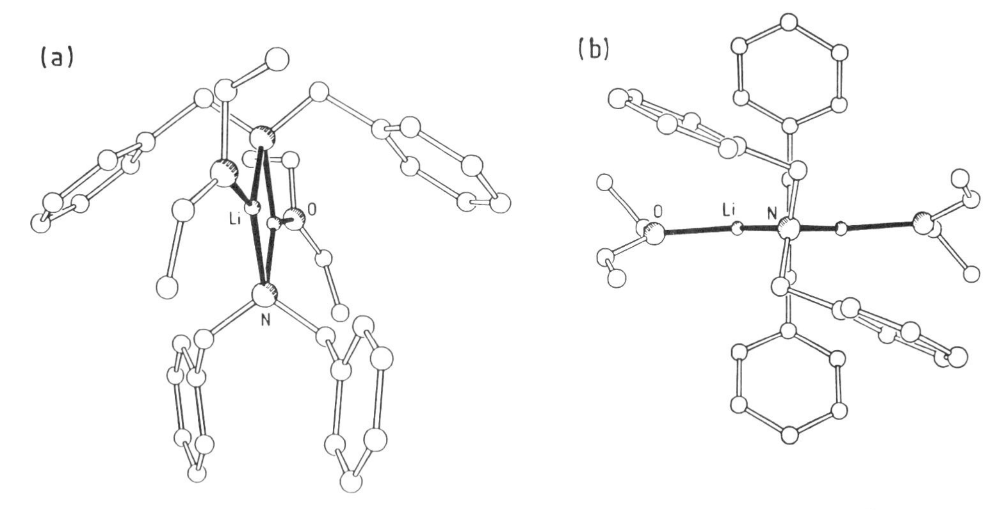

FIG. 37. Molecular structure of $[(PhCH_2)_2NLi \cdot OEt_2]_2$ (**61**): (a) view showing the perpendicular projection of $PhCH_2$ groups relative to the $(NLi)_2$ ring; (b) view showing the lateral extension of Et_2O ligands.

base, where stacking is not precluded, and cubic tetramers result (see Section II,B, Figs. 14 and 17, Section II,E, and Tables II–V).

Why are complexes (**61**)–(**68**) *dimers?* The same question was posed earlier regarding iminolithium species (Section II,B;Fig. 13). For the lithium amide complexes, the structures of (**61**), (**62**), (**63**), and (**64**) are particularly relevant here because the uncomplexed analogs, $[(PhCH_2)_2NLi]_3$ (**55**) and $[(Me_3Si)_2NLi]_3$ (**56**) (Section III,B,1; Table VI], are both trimers. As shown in Fig. 38a, the wide angle at Li in such trimers (~145°) leaves only a narrow coordination arc for an incoming complexant. In contrast, in a $(RR'NLi)_2$ dimer the ring angle at Li is expected to be ~100° (Fig. 38b): such is found for $[(Me_3Si)_2NLi]_2$ in the gas phase (Table VI). This leaves much more space for accommodation of an additional ligand. We will address this question in a more quantitative way in Section III,C,2.

Further comparison of the structures of (**61**) and (**62**) (Table XII) with that of (**55**) (Table VI) and of (**63**) and (**64**) with that of (**56**) shows that complexation of the Li centers (making them three coordinate) lengthens the N—Li distances slightly. Judging from Li—O bond lengths, HMPA and THF seem to be rather stronger complexants than Et_2O. A final point worth noting is that the unsymmetrically substituted lithium amides ($R \neq R'$), i.e., (**65**)–(**68**), all crystallize in the *transoid* form. This is illustrated for (**68**) in Fig. 39. This particular structure is the

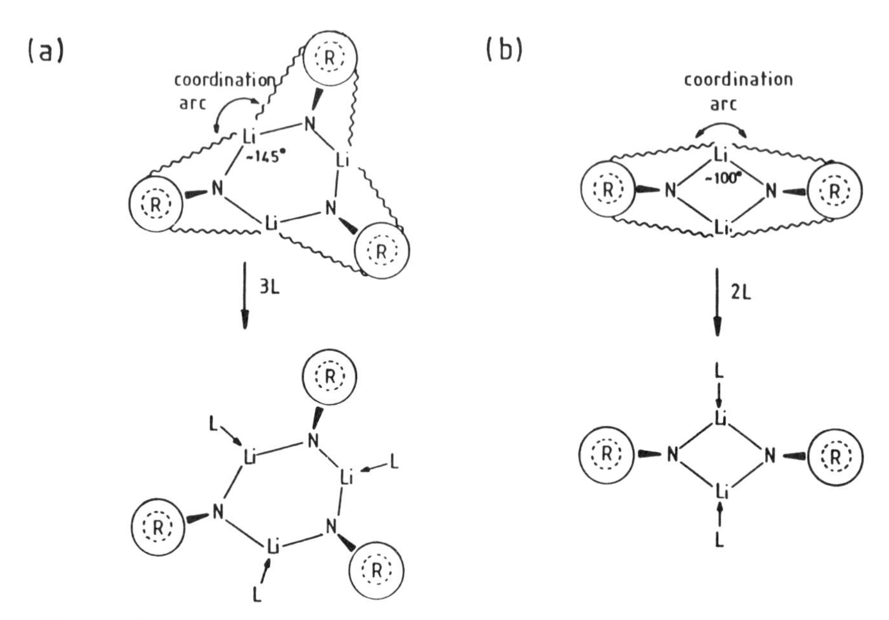

FIG. 38. Coordination arcs available for attachment of Lewis base molecules (a) to a trimeric lithium amide ring and (b) to a dimeric lithium amide ring.

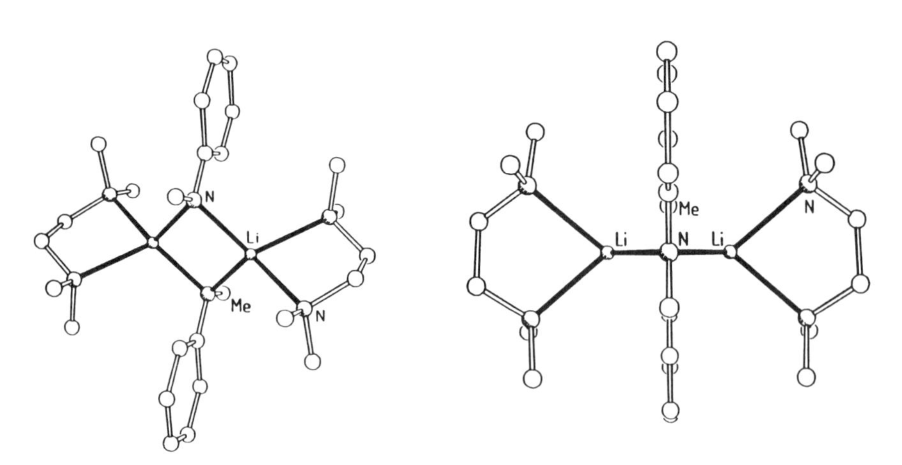

FIG. 39. Molecular structure of [PhN(Me)Li·TMEDA]$_2$ (**68**).

only one containing bidentate ligands. Each Li is four coordinate, leading to longer (amide)N—Li bonds and a smaller angle at Li.

Table XIII (*189–199*) gives details of solid-state lithium amide monomeric complexes (**69**)–(**87**). These include just three [(**79**), (**80**), and (**87**)] solvent-separated ion pairs. The remainder are contact-ion pairs, each with an (amido)N—Li bond. Association to dimers or higher oligomers is prevented sterically. The size of the R and/or R′ group in the $RR'N^-$ anions can lead to monomers even when Li^+ is complexed only by a single bidentate (e.g., TMEDA) or by two monodentate (e.g., THF or Et_2O) ligands. In such cases [(**69**), (**71**), (**72**), (**75**)–(**78**), and (**81**)–(**83**)], the lithium centers are only three coordinate. Electronic factors in the anion [notably, B⋯N multiple bonding in (**75**)–(**78**)] also may reduce the charge density at N, and lower the ability to bridge two

TABLE XIII

MONOMERIC LITHIUM AMIDE COMPLEXES: KEY STRUCTURAL PARAMETERS (X-RAY DIFFRACTION DATA)

Complex	Formula number[a]	N—Li distance (Å)	Ligand–Li distances (Å)	Ref.
Ph(naphthyl)NLi·TMEDA	(**69**)	1.97	2.12, 2.13	*163*
Ph(naphthyl)NLi·PMDETA	(**70**)	2.00	2.18, 2.21, 2.22	*163*
2,4,6-$Bu^t_3C_6H_2$(H)NLi·TMEDA	(**71**)	1.90	2.14, 2.16	*80*
2,4,6-$Bu_3^tC_6H_2[Si(CHMe_2)_2Cl]$NLi·2THF	(**72**)	1.99	—	*189*
$C_{12}H_8SNLi$·3THF	(**73**)	1.98	—	*190*
$C_{12}H_8ONLi$·3THF	(**74**)	2.03	—	*190*
$Mes_2B(Ph)NLi·2Et_2O$	(**75**)	1.94	av. 1.93	*191*
$Mes_2B(Mes)NLi·2Et_2O$	(**76**)	1.94	av. 1.95	*192*
$Bu_2^tBN(Bu^t)Li$·TMEDA	(**77**)	1.97	2.17, 2.24	*193*
$MeBN(Bu^t)BMe_2N(Bu^t)Li$·TMEDA	(**78**)	2.00	2.13, 2.20	*193*
$[Mes_2BNBMes_2]^-·[Li·3Et_2O]^+$	(**79**)	—	1.88, 1.90, 1.91	*187*
$[Mes_2BNSiPh_3]^-·[Li(12\text{-crown-}4)_2]^+$	(**80**)	—	—	*194*
$(Me_3Si)_2NLi$·TMEDA	(**81**)	1.79	2.12, 2.16	*195*
$(Ph_3Si)_2NLi$·2THF	(**82**)	2.00	1.95	*196*
$(Ph_2MeSi)_2NLi$·2THF	(**83**)	1.95	1.91	*196*
$(Me_3Si)_2NLi$·(12-crown-4)	(**84**)	1.97	2.09, 2.11, 2.33, 2.39	*194, 197*
Ph_2NLi·(12-crown-4)	(**85**)	2.01	2.11, 2.16 2.23, 2.24	*194, 198*
$(Ph_2MeSi)_2NLi$·(12-crown-4)	(**86**)	2.06	2.17, 2.23	*196*
$[(Ph_3Si)_2N]^-·[Li(12\text{-crown-}4)_2]^+$·THF	(**87**)	—	—	*194, 196, 199*

[a] See text.

Li^+ centers. Attachment of three monodentate ligands [THF; (**73**) and (**74**)], a tridentate one [PMDETA; (**70**)], or a single 12-crown-4 [(**84**)–(**86**)] to Li^+ also results in monomers. The bulk of these donors prevents association, even irrespective of the size of the amido R,R′ groups. Of course, Li^+ is four or five coordinate. These features are seen in the structure of (**70**) (Fig. 40); although the amido R,R′ groups (Ph, naphthyl) are two dimensional, the PMDETA ligand prohibits any further association of the monomer.

A further structural feature, namely "agostic" Li···H or Li···HC interactions (see Section I,B), occurs in several of these monomers. In the lithiated phenothiazin and phenoxazin complexes [(**73**) and (**74**), respectively], the annelated benzene rings are tilted, resulting in short Li^+—*o*-H distances. In (**73**) the four-coordinate Li^+ cation is placed unsymmetrically between the *ortho* hydrogens (Li—*o*-H, 2.54 and 2.82 Å), but in (**74**) the Li—*o*-H distances are identical (2.76 Å). More typically, such interactions are apparent when the metal centers are coordinatively unsaturated, i.e., when the steric bulk of the R,R′ amido groups and/or the bidentate neutral ligand (or two monodentate ones) prevents association. Note the short intramolecular Li—H(N) distance of 2.08 Å in (**71**) and the contacts with both aryl and *t*-butyl groups in (**72**) (Li—C, 2.47 and 2.75 Å, respectively). More unusual *inter*molecular interactions are found in the structure of (**69**) (Fig. 41a). In the monomer, the aryl groups adopt very different orientations, naphthyl being roughly perpendicular to the N—Li bond axis, but phenyl running alongside this axis. This arrangement allows these monomers to associate in "slipped" vertical pairs; each unit is displaced opposite to the other (Fig. 41b). Each Li^+ interacts with one *ortho*- and one *meta*-

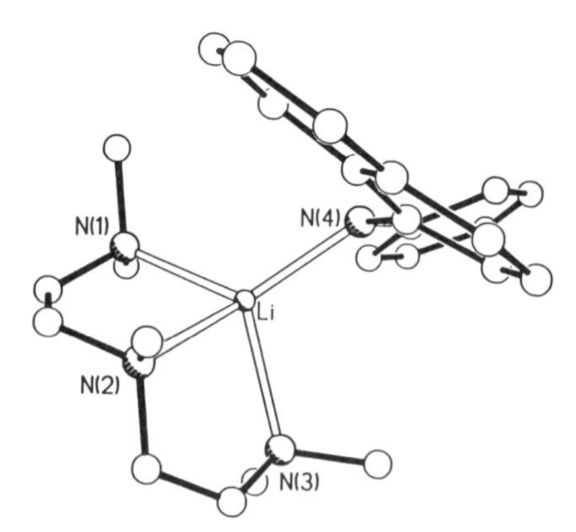

FIG. 40. Molecular structure of the monomer Ph(naphthyl)NLi·PMDETA (**70**).

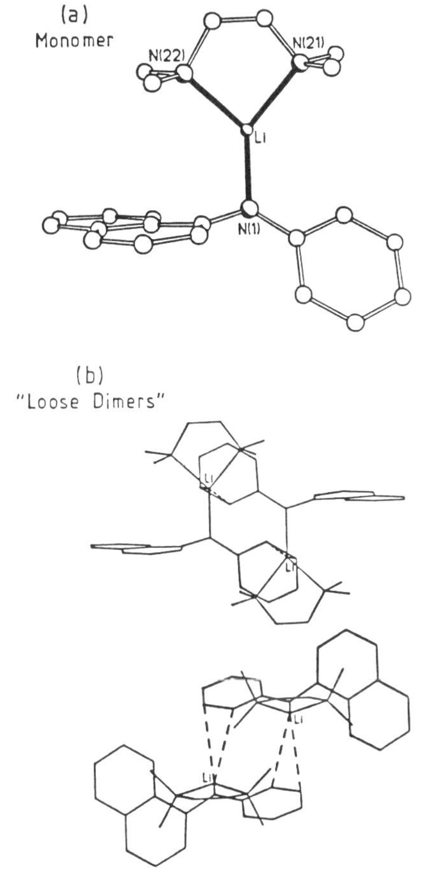

FIG. 41. Molecular structure of Ph(naphthyl)NLi·TMEDA (**69**): (a) the monomeric unit; (b) loose association of these monomers into dimers, via Li···HC interactions.

CH unit of the neighboring phenyl group (Li—C′, 3.12 and 3.15 Å, respectively) giving "loose dimers."

2. *Calculational Studies*

Section III,B,2 described how results from *ab initio* and MNDO calculations can explain many of the structural features found in the solid state for uncomplexed lithium amides $(RR'NLi)_n$. In particular, they explain (1) why rings formed experimentally, with $n = 2, 3$, and 4, have R,R′ groups perpendicular to the $(NLi)_n$ ring plane; (2) why a trimer ($n = 3$) is favored over a dimer ($n = 2$); and (3) why the only known tetramer ($n = 4$) is a ring, rather than a tetrahedral "stack" of two dimers. Here we discuss the results of calculations on complexed lithium amides. These also rationalize the experimentally found structures (Section III,C,1) of such species.

One marked structural feature is that the six-membered rings preferred for $(RR'NLi)_n$ species [e.g. (**55**) and (**56**) in Table VI, Section III,B,1] become four-membered rings on complexation with monodentate Lewis bases (base: Li ratio, 1 : 1) [e.g. (**61**)–(**67**) in Table XII, Section III,C,1]. A qualitative explanation for this diminution in ring size suggests that the dimer offers a wider coordination arc for attachment of the base (Fig. 38, Section III,C,1). However, more quantitative reasons have emerged from an *ab initio* study (6-31G basis set) on the model systems $(H_2NLi{\cdot}H_2O)_n$, with $n = 1, 2$, and 3 (*20*). The optimized geometries for these complexes are shown in Fig. 42. Comparing the structures of $H_2NLi{\cdot}OMe_2$ and $H_2NLi{\cdot}H_2O$ confirms that H_2O is a reasonable model (electronically) for the usual O ligands. In each of the aquo species, the H_2O molecule adopts a perpendicular (staggered) orientation to the plane of the H_2N groups; in the all-planar (eclipsed) conformations, the energies of the monomer, dimer, and trimer are higher (by 1.0, 5.8, and 10.4 kcal mol^{-1}, respectively). Expectedly, the N—Li bond lengths are longer than those in the uncomplexed analogs [(*165, 167*); see Table IX, Section III,B,2]; the largest increase (~0.05 Å) is for the trimer and this can be attributed to more pronounced steric effects at Li as the $O\hat{Li}N$ angle decreases with increased association. The Li—O bond lengths increase from monomer, through dimer, to trimer so that stabilization by solvation becomes less important. Thus, the calculated total hydration energies (−26.4, −38.0, and −40.7 kcal mol^{-1} for $n = 1, 2$, and 3, respectively), when expressed in terms of each additional H_2O attachment, become −26.4, −19.0, and −13.6 kcal mol^{-1}, respectively. The calculated association energies for $H_2NLi{\cdot}H_2O$ to $(H_2NLi{\cdot}H_2O)_n$ are −58.6 and −95.4 kcal mol^{-1} for $n = 2$ and $n = 3$, respectively. These values are much less than the association energies of the uncomplexed species (for $n = 2$, −73.7 kcal mol^{-1} and for $n = 3$, −133.9 kcal mol^{-1}; see also Table VIII, Section III,B,2).

The experimental preference for solvated dimers over solvated trimers (but the reverse for uncomplexed systems) can best be explained by considering association energies as sums of two components (Fig. 43). The first is the energy needed to reorganize the monomeric units, H_2NLi or $H_2NLi{\cdot}H_2O$, into the required geometry for the dimer or trimer, i.e., lengthening of the N—Li bond, and bending the H_2N group toward this bond. The second is the energy gained by association of these reorganized monomers to $(H_2NLi)_n$ or $(H_2NLi{\cdot}H_2O)_n$, $n = 2$ or 3. For the uncomplexed species (Fig. 43a), the reorganization energies are

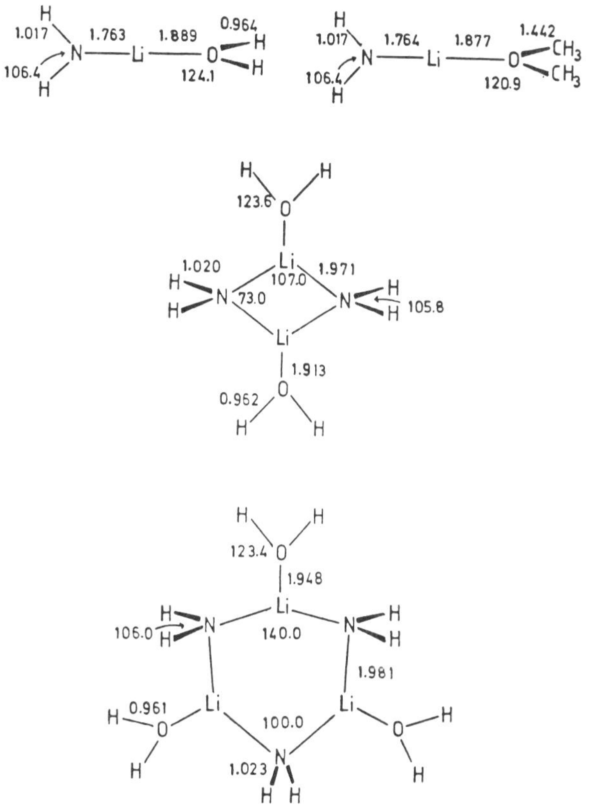

FIG. 42. Optimized structures (6-31G basis set) of $(H_2NLi{\cdot}H_2O)_n$ complexes, with $n = 1$, 2, and 3; the structure of $H_2NLi{\cdot}OMe_2$ is included for comparison with $H_2NLi{\cdot}H_2O$.

small, particularly that for the trimer. The dominant term is the association energy of the reorganized H_2NLi unit, and this is more favorable for the trimer. Thus, both components make the trimer $(H_2NLi)_3$ the preferred association product. In comparison, both reorganization energies are larger for $H_2NLi{\cdot}H_2O$ (Fig. 43b), but these energies are now considerably greater for the trimer than for the dimer. In the trimer, the H_2O ligand is moved from a linear OLiN arrangement to an angle of 110°. In the dimer, the ligand is only moved to 126.5° (Fig. 42). Though the total association energies show that $H_2NLi{\cdot}H_2O$ still prefers to form a trimer rather than a dimer (by 2.5 kcal mol^{-1} per unit), this preference is reduced (*cf.* 7.8 kcal mol^{-1} per unit for the uncomplexed H_2NLi species). With the much bulkier complexants used in practice (notably, Et_2O, THF, and HMPA), the reorganization term may become

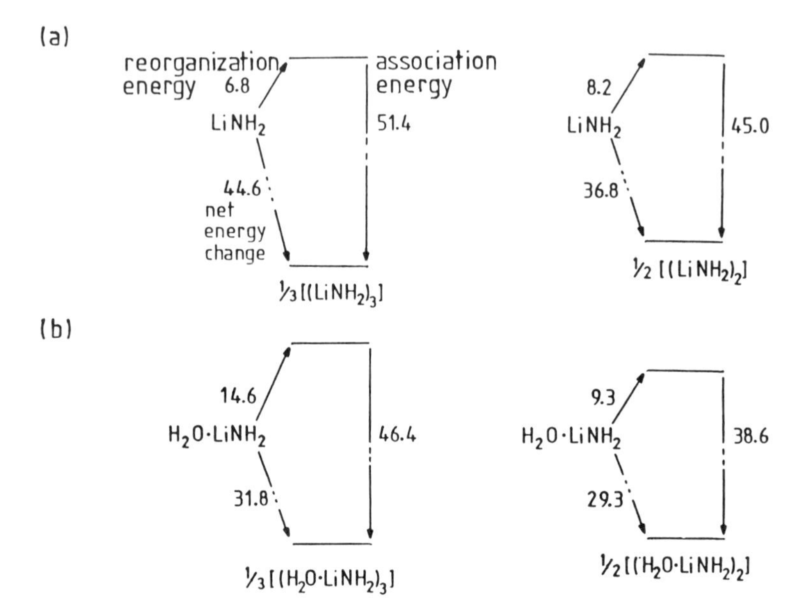

FIG. 43. Calculated energy profiles (6-31G basis set, energies in kcal mol^{-1}) for formation of (a) $(H_2NLi)_n$, $n = 2$ or 3, and (b) $(H_2NLi{\cdot}H_2O)_n$, $n = 2$ or 3.

dominant, so that association to a dimer would be expected, in line with experimental findings.

A further notable structural feature of complexed lithium amides is the isolation of ladder structures when ligands are attached to only *some* of the Li centers, i.e., when the base : Li ratio is $<1:1$ [(**59**) and (**60**) in Fig. 35, Section III,C,1]. Individual ring structures are not observed, even though experiment and calculation both show that dimeric rings are strongly favored when *each* Li is complexed. These structural preferences have been probed by *ab initio* calculations on $(H_2NLi)_4$, at the 6-31G and STO-3G (a smaller basis set) levels, and on $(H_2NLi)_4{\cdot}2H_2O$ and $(H_2NLi{\cdot}H_2O)_4$ (STO-3G level only) (*21*). The computations on $(H_2NLi)_4$ at 6-31G (see Fig. 31 and Tables X and XI, Section III,B,2) show that a ring is preferred, followed by a ladder (+7.4 kcal mol^{-1}), and then by a stack of two dimers (+11.4 kcal mol^{-1}). These preferences are reflected by the N—Li bond lengths, which are shortest in the ring and longest in the stack, whereas the N—N and Li—Li repulsive contacts are longest in the ring and shortest in the stack. In each instance, the ladder structure displays intermediate distances. To calibrate the results on the aquo complexes, $(H_2NLi)_4$ was reexamined at STO-3G (excluding the stacked structure). The ring also was found to be preferred to the ladder, and by a larger energy difference, 10.9 kcal mol^{-1}. For both, the STO-3G bond lengths and angles agree quite well (maximum changes of 0.08 Å and 3°, respectively) with those found at the 6-31G level.

For $(H_2NLi)_4{\cdot}2H_2O$, the optimized structure with the water ligands bound only to the outer Li atoms of the ladder is more stable (by 7.2 kcal mol^{-1}) than the ring structure with H_2O coordinated to the opposite Li atoms (Fig. 44). The large H_2O complexation energies (−85.2 and −67.1 kcal mol^{-1}, respectively) of both complexes are exaggerated by STO-3G, but the difference between them (18.1 kcal mol^{-1}) reverses the ring-over-ladder preference found for uncomplexed $(H_2NLi)_4$. This energy difference is reflected in changes in geometry at the coordination sites of the $(H_2NLi)_4$ ladder and ring structures. Thus, the $N\hat{Li}N$ angles at the outer rungs of the ladder change little (112.1–108.2°) upon coordination by H_2O, whereas in the ring model the angles at Li compress much more (162.4–138.4°) to accommodate these additional ligands. Significantly, attempts to optimize a ladder structure with H_2O attached to the inner Li atoms failed: the inner N—Li bonds of the ladder are cleaved on optimization and the ring geometry results.

Three possible structures of $(H_2NLi{\cdot}H_2O)_4$ were optimized (Fig. 45): a ring and a ladder, both with each Li complexed by H_2O, and a ladder with only the end-Li centers complexed. The last is the most stable structure, 17.1 kcal mol^{-1} lower in energy than the ring. Compared to

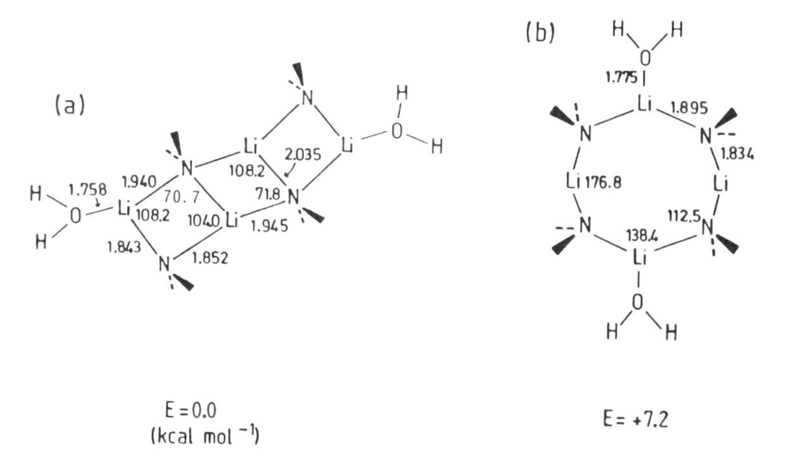

FIG. 44. Optimized geometries and relative energies (STO-3G basis set, energies in kcal mol^{-1}) of $(H_2NLi)_4{\cdot}2H_2O$ complexes: (a) a ladder, with H_2O on the outer Li atoms only; (b) a ring, with H_2O ligands on diagonally opposite Li atoms.

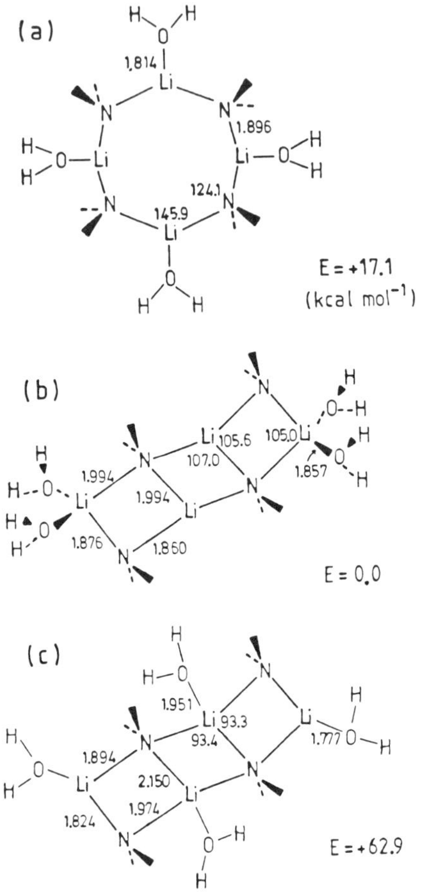

FIG. 45. Optimized geometries and relative energies (STO-3G, kcal mol^{-1}) of $(H_2NLi \cdot H_2O)_4$ complexes: (a) a ring, with each Li complexed by H_2O; (b) a ladder, with H_2O on the outer Li atoms only; (c) a ladder, with H_2O on each Li atom.

the $(H_2NLi)_4 \cdot 2H_2O$ species, further complexation increases the preference for a ladder over a ring. This arises because addition of four H_2O molecules to the ends of a ladder results in a gain in stability of 142.9 kcal mol^{-1} (*cf.* 114.9 kcal mol^{-1} for their addition to a ring). This reflects the extent of reorganization in each case: the outer $N\hat{Li}N$ angles of the ladder are reduced only by 7.1° on bis(solvation) (112.1–105.0°) whereas the $N\hat{Li}N$ angles in the ring change much more (by 16.5 to 145.9°). The third structure, a ladder with H_2O on each Li, is 62.9 kcal mol^{-1} less stable than the solely end-solvated ladder. This reflects the crowded nature of the geometry about the inner Li atoms if H_2O molecules are attached here. Such steric hindrance leads to long (inner) Li—O bonds, 1.951 Å [*cf.* (outer) Li—O bond lengths of 1.777 Å, and Li—O bond lengths in the end-complexed ladder and in the ring, 1.857 and 1.814 Å, respectively].

These calculational results concur extremely well with the experimental findings. Lithium amide complexes are dimeric rings (or monomers) if the complexant : Li ratio is 1 : 1 or greater, and if every Li bears (at least) one complexant. However, there is a switch of structural preference from ring to ladder if a base is not coordinated to each Li. In such cases, only the end-Li centers of the ladder are complexed, even though a 1 : 1 complexant : Li ratio may be present under the experimental conditions.

3. *Structures in Solution*

The structures of lithium amide complexes $(RR'NLi\cdot xL)_n$ in solution have been probed by colligative measurements (notably, cryoscopy and vapor-pressure barometry) and by combinations of ^{6}Li, ^{7}Li, ^{13}C, and ^{15}N NMR spectroscopy. In several cases, attempts have been made to correlate the results from the two kinds of experiments. When hydrocarbon solvents (notably, benzene, toluene, and their deuterated versions) are used, the preisolated complex is dissolved. With coordinating solvents (notably, Et_2O and THF), the "parent" amide $(RR'NLi)_n$ is dissolved and the complexes are formed in solution. These studies give information on the nature of the lithium amide reagents under the conditions in which they are employed. Solid-state structures are a guide. As shown in Section III,C,1, lithium amide complexes can be limited ladders (only two examples known), or, more commonly, dimers or monomers in the crystal. Similarly, dimers (n = 2) and monomers (n = 1) for $(RR'NLi\cdot xL)_n$ complexes dominate in solution. There are, however, three important caveats: (1) the degree of solvation or complexation (x) of these species can vary, *e.g.*, dimers can have a total of two, three, or four complexants (x = 1, 1.5, or 2) and monomers can have two, three, or four complexants; (2) there are frequently equilibria between dimeric solvates, between monomeric solvates, and between dimers and monomers (of various kinds); (3) all the variations described in (1) and (2) are solvent dependent, and for a particular solvent usually are temperature and concentration dependent.

These are the essential features. Hence, this section is not intended to be comprehensive but rather provides key references and gives illustrative examples. The section is organized in the same order as in Section

III,C,1, which discussed solid-state structures of lithium amide complexes.

The two limited-ladder structures known in the solid state, $\{[H_2\overline{C(CH_2)_3N}Li]_3\cdot PMDETA\}_2$ (**59**) and $\{[H_2\overline{C(CH_2)_3N}Li]_2\cdot TMEDA\}_2$ (**60**), behave very differently in solution (*21*). Complex (**60**) appears to retain its integrity (n = 2) on dissolution in hydrocarbon solvents. Cryoscopic relative molecular mass (crmm) measurements in benzene give fairly constant association state values (n) of 2.02 ± 0.06 and 2.10 ± 0.06 for solutions of 3.4×10^{-2} and 9.0×10^{-2} mol dm^{-3} concentrations, respectively. In addition, the ^{7}Li NMR spectra of d_8-toluene solutions of (**60**) at −95°C all reveal essentially just two signals of equal integral (*e.g.*, Fig. 46a and inset) corresponding to the two kinds of Li environment (ladder-end and inner) expected for intact dimeric (**60**) (see Fig. 35b). The natural-abundance ^{6}Li NMR spectra show these two 1 : 1 signals more clearly (e.g., Fig. 46b), because the low-quadrupole moment of ^{6}Li(*cf.* ^{7}Li) affords much sharper resonances. In contrast, complex (**59**) exhibits very complicated solution behavior. Cryoscopic rmm measurements on benzene solutions give n values of 1.77 ± 0.12 to 6.51 ± 0.31 (*cf.* n = 2 in the solid state) for concentrations 2.0×10^{-2} to 6.5×10^{-2} mol dm^{-3}, respectively. However, at yet higher concentrations, crmm, and hence n, values fall, e.g., n = 5.26 ± 0.32 at 8.0×10^{-2} mol dm^{-3}. Such results have been interpreted (*21*) in terms of PMDETA ligands amending their role from tridentate within each ladder to bidentate or even monodentate. Resultingly "free" Me_2N and/ or MeN groups might then complex to Li centers in other ladders, so joining ladders together and giving n values of 4, 6, . . . etc. These two processes should show different concentration dependence (ligand amendment being encouraged at low concentrations but association of resulting coordinatively unsaturated ladders being favored at high concentrations), and support for this interpretation has come from variable-concentration ^{7}Li NMR spectra of d_8-toluene solutions of (**59**) at −95°C. At very high concentrations (double that of solutions examined by cryoscopy), essentially three signals of equal integral are observed (Fig. 46c and inset). These correspond to the three Li environments of the intact dimeric ladder (see Fig. 35a); at this high concentration, ligand amendment is suppressed. Twofold dilution results in a more complicated ^{7}Li NMR spectrum (Fig. 46d); the three signals (asterisks) attributed to dimeric (**59**) are still apparent, but there are several other distinct resonances and probably many more beneath these broad peaks. Further dilution (to 6.0×10^{-2} mol dm^{-3}) gives one broad, ill-defined resonance. Crucially, though, for even more dilute solutions, the spectra begin to simplify again and the distinct

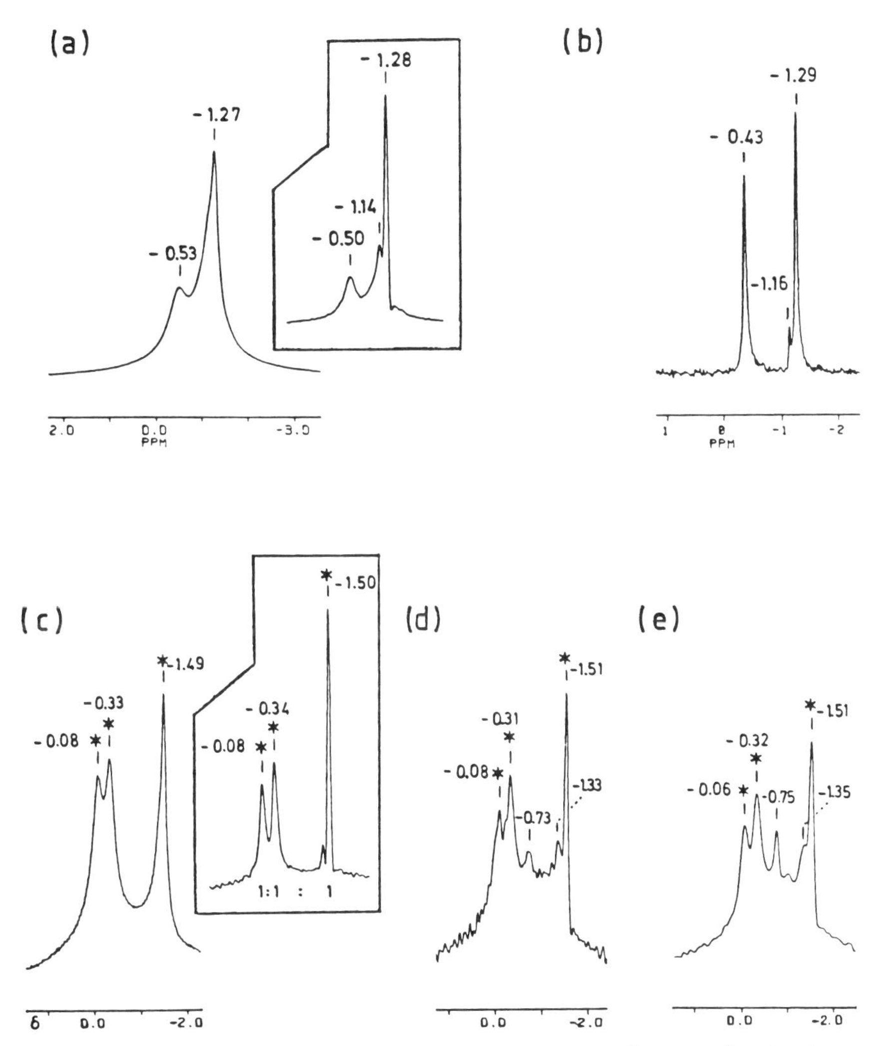

FIG. 46. (a) ^{7}Li NMR spectrum (139.96 MHz) of (**60**), 2.5×10^{-1} mol dm^{-3} in d_8-toluene at −95°C; inset, resolution-enhanced spectrum; (b) ^{6}Li NMR spectrum (52.99 MHz) of (**60**), 6.7×10^{-1} mol dm^{-3} in d_8-toluene at −95°C; (c, d, and e) ^{7}Li NMR spectra (139.96 MHz) of (**59**) in d_8-toluene at −95°C: (c) 1.8×10^{-1} mol dm^{-3}; inset, resolution-enhanced spectrum; (d) 9.0×10^{-2} mol dm^{-3}, resolution-enhanced spectrum; (e) 4.0×10^{-2} mol dm^{-3}, resolution-enhanced spectrum.

signals due to dimeric (**59**) reemerge (e.g., Fig. 46e); at these low concentrations, partial PMDETA amendment can occur, but only limited association of the resulting ladders is possible.

A similar strategy of linking variable-concentration rmm results to variable-concentration ^{7}Li NMR spectroscopic results has been used to ascertain the solution equilibria of amidolithium complexes having at least one complexant per Li (*cf.* the aforementioned ladder species, having a deficiency of complexant). For example, $[(PhCH_2)_2NLi \cdot OEt_2]_2$ (**61**), a dimer in the solid state (see Section III,C,1) affords cryoscopic n values of 1.20–1.07 (3.1×10^{-2} to 1.8×10^{-2} mol dm^{-3} concentrations in benzene, respectively) (*20, 94*). The ^{7}Li NMR spectra of benzene solutions of (**61**) exhibit only two resonances, at the same positions as those found in the spectra of the uncomplexed amide $[(PhCH_2)_2NLi]_3$ (**55**). Amide (**55**) was shown earlier (Section III,B,3) to engage in a trimer $\rightleftharpoons$ monomer equilibrium. Hence, these results prove that on dissolution (**61**) totally loses Et_2O (hence n values <2), the uncomplexed dimers so produced then rearranging to trimers and monomers. In a similar study, linking variable-concentration and -temperature ^{1}H and ^{7}Li NMR spectroscopic results to those from isopiestic and cryoscopic rmm values, $[(Me_3Si)_2NLi]_n$ (**56**), like (**55**), a trimer in the solid state, was proposed to exist as a dimer $\rightleftharpoons$ tetramer equilibrium in hydrocarbons. In THF, however, an equilibrium between a complexed dimer and a complexed monomer, both of uncertain degrees of solvation, was indicated (*200*). The experimental solution data for a complex of lithium dicyclohexylamide is even more illustrative. The parent species $(c\text{-hexyl}_2NLi)_n$ is known, but crystal twinning has so far prevented a solid-state structure determination (n probably equals 3 or 4; see Section III,B,1). This lithium amide is used widely in organic syntheses as a bulky reagent of somewhat greater basicity than the more familiar LDA, $(Pr^i{}_2NLi)_n$ (e.g., *6–8*). However, reported reactions of this reagent are carried out in complexing solvents, e.g., Et_2O and THF; in fact, this species is rather insoluble in hydrocarbons (*94, 157*). Thus the reagent is *not* $(c\text{-hexyl}_2NLi)_n$, but a complex of it. One such complex, $(c\text{-hexyl}_2NLi \cdot HMPA)_n$, has been isolated from lithiation of c-hexyl$_2$NH in hexane containing HMPA (*94*). Although its solid-state structure is as yet unknown (but probably $n = 2$), crmm measurements on benzene solutions give n values of 1.60–1.15 for 4.0×10^{-2} to 2.0×10^{-2} mol dm^{-3} concentrations, respectively. The 25°C ^{7}Li NMR spectra of benzene solutions of various concentrations (Fig. 47) each consist of two signals, neither of which can be due to $(c\text{-hexyl}_2NLi)_n$ species (not ruled out by crmm results) because this uncomplexed lithium amide affords one broad ^{7}Li signal at δ +0.75 ppm. Hence these results point to

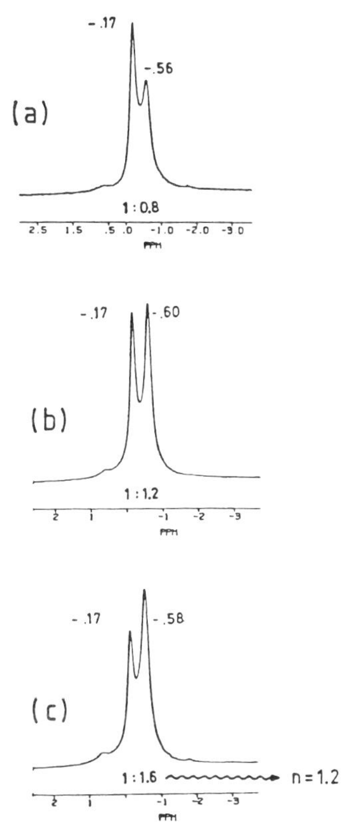

FIG. 47. ^{7}Li NMR spectrum (139.96 MHz) of (c-hexyl$_2$NLi·HMPA)$_n$ in d_6-benzene at 25°C: (a) 2.0×10^{-1} mol dm^{-3}; (b) 1.0×10^{-1} mol dm^{-3}; (c) 2.0×10^{-2} mol dm^{-3}.

a (remarkably slow) monomer ⇌ dimer solution equilibrium. The low-frequency resonance ($\sim\delta$ -0.6) becomes more pronounced on dilution, and so can be assigned to the monomeric complex, c-hexyl$_2$NLi·HMPA; the higher frequency signal must be due to an aggregate, almost certainly a dimer, (c-hexyl$_2$NLi·HMPA)$_2$. Assuming a dimer (as pointed out earlier, there are no known solid-state complexed trimers), there is fair agreement between n values measured by cryoscopy and n values implied by consideration of ^{7}Li signal integration ratios; for example, for the most dilute solution looked at both by spectroscopy (Fig. 47c) and by cryoscopy, n is calculated at 1.20 and 1.15, respectively. Clearly, studies such as these can pinpoint the actual species present, at various concentrations and temperatures, in solutions of lithium amide reagents; such information is crucial to the proper understanding of how these reagents operate in their regio- and stereospecific reactions with organics.

Several important studies have employed the somewhat different strategy of preparing a lithium amide complex $(RR'NLi\cdot xL)_n$ *in situ* in a complexing solvent (e.g., Et_2O and THF), and then identifying solution species present by $^{6,7}Li$, ^{13}C, and ^{15}N NMR spectroscopies (singly or in conjunction) with or without rmm measurements. With very few exceptions, species found are dimers or monomers (or both), although the degree of solvation (x) can vary considerably. These typical structures are illustrated in Fig. 48, structures (**I**)–(**VI**). A series of Li salts of aromatic secondary amines, including those with R = Ph, R = Me, Pr^i, Bu^n, Bu^t, and RR'N = indolide, $\overline{C_6H_4\cdot(CH_2)_2N}$, were studied in Et_2O and THF solutions, chiefly by ^{13}C NMR spectroscopy (*201*). In THF at 17°C, lithium indolide appears to be mainly dimeric; vapor pressure barometric measurements give n values of 1.8–1.9 for 0.09–0.49 mol dm^{-3} concentrations. In the ^{13}C NMR spectra, chemical shifts are con-

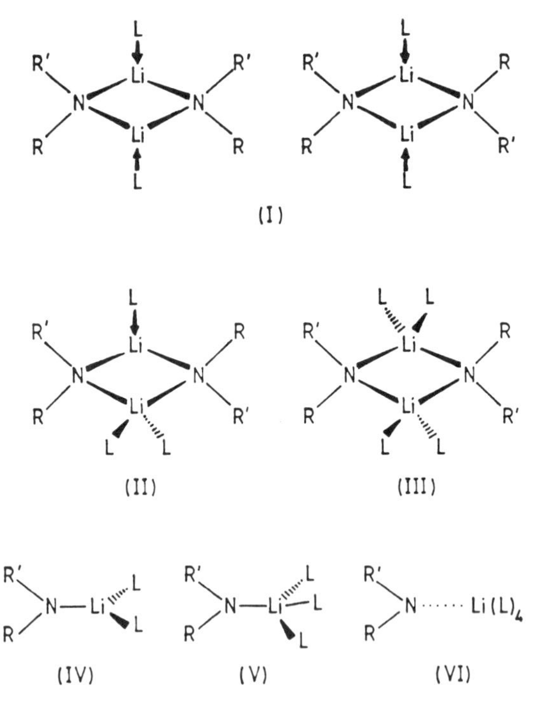

FIG. 48. Principal structures identified for lithium amides $(RR'NLi)_n$, in coordinating solvents L: (I) dimers with one molecule of L per Li (*cis*/*trans* or *E*/*Z* isomers); (II) a trisolvated dimer; (III) a tetrasolvated dimer; (IV)–(VI) monomers with two or three (contact-ion pairs) or four (solvent-separated ion pairs) L molecules per Li.

stant from −100 to 40°C, but then become temperature (but not concentration) dependent. The shift of C(4) ("para" to N) in particular decreases, implying decreased charge density at C(4) and increased localization of negative charge on N. Such results were interpreted in terms of a tetrasolvated dimer [Fig. 48, structure (**III**)] prevailing at very low temperatures, but a bisolvated dimer [(**I**)] being present at higher temperatures. In Et_2O solutions, only dimer (**I**) could be observed, although at −120°C it was possible to detect its *E* and *Z* isomers. Similar analysis of ^{13}C chemical shifts showed that $(PhMeNLi)_n$, $(PhPr^iNLi)_n$, and $(PhBu^nNLi)_n$ also exist in THF as dimers of types (**I**) and (**III**) together with a trisolvated monomer [Fig. 48, structure (**V**)]. In contrast, $(PhBu^tNLi)_n$ in THF is a monomer alone, (**V**) or the bisolvated species (**IV**), depending on temperature. All four lithium N-substituted anilides exist in Et_2O solutions only as the bisolvated dimer, (**I**). In the case of $(PhMeNLi{\cdot}Et_2O)_2$, the presence of a dimer was confirmed by observation of a ^{15}N pentuplet resonance at −100°C for a solution of the $^6Li,^{15}N$ isotopomer (J = 3.8 Hz), i.e., each N atom is attached to two Li atoms. Similarly, a ^{15}N triplet resonance (J = 7.5 Hz) was found at −80°C in the spectrum of $^6Li,^{15}N$-enriched $(PhPr^iNLi)_n$ in THF, in agreement with the presence of a monomer [probably (**V**)]. In a related study (*90*), $^{6,7}Li$, ^{13}C, and ^{15}N NMR spectroscopies were all employed to show that addition of HMPA (two to four equivalents) to an ether solution of $(PhPr^iNLi)_n$ converts the dimer [of type (**I**)] into a mixture of monomer $PhPr^iNLi{\cdot}(THF)_{3-n}(HMPA)_n$ and the triple-ion salt, $[(PhPr^iN)_2Li{\cdot}(HMPA)_n]^-{\cdot}Li(HMPA)_4^+$ [*cf.* the 'ate complex (**8**); see Section II,B].

Many of the above results on lithium arylamides have been backed up by determinations of ^7Li quadrupole splitting constants (QSCs) from ^7Li and ^{13}C spin-lattice relaxation data (*202*). The QSCs (in kHz) are remarkably sensitive to the degree of aggregation (n) and, especially, to the degree of solvation (x). For example, lithium indolide in Et_2O is believed to exist as a bisolvated dimer [Fig. 48, structure (**I**)] and the ^7Li QSC at 30°C is 317 ± 6 kHz, whereas in THF the tetrasolvated dimer (**III**) dominates (QSC at −80°C, 151 ± 5 kHz). Similarly, $(PhPr^iNLi)_n$ has been shown to exist as the trisolvated monomer (**V**) in THF at −80°C (QSC, 211 kHz) though it converts to the tetrasolvated dimer (**III**) (typical QSC ~ 155) and then to the bisolvated dimer (**I**) (typical QSC ~ 315 kHz) as the temperature increases. These empirical correlations between QSC values and solution structures have also been applied to lithium phenolates in solvents such in Et_2O, THF, pyridine, Me_3N, and dioxolane; as discussed in Section II,E, unlike lithium amides, rings of type $(ROLi{\cdot}xL)_n$ (n = 2 or 3) have the potential to stack once, so that tetramers and hexamers are also observed (*202–204*).

A final approach to unraveling lithium amide solution structures has been to employ ^6Li–^{15}N double labeling; ^6Li ($I = 1$) and ^{15}N ($I = \frac{1}{2}$) NMR resonance patterns reveal the number of N—Li connectivities, and this, particularly in conjunction with rmm measurements, often allows complete structural assignment. For example, in a study of lithiated cyclohexanone phenylimine [(Ph(1-*c*-hexenyl)NLi)]$_n$ in hydrocarbon (benzene or toluene)/THF solutions (*205*), cryoscopic measurements at 0°C implied the presence of a bisolvated dimer [Fig. 48, structure (**I**)] at low THF concentrations. The ^6Li NMR spectrum at −94°C of the doubly labeled amide in d_8-toluene containing THF (two equivalents per Li) showed two triplets (J = 3.3 Hz), and the ^{15}N NMR spectrum of the same solution consisted of two pentuplets (average J = 3.5 Hz). Thus, a cyclic structure is confirmed, with N—Li—N and Li—N—Li units; almost certainly the structure is the bisolvated dimer (**I**), existing in solution as the two (*cis* and *trans*) stereoisomers. At higher THF concentrations, the ^6Li NMR spectrum contains a doublet (J = 6.3 Hz) and the ^{15}N spectrum contains a triplet (J = 6.1 Hz), consistent with the presence of a monomeric, contact-ion pair species, probably trisolvated [Fig. 48, structure (**V**)]. Similar experiments have indicated that [Pr^i(*c*-hexyl)NLi]$_n$ in d_8-toluene/THF solutions exists as a mix of *cis* and *trans* stereoisomeric dimers, probably having one THF per Li [Fig. 48, structure (**I**)] (*206*). The N—Li—N and Li—N—Li connectivities proved by ^6Li and ^{15}N coupling patterns do not, of course, rule out higher oligomers; however, lower symmetry trimers and tetramers should exhibit increased spectral complexity (not observed), plus there are no known examples of solid-state complexed lithium amide oligomers higher than dimers (see Section III,C,1). The assignment of bisolvation is less clear-cut; the vast majority of complexed lithium amide dimers have one O complexant per Li (Table XII), although there is one example of a related trisolvated lithium dienamide dimer [Fig. 48, structure (**II**)] (*207*).

Recent studies have begun to explore the consequences of complexed lithium amide solution structures for reaction mechanisms and rates. Double-labeling (^6Li and ^{15}N) NMR experiments, allied to colligative measurements, show that lithium diphenylamide in THF/hydrocarbon solutions exists as a cyclic oligomer at low THF concentrations, probably a bi- or trisolvated dimer [Fig. 48, structure (**I**) or (**II**), respectively] (*208*). At intermediate THF concentration, ^{13}C NMR spectroscopy detects a second species, a monomer of uncertain solvation [(**IV**), (**V**), or (**VI**)]. In the presence of LiBr, a 1 : 1 adduct is formed at low THF concentrations, and NMR and cryoscopic measurements show this to be a trisolvated mixed dimer, $Ph_2NLi \cdot LiBr \cdot 3THF$ [*cf.* (**52**) and (**53**) in Table V, Section II,E]; at higher THF concentrations this mixed dimer dissociates to the monomeric lithium amide and free LiBr. Rate studies of the N-alkylation of Ph_2NLi with Bu^nBr in THF/hydrocarbon solutions imply that all the above three species are involved in the reaction mechanism, i.e., the dimer, the mixed dimer (formed after a considerable induction period), and, at intermediate THF concentrations, the monomer (*209*). The significance of which species are present in THF/hydrocarbon solutions of LDA, $(Pr^i_2NLi)_n$, regarding this reagent's reaction with an *N,N*-dimethylhydrazone, have also been examined (*210*). Earlier colligative measurements (*211*) had indicated that LDA in THF exists in a solvated dimer ⇌ monomer equilibrium. However, ^6Li and ^7Li NMR experiments in this most recent study could detect only an oligomer, probably a bisolvated dimer [Fig. 48, structure (**I**)], over a wide range of concentrations. Rate measurements on the lithiation of 2-methylcyclohexanone *N,N*-dimethylhydrazone, $MeCH(CH_2)_4C{=}N{\cdot}NMe_2$, using this LDA reagent suggest that the hydrazone itself acts as a further complexant toward this dimer, producing the monomer $Pr^i_2NLi{\cdot}THF{\cdot}$hydrazone. This kinetically active monomer than undergoes rate-determining proton transfer. These results spotlight a very important point. Reactions between organolithium reagents and organic substrates in general are represented conventionally in terms of nucleophilic attack by the organolithium residue (R^-, R_2N^-, etc.; the "carbanion") on the organic. It may well be that the opposite is true: that many organics act first as Lewis bases/nucleophiles toward the organolithium reagent, i.e., Li^+, not C^- or N^-, is the effective reactive and controlling center in these reagents.

IV. Conclusions

As is implied by the title of this review, the goal has been not only to describe and explain the known structures of organonitrogen–lithium compounds, but also to emphasize some key points about organolithium chemistry in general. The factual evidence and the arguments have been presented in the text, and we do not repeat these here. Rather, we give a checklist of the major conclusions we have reached. These conclusions (or, arguably, generalities) are presented in the following discussion.

A. Definitions and Nomenclature

We believe that the term "organolithium" can now in effect be taken to mean a lithiated derivative of *any* organic molecule. The stricture that a C—Li bond must be present is inappropriate and unwieldy. It seems best to link the term to the type of compound that has been lithiated: thus, suitable species with O—Li bonds (derived from alcohols, methyl ketones, esters, sulfones, etc.) and ones with N—Li bonds (derived from imines, amines, etc.) qualify. Claiming such species under the umbrella of "coordination chemistry," just because they contain no C—Li bonds, is somewhat grandiose: there *are* no "noncoordination" compounds. For the purist who insists on the presence of a C—Li bond to justify the epithet "organolithium," then "lithiated organic" seems a suitable compromise description.

B. Bonding

Organolithium compounds do not have any appreciable, or even significant, covalent character as regards the bonds to lithium. The central E—Li bonds (E = C, or N, or O, etc.) are essentially *ionic* ones. Many of the properties of these compounds might well be reminiscent of those of covalently bound discrete molecules, e.g., they are frequently crystalline materials, with low melting points and with good solubilities in weakly polar organic solvents. However, such properties are attributable to the *structures* of these compounds, which are often ones of limited association and dimensionality. It is these restricted aggregations, and the fact that the peripheries of the aggregates consist of organic groups, that lead to these properties. Accordingly, organolithiums are best termed "supramolecules" or "ionic molecules."

The short Li—Li distances (often much shorter than those in the metal and in the Li_2 molecule) found in the structures of many organolithium compounds do not imply that *lithium–lithium bonding* is of thermodynamic importance in these species. Close approach of Li centers is usually dictated by the need for acute angles at bridging atoms (E), e.g., of $(ELi)_n$ rings (n = 2 and 3 especially). Thus, the distances represent the optimization of maximized E—Li bridge bonding and minimized $Li^{\delta+}$—$Li^{\delta+}$ repulsions.

There is much evidence that low-coordinate lithium centers found within certain organolithiums $(RLi)_n$ (e.g., two-coordinate centers in rings with n = 2 or 3) will interact with C—H bonds within the R groups. Such "*agostic*" interactions can be intra- or intermolecular.

C. Implications for Organic Reaction Mechanisms

It is a gross and unjustified simplification to consider an organolithium as a monomer "RLi" or as a "*carbanion*" R^-. Most organolithiums are associated, $(RLi)_n$ (commonly, n = 2, 3, 4, 6, or ∞). Frequently, because many lithiations are carried out in polar solvents, they are complexed $(RLi{\cdot}xL)_n$, with L being a polar neutral ligand (Lewis base). For many such complexes, association persists (often n = 2 or 4). Monomers *are* known, but they are usually contact-ion pairs. Free carbanions are nebulous: even for supposedly solvent-separated ion pairs, $R^- {\cdots} Li^+(L)_x$, there is often a retained interaction between the ions. These points have implications for the mechanisms of the reactions of these organolithium reagents. It is not logical to ignore the lithium and to represent such reactions as proceeding *via* nucleophilic R^- attack on an electrophilic center of an added substrate. Indeed, all calculational investigations to date imply that it is the "nucleophilic" (coordinating) center of the substrate that first interacts with the Li^+ center of the intact organolithium $(RLi)_n$.

Experimental mechanistic studies of such reactions are in their early days. However, it is clear that proper understanding will rely on the prior isolation of the organolithium reagent, rather than on its production, then use, *in situ*. Thereafter, the reagent needs to be identified (e.g., RLi or $RLi{\cdot}xL$?) and characterized structurally, in the solid by X-ray diffraction and, perhaps more importantly, in solutions (by colligative molecular mass measurements and multinuclear NMR spectroscopic studies) in which it will be used.

D. Structures of Uncomplexed Organolithiums

The basic structural building block of any organolithium is an ion pair, R^-L^+. In the absence of a complexing agent, these will associate, first to give *rings*, $(RLi)_n$, with n = 2 or 3, or, less commonly, 4. The coordination numbers of the cations are raised thereby to two. Further association of these rings might (and usually does) occur. Whether this happens, and to what extent, depends only on the *steric* constraints imposed by the groups R. These associations, to rings and then possibly beyond rings, are a consequence of the *electrostatic nature* of the bonding. They are nothing to do with the supposed preference of lithium to reach four coordination, thereby (supposedly) using to the full its four valence orbitals: many (indeed, probably most) uncomplexed organolithiums have Li^+ centers in two or (particularly) three coordination. The R^- anions are *groups* (*cf.* point-charge anions such as Cl^-, which allow mutual six coordination of cations and anions). Such groups fall into two categories.

1. *For the majority of organolithiums,* RLi {e.g., R = aryl, R = alkoxy and aryloxy (R′O—), R = enolato [R′(=CH_2)CO—], R = im-

ino($R_2'C{=}N{-}$)}, the initially formed $(RLi)_n$ ring systems are quite flat. The central $(ELi)_n$ ring (E = C or O or N) is itself planar, and the substituents on E, up to and including the α atom of R′, lie in this same plane. Hence such rings can and do associate vertically, a process termed *stacking*. If the R′ groups are themselves small and/or especially flat (e.g., Me and/or Ph) this stacking can be extensive; amorphous materials result. If the R′ groups are bulkier, double stacking of $(RLi)_n$ rings (n = 2 or 3) is the norm: hence, most crystalline (by inference, oligomeric) uncomplexed organolithiums are tetrahedral tetramers or pseudooctahedral hexamers.

2. *Lithium amides* [R = amido ($R_2'N{-}$)] are the exception. The $(R_2'NLi)_n$ rings (n = 2, 3, or 4) cannot stack because the R′ groups project above and below the $(NLi)_n$ ring plane: the N^- center has a tetrahedral disposition of two R′ groups and two lone pairs. Further association of such rings must be lateral, joining N—Li ring edges in a process termed *laddering*. If R′ groups are small and/or tied together, such laddering is usually extensive, giving amorphous materials. The only generally found alternative is that, with large or floppy R′ groups, the rings themselves (n = 2, 3, or 4) are isolated.

E. Structures of Complexed Organolithiums

Complexation of $(RLi)_n$ rings (n = 2, 3, or 4) by added Lewis base molecules (L) or by base functions within R (e.g., NMe_2 or OMe groups) introduces a further steric constraint on their association. In addition, any ring forthcoming after complexation will be a dimeric one, i.e., $(RLi{\cdot}xL)_2$ complexes are produced, even if n = 3 or 4 for the parent $(RLi)_n$. Such dimers might stack (except when R = $R_2'N{-}$), but now they can do so only once. Hence, many complexed tetrahedral tetramers are known. If even such limited (twofold) stacking is precluded sterically, the dimers themselves are found. Dimers are also common for complexed lithium amides: stacking is excluded anyway (see Section IV,D,2) but now so too is laddering, because the base molecules occupy the lateral space around the ring. In the limit, of course, complexation can prevent R^-Li^+ ion pairs from associating even only as far as a single ring. Monomers result, $R^-Li^+{\cdot}xL$. If the R^- group is large, x can equal two (two monodentate ligands or a single bidentate one). More usually, x equals three or more (e.g., tridentate bases or crowns).

The above comments pertain to complexes in which each lithium center is ligated by at least one base function. Where there is a deficiency of the ligand L (less than one per Li), then these ligands are found only on the end-Li centers of aggregates. Such aggregates can be viewed as "intercepted" stacks or ladders.

The identities and structures of complexed organolithiums are of particular importance. These are the species that will be present in solutions of $(RLi)_n$ in polar solvents and most reactions of organolithiums are carried out using such solutions.

Acknowledgments

Many colleagues have provided us with reprints of their published work and/or with information prior to publication; we thank in particular Professors Gernot Boche (Marburg), David B. Collum (Cornell), Michael F. Lappert (Sussex), Philip P. Power (Davis, California), Dieter Seebach (ETH, Zurich), Erwin Weiss (Hamburg), and Paul G. Williard (Brown University). Thanks are due also to the many of our co-workers and collaborators who, through their efforts and ideas, have contributed greatly to the understanding of the structures of organonitrogen–lithium species. Regarding the United Kingdom, we mention particularly Drs. David R. Armstrong (Strathclyde), Donald Barr (Cambridge), William Clegg (Newcastle), Robert E. Mulvey (Strathclyde), and David Reed (Edinburgh), Professor Kenneth Wade (Durham), and Dr. Dominic S. Wright (Cambridge); in Erlangen, we mention especially Drs. Walter Bauer, Matthias Bremer, and Timothy Clark.

The research in this area in Cambridge (and earlier in Strathclyde) has been supported by the U.K. Science and Engineering Research Council and by the Associated Octel Company Ltd. Organo-alkali metal research in Erlangen is supported by the Deutscheforschungsgemeinschaft, the Fonds der Chemischen Industrie, Stiftung Volkswagenwerk, and the Convex Computer Corporation. Such financial input is essential, and it is appreciated.

Finally, we acknowledge with gratitude the provision of a Ciba–Geigy "Award for Collaboration in Europe." These funds allowed two-way visits to be made, during which this review was planned and partially written.

References

1. Dietrich, H., *Acta Crystallogr.* **16,** 681 (1963). The structure of $(EtLi)_4$ was later refined at low temperature: *J. Organomet. Chem.* **205,** 291 (1981).

2. Setzer, W., and Schleyer, P.v.R., *Adv. Organomet. Chem.* **24,** 353 (1985).

3. Schade, C., and Schleyer, P.v.R *Adv. Organomet. Chem.* **27,** 169 (1987). This review gives an extensive bibliography.

4. Seebach, D., *Angew. Chem.* **100,** 1685 (1988). *Angew, Chem., Int. Ed. Engl.* **27,** 1624 (1988).

4a. Boche, G., *Angew. Chem.* **101,** 286 (1989); *Angew. Chem., Int. Ed. Engl.* **28,** 277 (1989).

5. Williard, P. G., *In* "Comprehensive Organic Synthesis," Pergamon, Oxford (in press).

6. Fieser, M., "Reagents for Organic Synthesis," Vol. 15. Wiley (Interscience), New York, 1990. Vol. 15 and earlier volumes give specific uses of lithium amides arranged according to individual N—Li compounds.

7. Stowell, J. C., "Carbanions in Organic Synthesis." Wiley (Interscience), New York, 1979.

8. Wakefield, B. J., "The Chemistry of Organolithium Compounds." Pergamon, Oxford, 1974.
9. Barr, D., Berrisford, D. J., Jones, R. V. H., Slawin, A. M. Z., Snaith, R., Stoddart, J. F., and Williams, D. J., *Angew. Chem.* **101,** 1048 (1989); *Angew Chem., Int. Ed. Engl.* **28,** 1044 (1989).
10. Heathcock, C. H., *in* "Asymmetric Synthesis" (J. D. Morrison, ed.), Vol. 3B. Academic Press, New York, 1984.
11. Gill, G. B., and Whiting, D. A., *Aldrichim. Acta* **19,** 31 (1986).
12. Wardell, J. L., *in* "The Chemistry of the Metal-Carbon Bond" (F. R. Hartley, ed.), Vol. 4. Wiley, Chichester, 1987.
13. Brandsma, L., and Verkruijsse, H. D., "Preparative Polar Organometallic Chemistry," Vol. 1. Springer, Berlin, 1987.
14. King, R. B., and Eisch, J. J., eds., "Organometallic Syntheses," Vol. 3, pp. 352–406. Elsevier, Amsterdam, 1986.
15. Shriver, D. F., and Drezdzon, M. A., "The Manipulation of Air-Sensitive Compounds," 2nd ed. Wiley, New York, 1986.
16. Lappert, M. F., Power, P. P., Sanger, A. R., and Srivastava, R. C., "Metal and Metalloid Amides." Wiley, New York, 1980.
17. Rogers, R. D., Atwood, J. L., and Grüning, R., *J. Organomet. Chem.* **157,** 229 (1978).
18. Mootz, D., Zinnius, A., and Böttcher, B., *Angew. Chem.* **81,** 398 (1969); *Angew Chem., Int. Ed. Engl.* **8,** 378 (1969).
19. Lappert, M. F., Slade, M. J., Singh, A., Atwood, J. L., Rogers, R. D., and Shakir, R., *J. Am. Chem. Soc.* **105,** 302 (1983).
20. Armstrong, D. R., Mulvey, R. E., Walker, G. T., Barr, D., and Snaith, R., *J. Chem. Soc., Dalton Trans.* p. 617 (1988).
21. Armstrong, D. R., Barr, D., Clegg, W., Hodgson, S. M., Mulvey, R. E., Reed, D., Snaith, R., and Wright, D. S., *J. Am. Chem. Soc.* **111,** 4719 (1989).
22. Köster, H., Thoennes, D., and Weiss, E., *J. Organomet. Chem.* **160,** 1 (1978).
23. Schubert, B., and Weiss, E., *Angew. Chem.* **95,** 499 (1983); *Angew. Chem., Int. Ed. Engl.* **22,** 496 (1983).
24. Thoennes, D., and Weiss, E., *Chem. Ber.* **111,** 3157 (1978).
25. Schubert, B., and Weiss, E., *Chem. Ber.* **116,** 3212 (1983).
26. Seebach, D., Amstutz, R., Laube, T., Schweizer, W. B., and Dunitz, J. D., *J. Am. Chem. Soc.* **107,** 5403 (1985).
27. Schümann, U., Kopf, J., and Weiss, E., *Angew. Chem.* **97,** 222 (1985); *Angew. Chem., Int. Ed. Engl.* **24,** 215 (1985).
28. Klumpp, G. W., Vos, M., de Kanter, F. J. J., Slob, C., Krabbendam, H., and Spek, A. L., *J. Am. Chem. Soc.* **107,** 8292 (1985).
29. Lee, K.-S., Williard, P. G., and Suggs, J. W., *J. Organomet. Chem.* **299,** 311 (1986).
30. Jastrzebski, J. T. B. H., van Koten, G., Konijn, M., and Stam, C. H., *J. Am. Chem. Soc.* **104,** 5490 (1982).
31. Jastrzebski, J. T. B. H., van Koten, G., Christophersen, M. N., and Stam, C. H., *J. Organomet. Chem.* **292,** 319 (1985).
32. Moene, W., Vos, M., de Kanter, F. J. J., and Klumpp, G. W., *J. Am. Chem. Soc.* **111,** 3463 (1989).
33. van der Zeijden, A. A. H., and van Koten, G., *Recl. Trav. Chim. Pays-Bas* **107,** 431 (1988).
34. Klumpp, G. W., *Recl. Trav. Chim. Pays-Bas* **105,** 1 (1986).
34a. Snieckus, V., *Chem. Rev.* **90,** 879 (1990).
35. Wehmann, E., Jastrzebski, J. T. B. H., Ernsting, J.-M., Grove, D. M., and van Koten, G., *J. Organomet. Chem.* **353,** 145 (1988).

36. Schleyer, P.v.R., *Pure Appl. Chem.* **55,** 355 (1983); **56,** 151 (1984).
37. Streitwieser, A., *Acc. Chem. Res.* **17,** 353 (1984).
38. Reed, A. E., Weinstock, R. B., and Weinhold, F., *J. Chem. Phys.* **83,** 735 (1985).
39. Oliver, J. P., *Adv. Organomet. Chem.* **15,** 235 (1977).
40. O'Neill, M. E., and Wade, K., *in* "Comprehensive Organometallic Chemistry" (G. Wilkinson, F. G. A. Stone, and E. W. Abel, eds.), Vol. 1, p. 1. Pergamon, Oxford, 1982.
41. Wardell, J. L., *in* "Comprehensive Organometallic Chemistry" (G. Wilkinson, F. G. A. Stone, and E. W. Abel, eds.), Vol. 1, p. 43. Pergamon, Oxford, 1982.
42. Kaufmann, E., Schleyer, P.v.R., Houk, K. N., and Wu, Y.-D., *J. Am. Chem. Soc.* **107,** 5560 (1985).
43. Coates, G. E., Green, M. L. H., and Wade, K., "Organometallic Compounds: The Main Group Elements," 3rd ed., Vol. 1, pp. 35–38. Methuen, London, 1967.
44. Weiss, E., and Lucken, E. A. C., *J. Organomet. Chem.* **2,** 197 (1964); Weiss, E., and Hencken, G., *ibid.* **21,** 265 (1970).
45. Brockhart, M., and Green, M. L. H., *J. Organomet. Chem.* **250,** 395 (1983).
46. Rhine, W., Stucky, G. D., and Patterson, S. W., *J. Am. Chem. Soc.* **97,** 6401 (1975).
47. Zerger, R., Rhine, W., and Stucky, G. D., *J. Am. Chem. Soc.* **96,** 6048 (1974).
48. Tecle, B., Rahman, A. F. M. M., and Oliver, J. P., *J. Organomet. Chem.* **317,** 267 (1986).
49. Barr, D., Snaith, R., Mulvey, R. E., and Perkins, P. G., *Polyhedron* **7,** 2119 (1988).
50. Günther, H., Moskau, D., Bast, P., and Schmalz, D., *Angew. Chem.* **99,** 1242 (1987); *Angew. Chem., Int. Ed. Engl.* **26,** 1212 (1987).
51. Bauer, W., Clark, T., and Schleyer, P. v.R., *J. Am. Chem. Soc.* **109,** 970 (1987).
51a. Hoffmann, D., Bauer, W., and Schleyer, P. v.R., *J. Chem. Soc., Chem. Commun.* p. 208 (1990).
52. Ritchie, J. P., and Bachrach, S. M., *J. Am. Chem. Soc.* **109,** 5909 (1987).
53. Günther, H., Moskau, D., Dujardin, R., and Maercker, A., *Tetrahedron Lett.* **27,** 2251 (1986).
54. Hall, B., Farmer, J. B., Shearer, H. M. M., Sowerby, J. D., and Wade, K., *J. Chem. Soc., Dalton Trans.* p. 102 (1979).
55. Collier, M. R., Lappert, M. F., Snaith, R., and Wade, K., *J. Chem. Soc. D* p. 370 (1972).
56. Jennings, J. R., Snaith, R., Mahmoud, M. M., Wallwork, S. C., Bryan, S. J., Halfpenny, J., Petch, E. A., and Wade, K., *J. Organomet. Chem.* **249,** C1 (1983).
57. Snaith, R., Summerford, C., Wade, K., and Wyatt, B. K., *J. Chem. Soc. A* p. 2635 (1970).
58. Bryan, S. J., Clegg, W., Snaith, R., Wade, K., and Wong, E. H., *J. Chem. Soc., Chem. Commun.* p. 1223 (1987).
59. Farmer, J. B., Snaith, R., and Wade, K., *J. Chem. Soc. D* p. 1501 (1972).
60. Hall, B., Keable, J., Snaith, R., and Wade, K., *J. Chem. Soc., Dalton Trans.* p. 986 (1978).
61. Kilner, M., and Pinkney, J. N., *J. Chem. Soc. A* p. 2887 (1971).
62. Kilner, M., and Midcalf, C., *J. Chem. Soc., Dalton Trans.* p. 1620 (1974).
63. Chan, L.-H., and Rochow, E. G., *J. Organomet. Chem.* **9,** 231 (1967).
64. Anderson, H. J., Wang, N.-C., and Jwili, E. T. P., *Can. J. Chem.* **49,** 2315 (1971).
65. Pickard, P. L., and Talbot, T. L., *J. Org. Chem.* **26,** 4886 (1961).
66. Pattison, I., Wade, K., and Wyatt, B. K., *J. Chem. Soc. A* p. 837 (1968).
67. Samuel, B., Snaith, R., Summerford, C., and Wade, K., *J. Chem. Soc. A* p. 2019 (1970).
68. Sanger, A. R., *Inorg. Nucl. Chem. Lett.* **9,** 351 (1973).

69. Shearer, H. M. M., Wade, K., and Whitehead, G. *J. Chem. Soc., Chem. Commun.* p. 943 (1979).
70. Clegg, W., Snaith, R., Shearer, H. M. M., Wade, K., and Whitehead, G., *J. Chem. Soc., Dalton Trans.* p. 1309 (1983).
71. Barr, D., Clegg, W., Mulvey, R. E., Snaith, R., and Wade, K., *J. Chem. Soc., Chem. Commun.* p. 295 (1986).
72. Armstrong, D. R., Barr, D., Snaith, R., Clegg, W., Mulvey, R. E., Wade, K., and Reed, D., *J. Chem. Soc., Dalton Trans.* p. 1071 (1987).
73. Ilsley, W. H., Schaaf, T. F., Glick, M. D., and Oliver, J. P., *J. Am. Chem. Soc.* **102,** 3769 (1980).
74. Armstrong, D. R., and Walker, G. T., *J. Mol. Struct. (Theochem.)* **137,** 235 (1986).
75. Maercker, A., Bsata, M., Buchmeier, W., and Engelen, B., *Chem. Ber.* **117,** 2547 (1984).
76. Barr, D., Clegg, W., Mulvey, R. E., and Snaith, R., *J. Chem. Soc., Chem. Commun.* pp. 285, 287 (1984).
77. Barr, D., Clegg, W., Mulvey, R. E., and Snaith, R., *J. Chem. Soc., Chem. Commun.* p. 57 (1989).
78. Barr, D., Snaith, R., Clegg, W., Mulvey, R. E., and Wade, K., *J. Chem. Soc., Dalton Trans.* p. 2141 (1987).
79. Barr, D., Clegg, W., Mulvey, R. E., Reed, D., and Snaith, R., *Angew. Chem.* **97,** 322 (1985); *Angew. Chem., Int. Ed. Engl.* **24,** 328 (1985).
80. Fjeldberg, T., Hitchcock, P. B., Lappert, M. F., and Thorne, A. J., *J. Chem. Soc., Chem. Commun.* p. 822 (1984).
81. Engelhardt, L. M., May, A. S., Raston, C. L., and White, A. H., *J. Chem. Soc., Dalton Trans.* p. 1671 (1983).
82. Boche, G., Marsch, M., and Harms, K., *Angew. Chem.* **98,** 373 (1986); *Angew. Chem., Int. Ed. Engl.* **25,** 373 (1986).
83. Boche, G., Harms, K., and Marsch, M., *J. Am. Chem. Soc.* **110,** 6925 (1988).
84. Zarges, W., Marsch, M., Harms, K., and Boche, G., *Chem. Ber.* **122,** 1307 (1989).
85. Barr, D., Clegg, W., Mulvey, R. E., and Snaith, R., *J. Chem. Soc., Chem. Commun.* p. 79 (1984).
86. Barr, D., Clegg, W., Mulvey, R. E., and Snaith, R., *J. Chem. Soc., Chem. Commun.* p. 226 (1984).
87. Wittig, G., *Angew. Chem.* **70,** 65 (1958).
88. Eaborn, C., Hitchcock, P. B., Smith, J. D., and Sullivan, A. C., *J. Chem. Soc., Chem. Commun.* p. 827 (1983).
89. Schümann, U., and Weiss, E., *Angew. Chem.* **100,** 573 (1988); *Angew. Chem., Int. Ed. Engl.* **27,** 584 (1988).
90. Jackman, L. M., Scarmoutzos, L. M., and Porter, W., *J. Am. Chem. Soc.* **109,** 6524 (1987).
91. Dye, J. L., *Prog. Inorg. Chem.* **32,** 327 (1984).
92. Edmonds, R. N., Holton, D. M., and Edwards, P. P., *J. Chem. Soc., Dalton Trans.* p. 323 (1986).
93. Clegg, W., Mulvey, R. E., Snaith, R., Toogood, G. E., and Wade, K., *J. Chem. Soc., Chem. Commun.* p. 1740 (1986).
94. Reed, D., Barr, D., Mulvey, R. E., and Snaith, R., *J. Chem. Soc., Dalton Trans.* p. 557 (1986).
95. Barr, D., Snaith, R., Mulvey, R. E., Wade, K., and Reed, D., *Magn. Reson. Chem.* **24,** 713 (1986).
95a. Bauer, W., Winchester, W. R., and Schleyer, P. v. R., *Organometallics* **6,** 2371 (1987).

96. Fraenkel, G., Fraenkel, A. M., Geckle, M. J., and Schloss, F., *J. Am. Chem. Soc.* **101,** 4745 (1979).
97. Fraenkel, G., Henrichs, M., Hewitt, J. M., Su, B. M., and Geckle, M. J., *J. Am. Chem. Soc.* **102,** 3345 (1980).
98. Fraenkel, G., Henrichs, M., Hewitt, J. M., and Su, B. M., *J. Am. Chem. Soc.* **106,** 255 (1984).
99. Das, R., and Wilkie, C. A., *J. Am. Chem. Soc.* **94,** 4555 (1972).
100. Bauer, W., and Seebach, D., *Helv. Chim. Acta* **67,** 1972 (1984).
101. Armstrong, D. R., and Perkins, P. G., *Coord. Chem. Rev.* **38,** 139 (1981).
102. Kaufmann, E., Raghavachari, K., Reed, A. E., and Schleyer, P.v.R., *Organometallics* **7,** 1597 (1988).
103. Kaufmann, E., Gose, J., and Schleyer, P. v.R., *Organometallics* **8,** 2577 (1989).
103a. Würthwein, E.-U., Sen, K. D., Pople, J. A., and Schleyer, P. v.R., *Inorg. Chem.* **23,** 496 (1983).
104. Dewar, M. J. S., and Thiel, W., *J. Am. Chem. Soc.* **99,** 4899 (1977).
105. Clark, T., Rohde, C., and Schleyer, P. v.R., *Organometallics* **2,** 1344 (1983).
106. Thiel, W., and Clark, T., submitted for publication.
107. Glaser, R., and Streitwieser, A., *J. Mol. Struct.* **163,** 19 (1988).
108. Kaneti, J., Schleyer, P. v.R., Clark, T., Kos, A. J., Spitznagel, G. W., Andrade, J. G., and Moffat, J. B., *J. Am. Chem. Soc.* **108,** 1481 (1986).
109. Bachrach, S. M., and Streitwieser, A., *J. Am. Chem. Soc.* **106,** 2283, 5818 (1984).
110. Waterman, K. C., and Streitwieser, A., *J. Am. Chem. Soc.* **106,** 3138 (1984).
111. Sapse, A.-M., Raghavachari, K., Schleyer, P. v.R., and Kaufmann, E., *J. Am. Chem. Soc.* **107,** 6483 (1985).
112. Streitwieser, A., *J. Organomet. Chem.* **156,** 1 (1978).
113. Bushby, R. J., and Steel, H. L., *J. Organomet. Chem.* **336,** C25 (1987).
113a. Bushby, R. J., and Steel, H. L., *J. Chem. Soc., Perkin Trans. 2* p. 1143 (1990).
114. Kato, H., Hirao, K., and Akagi, K., *Inorg. Chem.* **20,** 3659 (1981).
115. Raghavachari, K., Sapse, A.-M., and Jain, D. C., *Inorg. Chem.* **26,** 2585 (1987).
116. Graham, G., Richtsmeier, S., and Dixon, D. A., *J. Am. Chem. Soc.* **102,** 5759 (1980).
117. Snow, M. R., *J. Am. Chem. Soc.* **92,** 3610 (1970).
118. Cremashi, P., Gamba, A., and Simonetta, M., *Theor. Chim. Acta* **40,** 303 (1975).
119. Wipff, G., Weiner, D., and Kollmann, P., *J. Am. Chem. Soc.* **104,** 3245 (1982).
120. Bushby, R. J., and Tytko, M. P., *J. Organomet. Chem.* **270,** 265 (1984).
121. Wipff, G., and Kollmann, P., *Nouv. J. Chem.* **9,** 457 (1985).
122. Hope, H., and Power, P. P., *J. Am. Chem. Soc.* **105,** 5320, (1983).
123. Spek, A. L., Duisenberg, A. J. M., Klumpp, G. W., and Geurink, P. J. A., *Acta Crystallogr., Sect. C* **C40,** 372 (1984).
124. Polt, R. L., Stork, G., Carpenter, G. B., and Williard, P. G., *J. Am. Chem. Soc.* **106,** 4276 (1984).
125. Jastrzebski, J. T. B. H., van Koten, G., Goubitz, K., Arlen, C., and Pfeffer, M., *J. Organomet. Chem.* **246,** C75 (1983).
126. Harder, S., Boersma, J., Brandsma, L., van Heteren, A., Kanters, J. A., Bauer, W., and Schleyer, P. v.R., *J. Am. Chem. Soc.* **110,** 7802 (1988).
127. Dietrich, H., Mahdi, W., and Storck, W., *J. Organomet. Chem.* **349,** 1 (1988).
128. Collins, J. B., Dill, J. D., Jemmis, E. D., Apeloig, Y., Schleyer, P. v.R., Seeger, R., and Pople, J. A., *J. Am. Chem. Soc.* **98,** 5419 (1976).
129. Nillsen, E. W., and Skancke, A., *J. Organomet. Chem.* **116,** 251 (1976).
130. Laidig, W. D., and Schäfer, H. F., *J. Am. Chem. Soc.* **100,** 5972 (1978).
131. Bachrach, S. M., and Streitwieser, A., *J. Am. Chem. Soc.* **106,** 5818 (1984).
132. Maercker, A., and Theis, M., *Top. Curr. Chem.* **138,** 1 (1987).

133. Harder, S., Boersma, J., Brandsma, L., Kanters, J. A., Bauer, W., and Schleyer, P.v.R., unpublished work.
134. Geissler, M., Schümann, U., and Weiss, E., *Proc. Int. Conf. Organomet. Chem. 12th, Vienna, 1985* Abstracts, p. 12 (1985).
135. Geissler, M., Kopf, J., Schubert, B., Weiss, E., Neugebauer, W., and Schleyer, P.v.R., *Angew. Chem.* **99,** 569 (1987); *Angew. Chem., Int. Ed. Engl.* **26,** 587 (1987).
136. Williard, P. G., and Carpenter, G. B., *J. Am. Chem. Soc.* **107,** 3345 (1985).
137. Williard, P. G., and Carpenter, G. B., *J. Am. Chem. Soc.* **108,** 462 (1986).
138. Jastrzebski, J. T. B. H., van Koten, G., and van der Mieroop, W. F., *Helv. Chim. Acta* **142,** 169 (1988).
139. Amstutz, R., Schweizer, W. B., Seebach, D., and Dunitz, J. D., *Inorg. Chim. Acta* **64,** 2617 (1981).
140. Williard, P. G., and Salvino, J. M., *Tetrahedron Lett.* **26,** 3931 (1985).
141. Graalmann, O., Klingebiel, U., Clegg, W., Haase, M., and Sheldrick, G. M., *Angew. Chem.* **96,** 904 (1984); *Angew. Chem., Int. Ed. Engl.* **23,** 891 (1984).
142. Laube, T., Dunitz, J. D., and Seebach, D., *Helv. Chim. Acta* **68,** 1373 (1985).
143. Bauer, W., Laube, T., and Seebach, D., *Chem. Ber.* **118,** 764 (1985).
144. Hvoslef, J., Hope, H., Murray, B. D., and Power, P. P., *J. Chem. Soc., Chem. Commun.* p. 1438 (1983).
145. Çetinkaya, B., Gümrükçü, I., Lappert, M. F., Atwood, J. L., and Shakir, R., *J. Am. Chem. Soc.* **102,** 2086 (1980).
146. Amstutz, R., Dunitz, J. D., Laube, T., Schweizer, W. B., and Seebach, D., *Chem. Ber.* **119,** 434 (1986).
147. Barr, D., Snaith, R., Wright, D. S., Mulvey, R. E., and Wade, K., *J. Am. Chem. Soc.* **109,** 7891 (1987).
148. Barr, D., Doyle, M. J., Mulvey, R. E., Raithby, P. R., Reed, D., Snaith, R., and Wright, D. S., *J. Chem. Soc., Chem. Commun.* p. 318 (1989).
149. Schmidbaur, H., Schier, A., and Schubert, U., *Chem. Ber.* **116,** 1938 (1983).
150. Armstrong, D. R., Barr, D., Clegg, W., Mulvey, R. E., Reed, D., Snaith, R., and Wade, K., *J. Chem. Soc., Chem. Commun.* p. 869 (1986).
151. Wittig, G., and Hesse, A., *Org. Synth.* **50,** 66 (1970).
152. Rathke, M. W., and Lindert, A., *J. Am. Chem. Soc.* **93,** 2318 (1971).
153. Olofson, R. A., and Dougherty, C. M., *J. Am. Chem. Soc.* **95,** 581, 582 (1973).
154. Corey, E. J., and Gross, W., *Tetrahedron Lett.* **25,** 495 (1984).
155. Cliffe, I. A., Crossley, R., and Shepherd, R. B., *Synthesis* p. 1138 (1985).
156. Rannenberg, M., Hausen, H.-D., and Weidlein, J., *J. Organomet. Chem.* **376,** C27 (1989).
157. Barr, D., Clegg, W., Mulvey, R. E., and Snaith, R. unpublished observations.
158. Hitchcock, P. B., Lappert, M. F., and Smith, S. J., *J. Organomet. Chem.* **320,** C27 (1987).
159. Hitchcock, P. B., Lappert, M. F., Power, P. P., and Smith, S. J., *J. Chem. Soc., Chem. Commun.* p. 1669 (1984).
160. Andrew, P. C., Armstrong, D. R., Mulvey, R. E., and Reed, D., *J. Am. Chem. Soc.* **110,** 5235 (1988).
161. Grüning, R., and Atwood, J. L., *J. Organomet. Chem.* **137,** 101 (1977).
162. Williard, P. G., *Acta Crystallogr., Sect. C* **C44,** 270 (1988).
163. Barr, D., Clegg, W., Mulvey, R. E., Snaith, R., and Wright, D. S., *J. Chem. Soc., Chem. Commun.* 716 (1987).
164. Snaith, R., Barr, D., Wright, D. S., Clegg, W., Hodgson, S. M., Lamming, G. R., Scott, A. J., and Mulvey, R. E., *Angew. Chem.* **101,** 1279 (1989); *Angew. Chem., Int. Ed. Engl.* **28,** 1241 (1989).

165. Sapse, A.-M., Kaufmann, E., Schleyer, P. v.R., and Gleiter, R., *Inorg. Chem.* **23,** 1569 (1984).
166. Hodoščck, M., and Šolmajer, T., *J. Am. Chem. Soc.* **106,** 1854 (1984).
166a. Kaufmann, E., Clark, T., and Schleyer, P. v.R., *J. Am. Chem. Soc.* **106,** 1856 (1984).
167. Armstrong, D. R., Perkins, P. G., and Walker, G. T., *J. Mol. Struct. (Theochem.)* **122,** 189 (1985).
168. Gregory, K., and Schleyer, P. v.R., unpublished results.
169. Raghavachari, K., Sapse, A.-M., and Jain, D. C., *Inorg. Chem.* **27,** 3862 (1988).
170. Kaufmann, E., Tidor, B., and Schleyer, P. v.R., *J. Comput. Chem.* **7,** 334 (1986).
171. Kolos, W., *Theor. Chim. Acta* **51,** 219 (1979).
172. Boys, S. F., and Bernardi, F., *Mol. Phys.* **19,** 553 (1970).
173. Clark, T., Chandrasekhar, J., Spitznagel, G., and Schleyer, P. v.R., *J. Comput. Chem.* **4,** 294 (1983).
174. Spitznagel, G., Clark, T., Chandrasekhar, J., and Schleyer, P.v.R., *J. Comput. Chem.* **3,** 363 (1982).
175. Chandrasekhar, J., Andrade, J. G., and Schleyer, P. v.R., *J. Am. Chem. Soc.* **103,** 5609 (1981).
176. Frisch, M. J., Pople, J. A., and Binkley, J. S., *J. Chem. Phys.* **80,** 3265 (1984).
177. Clegg, W., Snaith, R., and Wade, K., *Inorg. Chem.* **27,** 3861 (1988).
178. Armstrong, D. R., Perkins, P. G., and Stewart, J. J. P., *J. Chem. Soc., Dalton Trans.* p. 2277 (1973).
179. Durst, T., *in* "Comprehensive Carbanion Chemistry," Part B (E. Buncel and T. Durst, eds.) Elsevier, Amsterdam, 1984.
180. Normant, H., *Angew. Chem., Int. Ed. Engl.* **6,** 1046 (1967).
181. Mukhopadhyay, T., and Seebach, D., *Helv. Chim. Acta* **65,** 385 (1982).
182. Fraser, R. R., and Mansour, T. S., *Tetrahedron Lett.* **27,** 331 (1986).
183. Hey, E., Hitchcock, P. B., Lappert, M. F., and Rai, A. K., *J. Organomet. Chem.* **325,** 1 (1987).
184. Jones, R. A., Stuart, A. L., and Wright, T. C., *J. Am. Chem. Soc.* **105,** 7459 (1983).
185. Engelhardt, L. M., Jolly, B. S., Junk, P. C., Raston, C. L., Skelton, B. W., and White, A. H., *Aust. J. Chem.* **39,** 1337 (1986).
186. Çetinkaya, B., Hitchcock, P. B., Lappert, M. F., Misra, M. C., and Thorne, A. J., *J. Chem. Soc., Chem. Commun.* p. 148 (1984).
187. Bartlett, R. A., Chen, H., Dias, H. V. R., Olmstead, M. M., and Power, P. P., *J. Am. Chem. Soc.* **110,** 446 (1988).
188. Atwood, J. L., Bott, S. G., Hawkins, S. M., and Lappert, M. F., unpublished results.
189. Boese, R., and Klingebiel, U., *J. Organomet. Chem.* **315,** C17 (1986).
190. Bremer, M., Flock, R., and Schleyer, P.v.R., unpublished results.
191. Bartlett, R. A., Feng, X., Olmstead, M. M., Power, P. P., and Weese, K. J., *J. Am. Chem. Soc.* **109,** 4851 (1987).
192. Chen, H., Bartlett, R. A., Olmstead, M. M., Power, P. P., and Shoner, S. C., *J. Am. Chem. Soc.* **112,** 1048 (1990).
193. Paetzold, P., Pelzer, C., and Boese, R., *Chem. Ber.* **121,** 51 (1988).
194. Power, P. P., *Acc. Chem. Res.* **21,** 147 (1988).
195. Atwood, J. L., Lappert, M. F., Leung, W.-P., and Zhang, H., unpublished results.
196. Chen, H., Bartlett, R. A., Dias, H. V. R., Olmstead, M. M., and Power, P. P., *J. Am. Chem. Soc.* **111,** 4338 (1989).
197. Power, P. P., and Xiaojie, X., *J. Chem. Soc., Chem. Commun.* p. 358 (1984).
198. Bartlett, R. A., Dias, H. V. R., Hope, H., Murray, B. D., Olmstead, M. M., and Power, P. P., *J. Am. Chem. Soc.* **108,** 6921 (1986).
199. Bartlett, R. A., and Power, P. P., *J. Am. Chem. Soc.* **109,** 6509 (1987).

200. Kimura, B. Y., and Brown, T. L., *J. Organomet. Chem.* **26,** 57 (1971).
201. Jackman, L. M., and Scarmoutzos, L. M., *J. Am. Chem. Soc.* **109,** 5348 (1987).
202. Jackman, L. M., Scarmoutzos, L. M., and DeBrosse, C. W., *J. Am. Chem. Soc.* **109,** 5355 (1987).
203. Jackman, L. M., and DeBrosse, C. W., *J. Am. Chem. Soc.* **105,** 4177 (1983).
204. Jackman, L. M., and Smith, B. D., *J. Am. Chem. Soc.* **110,** 3829 (1988).
205. Kallman, N., and Collum, D. B., *J. Am. Chem. Soc.* **109,** 7466 (1987).
206. Galiano-Roth, A. S., Michaelides, E. M., and Collum, D. B., *J. Am. Chem. Soc.* **110,** 2658 (1988).
207. Seebach, D., Bauer, W., Hansen, J., Laube, T., Schweizer, W. B., and Dunitz, J. D., *J. Chem. Soc., Chem. Commun.* p. 853 (1984).
208. DePue, J. S., and Collum, D. B., *J. Am. Chem. Soc.* **110,** 5518 (1988).
209. DePue, J. S., and Collum, D. B., *J. Am. Chem. Soc.* **110,** 5524 (1988).
210. Galiano-Roth, A. S., and Collum, D. B., *J. Am. Chem. Soc.* **111,** 6772 (1989).
211. Seebach, D., and Bauer, W., *Helv. Chim. Acta* **67,** 1972 (1984).

ADVANCES IN INORGANIC CHEMISTRY, VOL. 37

CUBANE AND INCOMPLETE CUBANE-TYPE MOLYBDENUM AND TUNGSTEN OXO/SULFIDO CLUSTERS

TAKASHI SHIBAHARA

Department of Chemistry, Okayama University of Science, Okayama 700, Japan

I. Introduction
Previous Reviews
II. Incomplete Cubane-Type Clusters with $Mo_3O_{4-n}S_n$ Cores
A. Incomplete Cubane-Type Clusters with Mo_3O_4 Cores
B. Incomplete Cubane-Type Clusters with $Mo_3O_3S/Mo_3O_2S_2/Mo_3OS_3$ Cores
C. Incomplete Cubane-Type Clusters with Mo_3S_4 Cores
III. Cubane-Type Clusters with $Mo_4O_{4-n}S_n$ Cores
A. Cubane-Type Clusters with Mo_4O_4/Mo_4OS_3 Cores
B. Cubane-Type Clusters with Mo_4S_4 Cores
IV. Incomplete Cubane-Type Clusters with $W_3O_{4-n}S_n$ Cores
A. Incomplete Cubane-Type Clusters with $W_3O_4/W_3O_3S/W_3O_2S_2/W_3OS_3$ Cores
B. Incomplete Cubane-Type Clusters with W_3S_4 Cores
V. Cubane-Type Mixed-Metal Clusters with Mo_3MS_4 Cores
VI. Note on Preparation of Clusters
A. Compounds with Mo_3O_4 Cores
B. Compounds with $Mo_3O_{4-n}S_n$ (n = 1–4) Cores
C. Compounds with Mo_4S_4 Cores
D. Compounds with W_3S_4 Cores
E. Compounds with $MoMS_4$ Cores
Abbreviations
References

I. Introduction

This review surveys the structural aspect of cubane-type compounds with $Mo_4O_{4-n}S_n$ (n = 0, 3, or 4) cores and incomplete cubane-type compounds with $Mo_3O_{4-n}S_n$ and $W_3O_{4-n}S_n$ (n – 0–4) cores (Fig. 1). Compounds with mixed-metal cubane-type cores Mo_3MS_4 (M = metals) are also included. Notes on synthetic procedures are given in Section

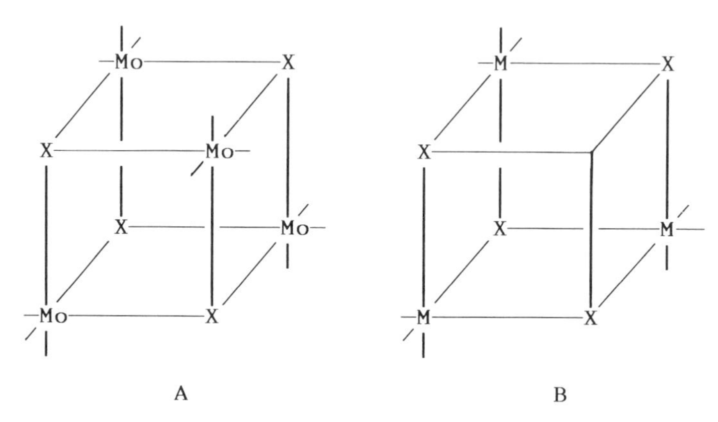

FIG. 1. Cubane-type (A) Mo_4X_4 and incomplete cubane-type (B) M_3X_4 cores. M = Mo, W; X = O, S.

VI, which, however, does not cover all of the available literature. Although the solid-state chemistry of molybdenum has been developed with, for example, infinite repeat Mo_3S_4 units, only discrete compounds with these cores are included in this review. Also, other related triangular Mo_3 and W_3 compounds such as $Mo_3(\mu_3\text{-S})(\mu\text{-S}_2)_3$ (*1*), $Mo_3(\mu_3\text{-H})(\mu_3\text{-I})(\mu\text{-I})_3$ (*2*), $Mo_3(\mu_3\text{-O})(\mu\text{-S}_2)_3$ (*3*) and $M_3(\mu_3\text{-O})_2$ (M = Mo, W) (*4*) are not described here, though they are interesting. This review has been organized primarily according to the number of Mo and W metals (three or four) and secondarily according to the number of bridging sulfur atoms.

PREVIOUS REVIEWS

The first systematic review of trinuclear clusters of the early transition elements was published in 1980 by Müller *et al.* (*5*), who classified the clusters into three categories according to the number of capping ligands (μ_3-type): 0, 1, and 2. Jiang *et al.* (*6*) reported a comprehensive review of trinuclear clusters in 1985, in which they gave a more detailed classification system. Thus in addition to the number of capping ligands, the number of bridging ligands (μ_2-type) was used for classification purposes. Both reviews considered the electronic structures of the clusters. The relationship between the structures of the clusters and the number of valence electrons has also been discussed by Gubin (*7*). Reviews focusing on more specialist fields have also appeared. Thus sulfur ligands have been considered by Dance (*8*), by Müller *et al.*, (*9, 10*), and by Zanello (*11*); Mo_3 clusters with μ_3-O and μ_3-OR (R = alkyl) have been considered by Chisholm *et al.* (*12*); and related electrochemical studies have been made by Zanello (*11*). Furthermore, Cotton has described the results of recent research on compounds with Mo—Mo bonds (*13*). Cubane-type clusters have been reviewed by Vahrenkamp (*14*), by Zanello (*11*), and by Harris (*15*). The latter in particular considers binding aspects. The binding properties of some of incomplete cubane-type molybdenum and tungsten compounds have been discussed by Cheng *et al.* (*16*). Chemical and physical properties of triangular bridged metal compounds have been reviewed by Cannon and White (*17*). A review of mixed-metal clusters has also been published (*18*). Earlier reviews by Stiefel (*19*) and by Spivack and Dori (*20*) have emphasized structural aspects of Mo compounds. A more recent review by Young (*21*) includes some description of mixed-valence compounds of both molybdenum and tungsten.

II. Incomplete Cubane-Type Clusters with $Mo_3O_{4-n}S_n$ Cores

In contrast to the situation a decade ago, many incomplete cubane-type clusters with $Mo_3O_{4-n}S_n$ cores have been prepared and the structures have been determined by X-ray structure analyses. The results obtained are summarized in Tables I–III. The formal oxidation state of molybdenum in the compounds cited here is in all cases IV. Unlike Mo(VI) and Mo(V) compounds, mononuclear oxo or dioxo compounds of the Mo(IV) state are relatively rare and all the incomplete cubane-type compounds cited here have no terminal oxo ligand. Three Mo atoms form an equilateral triangle, and three single bonds exist between each Mo. Except for the compounds **1**, **8**, and **31** (Table III), and excluding Mo—Mo bonds, each molybdenum is octahedrally coordinated.

A. Incomplete Cubane-Type Clusters with Mo_3O_4 Cores

Structural parameters for the compounds with Mo_3O_4 cores are found in Table I (*22-28*). Some of the compounds are shown in Figs. 2 and 3. The Mo—Mo distance is ~2.5 Å (consistent with metal–metal bonding), and the Mo-μ_3O distance is longer than Mo-μ_2O by ~0.1 Å. The angles in Mo—μ_3O—Mo and Mo—μ_2O—Mo fall in the narrow range 75.0–76.6 and 80.6–81.5, respectively. The following tendency of Mo—L distances was observed: Mo—F < Mo—O (RCO*O*) < Mo—N (*N*CS) ≃ Mo—O (H_2*O*) < Mo—N (R_3*N*).

Compound **1** (Fig. 2) was the first to be obtained from an aqueous Mo(IV) solution. In a related study at about the same time, the two Mo_3O_4 units in compound **4** (Fig. 3) were found to be bridged by three

TABLE I

STRUCTURAL PARAMETERS FOR INCOMPLETE CUBANE-TYPE COMPOUNDS WITH Mo_3O_4 CORES[a]

Compound	Mo — Mo	Mo — μ_3O	Mo — μ_2O	Mo — L[b]	Ref.
(1) $Cs_2[Mo_3O_4(ox)_3(H_2O)_3] \cdot 4H_2O \cdot 1/2H_2C_2O_4$	2.486[1]	2.019[6]	1.921[4]	2.091[4], O(ox) 2.154[7], O(H_2O)	*22,23*
(2) $Cs_2[Mo_3O_4(ox)_3(H_2O)_3] \cdot CsCF_3SO_3 \cdot 3H_2O$	2.491[1]	2.01[1]	1.908[7]	2.102[8], O(ox) 2.15[1], O(H_2O)	*22*
(3) $[Pt(en)_2][Mo_3O_4(ox)_3(H_2O)_3] \cdot 3H_2O$	2.494[6]	2.020[3]	1.915[6]	2.094[16], O(ox) 2.164(4), O(H_2O)	*24*
(4) $Na_4[(Mo_3O_4)_2(edta)_3] \cdot 14H_2O$	2.506[13]	2.042[17]	1.918[18]	2.090[21], O 2.274[42], N	*25*
(5) $(NH_4)_5[Mo_3O_4F_9] \cdot NH_4F \cdot H_2O$	2.505	2.032	1.920	2.034, F	*26*
(6) $((CH_3)_4N)_4[Mo_3O_4(NCS)_8(H_2O)] \cdot 3H_2O$	2.518[1]	2.07[4]	1.93[4]	2.26[2], O 2.14[4], N	*27*
(7) $Na_2[Mo_3O_4(mida)_3] \cdot 7H_2O$	2.495[10]	2.043[27]	1.918[10]	2.090[22], O 2.231[6], N	*28*

[a] Note: Numbers in parentheses in the all of the tables indicate the estimated standard deviation for that particular value; brackets indicate $\sqrt{\Sigma_i(x_i - \bar{x})^2/(n-1)}$ for several values whose unweighted arithmetic average is given, with $\bar{x}$ the mean of n values. In some cases, indicated by an asterisk, brackets are used to indicate $\sqrt{\Sigma_i(x_i - \bar{x})^2/n(n-1)}$.

[b] The ligating atom is noted after the numeral.

edta ligands. In compound **6,** it should be noted that the thiocyanate ligand coordinates to molybdenum through the nitrogen and not the sulfur atom.

B. INCOMPLETE CUBANE-TYPE CLUSTERS WITH $Mo_3O_3S/Mo_3O_2S_2/Mo_3OS_3$ CORES

Structural parameters for the compounds with Mo_3O_3S, $Mo_3O_2S_2$, or Mo_3OS_3 cores are found in Table II (*29–34*). Structures of some of the compounds are illustrated in Figs. 4–8. Introduction of μ_3-S causes an elongation of the Mo—Mo distance of 0.1 Å, or sometimes a little more. In compound **1** (Fig. 4) one CO_2 group of each $Hnta^{2-}$ ligand is left uncoordinated, the distinction being a long (av. 1.325 Å) and a short (1.205 Å) C—O distance in each case. There are three such —COOH groups in the trimeric anionic compound. In compound 2, the coordination of the cysteinato ligand is as indicated in Fig. 5. Compound **3** has one μ_3-S, one μ_2-S, and two μ_2-O bridging ligands in the trimeric core (Fig. 6). In compound **4** (Fig. 7), two nitrogen atoms occupy the δ position (*25*) [trans to μ_3-X (X = O or S); cf. Figs. 3 and 4] and the other

TABLE II

STRUCTURAL PARAMETERS FOR INCOMPLETE CUBANE-TYPE COMPOUNDS WITH $Mo_3O_{4-n}S_n$ (n = 1–3) CORES[a]

Compound	Mo — Mo[b]	Mo — μ_3X[c]	Mo — μ_2Y[c]	Mo — L	Ref.
(1) $Ba[Mo_3O_3S(Hnta)_3] \cdot 10H_2O$	2.589[6], SO	2.360[7], S	1.917[9], O	2.097[15], O 2.264[4], N	*29,30*
(2) $K_2[Mo_3O_3S(cys)_3] \cdot 6H_2O$	2.624[9], SO	2.367[13], S	1.933[52], O	2.190[22], O 2.219[26], N 2.440[20], S	*30*
(3) $(pyH)_5[Mo_3O_2S_2(NCS)_9] \cdot 2H_2O$	2.635[0], SO 2.715(4), SS	2.320[20], S	1.945[21], O 2.257[2], S	2.135[42], N	*31*
(4) $Ba[Mo_3OS_3(ida)_3] \cdot 7H_2O$	2.612(2), SO 2.725[12], SS	2.352[14], S	1.944[8], O 2.309[15], S	2.130[27], O 2.249[14], N	*32*
(5) $[Mo_3OS_3(dtp)_4(im)]$	2.651[6], OS	2.052[37], O	2.280[12], S	2.598, S[d] 2.622, S[e] 2.232, N	*33,34*
(6) $[Mo_3(\mu_3-O)S_3(dtp)_4(oxazole)]$	2.638[9], OS	2.027, O	2.282, S	2.590, S[d] 2.584, S[e]	*34*
(7) $[Mo_3(S,O)_{0.5}S_3(dtp)_4(SC(NH_2)_2)]$[f]	2.681[11]	2.202	2.282	2.568, S[d] 2.610, S[e] 2.633, S	*34*

[a] See footnote *a,* Table I.
[b] The μ_3X then μ_2Y atoms between the two Mo atoms follow the numeral.
[c] The μ_3X or μ_2Y atom follows the numeral.
[d] Terminal.
[e] Bridge.
[f] Disordered.

nitrogen atom resides in the γ position [trans to μ_2-X (X = O or S)], whereas all the nitrogen atoms occupy the δ position in the related compounds reported so far. Compounds **5** (Fig. 8) and **6** are rare cases in having μ_3-O alongside μ_2-S ligation, whereas all the other compounds exhibit a preference for μ_3-S occupancy on introducing sulfurs into the core. A statistical disorder of μ_3-X (X = O or S) has been reported for compound **7.**

C. INCOMPLETE CUBANE-TYPE CLUSTERS WITH Mo_3S_4 CORES

Structural parameters for the compounds with Mo_3S_4 cores are found in Table III (*34–52*). The structures of some of these compounds are illustrated in Fig. 9–13. Larger variations are observed in the Mo—Mo distances (2.731–2.830 Å) of the series of compounds as compared to

TABLE III

STRUCTURAL PARAMETERS FOR INCOMPLETE CUBANE-TYPE COMPOUNDS WITH Mo_3S_4 CORES[a]

Compound	Mo — Mo	Mo — μ_3S	Mo — μ_2S	Mo — L	Ref.
(1) $[Mo_3S_4Cp_3][Sn(CH_3)_3Cl_2]$	2.812[1]	2.314[6]	2.294[6]	2.030[7], Cp	*35*
(2) $K_5[Mo_3S_4(CN)_9] \cdot 3KCN \cdot 4H_2O$	2.775[8]	2.363[7]	2.322[10]	2.189[10], C[b] 2.220[3], C[c]	*36*
(3) $K_5[Mo_3S_4(CN)_9] \cdot 7H_2O$	2.765[7]	2.363[4]	2.312[5]	2.159[13], C[b] 2.194[17], C[d]	*37*
(4) $Ca[Mo_3S_4(ida)_3] \cdot 11.5H_2O$	2.754[11]	2.348[9]	2.294[8]	2.166[23], O 2.274[5], N	*38*
(5) $(NH_4)_3[Mo_3S_4(Hnta)_2(nta)] \cdot 3EtOH$	2.769(1)	2.344(5)	2.298(3)		*39*
(6) $[Mo_3S_4([9]aneN_3)_3](ZnCl_4)(ZnCl_3H_2O)_2 \cdot 3H_2O$	2.773[7]	2.361[10]	2.278[9]	2.27[5], N	*40*
(7) $[Mo_3S_4(H_2O)_9](pts)_4 \cdot 9H_2O$	2.732[7]	2.332[4]	2.286[6]	2.190[14], O	*41*
(8) $[(C_2H_5)_4N]_2[Mo_3S_4(SCH_2CH_2S)_3]$	2.783[12]	2.345[3]	2.293[22]	2.475[5], S[b] 2.347[8], S[d]	*42*
(9) $[Mo_3S_4(dtp)_4(H_2O)]$	2.754[18]	2.346[9]	2.281[19]	2.361(6), O 2.575[20], S	*34,43*
(10) $[Mo_3S_4(dtp)_4P(C_6H_5)_3] \cdot 0.86CH_2Cl_2$	2.744[12]	2.335[16]	2.288[15]	2.647(3), P 2.604[6], S[e] 2.588[9], S[f]	*34,44*
(11) $[Mo_3S_4(dtp)_4(PhCH_2CN)]$	2.756[9]	2.341[5]	2.282[6]	2.261(6), N 2.648[10], S[g] 2.574[17], S[h]	*34,45*
(12) $[Mo_3S_4Cl_3(dmpe)_3] \cdot PF_6 \cdot CH_3OH$	2.766(4)	2.360(9)	2.336(7)[j] 2.290(7)[k]	2.534(8), P[b] 2.605(8), P[d] 2.473(7), Cl	*46,47*
(13) $[Mo_3S_4Cl_4(PEt_3)_3(MeOH)_2]$	2.766[24]	2.355[4]	2.288[11]	2.330[5], O 2.599[6], P 2.537[6], Cl[k] 2.422[8], Cl[l]	*48*
(14) $[Mo_3S_4Cl_4(PEt_3)_4(MeOH)]$	2.790[5]	2.359[8]	2.289[23]	2.324(8), O 2.660[43], P[m] 2.600[23], P[n] 2.517[75], Cl[k] 2.452(2), Cl[l]	*48*
(15) $Cs_2[Mo_3S_4(ox)_3(H_2O)_3] \cdot 3H_2O$	2.738(5)	2.33(1)	2.28(1)	—	*46,49*
(16) $[\{Mo_3S_4(HB(pz)_3)_2\}_2(\mu - O)(\mu - pz)_2] \cdot 2THF$	2.830[24]	2.352[11]	2.292[22]	2.27[6], N 1.91[1], O	*50*
(17) $[Mo_3S_4Cl_4(PPh_3)_3(H_2O)_2] \cdot 3THF$	2.758[15]	2.345[8]	2.286[6]	2.503[3], Cl[k] 2.427[10], Cl[l] 2.621[18], P 2.273[1], O	*51*
(18) $[Mo_3S_4Cl_3(dmpe)_3]Cl \cdot 2MeOH$	2.772[6]	2.351[10]	2.309[23]	2.496[10], Cl 2.563[29], P	*51*
(19) $[Mo_3S_4(dtc)_4(dmf)] \cdot EtOH$	2.741[38]	2.336(3)(av.)	2.297(3)(av.)	2.278(7), O (DMF) 2.569(4), S[o] 2.524(3), S[p]	*52*
(20) $[Mo_3S_4(dtc)_4(py)] \cdot 2py \cdot H_2O$	2.741[36]	2.344(2)(av.)	2.289(2)(av.)	2.361(5), Py 2.593(2), S[o] 2.523(2), S[p]	*34,52*

TABLE III *Continued*

Compound	Mo — Mo	Mo — μ_3S	Mo — μ_2S	Mo — L	Ref.
(21) $[Mo_3S_4(dtp)_4(SC(NH_2)(NHC_3H_5))]$	2.755[10]	2.341	2.286	2.582, S^q 2.608, S^r 2.618, S	*34*
(22) $[Mo_3S_4(dtp)_4(py)]$	2.754[10]	2.339	2.282	2.573, S^q 2.623, S^r 2.360, N	*34*
(23) $[Mo_3S_4(dtp)_4(dmf)]$	2.741[38]	2.337	2.303	2.525, S^q 2.569, S^r 2.278, O	*34*
(24) $[Mo_3S_4(dtp)_4(PhCH_2SH)]$	2.755[11]	2.337	2.282	2.568, S^q 2.643, S^r 2.734, S	*34*
(25) $[Mo_3S_4(dtp)_4(oxazole)]$	2.760[8]	2.337	2.282	2.573, S^q 2.617, S^r 2.316, N	*34*
(26) $[Mo_3S_4(dtp)_4(CH_3CH_2CN)]$	2.754[10]	2.344	2.281	2.572, S^q 2.633, S^r 2.306, N	*34*
(27) $[Mo_3S_4(dtp)_3(im)_3](dtp)$	2.760[4]	2.339	2.287	2.571, S^s 2.266, N	*34*
(28) $[Mo_3S_4(dtp)_3(\mu_2-O_2CCH_2CH_3)(py)]$	2.731[42]	2.335	2.288	2.549, S^s 2.372, N	*34*
(29) $[Mo_3S_4(dtp)_3(\mu_2-O_2CH)(py)]$	2.740[37]	2.332	2.287	2.545, S^s 2.385, N	*34*
(30) $[Mo_3S_4(dtp)_3(\mu_2-O_2CCH_3)(py)]$	2.739[45]	2.334	2.293	2.549, S^s 2.385, N	*34*
(31) $[Mo_3S_4(dtp)_3(bipy)](dtp)$	2.751[20]	2.337	2.275	2.589, S^s 2.573, S	*34*

[a] See footnote *a*, Table I.
[b] Trans to μ_3S.
[c] Cis to μ_2S.
[d] Cis to μ_3S.
[e] One Pand two S ligating atoms to one Mo.
[f] Three ligating S atoms to each Mo.
[g] Bridging ligand.
[h] Nonbridging ligand.
[i] Trans to the Mo — P bond.
[j] Trans to the Mo — Cl bond.
[k] Two Cl atoms to one Mo atom.
[l] One Cl atom to each Mo atom.
[m] Two P atoms to one Mo atom.
[n] One P atom to each Mo atom.
[o] Bridging dtc.
[p] Terminal dtc.
[q] Terminal dtp.
[r] Bridging dtp.
[s] Terminal.

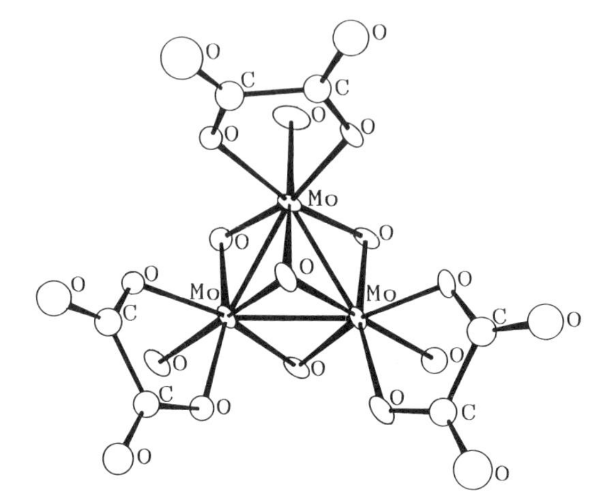

FIG. 2. Perspective view of $[Mo_3O_4(ox)_3(H_2O)_3]^{2-}$.

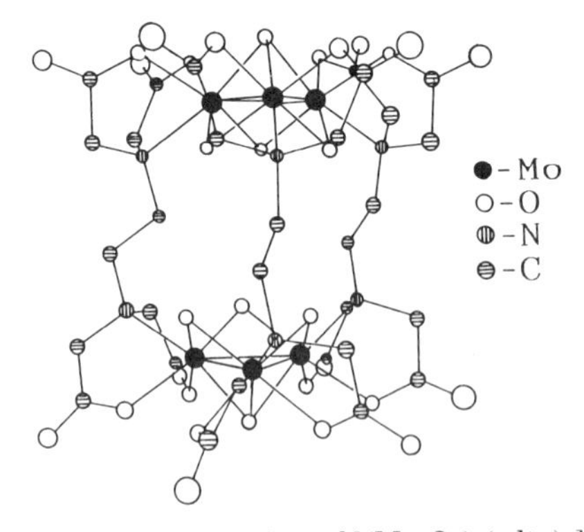

FIG. 3. Perspective view of $[(Mo_3O_4)_2(edta)_3]^{4-}$.

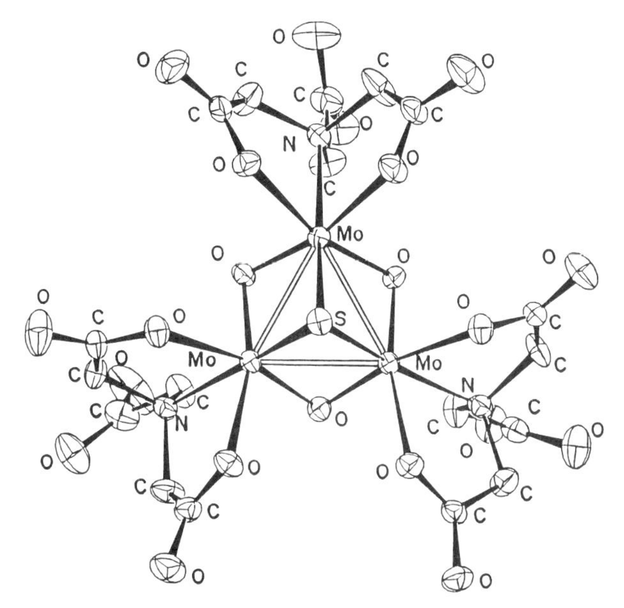

FIG. 4. Perspective view of $[Mo_3O_3S(Hnta)_3]^{2-}$.

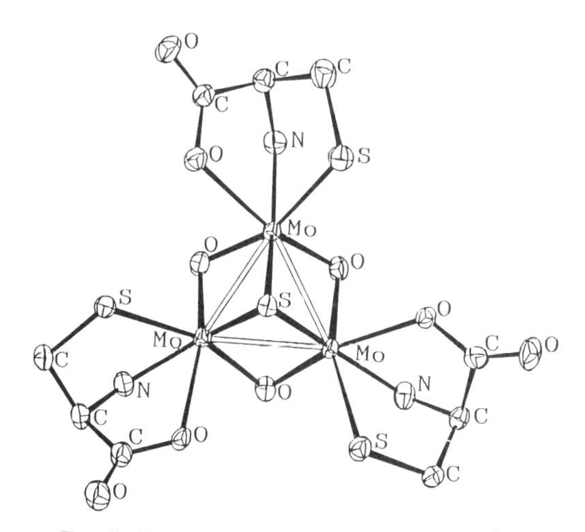

FIG. 5. Perspective view of $[Mo_3O_3S(cys)_3]^{2-}$.

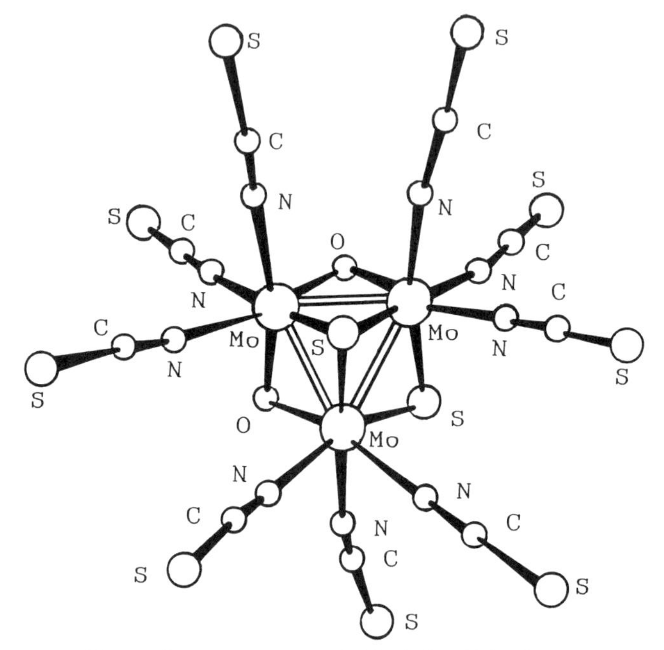

FIG. 6. Perspective view of $[Mo_3O_2S_2(NCS)_9]^{5-}$.

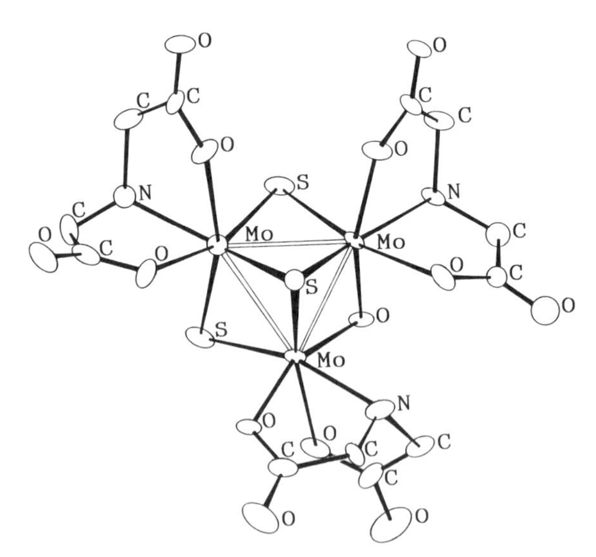

FIG. 7. Perspective view of $[Mo_3OS_3(ida)_3]^{2-}$.

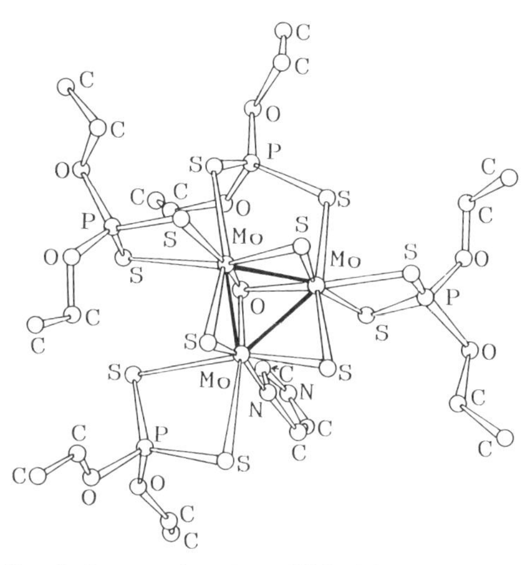

FIG. 8. Perspective view of $[Mo_3OS_3(dtp)_4(im)]$.

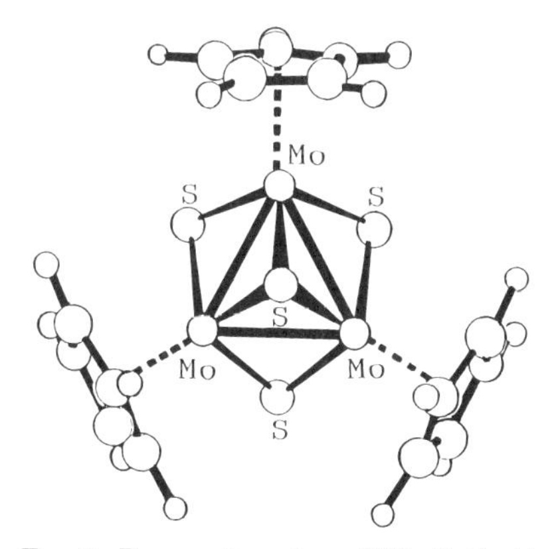

FIG. 9. Perspective view of $[Mo_3S_4Cp_3]^+$.

those having an Mo_3O_4 core (2.486–2.518 Å). However, this may be due to a greater variety of the ligands used in the Mo_3S_4 compounds. The cyclopentadienyl complex **1** (Table III) was the first compound reported to have an Mo_3S_4 core. The cation (Fig. 9) of compound **1** consists of three Mo(h^5-C_5H_5) centers positioned at the vertices of an equilateral triangle. In the cyano compounds **2** and **3,** the Mo—C bonds trans to μ_3-S are slightly shorter than those of Mo—C (cis to μ_3-S). The coordination modes of **4** and **5** are the same of those of **7** (Table I) and **1** (Table II). Each molybdenum in **6** (Fig. 10) has three ligating nitrogen atoms. Water molecules only are coordinated to the Mo_3S_4 core in **7** (Fig. 11). Each molybdenum atom in compound **8** (Fig. 12) is five coordinated, again ignoring the Mo—Mo bonds, and the mean planes defined by each ligand are nearly perpendicular to the equilateral triangle of Mo(IV) atoms. It is interesting that there are three terminal and one bridging $S_2P(OEt)_2$ ligands in **9** (Fig. 13), **10,** and **11.** This ligand has been used extensively and to good effect by the group in Fuzho, China. There are two kinds of Mo—μ_2S and Mo—P distances, respectively, in **12.** Structure **14** can be derived from that of **13** by replacing one of the methanol ligands with a PEt_3 and changing the coordination environment at Mo(2). In compound **16,** two incorporating Mo_3S_4 clusters are bridged by an oxygen atom (with the Mo—O—Mo angle close to 135°) and two bridging bidentate pyrazole $BH(pz)_3^-$ anions. The latter is potentially a tridentate anion. Compound **16** has the longest Mo—Mo distance. The molecular structure of **17** has an idealized C_s symmetry. Other compounds (**18–31**) in this series are examples containing either the dtc or dtp ligands.

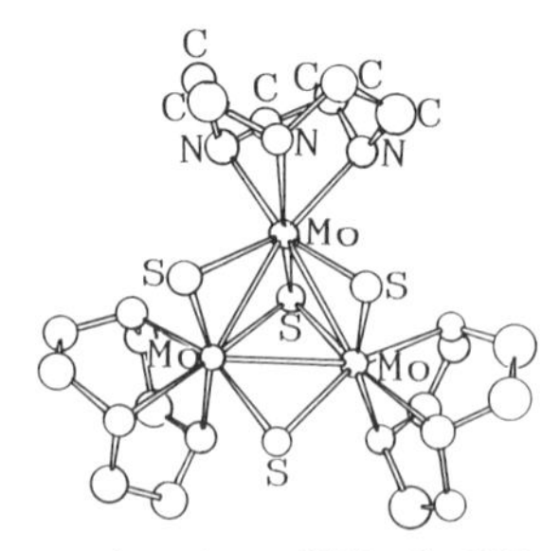

FIG. 10. Perspective view of $[Mo_3S_4([9]aneN_3)_3]^{4+}$.

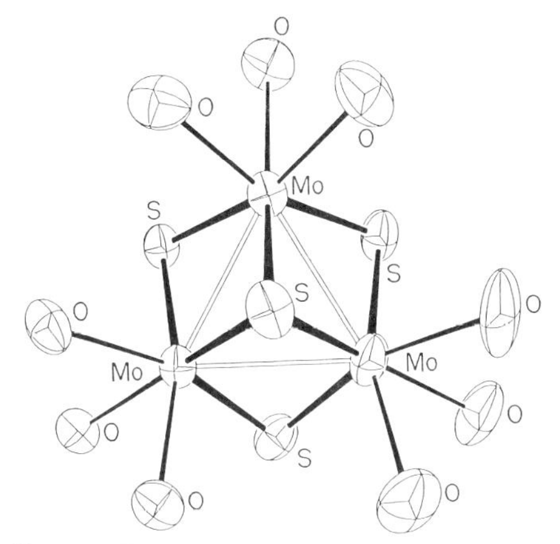

FIG. 11. Perspective view of $[Mo_3S_4(H_2O)_9]^{4+}$.

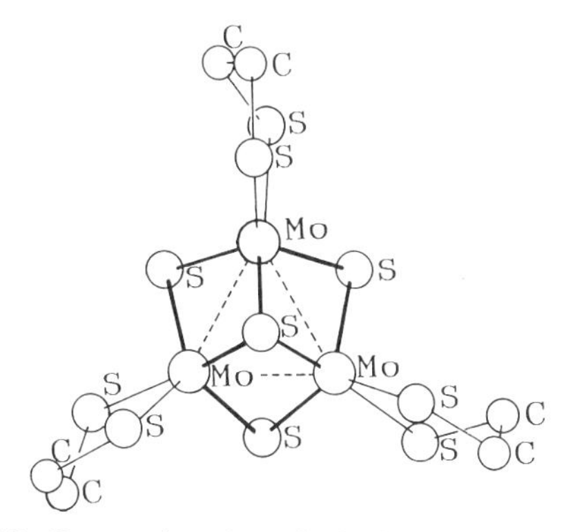

FIG. 12. Perspective view of $[Mo_3S_4(SCH_2CH_2S)_3]^{2-}$.

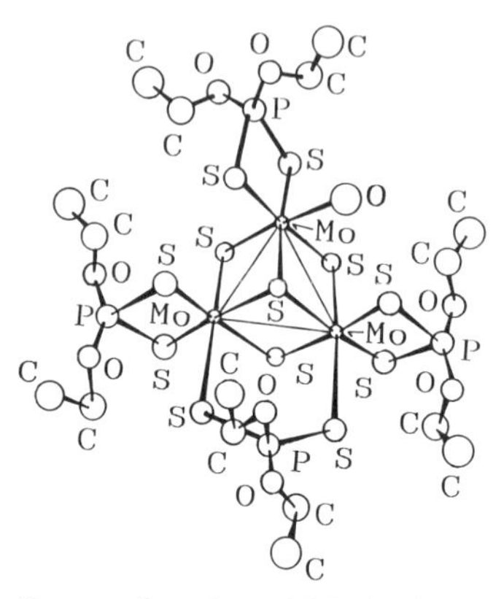

FIG. 13. Perspective view of $[Mo_3S_4(dtp)_4(H_2O)]$.

III. Cubane-Type Clusters with $Mo_4O_{4-n}S_n$ Cores

Although the series is incomplete, cubane-type clusters with $Mo_4O_{4-n}S_n$ ($n = 0, 3,$ or 4) cores have been prepared and structures have been determined by X-ray crystallography. The results are summarized in Tables IV (*53–57*) and V (*58–71*). Contrary to the case of incomplete cubane-type clusters, wherein only the IV oxidation state of molybdenum appears, the oxidation state of molybdenum varies from VI to III. The total oxidation number of each Mo_4 cluster is indicated in the tables.

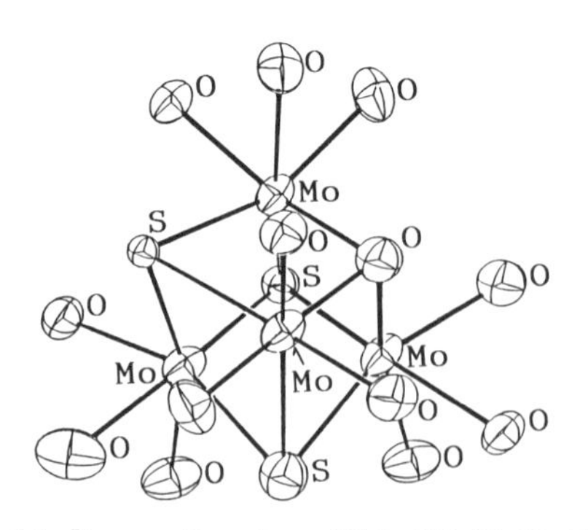

FIG. 14. Perspective view of $[Mo_4OS_3(H_2O)_{12}]^{5+}$.

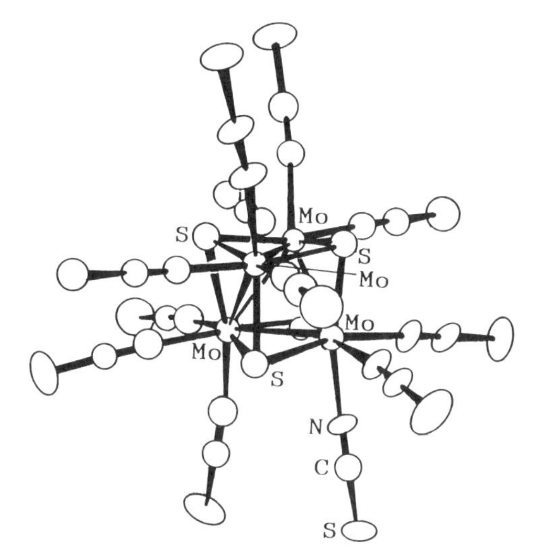

FIG. 15. Perspective view of $[Mo_4S_4(NCS)_{12}]^{6-}$.

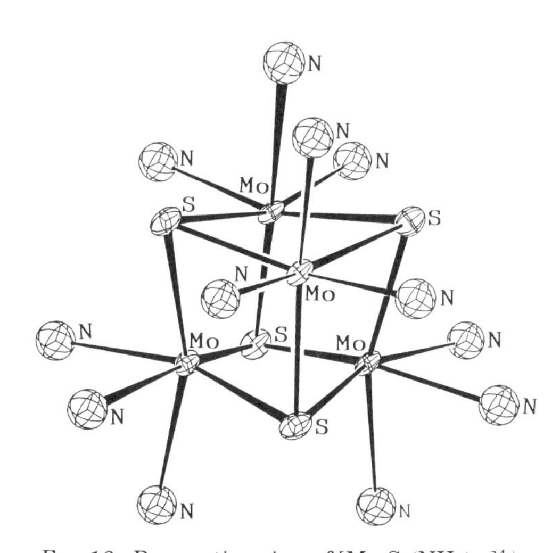

FIG. 16. Perspective view of $[Mo_4S_4(NH_3)_{12}]^{4+}$.

A. Cubane-Type Clusters with Mo_4O_4/Mo_4OS_3 Cores

Structural parameters for the compounds with Mo_4O_4 or Mo_4OS_3 cores are found in Table IV. No compounds with Mo_4O_3S or $Mo_4O_2S_2$ cores have been reported. Compared to the incomplete cubane-type compounds described above, only limited numbers of cubane-type compounds with oxo bridge(s) are known. Each molybdenum in compounds **1** and **2** has two terminal oxo ligands. X-Ray photoelectron spectroscopy (XPS) spectra show the oxidation states of the Mo, Rh, and Ir atoms to be VI, III, and III, respectively. The dimension of the cube in **3** may be compared with the corresponding distances for **1** and **2** in Table V. The Mo_4 units of **4** and **5** have crystallographic C_{3v} and approximate C_{3v} symmetry, respectively: the Mo_4 unit is a triangular pyramid and slant-edge lengths are shorter than basal-edge lengths, though an oxo ligand bridges the basal three molybdenum atoms. Compound **4** has three terminal and three bridging dtp ligands and **5** has four terminal and two bridging dtp ligands. The structure of the aqua complex of Mo_4OS_3, **6,** (Fig. 14) has been determined recently. Two groups of Mo—Mo distances are observed: the difference is large in compound **3** and small in compounds **4** and **5.**

B. Cubane-Type Clusters with Mo_4S_4 Cores

Structural parameters for the compounds with Mo_4S_4 cores are listed in Table V. In compounds **1, 2,** and **3,** which have Mo_4 (+20) cores, the two Mo_2S_2 units are separated by more than 3.6 Å. The Mo_4 unit of compound **4** (Fig. 15) is a triangular pyramid, with slant-edge lengths of

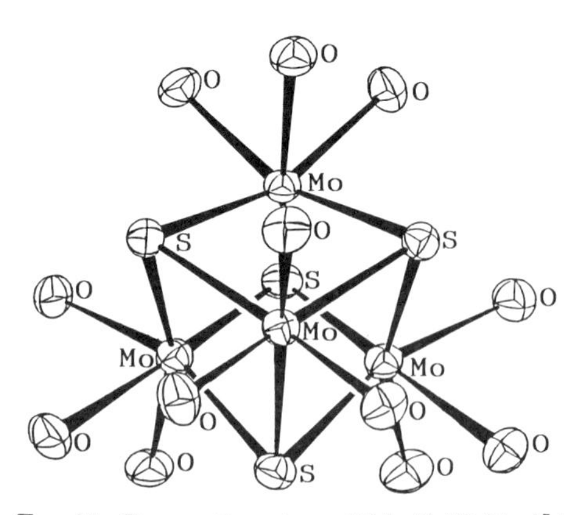

FIG. 17. Perspective view of $[Mo_4S_4(H_2O)_{12}]^{5+}$.

TABLE IV

STRUCTURAL PARAMETERS FOR CUBANE-TYPE COMPLEXES WITH $Mo_4O_{4-n}S_n$ CORES (n = 0 or 3)[a]

Compound	Mo — Mo	Mo — μ_3O	Mo — μ_3S	Mo — L	Ref.
(1) $[Mo_4O_4(O)_8(ORhC_5Me_5)_4] \cdot 2H_2O$, $Mo_4(+24)$	2.359(2)	1.967(8)[b] 2.340[2][c]	—	1.705[11], O_t 1.901(8), O (Rh) 3.225[7], Rh	*53*
(2) $[Mo_4O_4(O)_8(OIrC_5Me_5)_4]$, Mo_4 (+24)	2.343(5)	1.96(2)[b] 2.35[3][c]	—	1.71[1], O_t 1.89(2), O (Ir) 3.257(1), Ir	*53*
(3) $[Mo_4O_4(O)_4(OOPMe_2)_2(OSPMe_2)_2]$, $Mo_4(+20)$	2.635[d] 3.42[d]	1.976[b] 2.384[c]	—	1.636, O_t 2.081, O 2.490, S	*54*
(4) $[Mo_4OS_3(dtp)_6] \cdot 3CH_3CN$, $Mo_4(+14)$	2.831(1)[e] 2.700(1)[e]	2.036(5)	2.356[30]	2.556[19], S	*55*
(5) $[Mo_4OS_3(dtp)_6]$, $Mo_4(+14)$	2.817[5][e] 2.726[46][e]	2.005[11]	2.344[24]	2.562[29], S	*56*
(6) $[Mo_4OS_3(H_2O)_{12}](pts)_5 \cdot 14H_2O$, $Mo_4(+13)$	2.737[13]	1.96[7]	2.361[18]	2.184[15], O	*57*

[a] See footnote *a*, Table I.
[b] Cis to O_t.
[c] Trans to O_t.
[d] Two short and four long distances.
[e] Three long and three short distances.

2.791(1) Å and basal-edge lengths of 2.869(1) Å. The six Mo—Mo distances can be divided into two sets in the Mo_4 (+14) compounds: three long and three short for **4** and two short and four long for **5, 6, 11,** and **14,** though the difference between the sets is not so large as is found for the Mo_4 (+20) compounds. Four terminal and two bridging $S_2CN(C_2H_5)_2$ ligands exist in **5.** Likewise, the presence of four terminal and two bridging S_2PEt_2 ligands can be seen in **6.** The Mo—Mo distances of the Mo_4 (+12) compounds fall in a narrow range, as seen in compounds **8** (Fig. 16), **9, 13,** and **16.** The range of the Mo—Mo distance in **7** (Fig. 17) is smaller than that of compound **12,** but is a little larger than those of Mo_4 (+12) compounds. As far as compound **10** is concerned, the situation is complicated: if the charge of the nitrogen oxide is estimated (*72*) to be −1 from the observed ν(NO) = 1450 cm^{-1}, the oxidation number of Mo_4 is calculated to be +12; however, the observed Mo—Mo distance indicates that the oxidation number of Mo_4 is +20. If the +1 charge of NO is assigned to NO from the Mo—N—O linearity, the charge of Mo_4 is calculated to be +4, which is unlikely. The two sets of compounds, $[Mo_4S_4(edta)_2]^{2-/3-/4-}$ [**11, 12** (Fig. 18), and **13,** respectively] and $[Mo_4S_4(C_5H_4Pr^i)_4]^{0/+1/+2}$ (**14, 15,** and **16,** respectively), are the only cases so

TABLE V

STRUCTURAL PARAMETERS FOR CUBANE-TYPE COMPLEXES WITH Mo_4S_4 CORES[a]

Compound	Mo—Mo	Mo—μ_3S	Mo—L	Ref.
(1) $[Mo_4S_4(NC_6H_4CH_3)_4(dtp)_4]$, $Mo_4(+20)$	2.862[1][b] 3.69[4][b]	2.366[5][c] 2.704[12][c]	1.723[19], N[d] 2.552[13], S	*58*
(2) $[Mo_4S_4(O)_2(NC_6H_5)_2(dtp)_4]$, $Mo_4(+20)$	2.844[2][b] 3.85[4][b]	2.359[7][c] 2.843[55][c]	1.71[2], N 1.68[2], O 2.532[9], S	*59*
(3) $[Mo_4S_4(NC_6H_4CH_3)_4(S_2CN(i-C_4H_9)_2)_4]$, $Mo_4(+20)$	2.881[3][b] 3.66[4][b]	2.382[15][c] 2.687[25][c]	1.72[1], N 2.524[24], S	*60*
(4) $(NH_4)_6[Mo_4S_4(NCS)_{12}] \cdot 10H_2O$, $Mo_4(+14)$	2.791(1)[e] 2.869(1)[e]	—	—	*61*
(5) $[Mo_4S_4(Et_2dtc)_6] \cdot 2CHCl_3$, $Mo_4(+14)$	2.732(5)[b] 2.861[6][b]	2.35[2]	2.52[3], S	*62, 63*
(6) $[Mo_4S_4(S_2PEt_2)_6]$, $Mo_4(+14)$	2.786[1][b] 2.879[1][b]	2.354[12]	2.575[12], S	*64*
(7) $[Mo_4S_4(H_2O)_{12}](pts)_5 \cdot 14H_2O$, $Mo_4(+13)$	2.802[18]	2.349[4]	2.188[11], O	*57*
(8) $[Mo_4S_4(NH_3)_{12}]Cl_4 \cdot 7H_2O$, $Mo_4(+12)$	2.797[7]	2.370[5]	2.336[15], N	*65*
(9) $K_8[Mo_4S_4(CN)_{12}] \cdot 4H_2O$, $Mo_4(+12)$	2.854[1]	2.381[1]	2.190[1], C	*66*
(10) $K_8[Mo_4S_4(NO)_4(CN)_8] \cdot 4H_2O$, $Mo_4(+12$ or $+20)$	2.99[3][b] 3.67[4][b]	2.76[2] 2.35[2]	2.21[2], CN 1.76[2], NO[f]	*67*
(11) $Na_2[Mo_4S_4(edta)_2] \cdot 6.5H_2O$, $Mo_4(+14)$	2.761[28][b] 2.860[14][b]	2.355[24]	2.273[14], N 2.095[11], O	*68*
(12) $Ca_{1.5}[Mo_4S_4(edta)_2] \cdot 13H_2O$, $Mo_4(+13)$	2.780[19][g] 2.863[25][g]	2.356[6]	2.289[8], N 2.141[9], O	*69*
(13) $Mg_2[Mo_4S_4(edta)_2] \cdot 22H_2O$, Mo4(+12)	2.783[11] —	2.355[4]	2.279[3], N 2.179[16], O	*68*
(14) $[Mo_4S_4(C_5H_4Pr^i)_4](I_3^{-1})_2$, $Mo_4(+14)$	2.805[21][g] 2.885[19][g]	2.343[2]	2.331[6], Cp	*70*
(15) $[Mo_4S_4(C_5H_4Pr^i)_4]BF_4$, $Mo_4(+13)$	2.894[8]	2.343[3]	2.348[6], Cp	*70*
(16) $[Mo_4S_4(C_5H_4Pr^i)_4]$, $Mo_4(+12)$	2.904[7]	2.344[2]	2.356[1], Cp	*70*
(17) $[(H_2O)_9Mo_3S_4MoS_4Mo_3(H_2O)_9](pts)_8 \cdot 18H_2O$	2.770[4][h] 3.046[18][h]	2.334[13][i] 2.452[9][i]	2.178[18], O	*71*

[a] See footnote *a*, Table I.
[b] Two short and four long distances.
[c] Eight short and four long distances.
[d] Mo≡N (triple bond).
[e] Three short and three long distances.
[f] The angles Mo—N—O = 166.3(5)° and 178.2(6)°.
[g] Four short and two long distances.
[h] Three short and three long distances per half of the core.
[i] Nine short and three long distances per half of the core.

far reported to form a set of consecutive oxidation numbers of molybdenum. The mean Mo—Mo distance in the former decreases as the oxidation number of Mo_4 decreases, whereas that in the latter increases. Data on the double-cubane-type compound, **17** (Fig. 19) are also included in the table.

IV. Incomplete Cubane-Type $W_3O_{4-n}S_n$ Cores

As seen in Table VI, the incomplete cubane-type compounds with $W_3O_{4-n}S_n$ cores are not as well known as is the case for the corresponding molybdenum analogs. The oxidation state of tungsten in the table is in all cases IV. A particularly important finding is that the dimensions of the tungsten compounds are nearly the same as those of the molybdenum analogs.

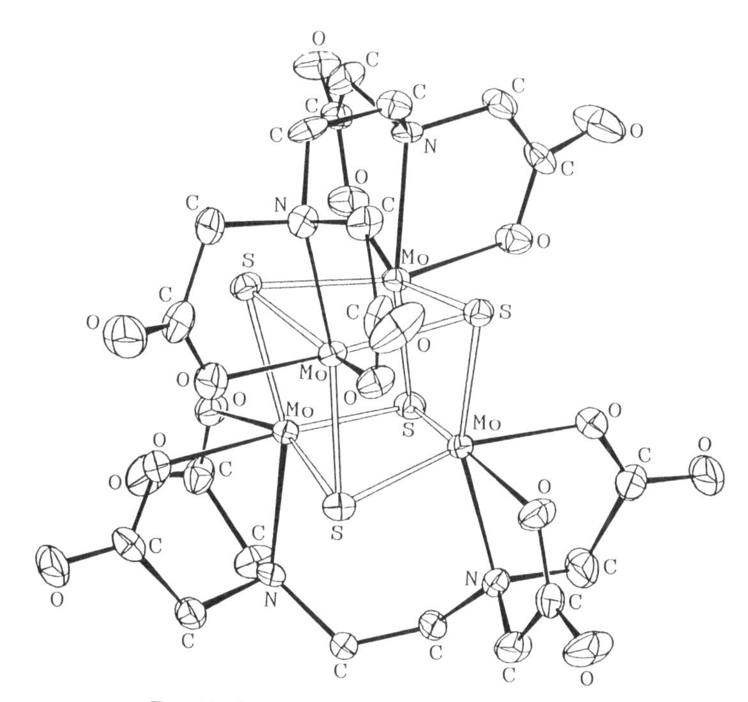

FIG. 18. Perspective view of $[Mo_4S_4(edta)_2]^{3-}$.

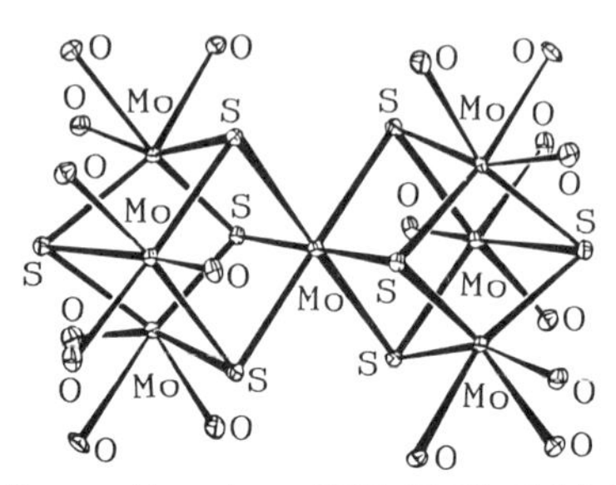

FIG. 19. Perspective view of $[(H_2O)_9Mo_3S_4MoS_4Mo_3(H_2O)_9]^{8+}$.

A. INCOMPLETE CUBANE-TYPE CLUSTERS WITH W_3O_4/W_3O_3S/$W_3O_2S_2$/W_3OS_3 CORES

Structural parameters for the compounds with W_3O_4, W_3O_3S, $W_3O_2S_2$, and W_3OS_3 cores are found in Table VI (*73–78*). The W—W distance of compounds incorporating the W_3O_4 cluster appears slightly longer than the corresponding distance in the molybdenum analogs. Compound **1** was the first compound having a W_3O_4 core for which the structure was determined. The structure of **2** (Fig. 20) can be compared with compound **6** in Table III. Compounds **4** (Fig. 21), **5, 6,** and **7** have a μ_3-S group. No μ_3-O trimeric compounds of tungsten with μ_2-S ligands have been reported, in contrast to the reported cases with Mo_3OS_3 cores.

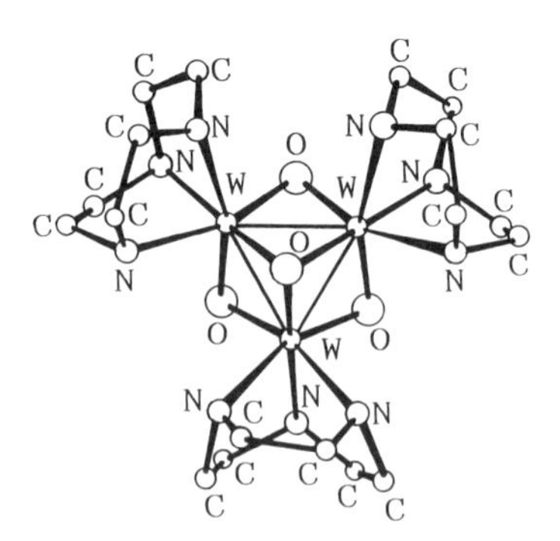

FIG. 20. Perspective view of $[W_3O_4([9]aneN_3)_3]^{4+}$.

TABLE VI

STRUCTURAL PARAMETERS FOR INCOMPLETE CUBANE-TYPE COMPLEXES WITH $W_3O_{4-n}S_n$ CORES (n = 0 – 3)[a]

Compound	W—W[b]	W—μ_3X[c]	W—μ_2Y[c]	W—L	Ref.
(1) $(NH_4)_5[W_3O_4F_9] \cdot NH_4F \cdot H_2O$	2.514[1]	2.07[1]	1.95[3]	2.03[4], F	*73*
(2) $[W_3O_4([9]aneN_3)_3][ZnBr_4]_2$	2.528[12]	2.10[4]	1.92[3]	2.22[4], N	*74*
(3) $(NH_4)_2((C_2H_5)_4N)_3[W_3O_4(NCS)_9] \cdot nH_2O$	2.534	2.039	1.911	2.110, N	*74a*
(4) $[W_3O_3S(NCS)_9]^{5-}$	2.612(6), SO	2.34(2), S	1.98(2), S	—	*75*
(5) $K_2[W_3O_3S(Hnta)_3] \cdot 9H_2O$	2.596(2), SO	2.380(8), S	1.949[3], S	2.096[32], O 2.236(20), N	*76*
(6) $Ba[W_3O_2S_2(Hnta)_3] \cdot 9H_2O$	2.612[7], SO 2.684(2), SS	2.375[8]	2.005[30], O 2.275[16], S	2.274[12], N 2.093[21], O	*77*
(7) $K_2[W_3OS_3(Hnta)_3] \cdot KCl \cdot 7H_2O$	2.620(1), SO 2.724[6], SS	2.344[8]	1.952[6], O 2.315[5], S	2.089[24], O 2.273[20], N	*78*

[a] See footnote *a*, Table I.
[b] The μ_3X then the μ_2Y atoms between the two W atoms follow the numeral.
[c] The μ_3X or μ_2Y atom follows the numeral.

B. INCOMPLETE CUBANE-TYPE CLUSTERS WITH W_3S_4 CORES

Structural parameters for the compounds with W_3S_4 are found in Table VII (*47, 79–84*). Compound **1** (Fig. 22) was the first to be prepared having a sulfur bridge in the triangle W_3 core. Compounds **2** (Fig. 23) and **3** can be compared with the corresponding molybdenum compounds, **7** and **12,** in Table III, respectively. In the case of compound **6,** each tungsten atom is coordinated by two μ_2-S, a μ_3-S, and a dtp. There are some differences among the W—W bond lengths and the W—μ_3S—W and W—μ_2S—W angles; these may be due to the effect of the acetate bridging ligand. Structure **7** is similar to structure **6.**

V. Cubane-Type Mixed-Metal Clusters with Mo_3MS_4 Cores

There are three kinds of cubane-type cores in this series of complexes (Fig. 24). There are the single cubane-type, Mo_3MS_4, the double cubane-type, $Mo_3S_4MMS_4Mo_3$, and the sandwich cubane-type, $Mo_3S_4MS_4Mo_3$. Structural parameters are collected in Table VIII (*34, 85–92*). Formal charges of the cores are noted after each compound in the table.

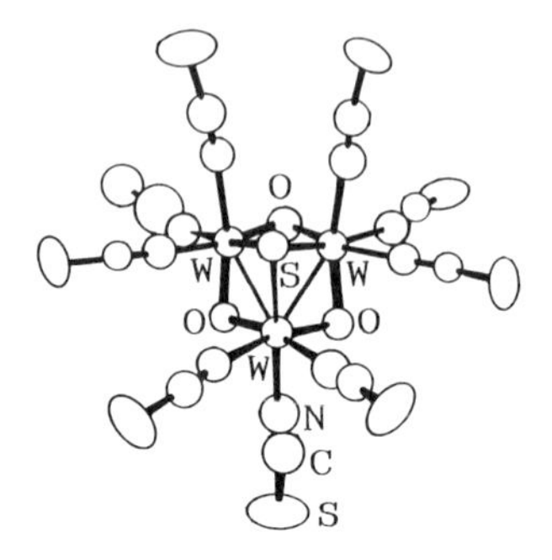

FIG. 21. Perspective view of $[W_3O_3S(NCS)_9]^{5-}$.

Compounds **1** (Fig. 25) to **7** are of the single cubane-type, and compound **8** (Fig. 26) is of the double cubane-type. Compounds **9** and **10** (Fig. 27) are of the sandwich cubane-type, wherein a tin or mercury atom is bonded to two incomplete cubane-type Mo_3S_4 corese, and are common to both cubes that are generated.

Although no distinct differences are found for the Mo—Mo distances in the mixed-metal clusters except for **5,** short (**1, 2, 3,** and **8**), and long (**6, 7, 9,** and **10**) Mo—M distances are observed. The first group has transition elements M in the cores, whereas the second group has nontransition elements M in the cores. The Mo—Cu distances in **3** and **8** are slightly longer than those of Mo—Fe and Mo—Ni in **1** and **2,** respectively.

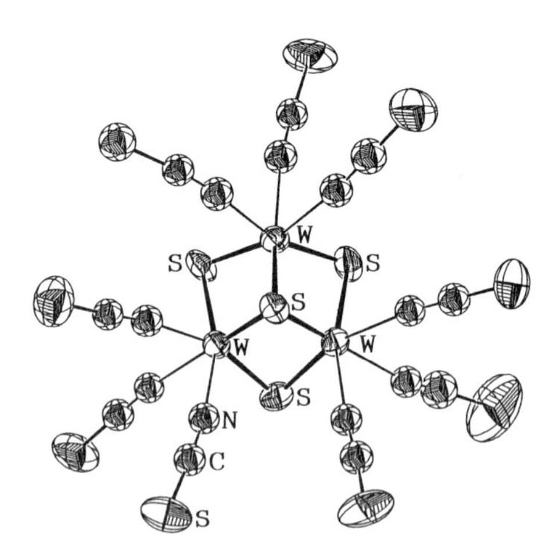

FIG. 22. Perspective view of $[W_3S_4(NCS)_9]^{5-}$.

TABLE VII

STRUCTURAL PARAMETERS FOR INCOMPLETE CUBANE-TYPE COMPLEXES WITH W_3S_4 CORES[a]

Compound	W—W	W—μ_3S	W—μ_2S	W—L	Ref.
(1) $(bpyH)_5[W_3S_4(NCS)_9] \cdot 3H_2O$	2.767[4]	2.363[12]	2.310[7]	2.153[32], N	*79*
(2) $[W_3S_4(H_2O)_9](pts)_4 \cdot 7H_2O$	2.723[15]	2.351[3]	2.283[5]	2.176[17], O	*80*
(3) $[W_3S_4Cl_3(dmpe)_3]PF_6 \cdot H_2O$	2.755(1)	2.382(5)	2.308[14][b]	2.595(3), P^b	*47, 81*
				2.520(4), P^c	
				2.488(4), Cl	
(4) $[W_3S_4Cl_3(depe)_3][PF_6]$	2.776[3]	2.368[15]	2.305[17]	2.468[7], Cl	*47*
				2.617[6], P^b	
				2.528[6], P^c	
(5) $[W_3S_4H_3(dmpe)_3][BPh_4]$	2.751[4]	2.354[2]	2.335[9]	2.516[6], P^b	*47*
				2.476[9], P^c	
(6) $[W_3S_4(dtp)_3(\mu_2-O_2CCH_3)(py)] \cdot (0.5HCON(CH_3)_2)$	2.6835(5)[d]	2.353[2]	2.306[11]	2.541[24], S	*82*
	2.748[7]			2.209[0], O	
				2.329(6), N	
(7) $[W_3S_4(dtp)_3(\mu_2\text{-}O_2CCH_3)(py)]$	2.6728(6)[d]	2.343[3]	2.298[6]	2.538[26], S	*83*
	2.745[7]			2.174[33], O	
				2.390[8], N	
(8) $[(NH_4)_3(HSO_3)][W_3S_4(S_4)_3(H_2O)_3]$	2.784(1)	—	2.283[30]	2.503[51], S	*84*
				2.335(7), O	

[a] See footnote *a*, Table I.
[b] Trans to μ_2S.
[c] Trans to μ_3S.
[d] Bridging CH_3CO_2 between two W atoms.

VI. Note on Preparation of Clusters

Preparation of the clusters described above is treated briefly here, although a thorough coverage of the literature is not intended.

A. COMPOUNDS WITH Mo_3O_4 CORES

The Mo(IV) aqua ion was first reported by Souchay *et al.* as a monomeric species in 1966 (*93*), and the $Mo_3O_4^{4+}$ core structure has been confirmed by single-crystal X-ray structure analyses, as described in this review, ^{95}Mo NMR (*94*), EXAFS structure analysis (*95*), and ^{18}O-labeling experiments (*96*), after the appearance of many contradictory reports (*97*). Information on the reduction products of the Mo(IV) aqua ion is also available (*98*), Richens and Sykes have summarized the preparation of the different aqua ions of molybdenum in oxidation states II to V (*99*).

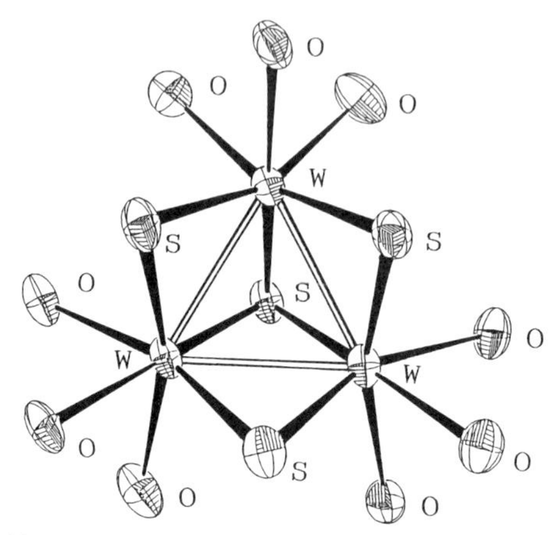

FIG. 23. Perspective view of $[W_3S_4(H_2O)_9]^{4+}$.

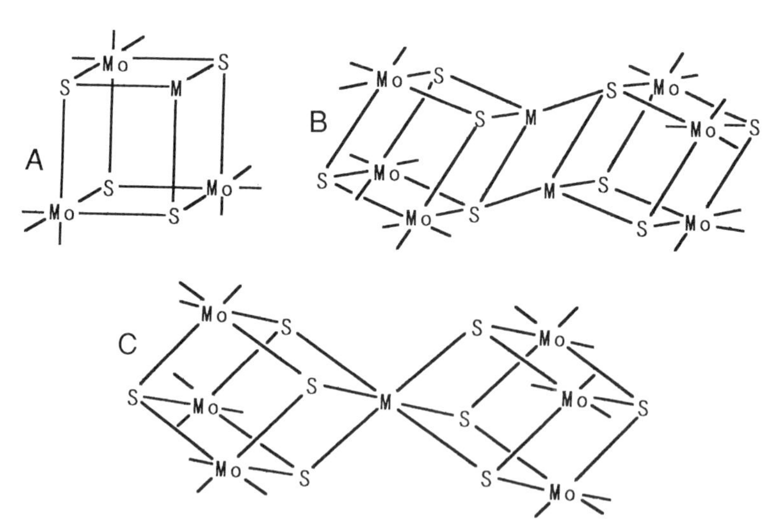

FIG. 24. Mixed-metal cubane-type cores. (A) Single cubane-type. (B) Double cubane-type. (C) Sandwich cubane-type.

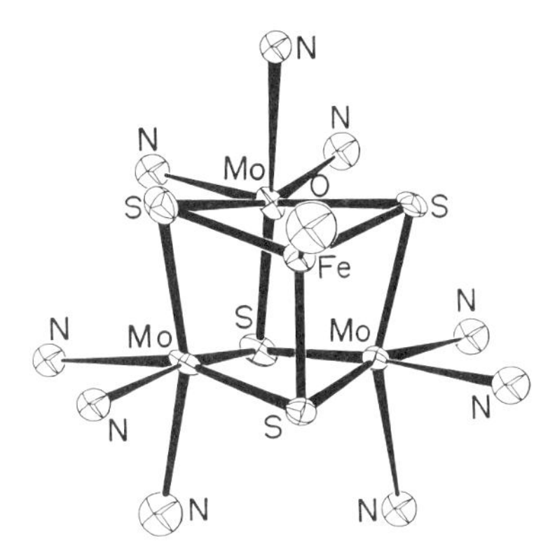

FIG. 25. Perspective view of $[Mo_3FeS_4(NH_3)_9(H_2O)]^{4+}$.

B. COMPOUNDS WITH $Mo_3O_{4-n}S_n$ (n = 1–4) CORES

Two methods are useful for the preparation of compounds with Mo_3O_3S: (1) reduction of $Mo_2O_3S^{2+}$ with $NaBH_4$ (*30*) and (2) reduction of $Mo_2O_3S^{2+}$ with $[MoCl_6]^{3-}$ (*100*). Compounds **1** and **2** in Table II were prepared from the aqua ion $Mo_3O_3S(aq)^{4+}$ and the corresponding ligands.

Some preparative methods of the sulfur-bridged incomplete cubane-type aqua ion $Mo_3S_4(aq)^{4+}$ have been reported: (1) reduction of $[Mo_2O_2S_2(cys)_2]^{2-}$ with $NaBH_4$ (*38*), (2) reaction of $Mo(CO)_6$ with Na_2S (*49*), and (3) electrolysis of $[Mo_2O_2S_2(cys)_2]^{2-}$ (*100*).

Compound **9** in Table III is prepared from $MoCl_3{\cdot}3H_2O$, P_2O_5, and H_2S (*34*).

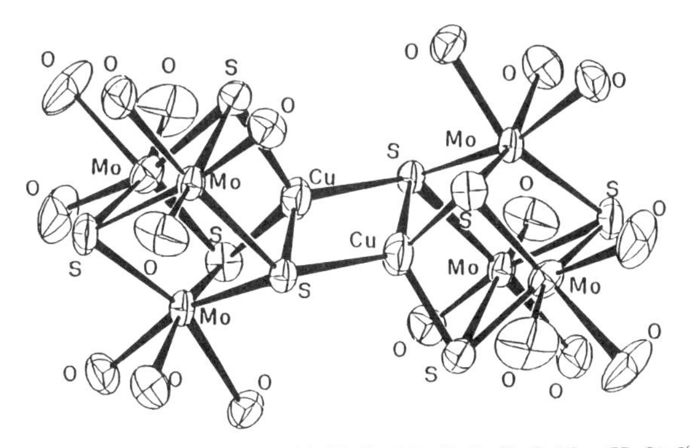

FIG. 26. Perspective view of $[(H_2O)_9Mo_3S_4CuCuS_4Mo_3(H_2O)_9]^{8+}$.

TABLE VIII

CUBANE-TYPE MIXED-METAL CLUSTERS WITH Mo_3MS_4 CORES[a]

Compound	Mo—Mo	Mo—M	M—μ_3S	Mo—L	M—L	Ref.
(1) $[Mo_3FeS_4(NH_3)_9(H_2O)]Cl_4$ (4+)	2.794[18]	2.683[10]	2.249[7]	2.269[34], N	2.039[0], O	*85*
(2) $[Mo_3NiS_4(H_2O)_{10}](pts)_4 \cdot 7H_2O$ (4+)	2.755[10]	2.641[9]	2.204[5]	2.21[1], O	1.97(1), O	*86*
(3) $[Mo_3CuS_4(dtp)_3(CH_3COO)(I)(dmf)]$ (5+)	2.770(2) 2.749(2) 2.679(2)[b]	2.849[39]	2.294[15]	2.538[29], S 2.214(13), O[c] 2.182[1], O[d]	2.454, I	*87*
(4) $[Mo_3CuS_4(dtp)_3(CH_3COO)(I)(H_2O)]$ (5+)	2.741[43]	—	—	2.527, S[e] 2.295,O	—	*34*
(5) $[Mo_3WS_4(Et_2PS_2)_6]$[f] (6+)	2.752(1)[g] 2.961(1)	—	2.357[18]	2.576[13], S	—	*88*
(6) $[Mo_3SbS_4(dtp)_4(Cl_3)(EtOH)]EtOH$[h] (7+)	2.735[12]	3.822[9]	2.775[36]	2.560[21], S[i] 2.576[9], S[j] 2.243(20), O	2.775[36], C1	*89*
(7) $[Mo_3SbS_4(dtp)_4(Cl_3)(oxazole)]$[h] (7+)	2.741[12]	3.825[19]	2.780[27]	2.553[37], S[h] 2.551(6), S[j] 2.271(12), N	2.383[18], C1	*89*
(8) $[H_2O)_9Mo_3S_4CuCuS_4Mo_3(H_2O)_9](pts)_8 \cdot 20H_2O$[k] (8+)	2.730[8]	2.886[93]	2.30[13]	2.178[22], O	—	*90*
(9) $[(H_2O)_9Mo_3S_4SnS_4Mo_3(H_2O)_9](pts)_8 \cdot 26H_2O$ (8+)	2.689[7]	3.712[37]	2.626[28]	2.177[21], O	—	*91*
(10) $[(H_2O)_9Mo_3S_4HgS_4Mo_3(H_2O)_9](pts)_8 \cdot 20H_2O$ (8+)	2.713[9]	3.833[93]	2.84[12]	2.172[27], O	—	*92*

[a] M represents metals; see footnote *a*, Table I.
[b] Acetato bridge between two Mo atoms.
[c] O atom of dmf.
[d] O atom of CH_3COO.
[e] Terminal.
[f] Three Mo atoms and one W atom are statistically disordered; three bridging and three terminal $Et_2PS_2^-$ exist.
[g] A bridging $Et_2PS_2^-$ exists between metals.
[h] Three terminal and a bridging dtp ligand exist.
[i] S atoms of the terminal dtp ligands.
[j] S atoms of the bridging dtp ligands.
[k] Cu—Cu′ = 2.424(3) Å.

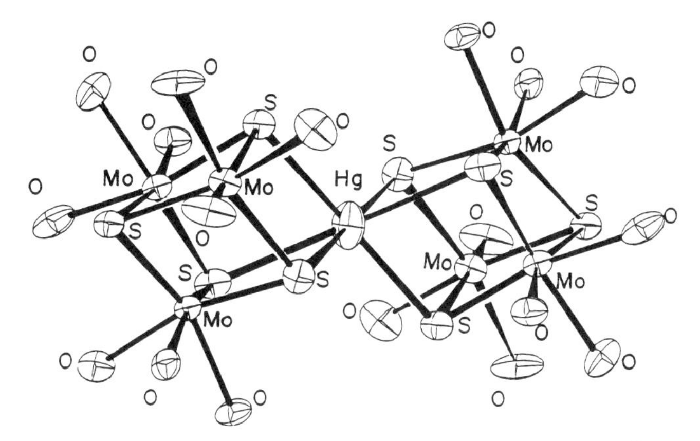

FIG. 27. Perspective view of $[(H_2O)_9Mo_3S_4HgS_4Mo_3(H_2O)_9]^{8+}$.

C. COMPOUNDS WITH Mo_4S_4 CORES

Compounds **11, 12,** and **13** in Table V were prepared by the reduction of $[Mo_2O_2S_2(edta)_2]^{2-}$ with $NaBH_4$ (*101*). Compounds **14, 15,** and **16** were prepared from $[Mo_2(\mu\text{-}Cl)_4(\eta\text{-}C_5H_4Pr^i)_2]$.

D. COMPOUNDS WITH W_3S_4 CORES

Compound **1** in Table VII was prepared by the reduction of $(NH_4)_2WS_4$ with $NaBH_4$. Compounds **3** and **4** in Table VII were prepared from WCl_4, NaHS, and the corresponding ligand, respectively. Compound **8** in Table VII was prepared from tungsten metal, sulfur, bromine, and $(NH_4)_2S_2$. Reaction of **8** with concentrated HCl and HBr gave $W_3S_4(aq)^{4+}$ by the elimination of the S_4^{2-} ligand.

E. COMPOUNDS WITH Mo_3MS_4 CORES

Compounds **1, 2, 8, 9,** and **10** in Table VIII were obtained by the reaction of the aqua ion $Mo_3S_4(aq)^{4+}$ in dilute HCl with corresponding metals. Compound **5** in Table VIII was prepared by the reaction of $Mo_3S_4(Et_2PS_2)_4$ with $W(CO)_3(CH_3CN)_3$. Compounds **3, 4, 6,** and **7** in Table VIII were prepared from $[Mo_3S_4(dtp)_4L]$ (L = H_2O or C_3H_3ON) and CuI (for **3** and **4**) or $SbCl_3$ (for **6** and **7**).

Abbreviations

ox	oxalate anion (2 −), $C_2O_4^{2-}$
nta	nitrilotriacetate anion (3 −), $N(CH_2CO_2)_3^{3-}$
edta	ethylenediaminetatraactate anion (4 −), $(O_2C)_2N(CH_2)_2N(CO_2)_2^{4-}$
ida	iminodiacetate anion (2 −), $NH(CO_2)_2^{2-}$
cys	L-cysteine anion (2 −), $SCH_2CH(NH_2)CO_2^{2-}$
mida	methyliminodiacetate anion (2 −), $NCH_3(CO_2)_2^{2-}$
pts	*para*-toluensulfonate anion (1 −), $CH_3C_6H_4SO_3^-$
[9]aneN_3	1,4,7-triazacyclononane, $\{NH(CH_2CH_2)\}_3$
dtc	diethyl dithiocarbamate anion (1 −), $Et_2NCS_2^-$
dtp	diethyl dithiophospahte anion (1 −), $S_2P(OEt)_2^-$
dmf	dimethylformamide, $HCON(CH_3)_2$
Him	imidazole, $C_3H_4N_2$
oxazole	C_3H_3NO
$HB(pz)_3$	tris (pyrazoborate)(1 −), $BH(C_3H_3N_2)_3^-$
Hpz	pyrazole, $C_3H_4N_2$
dmpe	1,2-bis (dimethylphosphine) ethane, $Me_2PCH_2CH_2PMe_2$
depe	1,2-bis (diethylphosphine) ethane, $Et_2PCH_2CH_2PEt_2$

References

1. Fedin, V. P., Kolesov, B. A., Mironov, Yu. V., and Fedorov, V. Ye., *Polyhedron* **8,** 2419–2423 (1989).
2. Cotton, F. A., and Poli, R., *J. Am. Chem. Soc.* **110,** 830–841 (1988).
3. Shang, M., Huang, J., and Lu, J., *Jiegou Huaxue* **3,** 17–20 (1984).
4. Ardon, M., and Bino, A., *Struct. Bonding (Berlin)* **65,** 1–28 (1987).
5. Müller, A., Jostes, R., and Cotton, F. A., *Angew. Chem., Int. Ed. Engl.* **19,** 875–882 (1980).
6. Jiang, Y., Tang, A., Hoffmann, R., Huang, J., and Lu, J., *Organometallics* **4,** 27–34 (1985).
7. Gubin, S. P., *Russ. Chem. Rev. (Engl. Transl.)* **54,** 305–322 (1985).
8. Dance, I. G., *Polyhedron* **5,** 1037–1104 (1986).
9. Müller, A., Diemann, E., Jostes, R., and Bogge, H., *Angew. Chem., Int. Ed. Engl.* **20,** 934–955 (1981).
10. Muller, A., Jaegermann, W., and Enemark, J. H., *Coord. Chem. Rev.* **46,** 245–280 (1982).
11. Zanello, P., *Coord. Chem. Rev.* **83,** 199–275 (1988).
12. Chisholm, M. H., Cotton, F. A., Fang, A., and Kober, E. M., *Inorg. Chem.* **23,** 749–754 (1984).
13. Cotton, F. A., *Polyhedron* **5,** 3–14 (1986).
14. Vahrenkamp, H., *Angew. Chem. Int. Ed. Engl.* **14,** 322–329 (1975).
15. Harris, S., *Polyhedron* **8,** 2843–2882 (1989).

16. Cheng, W., Zhang, Q., Huang, J., and Lu, J., *Polyhedron* **8,** 2785–2789 (1989).
17. Cannon, R. D., and White R. P., *Prog. Inorg. Chem.* **36,** 195–298 (1988).
18. Gladfelter, W. L., and Geoffroy, G. L., *Adv. Organomet. Chem.* **18,** 207–273 (1980).
19. Stiefel, E. I., *Prog. Inorg. Chem.* **22,** 1–223 (1977).
20. Spivack, B., and Dori, Z., *Coord. Chem. Rev.* **17,** 99–136 (1975).
21. Young, C. G., *Coord. Chem. Rev.* **96,** 89–251 (1989).
22. Benory, E., Bino, A., Gibson, D., Cotton, F. A., and Dori, Z., *Inorg. Chim. Acta* **99,** 137–142 (1985).
23. Bino, A., Cotton, F. A., and Dori, Z., *J. Am. Chem. Soc.* **100,** 5252–5253 (1978).
24. Rodgers, K. R., Murmann, R. K., Schlemper, E. O., and Schelton, M. E., *Inorg. Chem.* **24,** 1313–1322 (1985).
25. Bino, A., Cotton, F. A., and Dori, Z., *J. Am. Chem. Soc.* **101,** 3842–3847 (1979).
26. Müller, A., Ruck, A., Dartmann, M., and Reinsch-Vogell, U., *Angew. Chem.* **93,** 493–494 (1981).
27. Schlemper, E. O., Hussain, M. S., and Murmann, R. K., *Cryst. Struct. Commun.* **11,** 89–94 (1982).
28. Gheller, S. F., Hambley, T. W., Brownlee, R. T. C., O'Connor, M. J., Snow, M. R., and Wedd, A. G., *J. Am. Chem. Soc.* **105,** 1527–1532 (1983).
29. Shibahara, T., Hattori, H., and Kuroya, H., *J. Am. Chem. Soc.* **106,** 2710–2711 (1984).
30. Shibahara, T., Akashi, H., Nagahata, S., Hattori, H., and Kuroya, H., *Inorg. Chem.* **28,** 362–370 (1989).
31. Shibahara, T., Yamada, T., Kuroya, H., Hills, E. F., Kathirgamanathan, P., and Sykes, A. G., *Inorg. Chim. Acta* **113,** L19–L21 (1986).
32. Shibahara, T., Miyake, H., Kobayashi, K., and Kuroya, H., *Chem. Lett.* pp. 139–142 (1986).
33. Huang, M., Lu, S., Huang, J., and Huang, J., *Jiegou Huaxue* **6,** 29–33 (1987).
34. Huang, J. Q., Huang, J. L., Shang, M. Y., Lu, S. F., Lin, X. T., Lin, Y. H., Huang, M. D., Zhuang, H. H., and Lu, J. X., *Pure Appl. Chem.* **60,** 1185–1192 (1988).
35. Vergarmini, P. J., Vahrenkamp, H., and Dahl, L. F., *J. Am. Chem. Soc.* **93,** 6327–6329 (1971).
36. Müller, A., and Reinsch, U., *Angew. Chem., Int. Ed. Engl.* **19,** 72–73 (1980); Müller, A., Jostes, R., Eltzner, W., Nie, C.-S., Diemann, E., Bogge, H., Zimmermann, M., Dartmann, M., Reinsch-Vogell, U., Che, S., Cyvin, S. J., and Cyvin, B. N., *Inorg. Chem.* **24,** 2872–2884 (1985).
37. Howlader, N. C., Haight, G. P., Jr., Hambley, T. W., Lawrance, G. A., Rahmoeller, K. M., and Snow, M. R., *Aust. J. Chem.* **36,** 377–383 (1983).
38. Shibahara, T., and Kuroya, H., *Polyhedron* **5,** 357–361 (1986).
39. Cotton, F. A., Llusar, R., Marler, D. O., and Schwotzer, W., *Inorg. Chim. Acta* **102,** L25–L27 (1985).
40. Cotton, F. A., Dori, Z., Llusar, R., and Schwotzer, W., *Inorg. Chem.* **25,** 3654–3658 (1986).
41. Akashi, H., Shibahara, T., and Kuroya, H., *Polyhedron* **9,** 1671–1676 (1990).
42. Halbert, T. R., McGauley, K., Pan, W. H., Czernuszewicz, R. S., and Stiefel, E. I., *J. Am. Chem. Soc.,* **106,** 1849–1851 (1984).
43. Lin, X., Lin, Y., Huang, J., and Huang, J., *Kekue Tongbao* **32,** 810–815 (1987).
44. Huang, J., Lu, S., Lin, Y., Huang, J., and Lu, J., *Acta Chim. Sinica* **45,** 213–219 (1987).
45. Lu, S., Huang, J., Lin, Y., and Huang, J., *Jiegou Huaxue* **6,** 154–159 (1987).
46. Cotton, F. A., and Llusar, R., *Polyhedron* **6,** 1741–1745 (1987).

47. Cotton, F. A., Llusar, R., and Eagle, C. T., *J. Am. Chem. Soc.* **111,** 4332–4338 (1989).
48. Saito, T., Yamamoto, N., Yamagata, T., and Imoto, H., *Chem. Lett.* pp. 2025–2028 (1987).
49. Cotton, F. A., Dori, Z., Llusar, R., and Schwotzer, W., *J. Am. Chem. Soc.* **107,** 6734–6735 (1985).
50. Cotton, F. A., Llusar, R., and Schwotzer, W., *Inorg. Chim. Acta* **155,** 231–236 (1989).
51. Cotton, F. A., Kibala, P. A., Matusz, M., McCaleb, C. S., and Sandor, R. B. W., *Inorg. Chem.* **28,** 2623–2630 (1989).
52. Huang, M.-D., Lu,S.-F., Huang, J.-Q., and Huang, J.-L., *Acta Chim. Sin.* **47,** 121–127 (1989).
53. Hayashi, Y., Toriumi, K., and Isobe, K., *J. Am. Chem. Soc.* **110,** 3666–3668 (1988).
54. Mattes, R., and Muehlsiepen, K., *Z. Naturforsch., B: Anorg. Chem., Org. Chem.* **35B,** 265–268 (1980).
55. Lu, S., Huang, J., and Huang, J., *Acta Chim. Sin.* **43,** 918–922 (1985).
56. Lu, S., Lin, Y., Huang, J., and Huang, J., *Kexue Tongbao* **31,** 1609–1614 (1986).
57. Akashi, H., Shibahara, T., Narahara, T., Tsuru, H.,and Kuroya, H., *Chem. Lett.* pp. 129–132 (1989).
58. Edelblut, A. W., Folting, K., Huffman, J. C., and Wentworth, R. A. D., *J. Am. Chem. Soc.* **103,** 1927–1931 (1981).
59. Noble, M. E., Folting, K., Huffman, J. C., and Wentworth, R. A. D., *Inorg. Chem.* **22,** 3671–3676 (1983).
60. Wall, K. L., Folting, K., Huffman, J. C., and Wentworth, R. A. D., *Inorg. Chem.* **22,** 2366–2371 (1983).
61. Cotton, F. A., Diebold, M. P., Dori, Z., Llusar, R., and Schwotzer, W., *J. Am. Chem. Soc.* **107,** 6735–6736 (1985).
62. Mak, T. C. W., Jasim, K. S., and Chieh, C., *Inorg. Chem.* **24,** 1587–1591 (1985).
63. Mak, T. C. W., Jasim, K. S., and Chieh, C., *Angew. Chem., Int. Ed. Engl.* **23,** 391–392 (1984).
64. Keck, H., Kruse, A., Kuchen, W., Mathow, J., and Wunderlich, H., *Z. Naturforsch. B: Chem. Sci.* **42,** 1373–1378 (1987).
65. Shibahara, T., Kawano, E., Okano, M., Nishi, M., and Kuroya, H., *Chem. Lett.* pp. 827–828 (1986).
66. Müller, A., Eltzner, W., Bogge, H., and Jostes, R., *Angew. Chem., Int. Ed. Engl.* **21,** 795–796 (1982).
67. Müller, A., Eltzner, W., Clegg, W., and Sheldrick, G. M., *Angew. Chem.* **94,** 555–556 (1982).
68. Shibahara, T., Kuroya, H., Matsumoto, K., and Ooi, S., *Inorg. Chim. Acta* **116,** L25–L27 (1986).
69. Shibahara, T., Kuroya, H., Matsumoto, K., and Ooi, S., *J. Am. Chem. Soc.* **106,** 789–791 (1984).
70. Bandy, J. A., Davies, C. E., Green, J. C., Green, M. L. H., Prout, K., and Rodgers, D. P. S., *J. Chem. Soc., Chem. Commun.* pp. 1395–1397 (1983).
71. Shibahara, T., Yamamoto, T., Kanadani, H., and Kuroya, H., *J. Am. Chem. Soc.* **109,** 3495–3496 (1987).
72. Nakamoto, K., "Infrared and Raman Spectra of Inorganic and Coordination Compounds," pp. 295–297. Wiley, New York, 1978.
73. Mattes, R., and Mennemann, K., *Z. Anorg. Allg. Chem.* **437,** 175–182 (1977); *Angew. Chem.* **88,** 92–92 (1976).
74. Chaudhuri, P., Wieghardt, K., Gebert, W., Jibril, I., and Huttner, G., *Z. Anorg. Allg. Chem.* **521,** 23–36 (1985).

74a. Segawa, M., and Sasaki, Y., *J. Am. Chem. Soc.* **107,** 5565–5566 (1985).
75. Dori, Z., Cotton, F. A., Llusar, R., and Schwotzer, W., *Polyhedron* **5,** 907–909 (1986).
76. Shibahara, T., Takeuchi, A., Nakajima, M., and Kuroya, H., *Inorg. Chim. Acta* **143,** 147–148 (1988).
77. Shibahara, T., Takeuchi, A., Kunimoto, T., and Kuroya, H., *Chem. Lett.* pp. 867–870 (1987).
78. Shibahara, T., Takeuchi, A., and Kuroya, H., *Inorg. Chim. Acta* **127,** L39–L40 (1987).
79. Shibahara, T., Kohda, K., Ohtsuji, A., Yasuda, K., and Kuroya, H., *J. Am. Chem. Soc.* **108,** 2757–2758 (1986).
80. Shibahara, T., Takeuchi, A., Ohtsuji, A., Kohda, K., and Kuroya, H., *Inorg. Chim. Acta* **127,** L45–L46 (1987).
81. Cotton, F. A., and Llusar, R., *Inorg. Chem.* **27,** 1303–1305 (1988).
82. Zhan, H., Zheng, Y., Wu, X., and Lu, J., *J. Mol. Struct.* **196,** 241–247 (1989).
83. Zheng, Y., Zhan, H., and Wu, X., *Acta Crystallogr., Sect. C* **C45,** 1424–1426 (1989).
84. Fedin, V. P., Sokolov, M. N., Geras'ko, O. A., Sheer, M., Fedorov, V. Ye., Mironov, A. V., Slovokhotov, Yu. L., and Strutchkov, Yu. T., *Inorg. Chim. Acta* **165,** 25–26 (1989).
85. Shibahara, T., Akashi, H., and Kuroya, H., *J. Am. Chem. Soc.* **108,** 1342–1343 (1986).
86. Shibahara, T., Yamasaki, M., Akashi, H., and Katayama, T., *Inorg. Chem.* **30,** 2693–2699 (1991).
87. Wu, X., Lu, S., Zu, L., Wu, Q., Lu, J., *Inorg. Chim. Acta* **133,** 39–42 (1987).
88. Deeg, A., Keck, H., Kruse, A., Kuchen, W., and Wunderlich, H., *Z. Naturforsch., B: Chem. Sci.* **43,** 1541–1546 (1988).
89. Lu, S., Huang, J., Lin, Y., and Huang, J., *Acta Chim. Sin.* **45,** 666–675(Ch) (1987).
90. Shibahara, T., Akashi, H., and Kuroya, H., *J. Am. Chem. Soc.* **110,** 3313–3314 (1988).
91. Akashi, H., and Shibahara, T., *Inorg. Chem.* **28,** 2906–2907 (1989).
92. Shibahara, T., Akashi, H., Yamasaki, M., and Hashimoto, K. *Chem. Lett.* pp. 689–692 (1991).
93. Souchay, P., Cadiot, M., and Duhameaux, M., *C. R. Hebd. Seances Acad. Sci.* **262,** 1524–1527 (1966).
94. Gheller, S. F., Hambley, T. W., Brownlee, R. T. C., O'Conner, M. J., Snow, M. R., and Wedd, A. J., *J. Am. Chem. Soc.* **105,** 1527–1532 (1983).
95. Cramer, S. P., Eidem, P. K., Paffet, T. M., Winkler, J. R., Dori, Z., and Gray, H. B., *J. Am. Chem. Soc.* **105,** 799–802 (1983).
96. Murmann, R. K., and Shelton, M. E., *J. Am. Chem. Soc.* **102,** 3984–3985 (1980).
97. Ardon, M., Bino, A., and Yahao, G., *J. Am. Chem. Soc.* **98,** 2338–2339 (1976).
98. Richens, D. T., and Sykes, A. G., *Inorg. Chem.* **21,** 418–422 (1982); Paffet, M. T., and Anson, F. C., *ibid.* **22,** 1347–1355 (1983).
99. Richens, D. T., and Sykes, A. G., *Inorg. Synth.* **23,** 130–140 (1985).
100. Martinez, M., Ooi, B.-L., and Sykes, A. G., *J. Am. Chem. Soc.* **109,** 4615–4619 (1987).
101. See Refs. 68 and 69, Dimmock, P. W., McGinnis, J., Ooi, B.-L., and Sykes, A. G., *Inorg. Chem.* **29,** 1085–1089 (1990).

INTERACTIONS OF PLATINUM AMINE COMPOUNDS WITH SULFUR-CONTAINING BIOMOLECULES AND DNA FRAGMENTS

EDWIN L. M. LEMPERS and JAN REEDIJK

Department of Chemistry, Gorlaeus Laboratories, Leiden University, Leiden, The Netherlands

I. History of *cis*-Pt as an Antitumor Drug
II. Aqueous Solution Chemistry of *cis*-Pt
III. Antitumor Activity and DNA as the Target
 A. General
 B. Pt Binding to Nucleobases
 C. Binding of Pt to DNA
 D. Study of Pt Binding to Oligonucleotides
 E. Binding of Other Pt Compounds to Nucleic Acids
IV. Platinum–Sulfur Interactions
 A. General
 B. Inactivation of Platinum Amine Compounds
 C. Resistance to Platinum Antitumor Compounds
 D. Nephrotoxicity and Rescue Agents
 E. Rate-Determining Step of Pt—S Binding
 F. Reactivation of Pt(II) Compounds with Antitumor Activity
 G. Reduction of Pt(IV) Compounds Exhibiting Antitumor Activity
 H. Reaction Products of Pt–Amine Compounds and Sulfur-Containing Biomolecules
V. Prospect for Future Studies of Pt Antitumor Compounds
 Abbreviations
 References

I. History of *cis*-Pt as an Antitumor Drug

Over the past 20 years, *cis*-diamminedichloroplatinum(II), known for more than 145 years, has emerged as the classic compound in the context of antitumor drug therapy. Because it is generally accepted that binding of the compound to DNA is a major requirement for its biological activity, scientists have focused their attention especially on platinum–DNA interactions. In the present review, the latest results concerning these interactions are briefly described. However, most attention will be given to the reactions of this and related platinum compounds with S-containing biomolecules. Although not directly relevant in antitumor activity, such interactions are considered to have an overall inhibitory effect on the drug action, and are therefore of considerable importance.

The compound [*cis*-$PtCl_2(NH_3)_2$], here referred to as cisplatin, or *cis*-Pt,[1] and its trans isomer were known in the last century. Renewed interest in these compounds was evoked by the results of Rosenberg. As a physicist, he was impressed by the similarity between the mitotic spindle figures and the field lines of a dipole. This led him to investigate—in 1962—how electric fields may interfere with cell division of cultured bacterial cells. The field generated between platinum electrodes seemed to stop cell division without hampering cell growth, in this way inducing filamentous growth (*1*). Subsequent experiments by the same research group soon made clear that this curious phenomenon was caused by small amounts of compounds such as [*cis*-$PtCl_4(NH_3)_2$], formed during electrolysis by interaction of the electrolyte NH_4Cl and the "inert" Pt electrodes (*2*). After this important discovery, numerous Pt (II) and Pt (IV) compounds were found to show similar effects on bacterial growth (*3*). Surprisingly, only the cis and the trans isomer appeared to be effective. Subsequently, the antitumor activity of these and other platinum compounds have been studied. In particular, regarding the effect on tumors induced in animals, such as sarcoma, 180 and leukemia L1210 in mice (*4, 5*) [*cis*-$PtCl2_2(NH_3)_2$] (Fig. 1) turned out to be a very active compound (*6*). In many cases a total regression of the tumors was observed. Clinical trials of *cis*-Pt started in 1971 (*7*). Currently, *cis*-Pt is routinely used and has been particularly successful in the treatment of testicular and ovarian cancers (*8, 9*). The severe renal toxicity exhibited by the compound can be largely circumvented by the use of diuretics and prehydration. In 1978, *cis*-Pt was officially approved as a drug in the United States, and by 1983 this first metal antitumor compound had become the biggest selling antitumor drug. Moreover, the use of *cis*-Pt led to major improvements in response rates for head, neck, and lung tumors, even though responses to chemotherapy are often of limited duration for these tumors (*9, 10*). *cis*-Pt is also being used in combination therapy with antitumor drugs such as vinblastine (*11*), bleomycin (*11*), and adriamycin (*12*), or in combination with radiotherapy (*13*). For testicular and ovarian cancer the progress in the curing of these tumor types, effected by the use of *cis*-Pt, is

[1] See the complete list of abbreviations at the end of this review.

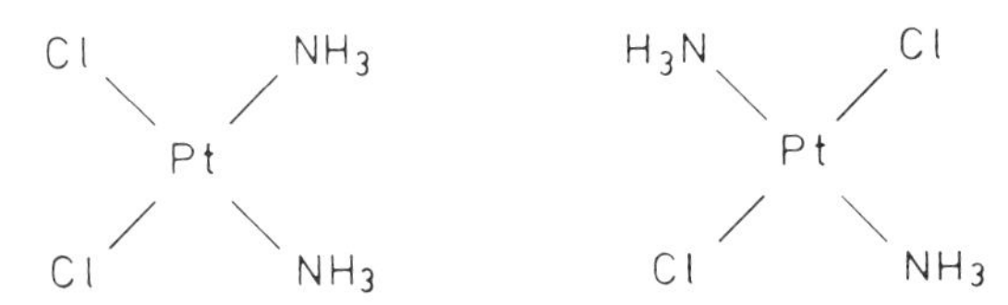

FIG. 1. Structure of *cis*-Pt (left) and *trans*-Pt (right. The trans isomer is not antitumor active.

spectacular (*10*), and especially for early recognized testicular cancer the cure rate is approaching 100%.

Besides *cis*-Pt other platinum compounds have been synthesized and tested for antitumor activity (*14–16*), with the major aim of obtaining higher solubility, better antitumor activity, and lower toxicity. The synthesis of most of these platinum compounds is largely governed by two empirical rules, concerning the trans-labilizing power of the coordinated ligands and the difference in platinum–ligand bond strength. All studies on these derivatives have made clear that there are a number of structural requirements that have to be fulfilled for a complex to show antitumor activity. These are as follows:

1. All complexes should have the cis geometry, with the general formulas [*cis*-$PtX_2(Am)_2$] for Pt(II) and [*cis*-$PtY_2X_2(Am)_2$] for Pt(IV) compounds (*14*). This geometric restriction is automatically answered for didentate amines such as ethylenediamine.

2. The active Pt(IV) compounds are octahedrally coordinated and possess axial bound chloride or—to improve the solubility—hydroxo ligands, i.e., two Y ligands in the trans orientation. These compounds are far more inert than the corresponding Pt(II) compounds that lack these axial ligands. Most likely the Pt(IV) compexes are reduced *in vivo* to the corresponding Pt(II) complexes, which are in fact the active species (*17–19*). They can therefore be considered as a type of "prodrug" that requires *in vivo* activation (substitution and reduction) to the square-planar Pt(II) compounds to exhibit antineoplastic activity. This hypothesis is supported by the observation that platinum(IV) compounds are unable to react with DNA under ambient conditions (*19*), and that appreciable amounts of Pt(II) derivatives can be detected in the urine of Pt(IV)-treated patients(*18*).

3. The ligand X is an anion that should have intermediate binding strength to Pt(II). Complexes with labile anions such as NO_3^- or ClO_4^- are usually highly toxic (*14–20*). Complexes with strongly bound anions are inert, although in some cases the dissociation of these anions appears to be activated *in vivo* (*15,21*). Examples of effective anions are

Cl^-, SO_4^{2-}, citrate(3−), oxalate(2−), and other carboxylic acid residues.

4. At least one of the two cis amine ligands should have an N—H group (*14*). All compounds lacking this property have been found to be inactive; however, it cannot be excluded that other compounds, acting through a different mechanism, will show antitumor activity. This N—H group is likely required for a hydrogen bond donor function, although steric effects cannot be excluded *a priori*.

A series of very promising "second-generation" platinum antitumor drugs that answer the above-mentioned characteristics is now known and several of these are ready for use in many clinics. A few examples of these compounds, all of which fulfill the above requirements, are shown in Fig. 2 (*22*). There are a few exceptions among antitumor-active Pt compounds that seem to deviate from these empirical rules in having only one anionic ligand. As examples, compounds of the general formulas [Pt(diam)(R′R″SO)Cl]NO_3 (*23*) and [*cis*-Pt$(NH_3)_2$(N-het)Cl]Cl (*24*) can be mentioned. Both classes of compounds are cationic in nature, water soluble, and have promising antitumor activities. Figure 3 contains two examples of these new compounds.

Currently, there are interesting new approaches in the design of antitumor drugs; a variety of platinum complexes are being designed

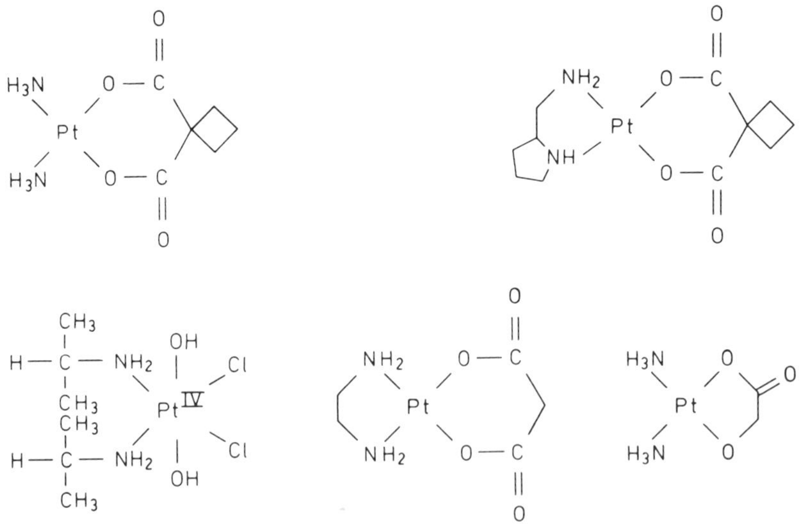

FIG. 2. Structure of some promising platinum antitumor drugs. CBDCA (upper left) is already routinely used in the clinic (*22*).

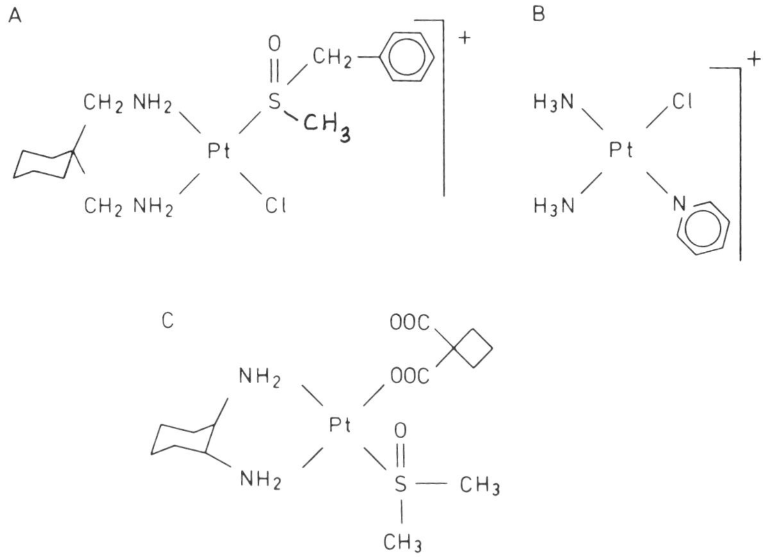

FIG. 3. Structure of [*cis*-Pt(damch)(MeBzSO)Cl]$^+$ (A), [*cis*-Pt$(NH_3)_2$(*N*-pyridine)Cl]$^+$ (B), and [*cis*-Pt(DACH)(Me_2SO)(CBDCA)] (C).

having the following characteristics: (1) containing carrier molecules as ligands for achieving higher drug concentrations, or slower release in tumor tissues (*25*); (2) containing other chemotherapeutic agents, such as intercalators (*26, 27*) and phosphono carboxylates (*28*) as coligands in the hope of obtaining some sort of synergistic effect; (3) containing more than one platinum atom (*29, 30*); and (4) containing radiosensitizers as ligands (*13, 31, 32*) for use in radiation therapy.

II. Aqueous Solution Chemistry of *cis*-Pt

When *cis*-Pt is dissolved in water, the labile chloride ions are slowly replaced by water molecules (hydrolysis) in a stepwise manner as shown in Eqs. (1)–(7) (*33–35*). The whole process of hydrolysis takes a few hours at 37°C.

$$[cis\text{-Pt}(NH_3)_2Cl_2] + H_2O \longrightarrow [cis\text{-Pt}(NH_3)_2Cl(H_2O)]^+ + Cl^- \quad (1)$$
$$[cis\text{-Pt}(NH_3)_2Cl(H_2O)]^+ + H_2O \longrightarrow [cis\text{-Pt}(NH_3)_2(H_2O)_2]^{2+} + Cl^- \quad (2)$$
$$[cis\text{-Pt}(NH_3)_2Cl_2 + OH^- \longrightarrow [cis\text{-Pt}(NH_3)_2Cl(OH)] + Cl^- \quad (3)$$
$$[cis\text{-Pt}(NH_3)_2Cl(OH)] + OH^- \longrightarrow [cis\text{-Pt}(NH_3)_2(OH)_2] + Cl^- \quad (4)$$
$$[cis\text{-Pt}(NH_3)_2Cl(H_2O)]^+ \longrightarrow [cis\text{-Pt}(NH_3)_2Cl(OH)] + H^+ \quad pK_a = 6.85 \quad (5)$$
$$[cis\text{-Pt}(NH_3)_2(H_2O)_2]^{2+} \longrightarrow [cis\text{-Pt}(NH_3)_2(OH)(H_2O)]^+ + H^+ \quad pK_a = 5.93 \quad (6)$$
$$[cis\text{-Pt}(NH_3)_2(OH)(H_2O)]^+ \longrightarrow [cis\text{-Pt}(NH_3)_2(OH)_2] + H^+ \quad pK_a = 7.87 \quad (7)$$

The relative amounts of all these Pt species vary as a function of the pH and the chloride concentration. Only platinum species with a coordinated water molecule are regarded to be reactive, because, in contrast to coordinated chloride or hydroxide, this ligand can be easily substituted by other donor molecules. Hydroxo species are formed as indicated in Eqs. (3)–(7) (*34, 35*), with [*cis*-$Pt(NH_3)_2(OH)_2$] as the stable end product in basic solution (*36*). It should be noted, however, that this species very easily dimerizes and trimerizes at higher concentrations, producing ions such as [*cis*-$Pt(NH_3)_2]_2(\mu\text{-}OH)_2$ and [*cis*-$Pt(NH_3)_2]_3(\mu\text{-}OH)_3$, as has been proved with, e.g., ^{195}Pt NMR spectroscopy (*36a, b*). Very recently, accurate pK_a values have been presented for the (de)hydronation equilibria (*36b*); the pK_a values have been added to Eqs. (5)–(7). Miller and House (*36c*) have accurately determined the kinetic parameters for the several hydrolysis reactions. They concluded that acid hydrolysis of *cis*-Pt *in vivo* is unlikely to proceed beyond [*cis*-$Pt(NH_3)_2Cl(H_2O)]^+$.

In blood plasma, the chloride ion concentration is sufficiently large (about 100 m*M*) to prevent *cis*-Pt hydrolysis, and the neutral platinum species most likely crosses the cell membrane. Inside the cell the chloride ion concentration is much lower (about 4m*M*), which allows for hydrolysis (*35, 37*). Because water is a far better leaving group than chloride or hydroxide (*38, 39*), the aqua species are most likely the reactive form of *cis*-Pt *in vivo*. Thus hydrolysis is the rate-limiting step in the reaction of *cis*-Pt with biomolecules such as proteins, RNA, and DNA (*40*).

In fact, the very recent ^{195}Pt NMR results of Bancroft *et al.* (*41*) indicate that, in agreement with Miller and House (*36c*), most likely [*cis*-$Pt(NH_3)_2Cl(H_2O)]^+$ is the predominant species that reacts with biomolecules (at least with DNA). Other Pt amine compounds that are antitumor active have different kinetics of the hydrolysis reactions, and usually react much slower. The second-generation drug CBDCA (Fig. 2) is known to hydrolyze (in a 1 m*M* solution) with a half-life at 37°C of a few days (*41a*) (compared to only 1 hour for *cis*-Pt).

III. Antitumor Activity and DNA as the Target

A. General

During the past decades a vast amount of evidence has been obtained that points to interactions of platinum compounds with DNA in the tumor cell as the origin for cytotoxic action. The most important observations are as follows:

1. Treatment of *Escherichia coli* bacteria with *cis*-Pt leads to filamentous growth, due to hampered cell division (*1*).

2. *cis*-Pt induces lysis in lysogenic bacteria (*42*) and a good correlation between the antitumor activity and prophage induction has been found.

3. Inhibition of mainly DNA synthesis has been shown for Ehrlich ascites cells (*43*) and human HV_3 cells (*43a*), as studied by the uptake of radiolabeled DNA, RNA, and protein precursors. However, RNA and protein inhibition are still being examined (*44*).

4. The mutagenic properties of *cis*-Pt result in frameshift and base substitution mutations (*45, 45a*).

5. Bacteria (*45, 46*) and human cell lines (*47, 48*) that are DNA repair deficient are far more sensitive toward *cis*-Pt than is the DNA repair-proficient organism.

Initially these hypotheses directed almost all attention to platinum–DNA interactions in the studies concerning the working mechanism of *cis*-Pt. However, it cannot be excluded that other *cis*-Pt-induced processes at the cellular level might attribute to the ultimate cell killing. In this respect the natural immune response should be mentioned. It appears, however, that an ever-increasing amount of research is focused on the interactions of platinum compounds with DNA.

In this section a brief summary of the most important Pt—DNA interactions, including *cis*-Pt, *trans*-Pt, and other related platinum compounds, and their relevance for antitumor activity will be presented. For more detailed information the reader is referred to the several reviews that have appeared on this subject during the last 5 years (*49–53*).

B. Pt Binding to Nucleobases

Knowledge that DNA is an important target for *cis*-Pt binding raises the question of which DNA sites are preferred by the platinum antitumor compounds. This will be briefly summarized herein. Because platinum is a class B metal, it should be expected that platinum compounds show a relatively high affinity for nitrogen donor sites in DNA, i.e., the nucleobases, rather than for the phosphate–deoxyribose moiety. Indeed, Pt compounds have a high affinity for nitrogen donor atoms. At physiological pH, the potential DNA donor atoms are guanine N7, cytosine N3, and adenine N1 and N7. No coordination to other atoms has been observed at physiological pH (*51a*) (see Fig. 4). In fact, early competition experiments have shown that platinum binding to DNA monomers is largely determined by kinetic factors, and these strongly favor guanine N7 binding (*50*).

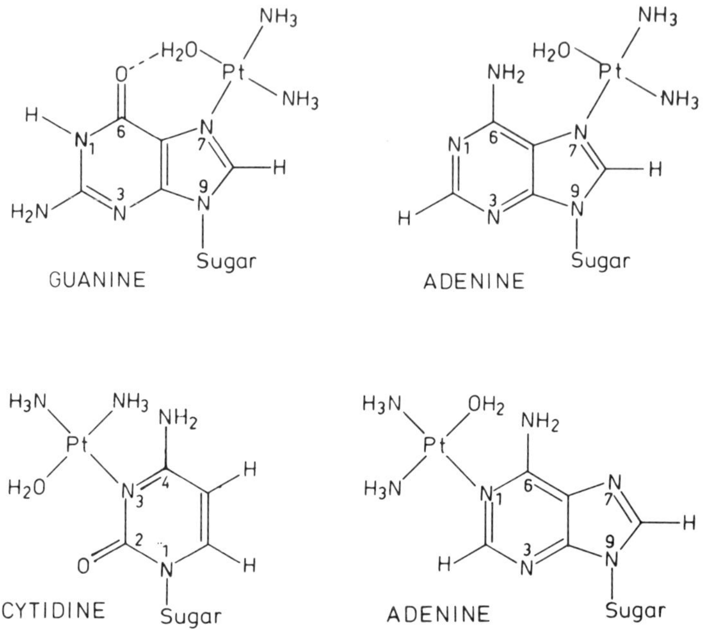

FIG. 4. Schematic structure of some nucleobases and their adducts with platinum amines.

Proton NMR has been widely applied and has been proved to be a powerful method to determine the details of these binding sites. The major conclusions can be summarized as follows:

1. The resonances of nucleobase protons that are present in the near vicinity of the platinated site are strongly shifted downfield, i.e., the purine H8 signal after N7 platination.

2. Because the chemical shifts of the nonexchangeable nucleobase proton signals are sensitive to (de)protonations of the aromatic structure, the absence of certain "protonation shifts," or alterations in the expected pK_a values, can give valuable information about the sites where platinum is bound. So, in N7-platinated guanine, no N7 protonation shift around pH 2 is observed, whereas the pK_a of the N1 protonation decreases from 9.5 to 8.5 (*54*).

3. A characteristic $^3J(^{195}Pt-^1H)$ coupling, manifested as "satellites" around the H8 resonance, is often observed in proton NMR spectra of N7-platinated guanine residues, when recorded at low magnetic fields, for instance, 60 MHz. This coupling, caused by ^{195}Pt (34% abundance),

disappears at higher fields due to efficient chemical shift anisotropy (CSA) relaxation (*54a*). Interestingly, this coupling remains visible in the spectra of Pt(IV)–nucleobase compounds when recorded at higher fields. This observation can be used as a tool to discriminate between Pt(II) and Pt(IV) adducts (*54b*).

It has also been established that in oligonucleotides, from di- to dodecanucleotides, reactions with *cis*-Pt or related Pt(II) compounds yield largely guanine N7 adducts (*55*). Several detailed NMR structures for such adducts have been reported in the last 5 years. The major conclusions are compatible with those described below for DNA, and will be reviewed in Section III,D.

C. Binding of Pt to DNA

A very interesting aspect of platinum–DNA interactions concerns the nature of the resulting adducts and their relative quantities. Due to the bifunctional nature of *cis*-Pt, several types of adducts in the DNA can be expected to be formed, to be distinguished in (1) interstrand chelates (binding of two nucleobases that are each positioned in one of the complementary DNA strands), (2) intrastrand chelates (binding of two nucleobases within the same DNA strands), (3) intrabase chelates (binding to two different atoms in one base), and (4) DNA–protein cross-links.

Once platinum is bound to DNA, the products appear to be very stable. Only strong nucleophiles such as thiourea (*56*) and cyanide (*57*) can relatively rapidly reverse the Pt–DNA bond. The binding of *cis*-Pt to DNA perturbs the DNA structure, which results in a decrease in melting temperature (*58, 59*), shortening (*60*), unwinding (*61*), and local denaturation (*62*) of the DNA helix. However, the degree of these distortions strongly depends on the used platinum levels. The distortion of the DNA structure as a result of the binding of one *cis*-Pt molecule proved to be only small (a few base pairs) (*57*).

A useful method (*63*) to study the binding positions of *cis*-Pt in DNA is based on the digestion of the high-molecular platinated DNA by enzymes, resulting in mononucleotides and platinum-containing mono- and dinucleotides. After degradation, the digestion mixture can be separated on the basis of charge by anion-exchange chromatography, and subsequently the platinated fragments can be identified and quantitated, e.g., by spectroscopic methods or immunochemical techniques. In this way, four reaction products of *cis*-Pt with salmon sperm DNA have been isolated, together comprising at least 90% of the platinum input (*63*). The structures of these adducts have been unambiguously assigned by proton NMR and Pt analyses (atomic absorption spectroscopy; AAS). In all cases, *cis*-Pt is linked to the N7 atom of guanine or adenine, and no indications have been found for binding at adenine N1, or at cytosine N3. The quantitation results of this study have demonstrated that platinum chelates are preferably formed on neighboring guanines, the so-called GG adduct (65%). To a lesser extent, also the AG chelate (25%) was found, but surprisingly no GA chelate. Also, an adduct originating from *cis*-Pt bound to two nonneighboring guanines (10%) could be isolated as [*cis*-$Pt(NH_3)_2(GMP\text{-}N7)_2$]. This product is thought to originate both from interstrand chelates and so-called GNG intrastrand chelates, i.e., adducts in which *cis*-Pt is bound to two next-neighboring guanines. Finally, also a minute amount of a monofunctional adduct (5%) with guanine could be identified and quantitated. At this point it should be noted that it is not known yet which of these adducts may be the "critical lesion(s)" that lead to cell mortality. The large abundance of the GG chelate has resulted in several studies by a variety of groups (*50–55*) on this adduct.

The six different kinds of major reaction products between *cis*-Pt and DNA that have been found are depicted in Fig. 5. Two of them (4 and 6)

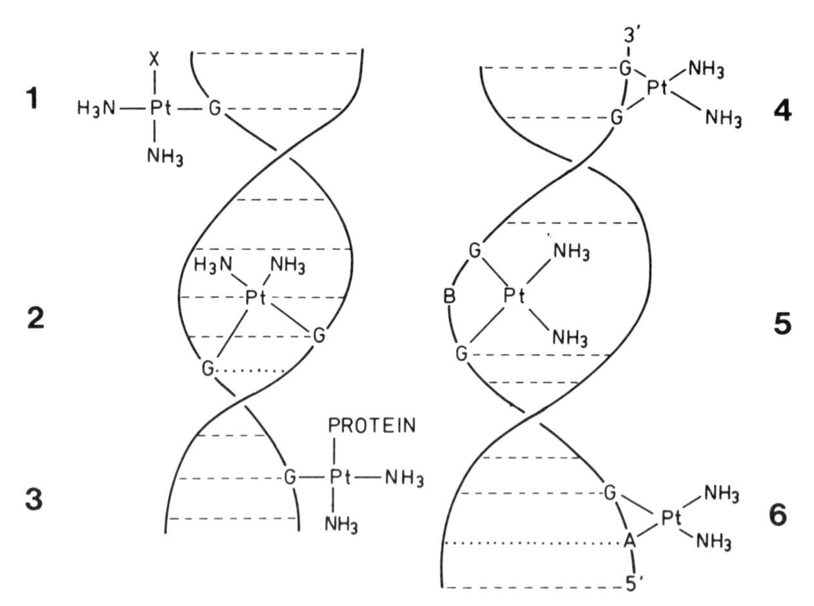

FIG. 5. A schematic representation of the known *cis*-Pt–DNA interactions. X = Cl^-, OH^-, and H_2O. See text for discussion.

are specific for *cis*-Pt and analogs. The other four can also be formed by trans isomers of platinum amine compounds. From the observation that the adducts 4 and 6 are specifically formed by *cis*-Pt and not by the inactive trans isomer, it seems likely that at least one of them is important for antitumor activity.

The intrastrand cross-link between two adjacent guanines (60–65%) (*63*) is higher than the statistically expected value of 44%, assuming a random distribution and an equal reactivity of all guanines. Therefore some kind of directing effect must be present, leading to binding of *cis*-Pt preferably to GG sequences in the DNA.

During *in vivo* studies under biologically relevant conditions, the *cis*-Pt loading of the DNA is much lower than for the above-mentioned *in vitro* studies. It has been calculated that mortality of HeLa cells occurs at an r_b value of 10^{-5} (i.e., one bound *cis*-Pt molecule per 10^5 nucleotides) (*64a*). This excludes atomic absorption spectroscopy for identification of the *in vivo* adducts. Immunochemical techniques, however, have shown to be very promising, and high sensitivity and selectivity levels have been reached. At the moment, only a few studies in which antibodies are raised against *cis*-Pt-treated DNA (*64*) or against synthetic *cis*-Pt adducts with mono- or dinucleotides are available (*64a*). With the latter method, quantitation of the different platinum–DNA adducts formed under *in vivo* conditions is possible. At the moment, femtomole (10^{-15} mol) amounts of the adducts can be detected with competitive enzyme-linked immunosorbent assay (ELISA) techniques. It has been demonstrated in this manner that the GG–Pt adduct is also the predominant adduct under *in vivo* conditions.

D. Study of Pt Binding to Oligonucleotides

As said above, only a summary of the many reported studies can be given here, and for details the reader is referred to the reviews (*49–53*) and the papers of the groups mentioned in Section III, B (*55*). A complete conformational NMR analysis of the solution structure of the adduct with a dinucleotide, i.e. [*cis*-Pt$(NH_3)_2${d(GpG)-N7(1),N7(2)}], has been available for some time (*50b*). The main features of this structure can be summarized as follows:

1. The two bases are coordinated through N7 to *cis*-Pt in a "head-to-head" orientation with a dihedral angle of about 60(20)°.
2. The deoxyribose moiety of the 5′ guanine has adopted an almost pure N-type conformation, compared to the S-conformation in free d(GpG).

3. The other conformational characteristics of d(GpG) are hardly changed upon the platination. The crystallization of *cis*-Pt adducts with oligonucleotides has been proved to be very difficult and only recently a few X-ray structures have become available (see below).

The observation that *cis*-Pt chelates preferably neighboring guanines led to studies (*55*) in which larger oligonucleotides with this sequence were reacted with the platinum compound. In addition, studies with the *trans*-Pt and with monofunctional Pt compounds have been performed. Again, only a brief summary can be given here. The distortion of DNA by *cis*-Pt appeared to be so large that up to self-complementary hexamers, no duplex formation occurs after platination. However, the GG-platinated non-selfcomplementary decanucleotide d(TCTCGGTCTC) forms a duplex with its complementary strand d(GAGACCGAGA). Detailed analysis of proton and phosphorus NMR spectra and consideration of the circular dichroism (CD) spectra led to the following conclusions (*64b*), which have been confirmed by others for related oligonucleotides (*55*).

1. The double helix is somewhat destabilized after the platination, as reflected by the decrease of the melting temperature of the duplex by 10–20°C at NMR concentrations (3 m*M*).
2. Base pair formation by hydrogen bonding seems also possible after the platination, as reflected by the appearance of iminoproton resonances. For the central GG sequence, these signals are only observed at low temperature, although shifted to lower field and broadened.
3. A careful study of the NMR spectra followed by conformational analysis suggests a rather small distortion of the double helix of this decanucleotide upon platination, which has been described as a "kink" of about 40° in the helical axis at or around the GG lesion. Later studies of Lippard (*72*) have determined the kink more accurately at 33°, with an unwinding at 13°.
4. Comparison of CD spectra and ^{31}P NMR spectra of both platinated DNA from several sources and the platinated ds decanucleotide, strongly suggets similar distortions in both cases. Molecular dynamics calculations by Kozelka *et al.* (*64c*), based on the above-mentioned decanucleotide, indicated that this kinklike structure can indeed exist. Just as in the crystals of the *cis*-Pt adduct of d(pGpG), an NH_3—OPO_3 hydrogen bond appears to be present in the calculated ds decanucleotide structure.

The recognition that the DNA structure distortion, resulting from platinum binding at GG, is rather small and localized and that the binding does not severely unwind the double helix, has directed research to detailed three-dimensional studies of small GG-containing DNA fragments.

Lippard (*64d*) succeeded in solving the X-ray structure of the *cis*-Pt adduct with d(pGpG). It turned out that the solid-state geometry of this compound is roughly the same compared to the above-mentioned solution structure of the GG adduct (*50, 50b*). The role of the 5′-phosphate group seems to be a stabilizing one, i.e., it is involved in a hydrogen bond with an NH_3 ligand of platinum. This could be an important explanation for the observation that platinum antitumor drugs need an acid N—H group to donate a hydrogen bond to a phosphate and/or a guanosine—O6. The phosphate–ammonia interaction could induce and/or stabilize DNA distortions, thereby interfering with the replication process.

A very similar structure was found subsequently for the adduct with d(CpGpG) (*64e*). The solution structure of the *cis*-Pt adduct of d(CpGpG) has been determined earlier by NMR (*50b*), and shows that in spite of the *cis*-Pt moiety, the cytosine base can still stack rather well on the central platinated G, whereas the structure of the platinated GG part is similar to the above-mentioned structure of the platinated d(GpG). Subsequently the crystal structure (*64e*) of the d(CpGpG) adduct could be determined by using X-ray diffraction. In the three independent molecules present in the crystal, no stacking interactions of the cytosine on the central guanine are present. On the other hand, extensive intermolecular stacking interactions and hydrogen bonding in the solid state would explain this discrepancy with the solution structure. Apparently, in the solid state the intermolecular forces are stronger than the stacking interactions in solution. We now have to wait till detailed three-dimensional structures, based on X-ray diffraction, will become available on platinated, double-stranded oligonucleotides.

E. Binding of Other Pt Compounds to Nucleic Acids

Many other active and inactive compounds have been reacted with oligonucleotides and with DNA. It appears that the compounds with a structure related to *cis*-Pt bind to DNA, forming very similar products, although the kinetics differ (*64f*).

The two classes of antitumor-active compounds [Pt(diam)(R′R″SO)Cl](NO_3) (*23*) and [*cis*-PT(NH_3)$_2$(N-het)Cl]Cl(*24*) are in principle "monofunctional" platinum compounds (see also Section I and Fig. 3) and their antitumor activity cannot be simply explained by the formation of products such as 4 and 6 in Fig. 5, especially because other "monofunctional" cationic complexes such as $[Pt(NH_3)_3Cl]^+$ (*65*) and $[Pt(dien)Cl]^+$ (*14*) were found to be antitumor inactive. These complexes are often used to model the first binding step of platinum antitumor compounds to DNA (*66–68*).

$$[Pt(diam)(R'R''SO)Cl] \xrightarrow[-Cl^-]{+DNA} [Pt(diam)(R'R''SO)(DNA)] \xrightarrow[-R'R''SO]{} [Pt(diam)(DNA)] \quad (8)$$

$$[Pt(diam)(R'R''SO)Cl] \xrightarrow[-R'R''SO]{+DNA} [Pt(diam)(DNA)Cl] \xrightarrow[-Cl^-]{} [Pt(diam)(DNA)] \quad (9)$$

For $[Pt(diam)(R'R''SO)Cl]^+$, the formation of a sulfoxide–Pt–DNA [Eq. (8)] or a chloride–Pt–DNA intermediate [Eq. (9)] with subsequent activation by *in vivo* displacement of sulfoxide or chloride, respectively, by a nucleic acid base (i.e., guanine) can explain the antitumor activity (*23*). The inertness of R′R″SO with respect to displacement by both chloride and water precludes the possibility of these complexes acting by simple loss of the sulfoxide ligand (*23*). Therefore the mechanism presented in Eq. (8) is the most likely one.

Recent ^{1}H NMR results of binding studies of the relatively simple compound $[Pt(en)(Me_2SO)Cl]Cl$ with the dinucleotide d(GpG) and with 5′-GMP are supportive for such a mechanism (*69*). Initially, relatively long-lived species of the type $[Pt(en)(Me_2SO)G]$ are formed. Eventually, Me_2SO hydrolysis occurs while forming [Pt(en){d(GpG)-N7(1),N7(2)}] and $[Pt(en)(5'GMP\text{-}N7)_2]$, the same as formed with the classical antitumor drug $[Pt(en)Cl_2]$. Recently, a novel synthesis has been described to prepare symmetrical and unsymmetrical (malonato) platinum(II) complexes (*70*). The intermediate species in this synthesis {[*cis*-Pt(DACH)(Me_2SO)(CBDCA)]; Fig. 3, compound C} might, in view of the above, exhibit interesting antitumor activities.

For [*cis*-$Pt(NH_3)_2$(N-het)Cl]$^+$ compounds an ammonia-loss pathway, as a result of the trans labilizing effect of a N-het, to achieve didentate binding to DNA has been considered as a possible mechanism of action (*24*). It is equally possible that [*cis*-$Pt(NH_3)_2$(N-het)Cl]$^+$ is monofunctionally bound to DNA, followed by a subsequent interaction of the N-het and DNA. Such an intercalative interaction must be located in a way to produce a nonrepairable lesion of DNA (*24*). It should be stated that these mechanisms of action form part of the discussion and are not definite hypotheses.

Reactions of the triamine complexes such as [*cis*-$Pt(NH_3)_2$(4-mepy)Cl]Cl with d(GpG) proved that there is no release of amine ligands and therefore the formation of bifunctional adducts, comparable to those induced by *cis*-Pt, is highly unlikely (*71, 72*). Only two mononuclear complexes, [*cis*-Pt$(NH_3)_2$(4-mepy){d(GpG)-N7(1)}] (**1**) and [*cis*-Pt$(NH_3)_2$(4-mepy){d(GpG)-N7(2)}] (**2**), and one dinuclear complex, [*cis*-Pt$(NH_3)_2$(4-mepy)$]_2${μ-d(GpG)-N7(1),N7(2)}] (**3**), are formed (see Scheme 1 (*71*). Even extreme conditions, such as the addition of nucleophilic sulfur-containing molecules (*71*) or raising the temperature (*72*), did not induce the formation of a chelate. In addition, the results (*71*) of bacterial survival and mutagenesis experiments with *E. coli* strains and data (*72*) obtained with monoclonal antibodies that bind to *cis*-Pt–DNA lesions showed that the *in vivo* formation of bifunctional adducts in DNA and a mechanism of both GN7 binding and intercalation are unlikely. Whether the Pt–DNA binding in general or other mechanisms are important for the antitumor activity of [*cis*-Pt$(NH_3)_2$(N-het)Cl]Cl remains unclear and is a subject of further studies. The fact that these compounds inhibit DNA replication at individual guanine residues is supportive for the importance of Pt–DNA binding (*72*).

IV. Platinum–Sulfur Interactions

A. General

Since the discovery of *cis*-Pt as an antitumor drug, the research on the mechanism of action has mainly been focused on the interactions with DNA, as summarized in the preceding section. Although such interac-

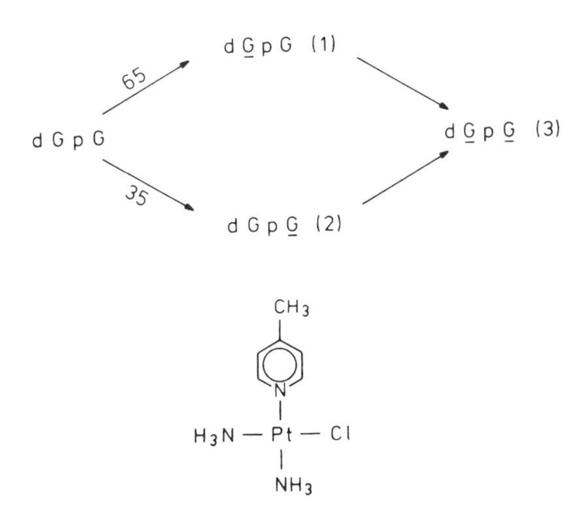

SCHEME 1. Summary of the two-step reaction of d(GpG) with [*cis*-Pt$(NH_3)_2$(4-mepy)Cl]Cl. The indicated values denote relative amounts (%).

tions are generally accepted to be ultimately responsible for antitumor activity, there are many other important biomolecules that can react with platinum amine compounds. Especially sulfur-containing molecules have a high affinity for platinum (*73*). Possible reactive biomolecules are cysteine, methionine, *S*-adenosyl-L-homocysteine, *S*-adenosyl-L-methionine, glutathione, metallothionein, and other proteins. Only in the last decade has information about (bio)chemically and medically interesting Pt–sulfur interactions become available. In general, interactions of platinum antitumor compounds with sulfur-containing biomolecules are considered to have an overall negative effect on the antitumor activity. Such interactions can be responsible for inactivation of Pt(II) species, for development of resistance, and for toxic side effects such as nephrotoxicity. In this section these interactions and their relevance in the mechanism of action of *cis*-Pt, as an antitumor drug, will be summarized. In addition, the mechanism of action of some nephrotoxicity-reducing agents, i.e., those with sulfur, such as sodium diethyldithiocarbamate (Naddtc) and sodium thiosulfate (STS), will be discussed.

B. Inactivation of Platinum Amine Compounds

The reaction of sulfur-containing biomolecules with platinum antitumor compounds, thereby preventing binding to the critical DNA target, is a possible mechanism of inactivation and is supported by numerous studies. Thus, glutathione (GSH, a cysteine-containing tripeptide; see also Fig. 6), which is the predominant intracellular thiol and is present in concentrations varying from 0.5 to 10 m*M*, is able to inhibit the reaction of DNA with $[Pt(en)Cl_2]$ (*74*) and with *cis*-Pt (*75, 76*). It has also been observed that the presence of cysteine can inhibit the reaction between *cis*-Pt and d-Guo (*77*). Furthermore, the antitumor activity of *cis*-Pt was proved to be inhibited by coadministered methionine (*78, 79*) and even a bis-adduct between *cis*-Pt and methionine has been isolated from the urine of patients (*80*).

The reactions of *cis*- and *trans*-Pt have been recently investigated with the use of ^1H NMR by Berners-Price and Kuchel (*81*) in intact

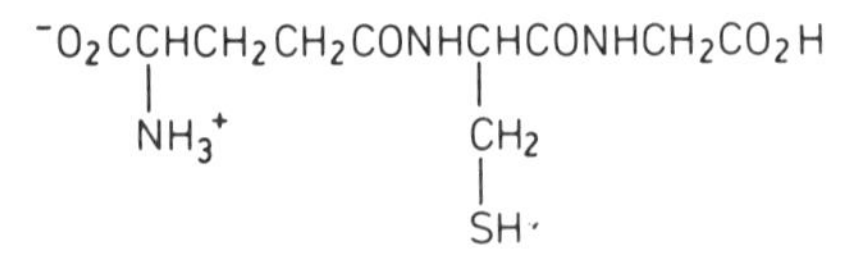

FIG. 6. Structure of glutathione.

human red blood cells. As expected, they observed that both Pt compounds react with GSH, leading to binding of the Pt species, as shown in Fig. 7. In addition to the binding to GSH, binding to hemoglobin was also observed. *trans*-Pt reacts more rapidly with GSH, which would suggest that in the case of *cis*-Pt higher percentages of the drug dose may reach the cell nucleus, before inactivation with GSH takes place. The more rapid reaction of *trans*-Pt compared to *cis*-Pt can be explained by the more labile chloride atoms in *trans*-Pt due to the trans effect; i.e., the chloride in *trans*-Pt is a better labilizer than is the amino group in *cis*-Pt (*73*).

A second mechanism of inactivation might be the reaction of sulfur-containing biomolecules with the *cis*-Pt–DNA monoadducts (product 1 in Fig. 4), which prevents those from rearranging to toxic bifunctional adducts. Supportive for such a mechanism is the observation that GSH can be cross-linked to DNA by *cis*-Pt (*41, 41a*) and [Pt(en)Cl_2] (*74*), and that cysteine can be cross-linked to d-Guo by *cis*-Pt (*77*). Furthermore, *cis*-Pt–DNA monoadducts can be experimentally quenched with thiourea, which reduces drug toxicity (*82, 83*). *trans*-Pt also yields monofunctional adducts after reaction with DNA, and these rearrange somewhat slower than does *cis*-Pt into bifunctional adducts (*41, 84*), clearly for sterical reasons. The relatively long-living monofunctional adducts react efficiently with GSH and proteins (*41a, 84–86*).

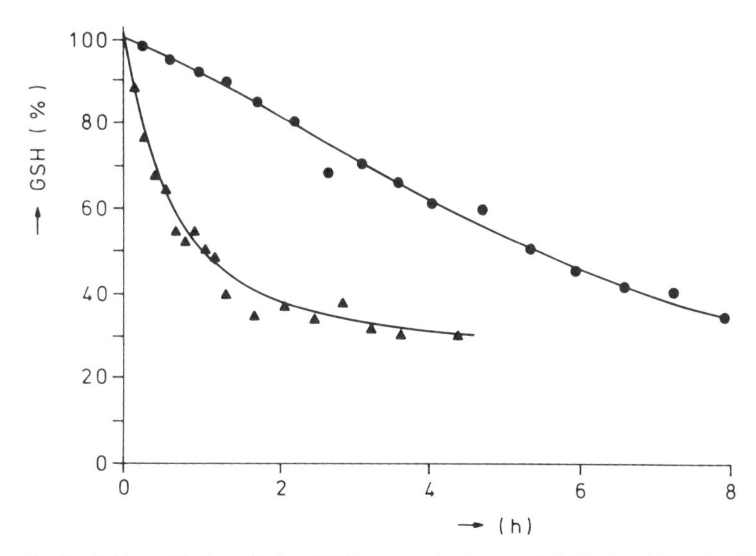

FIG. 7. Depletion of intracellular GSH on incubation at 310 K of red blood cells with *cis*-Pt (●) and *trans*-Pt (▲).

It can be stated now that in principle all S-containing biomolecules can inactivate Pt amine compounds by the two mechanisms and it is probable that the inactivation is not limited to the above described binding by GSH, methionine, and cysteine. For instance, it has been reported that high percentages of *cis*-Pt are bound to metallothionein (MT; a protein with a molecular weight of 6000–7000, containing 33% cysteine) in Ehrlich tumor cells (30%) (*87*), in *cis*-Pt-resistant cells (70%) (*88*), and in rat tissues (25–40%) (*89, 90*). In comparing *cis*-Pt and *trans*-Pt it has been shown that both inactivation mechanisms [i.e., the reaction with free Pt(II) species and with Pt(II)–DNA monoadducts] are more pronounced for *trans*-Pt, which could—at least partly—explain the reduced toxicity of this antitumor-inactive compound. This hypothesis is in agreement with studies in which especially the cytotoxicity of *trans*-Pt (*91*), but not of *cis*-Pt (*75, 91*), to a human ovarian carcinoma cell line was found to be increased markedly by depleting GSH levels.

C. Resistance to Platinum Antitumor Compounds

The clinical usefulness of *cis*-Pt is often limited by the development of resistance. The development of resistance to *cis*-Pt has been explained by several factors, including reduced drug accumulation, increased DNA repair processes, and an increase in the amount of inactivation proteins. In most *cis*-Pt-resistant tumors probably a combination of such factors plays a role. In this section, only an increase in cellular thiols in relation to resistance will be discussed. Such inactivation processes have already been discussed (Section IV,B), but attention will be directed now to the mechanism of resistance. These mechanisms are probably also of importance for the inactivation of *cis*-Pt in nonresistant tumors, though to a lesser extent.

It has been shown that many cancer cell lines that are resistant to *cis*-Pt have elevated levels of gluthathione (Table I) (*92–101*); on the other hand, depletion of GSH with buthionine sulfoxime reverses resistance only for the A2780 ovarian carcinoma cell line (*93*). Resistant cancer cell lines, which have no elevated levels of GSH, could also not be sensitized by depletion of GSH (*75, 91, 102*). These results are rather controversial and therefore the precise role of GSH in mediating *cis*-Pt resistance is still unclear at present. More research is necessary to explain the rather low increased levels of GSH with respect to the sometimes significant increase in resistance.

It has recently been shown by Kelley *et al.* (*103*) that tumor cell lines resistant to *cis*-Pt have significant elevated levels of MT (Table II). In contrast, the results of Schilder *et al.* (*104*) did not show a correlation

TABLE I

FUNCTION OF GLUTATHIONE IN *cis*-Pt RESISTANCE

Cell line	Tumor type	Increase factor in resistance	Increase factor in GSH	Ref.
A2780	Human ovarian carcinoma	14	3.2	*92,93*
P388	Mouse leukemia	24	1.2	*94*
L1210	Mouse leukemia	100	1.7	*95*
COLO 316	Human ovarian carcinoma	13	2.3	*96*
GLC_4	Human lung carcinoma	6.4	3.4	*97*
O-342	Rat ovarian carcinoma	?	1.9	*98*
ROT 68/C1	Rat ovarian carcinoma	20	1.4	*99*
RIF-1	Mouse fibrosarcoma	2.3	1.4	*100*
BE	Human colon carcinoma	5	3	*101*

between MT expression and *cis*-Pt resistance. The cell lines in Table II have been exposed to high concentrations of *cis*-Pt for long periods of time, which raises the question whether such a mechanism of development of resistance also occurs in clinically relevant situations. Again, much more work is needed to establish the exact role of MT in the development of resistance. It also remains to be determined whether long-term exposure to *cis*-Pt induces MT synthesis either directly (as $CdCl_2$ is known to do), or indirectly. Recent results showed that *cis*-Pt doubles the amount of MT in the liver and kidney of mice and it was concluded that *cis*-Pt can directly induce the synthesis of MT, although only when present in the hydrolyzed form (*105*).

TABLE II

FUNCTION OF METALLOTHIONEIN IN *cis*-Pt RESISTANCE[a]

Cell line	Tumor type	Increase factor in resistance	Increase factor in MT
SCC25	Human head and neck carcinoma	7.1	4.4
G3361	Human melanoma	6.7	2.0
SW2	Human small cell carcinoma	4.5	5.1
SL6	Human large cell carcinoma	2.5	3.4
L1210	Mouse leukemia	44	13.3

[a] From Ref. *103*.

D. Nephrotoxicity and Rescue Agents

A limitation of *cis*-Pt in its use as an antitumor drug is its concentration-dependent nephrotoxicity (*106, 107*), besides a variety of other side effects (*108*). Currently, the nephrotoxicity effects can be reduced by mannitol-induced diuresis (*109*) and hypersalination (*110*). On the basis of a similarity of histopathology of the kidney after Pt(II) or Hg(II) exposure in the rat, it has been suggested by Borch and Pleasants (*111*) that a similar mechanism might play a role in the nephrotoxicity of these metals [i.e., inactivation of enzymes by the coordination of Pt(II) or Hg(II) to thiol residues]. Supportive of this mechanism is that the total number of protein-bound sulfhydryl groups is depleted (14%) in kidneys after *cis*-Pt administration, especially in the mitochondrial fraction (*112, 113*). The enzyme adenosine triphosphatase, which is critical for kidney function, has been proposed as the site of action (*114*), although the high concentrations necessary for inhibition are unlikely to be achieved *in vivo*. Recently, it was demonstrated that the activity of gluthathione peroxidase was decreased significantly in *cis*-Pt-treated kidney mitochondria resulting in dysfunction (*115*). However, which enzymes are important in causing the nephrotoxicity is still controversial. Also, other mechanisms, not based on inactivation of enzymes, may play a role. Recently, it was postulated that the mitochondrial DNA damage induced by *cis*-Pt causes nephrotoxicity (*116*).

The affinity of sulfur for platinum complexes has led to investigations of numerous sulfur nucleophiles as inhibitors of *cis*-Pt nephrotoxicity, including Naddtc, STS, WR-2721, mesna, methionine, thiourea, cysteine, *N*-acetylcysteine, penicillamine, and GSH. Of these, Naddtc, STS, and WR-2721 are undergoing preclinical and/or clinical evaluation (Fig. 8). Some of the more promising compounds will be discussed here.

Naddtc has proved to be a very effective inhibitor of nephrotoxicity [it is also effective against bone marrow toxicity (*117–119*)] and should be administered 1–4 hours after *cis*-Pt, without interfering with the anti-

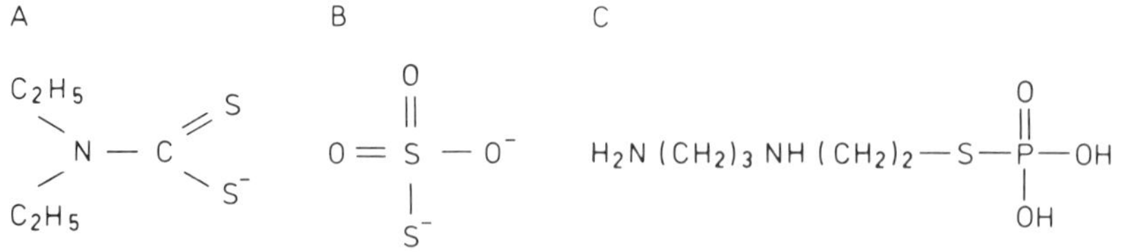

A: $(C_2H_5)_2N{-}C(=S){-}S^-$

B: $O{=}S(=O)(-O^-){-}S^-$

C: $H_2N(CH_2)_3NH(CH_2)_2{-}S{-}P(=O)(OH){-}OH$

FIG. 8. Structure of the nephrotoxicity inhibitors ddtc$^-$ (A), TS$^-$ (B), and WR-2721 (C).

tumor properties of *cis*-Pt (*111, 117, 120–123*). The noninterference with the antitumor properties of *cis*-Pt is in agreement with the observation that Naddtc could not reverse Pt–DNA cross-links (*123*), except for the *cis*-Pt–adenosine 1:2 complex and the *cis*-Pt–guanosine 1:1 complex (*124*), which are considered to be not important for the antitumor activity of *cis*-Pt. Naddtc and thiourea (*vide infra*) are the only rescue agents that provide protection from nephrotoxicity when administered after *cis*-Pt, at a time when most of the reactive platinum species have been taken up by cells or are already being excreted in the urine. Although pretreatment with Naddtc is inefficient in inhibiting nephrotoxicity, probably because it inactivates *cis*-Pt, one study is known in which treatment with Naddtc 12 hours prior to *cis*-Pt was effective in protecting kidney damage (*125*). This might be related to the fact that Naddtc is a powerful inducer of MT synthesis (*126*), which could be the actual inhibitor of the nephrotoxicity (i.e., by reacting with free *cis*-Pt in the kidney). Especially the time of administration of Naddtc, i.e., after *cis*-Pt treatment, agrees with the hypothesis that Naddtc acts as a real rescue agent. Thus it reduces the nephrotoxicity by removing the platinum from certain proteins, thereby restoring the original structure of the protein (111). Evidence for the Naddtc-induced dissociation of Pt–protein adducts has been reported. Thus α_2-macroglobulin (*127*), γ-glutamyltranspeptidase (*76, 124, 128*), and fumarase (*129*) are inactivated by *cis*-Pt and reactivated by Naddtc (e.g., see the reactivation experiments with fumarase in Fig. 9). Unfortunately, only for α_2-macroglobulin is the actual interaction of the protein with *cis*-Pt known, i.e., it is cross-linked through its methionine residues (*127*).

In order to investigate what kind of Pt–enzyme interactions can be

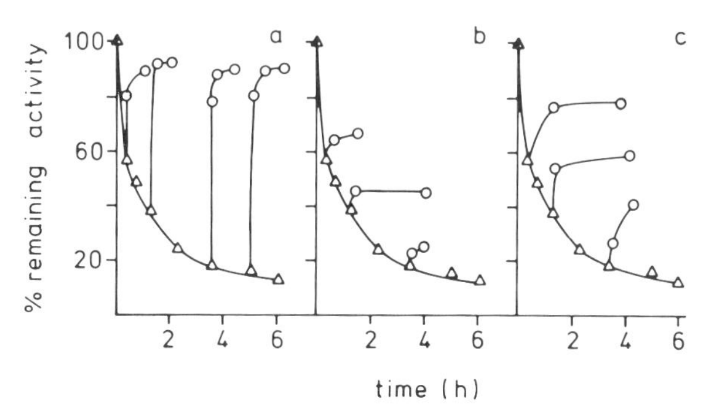

FIG. 9. Inhibition of fumarase by *cis*-Pt (△) and regeneration (○) by Naddtc (a), STS (b), and thiourea (c). The values for Naddtc and STS remained constant for 24 hours, but those for thiourea increased to 80%.

reversed, exchange reactions of model compounds with Naddtc were performed. The well-identified model compounds (*130*) (Fig. 10) can be considered to mimic Pt–cysteine and Pt–methionine adducts within a protein (*131*).

The results (*131*) have shown (Table III) that particularly the Pt–methionine type of bond in $[Pt(dien)GS\text{-}Me]^{2+}$ reacts fast with Naddtc, thereby restoring the original structure of the thioether linkage. The Pt–cysteine type of bond in $[Pt(dien)GS]^{+}$ appears to be inert toward Naddtc treatment. This is in agreement with the results of exchange reactions of Naddtc with the not well-characterized, but biologically more relevant, *cis*-Pt–methionine and —GSH system, which also points to the lability of only the Pt–methionine types of interactions (*131, 132*). The inert Pt—S bond in Pt–cysteine complexes under neutral conditions is probably the result of a very strong negatively charged $Pt—S^{-}$ binding compared to rather weak neutral Pt—S binding in Pt–methionine complexes. Extrapolating the results obtained with the model adducts to a protein leads to the following hypothesis (*131*) for the mechanism of reducing the nephrotoxicity by Naddtc: Naddtc would only reduce the nephrotoxicity when the Pt binding to proteins is caused by Pt coordination to methionine residues and not to cysteine residues. This is in contrast with the initial hypothesis (*111*), which suggests that Naddtc reverts the Pt–cysteine type bonds in proteins.

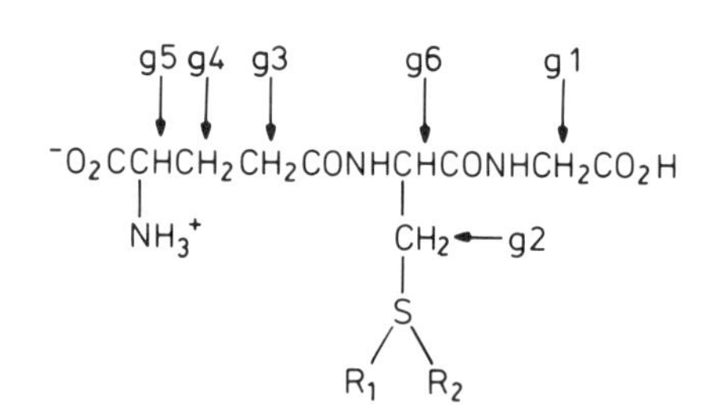

Model Adduct	R_1	R_2
Pt-methionine	CH_3	$Pt(dien)^{2+}$
	$Pt(dien)^{2+}$	$Pt(dien)^{2+}$
Pt-cysteine	$Pt(dien)^{2+}$	-------

FIG. 10. Schematic representation of the products formed between GSH, GS-Me, and [Pt(dien)Cl]Cl. The complexes are considered as models for Pt–protein interactions.

TABLE III

HALF-LIVES AND PRODUCTS OF EXCHANGE REACTIONS BETWEEN NADDTC, THIOUREA, OR STS AND PLATINATED SULFUR COMPOUNDS[a]

Compound	Naddtc	Thiourea	STS	Product
$[Pt(dien)GS]^{+}$	—	—	—	No reactions
$[Pt(dien)GS\text{-}Me]^{2+}$	<2 minutes	30 min	<2 minutes	GS-Me
$[\{Pt(dien)\}_2GS]^{3+}$	40 minutes	7 hours	20 minutes	$[Pt(dien)GS]^{+}$

[a] From Ref. *131*.

Naddtc does not react with all the platinated proteins, as was observed by Hegedus *et al.* (*133*). Therefore, the nephrotoxicity of *cis*-Pt is likely to be the result of inactivation of certain enzymes through binding at cysteines and/or methionines and Naddtc can reduce the nephrotoxicity only by reversing the Pt—S bonds of the methionine type.

Unfortunately, Naddtc is known to cause severe side effects, including burning of the mouth, chest tightness, and extreme anxiety (*134*). However, recently it has been demonstrated that the use of other dithiocarbamates, such as sodium di(hydroxyethyl)dithiocarbamate (Fig. 11), possess certain promising advantages (*125, 135, 136*). The polar groups on these derivatives of Naddtc can be expected to result in *cis*-Pt complexes with a character different than those of complexes formed with Naddtc (*136*). For example, the solubility will be much higher, which may result in lower toxic side effects.

Thiourea probably acts in a manner comparable to that of Naddtc and should also be administered after *cis*-Pt treatment (*78*). Like Naddtc, thiourea is able to remove platinum from platinated enzymes, such as leucine aminopeptidase (*76, 128*), γ-glutamyltranspeptidase (*76, 128*), and fumarase (*129*) (Fig. 9), and from Pt–methionine model adducts (Table III) (*131*). However, thiourea appears to be less useful as an inhibitor of nephrotoxicity; it also reacts quite rapidly with platinum–DNA cross-links (*56*).

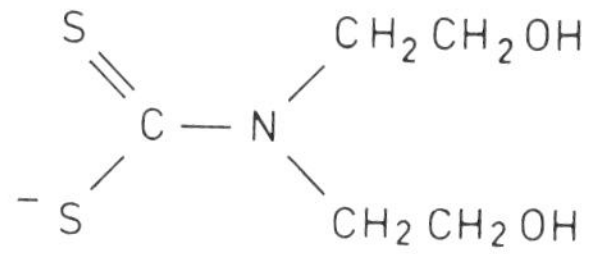

S CH_2CH_2OH
C — N
$^-$S CH_2CH_2OH

FIG. 11. Structure of the nephrotoxicity inhibitor di(hydroxyethyl)dithiocarbamate.

The rescue agent sodium thiosulfate (STS) (Fig. 8), which as such is nontoxic (*137*), should be injected during the period 1 hour before and 0.5 hour after *cis*-Pt treatment (*138*). The optimal protocol in fact uses STS intravenously (i.v.) in conjunction with intraperitoneal (i.p.) *cis*-Pt (*139–141*). Concurrent injection of STS and *cis*-Pt, i.p. (*138*) or i.v. (*142*), partly reduces the antitumor activity, probably by inactivation of *cis*-Pt. A likely explanation for the nephrotoxicity protecting effect is that STS is concentrated extensively in the kidney (*137*), where it has been proved to react rapidly with *cis*-Pt (*143*).

Numerous studies have shown that protein-bound *cis*-Pt cannot be released by STS (*76, 128, 129, 143*) (Fig. 9), although STS is able to break the Pt—S bond of the methionine type in the model system (Table III) (*131*). Therefore, this model system, does not mimic enzymes in every detail for the reaction with STS. This difference might originate from the 2^- charge of STS compared to the 1^- charge in Naddtc (*129*), keeping STS more separated from the active site.

Thus, though Naddtc acts as a "true" rescue agent, i.e., by reversing Pt–biomolecule interactions, STS most likely acts by local inactivation of *cis*-Pt and its concentration in the kidney. This may also explain why STS is not effective when administered after *cis*-Pt treatment.

The mechanism by which the more recently used WR-2721 (Fig. 8) reduces nephrotoxicity is not very well understood. WR-2721 protects against nephrotoxicity when administered just prior to *cis*-Pt (*144, 145*) and it is known that WR-2721 is preferentially taken up by normal cells and not by tumor cells (*146*). Recently, it was concluded that the uncharged form of the dephosphorylated WR-2721 (known as WE-1065) is the actual species taken up by both normal and tumor cells (*147*). It has been proposed that the conversion of WR-2721 to WR-1065 is slower in tumors, compared with normal tissues (*147*), possibly because tumors generally have lower levels of alkaline phosphatase (*148*). Furthermore, it has been proposed (*147*) that once formed, WR-1065 will have a decreased uptake rate in tumors, probably as a consequence of their lower pH (*149*) as compared with normal tissues; i.e., the neutral form of WR-1065 will only constitute 0.1% of the total drug present at pH 7 and 1% of the total at pH 8. The reactive WR-1065 is likely to bind directly to *cis*-Pt, thereby preventing side reactions of *cis*-Pt.

Pretreatment of rat with GSH also reduces kidney toxicity, which is particularly attractive because GSH is nontoxic and does not interfere with the antitumor effectiveness of *cis*-Pt (*150–153*). It has been proved that extracellular GSH accumulates in the kidney (*154*) and therefore it seems probable that the inactivation of *cis*-Pt also takes place in the kidney. Supportive for such a protective role is that depletion of GSH levels by buthionine sulfoxime or diethylmaleate has been shown (*155, 156*) to increase the nephrotoxicity of *cis*-Pt. A recently reported observation by Mayer *et al.* (*157*) that depletion of GSH levels by buthionine sulfoxime reduces the *cis*-Pt-induced nephrotoxicity is, apparently, in contradiction to these results. A possible explanation may be that the buthionine sulfoxime-induced protection is mediated by a mechanism that is independent of GSH depletion (*157*).

Finally, it should be mentioned that a new interesting method has been developed to protect cells from *cis*-Pt nephrotoxicity. Pretreatment of mice by administration of bismuth salts induces MT synthesis, especially in the kidney, but not in the tumor (*158*). This induction of MT was found to decrease the renal toxicity of *cis*-Pt, whereas the antitumor activity was not affected (*158, 159*); these observations can be rationalized by assuming that *cis*-Pt binds to MT in the kidney, as has been previously observed by other investigators (*89, 90*).

E. Rate-Determining Step of Pt—S Binding

It has now been generally accepted that the hydrolysis of the chloro ligand is the rate-determining step in the reaction of *cis*-Pt with DNA (*40*). Concerning the rate-determining step of platinum amine compounds with sulfur-containing biomolecules, the available data are quite controversial. It has been reported that the chloro hydrolysis is the rate-determining step in reactions of *cis*-Pt with leucine aminopeptidase (*76*), with γ-glutamyl transpeptidase (*76, 124*), and also with albumin (*160*). However, it has also been suggested that there may be a direct binding to proteins without prior aquation (*161*), and this has been observed with cysteine (*162*), GSH (*75, 162*), adenosine triphosphatase (*76*), and with MT (*163*).

To investigate the kinetics in more detail, the reaction rates of a simple pair of model compounds, [Pt(dien)X] [(X = Cl^-, H_2O) with GSH, GS-Me and 5′-GMP have been investigated and compared (*164*). The reaction products with GSH and GS-Me are shown in Fig. 10; an overview of the reactions between [Pt(dien)Cl] and GSH is presented in Scheme 2. These products are the first well-identified complexes between S-containing biomolecules and platinum amine compounds (*130*) and therefore are ideally suited as model compounds for kinetic studies. The results of the reactions are summarized in Table IV. In agreement with the above-mentioned hypothesis, the chloride hydrolysis is the rate-determining step in the reaction of $[Pt(dien)Cl]^+$ with 5-GMP,

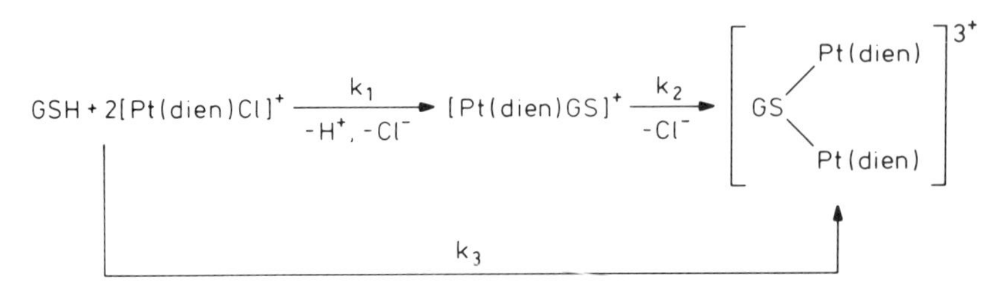

SCHEME 2. Overview of the reactions between [Pt(dien)Cl]Cl with GSH.

forming [Pt(dien)(5′-GMP-N7)]. This reaction can be completely inhibited in saturated NaCl. On the contrary, the reaction of Pt(dien)X with GSH and G*S*-Me is second order in Pt and S and the rate constant is nearly independent of the Cl^- concentration. In fact, the slightly slower reaction at higher $[Cl^-]$ indicates that only a small fraction proceeds through the hydrolysis pathway.

These results are clear evidence that sulfur reacts mainly directly with Pt amine compounds, substituting Cl^-, without prior aquation. As is evident from Table IV, the hydrolyzed species $[Pt(dien)(H_2O)]^{2+}$ will almost selectively react with 5′-GMP (3.6 M^{-1} sec^{-1} versus 0.51 M^{-1} sec^{-1} and 0.18 M^{-1} sec^{-1}), whereas the chloro species $[Pt(dien)Cl]^+$ will only react with sulfur. This information is of extreme importance in the strategy of the development of new Pt drugs. If it would be possible to develop a compound with structural properties such that the direct attack by sulfur is inhibited, but with a similar rate of chloro hydrolysis compared to *cis*-Pt, this would lead to compounds with improved antitumor properties and lower toxicities. The data depicted in Table IV were obtained at pH 5. However, it has been proved that GS^- reacts remark-

TABLE IV

RATE CONSTANTS FOR THE FORMATION OF THE 1:1 COMPOUNDS FOR 5′-GMP AND G*S*-Me AND OF THE 2:1 COMPOUND FOR GSH (pH 5)[a]

Compound	5′-GMP	G*S*-Me	GSH (k_3)
$[Pt(dien)H_2O]^{2+}$	3.6 M^{-1} sec^{-1}	0.51 M^{-1} sec^{-1}	0.18 M^{-1} sec^{-1}
$[Pt(dien)Cl]^+$	6.2×10^{-5} sec^{-1}	0.03 M^{-1} sec^{-1}	0.006 M^{-1} sec^{-1}

[a] From Ref. 164.

ably faster than GSH does with [Pt(dien)Cl]Cl (*130*), which is indicative for a slightly higher reactivity of glutathione at physiological pH. This would suggest a less effective mechanism of inactivation of *cis*-Pt in tumors, as these generally have a lower pH compared to normal tissues (*149*).

F. Reactivation of Pt(II) Compounds with Antitumor Activity

The high affinity of many platinum compounds for sulfur and the availability of many sulfur-containing biomolecules have raised the question whether Pt–sulfur biomolecule interactions could serve as a drug reservoir for platination at DNA, necessary for the antitumor activity of *cis*-Pt. Two reaction paths are possible, i.e., spontaneous release of plantinum from the sulfur, or nucleophilic displacement of platinum from sulfur by guanine (N7), for example. At the moment, there is no real evidence for the existence of such reactivation mechanisms. In fact, it has been reported that Pt–protein interactions in the plasma (albumin) are not reversible under normal conditions (*161, 165*). Further, a mixture of *cis*-Pt–methionine products does not show antitumor properties (*166*), indicating no induced platination of DNA. More research is required to investigate the existence of a reactivation mechanism. However, it is predicted that if such a reactivation phenomenon is operational, the most likely candidate is the labile Pt–methionine bond, as has been shown by its rapid reaction with Naddtc, STS, and thiourea (*vide supra*) (*131*).

G. Reduction of Pt(IV) Compounds Exhibiting Antitumor Activity

As was indicated in Section I, Pt(IV) complexes are most likely (*in vivo*) reduced to the corresponding active Pt(II) complexes. Evidence for such a mechanism comes from the fact that Pt(IV) compounds only bind to 5′-GMP in the presence of ascorbic acid (*19*). Other possible reducing agents are sulfhydryl-containing biomolecules, of which GSH is the most abundant compound in a cell. Indeed, it has been shown that in the presence of GSH, tetraplatin ([Pt(DACH)Cl_4]) reacts readily with DNA (*167*). At high concentrations of GSH, inhibition of DNA binding was observed due to a "simple" inactivation reaction between GSH and the formed Pt(II) species (*167*). In accordance with these observations, [*cis*-Pt(*i*-PrNH_2)$_2$$Cl_2$] has been identified as one of the metabolites of the corresponding Pt(IV) complex (*18*).

H. Reaction Products of Pt–Amine Compounds and Sulfur-Containing Biomolecules

In the last 5 years there has been a growing interest in the Pt—S interactions from a chemical point of view. The main reason for this is to understand the exact role (*vide supra*) of such interactions in the mechanism of the antitumor activity of *cis*-Pt. In this section the currently known products between Pt amine compounds and sulfur-containing biomolecules will be evaluated, with special attention to *cis*-Pt and the antitumor-inactive complexes *trans*-Pt and [Pt(dien)Cl]Cl. [Pt(dien)Cl]Cl is monofunctional in nature and is often used to model the first binding step of *cis*)-Pt (*66*). In addition, stable complexes can be expected (originating from the chelate effect of dien), whereas *cis*-Pt often gives complicated mixtures of products with all kinds of degradation products (*vide infra*).

Numerous *in vitro* studies have reported that both *trans*-Pt and *cis*-Pt bind to enzymes and proteins (*76, 124, 128, 129, 168–179*). In all these studies it has been suggested that cysteines and methionines are the binding sites and that, where reported, in general, *trans*-Pt reacts more readily (see also the above-described reaction with GSH) than *cis*-Pt. Only ribonucleotide reductase (*179*) appears to be an exception. Only in a few studies the actual binding site has been investigated. For thymidylate synthetase (*169*), albumin (*173, 177*), and ribonucleotide reductase (*179*), the reaction with *cis*-Pt takes place at cysteines, whereas for α_2-macroglobulin (*175, 177*) and α_1-antitrypsin (*176, 177*) the reactions take place at methionines. In general, high ratios of Pt:protein are required (at least >100) in order to observe detectable amounts of binding within a reasonable time. The presence of such interactions *in vivo* is not known. However, *cis*-Pt and [Pt(en)Cl_2] are known to react *in vivo* with plasma proteins (*165, 180, 181*), which makes it reasonable that reaction takes place, at least in those cases with albumin, α_2-macroglobulin, and/or α_1-antitrypsin.

A recent study by Bongers *et al.* (*182*) describes the interaction of MT with an excess of K_2PtCl_4. A product is formed containing 7 mol Pt/mol MT (at neutral pH) with coordination only through sulfur. Upon reaction of [*cis*-Pt(Am)$_2$Cl$_2$] with Mt, release of the amine ligands has been observed (*163, 182*). Therefore, products similar to those formed with K_2PtCl_4 can be expected. Such release of amine ligands upon coordination of [*cis*-Pt(Am)$_2$Cl$_2$] to methionine (*80, 183–186*), cysteine (*187, 188*), GSH (*76, 187–189*), and proteins (*190*) have already been observed and is rationalized by the large trans effect of a coordinated sulfur atom, which labilizes the amine ligand trans to the sulfur *(73)*. For this reason, amine release upon coordination of sulfur to [*trans*-$Pt(Am)_2Cl_2$] is not expected. It has even been speculated that the differing biological activities of *cis*-Pt and *trans*-Pt might be related to S-induced amine release only from *cis*-Pt *(190)*.

A number of investigations of the reaction of *cis*-Pt with GSH have been published *(76, 187–189, 191)*, but due to the complexity of the system, i.e., because of amine release, the reaction products still have not been characterized unambiguously. It has been suggested that [$Pt(GS)_2$] *(187)* with coordination via S and dehydronated peptide N atoms, or [*cis*-$PT(NH_3)_2(GS)(H_2O)$] *(191)*, is formed. On the other hand, it has been proved recently that eventually a polymeric structure is formed with formula $[Pt(GS)_2]_n$ *(76, 188, 189)*, involving loss of NH_3. Combining the results of the two most detailed studies *(188, 189)*, it is likely that initially intermediate species such as [*cis*-$Pt(NH_3)_2(GS)Cl$] and [$Pt_2(NH_3)_4(GS)_2$] (see Fig. 12) can indeed be formed. These unstable products lose NH_3 upon standing, eventually forming the polymeric $[Pt(GS)_2]_n$ with coordination exclusively via the S atom, but with several different Pt—S and Pt—S—Pt environments.

The reaction of *trans*-Pt with two equivalents of GSH results in a monomeric complex in which two GS^- units are coordinated via the cys-S atom [Eq. (10)] *(76, 131, 189)*. The unstable intermediate species, [*trans*-$Pt(NH_3)_2(GS)Cl$] and [*trans*-$\{Pt(NH_3)_2Cl\}_2GS$], could also be detected by Berners-Price *(189)*. The reaction of [$Pt(NH_3)_3Cl$]Cl with two equivalents of GS^- initially yields [$Pt(NH_3)_3GS$], which reacts further, eventually forming [*trans*-$Pt(NH_3)_2(GS)_2$] and free NH_3 [Eq. (11) *(131)*. This is rationalized by the large trans effect of the coordinated sulfur and the presence of a second nucleophilic sulfur. Therefore, the amine ligand is easily substituted by a second GS^- unit.

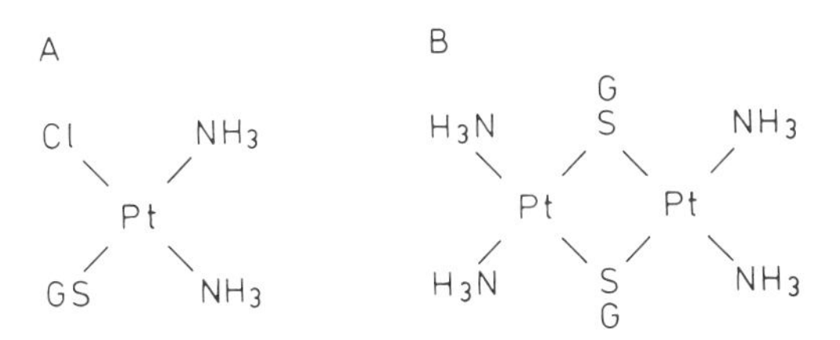

FIG. 12. Structure of the intermediate species [cis-$Pt(NH_3)_2(GS)Cl$] (A) and [$Pt_2(NH_3)_4(GS)_2$] (B) formed during reaction of *cis*-Pt with GSH.

$$[trans\text{-}Pt(NH_3)_2Cl_2] \xrightarrow[-\,Cl^-]{+\,GS^-} [trans\text{-}Pt(NH_3)_2(GS)Cl] \xrightarrow[-\,Cl^-]{+\,GS^-} [trans\text{-}Pt(NH_3)_2(GS)_2] \quad (10)$$

$$[Pt(NH_3)_3Cl]+ \xrightarrow[-\,Cl^-]{+\,GS^-} [Pt(NH_3)_3GS]^+ \xrightarrow[-\,NH_3]{+\,GS^-} [trans\text{-}Pt(NH_3)_2(GS_2] \quad (11)$$

For the complexes formed between GSH and [Pt(dien)Cl]Cl, the reader is referred to Fig. 10 and Scheme 2. In contrast with *cis*-Pt and $[Pt(NH_3)_3Cl]Cl$, no amine release is observed, which is rationalized by the strong chelate effect of the dien ring.

Cysteine yields a similar complex with *cis*-Pt, as shown in Fig. 12B (*188*). Other species that have been detected are [*cis*-$Pt(NH_3)_2$(Cys-S_2] and [*cis*-$Pt(NH_3)_2$(Cys-*N,S*)] (see Fig. 13) (*188*).

A number of complexes have been reported as products from the reaction of *cis*-Pt and methionine, i.e., [*cis*-$Pt(NH_3)_2$(Met-$S)_2$] (similar to that shown in Fig. 13A) (*185*), [*cis*-$Pt(NH_3)_2$(Met-*N,S*)] (similar to that shown in Fig. 13B) (*183, 185*), [*trans*-$Pt(NH_3)$(Met-*S*)(Met-*N,S*)] (Fig. 14A) (*183*), and [*trans*-Pt(Met-*N,S*$)_2$] (Fig. 14B) (*80, 184*). From these, [*cis*-$Pt(NH_3)_2$(Met-*N,S*)] and [*cis*-$Pt(NH_3)_2$(Met-$S)_2$] lose NH_3 on standing (*185*). Methionine yields a similar complex after reaction with *trans*-Pt, as formed between GSH and *trans*-Pt (i.e., [*trans*-$Pt(NH_3)_2$(Met-$S)_2$] (*192*).

The reaction of [Pt(dien)Cl]Cl with *S*-adenosyl-L-homocysteine (SAH, a biologically relevant molecule, as it is the coproduct of the methyl transfer reaction by *S*-adenosyl-L-methionine; see Fig. 15) results in a mixture of complexes (*193*), i.e., the mononuclear complex (1) $[Pt(dien)(SAH\text{-}S)]^{2+}$, with platination of SAH at the sulfur atom, the mononuclear complex (2) $[Pt(dien)(SAH\text{-}N)]^+$, which has a $Pt(dien)^{2+}$ unit coordinated to the amino group of the homocysteine unit, and the dinuclear complex (3) $[\{Pt(dien)\}_2(SAH\text{-}S,N)]^{3+}$, which has a $Pt(dien)^{2+}$ unit coordinated to the sulfur atom as well as a $Pt(dien)^{2+}$

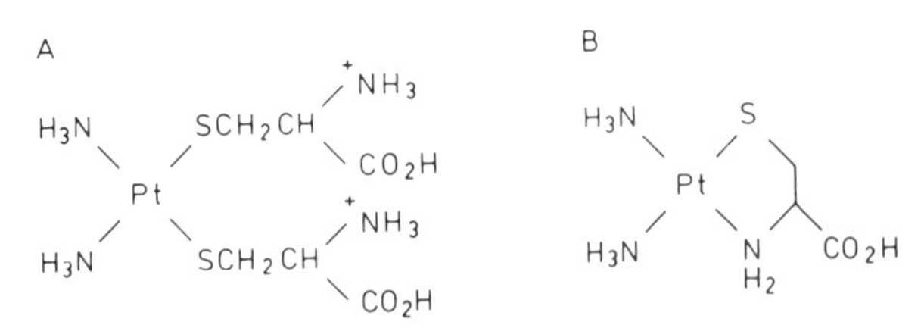

FIG. 13. Structure of [*cis*-$Pt(NH_3)_2$(Cys-$S)_2$] (A) and [*cis*-$Pt(NH_3)_2$(cys-*N,S*)] (B) formed during reaction of *cis*-Pt with cysteine.

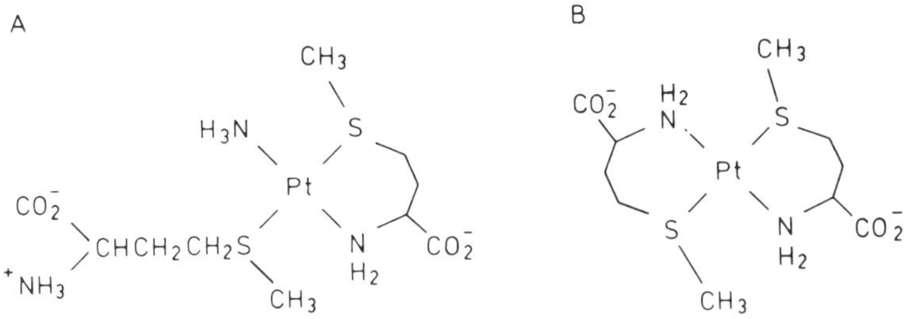

FIG. 14. Structure of [*trans*-Pt(NH_3)(Met-*S*)(Met-*N,S*)] (A) and [*trans*-Pt(Met-*N,S*)$_2$] (B) formed during reaction of *cis*-Pt with methionine.

coordinated to the amino group of the cysteine unit. The formation of these complexes and their interconversions has been depicted schematically in Fig. 16. As can be seen, the Pt–methionine type of bond in (1) is labile in the presence of the dehydronated amino group, resulting in the formation of (2). Such migrations of platinum are quite uncommon and this is the first case in which a Pt migrates from a sulfur to a nitrogen and *vice versa*.

In all Pt complexes described in this section, coordination to the sulfur atom is observed (i.e., with SAH, cysteine, methionine, GSH, MT, and other proteins), which is consistent with a high reactivity of sulfur. It is therefore reasonable to expect that significant amounts of platinum antitumor drugs will bind *in vivo* to sulfur-containing biomolecules. Although the kinetic reactivity of sulfur is high, the Pt—S bond of the methionine type is labile in the presence of other nucleophiles (*vide supra* for reactions with Naddtc and the interconversion of the Pt–SAH adducts). The relatively labile Pt—S methionine bond, but not the Pt—S cysteine bond, is likely to be relevant for the following cases:

1. The mechanism of suppressing the nephrotoxicity, i.e., Naddtc

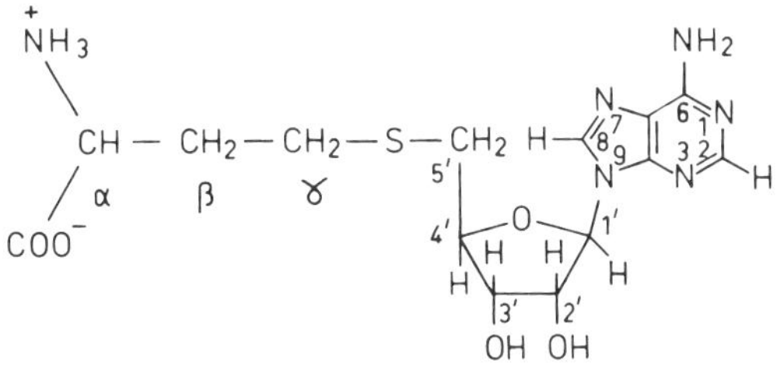

FIG. 15. Structure of *S*-adenosyl-L-homocysteine.

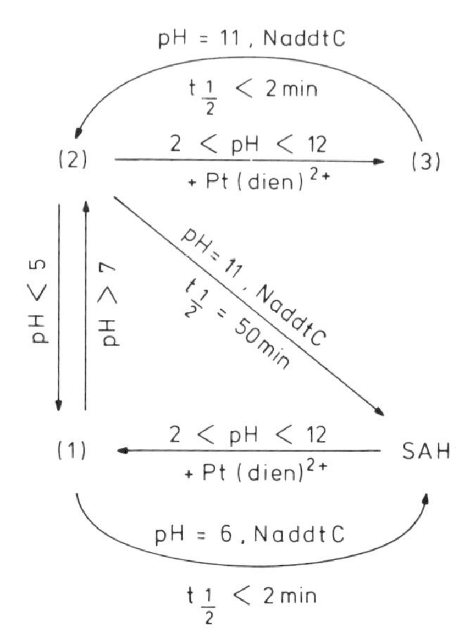

FIG. 16. Formation scheme of the complexes of SAH and their interconversions as a function of pH as well as decomposition reactions with Naddtc. Values for $t_{\frac{1}{2}}$ are also indicated.

reduces the nephrotoxicity by reversing Pt—S bonds only of the methionine type.

2. The development of permanent nephrotoxicity. Nephrotoxicity cannot be cured by Naddtc and thiourea when administered more than 4 hours after *cis*-Pt treatment. A possible mechanism for the development of permanent nephrotoxicity might be that the platinum migrates from a methionine to a cysteine within a protein, resulting in irreversible binding.

3. Acting as a drug reservoir for platination at DNA. Although there is no evidence for the existence of such a phenomenon, in principle Pt–methionine interactions are labile and therefore are likely candidates to serve as a drug reservoir for platination at DNA, i.e., spontaneous release and binding to DNA or nucleophilic displacement of platinum by guanine, for example.

V. Prospects for Future Studies on Pt Antitumor Compounds

Despite many intensive efforts to elucidate the detailed working mechanism of *cis*-Pt, as yet there is no conclusive proposal. It has been shown that both the cis and the transisomer can bind to DNA *in vivo*.

The difference in antitumor properties is thought to originate from differences in platinum–DNA adduct induction. As intrastrand chelation of neighboring purines is possible with *cis*-Pt, but—for steric reasons—not for *trans*-Pt, the AG and the GG chelates are likely candidates for the crucial lesion. A reasonable hypothesis states that these lesions, contrary to the GNG chelate and the interstrand cross-links, which can be formed by *trans*-Pt, also, are less or not at all recognized by cellular repair mechanisms, but on the other hand do interfere with DNA replication. Also, other important questions concerning the uptake of *cis*-Pt into cells (are certain cell walls selective for uptake of *cis*-Pt?) and the cellular transport (i.e., resulting in toxic side effects) and degradation are far from being answered. These questions promise a very interesting future for interdisciplinary research by chemists, biologists, pharmacologists, and physicians.

It is quite likely that the molecular basis of all kinds of toxicity, of resistance, and of repair will gain increasing interests in the coming years, and the first results are becoming available. Very recently, a factor in mammalian cells has been identified by Lippard, Donahue, and Chang *et al.* (*194–196*) that binds to *cis*-Pt-damaged DNA, but *not* to DNA modified with *trans*-Pt. This so-called DRP (damage recognition protein) could act in one of the following possibilities (*72*):

1. It could be a damage-repair protein, which would not work so well in (certain) tumor cells.
2. It could be a tumor-regulating protein, i.e., specific for tumor cells, and Pt-binding could perhaps deregulate the protein.
3. It could be a protein that would prevent Pt-damage repair by (other) proteins.

For an overview in the area of resistance, the reader is referred to recent contributions (*197–199*) dealing with suggestions to circumvent resistance.

Nephrotoxicity has also been reviewed recently (*200–202*), whereas Borch and Markman (*203*) have published an interesting paper on the modulation of *cis*-Pt toxicity, by using a combination of Naddtc and hypertonic saline, for example.

Finally, considerable progress has been reported about the possible use of Pt(IV) compounds, as in Fig. 2c, but with carboxylates as axial ligands, as oral anti-tumor drugs. (*204*).

Very recently, it has been observed that *E. coli* produces UvrABC excision repair proteins, and that the UvrAB complex binds to the convex side of a cisplatin-induced kink in DNA (*205*) It would be of great interest to study the similarities between this complex and the DRP protein mentioned above.

The results discussed in Section IV have clearly shown that there is increasing evidence for the importance of Pt–sulfur interactions from a bio(chemical) and medical point of view. Although such interactions are not responsible for the antitumor activity of *cis*-Pt, they probably contribute to some of the mentioned overall negative effects (inactivation, resistance, and nephrotoxicity). Much more research is needed to study these negative effects in detail. This could ultimately lead to compounds with better antitumor properties and lower side effects.

Some prospects for future studies from a chemical point of view are as follows:

1. Detailed kinetic studies to determine the rate-limiting step in the binding of Pt to S. This will be important in understanding the mechanisms of inactivation and resistance in normal tissues and tumors.
2. Systematic binding studies of a variety of Pt compounds with sulfur-containing molecules, which would eventually lead to the development of new antitumor drugs with structural properties such that the ratio of Pt binding at DNA compared to that of RSR is increased.
3. Extensive exchange reactions of platinated sulfurs of the methionine type with mononucleotides and longer DNA fragments. This is with the aim to investigate the possibility that such Pt—S methionine bonds could act as drug reservoirs for platination at DNA.
4. Exchange reactions of rescue agents with platinated proteins, which are known to form Pt—S cysteine bonds, to test the hypothesis (*131*) that the Pt—S cysteine bond is inert; this is important in the unraveling of the mechanism of the nephrotoxicity.
5. Ongoing Pt-binding studies with S-containing biomolecules and derivatives with the purpose to investigate in detail Pt migration reactions and competitions between a sulfur and the reactive guanine N7.

Abbreviations

cis-Pt	[*cis*-$PtCl_2(NH_3)_2$]
trans-Pt	[*trans*-$PtCl_2(NH_3)_2$]
CBDCA	[*cis*-$Pt(C_6H_6O_4)(NH_3)_2$]
DNA	deoxyribonucleic acid
RNA	ribonucleic acid
i.v.	intravenously
i.p.	intraperitoneally
Am	amine

diam	didentate amine
en	ethylenediamine
dien	diethylenetriamine
dach	1,2-diaminocyclohexane
damch	1,1-bis(aminomethyl)cyclohexane
i-$PrNH_2$	i-propylamine
N-het	heterocyclic amine
mepy	methylpyridine
R′R″SO	substituted sulfoxide
Me_2SO	dimethyl sulfoxide
MeBzSO	methyl benzyl sulfoxide
GSH	γ-L-glutamyl-L-cysteinylglycine
GS^-	GSH, dehydronated at the thiol group
G*S*-Me	*S*-methyl glutathione
MT	metallothionein
cys	cysteine
met	methionine
SAH	*S*-adenosyl-L-homocysteine
Naddtc	sodium diethyldithiocarbamate
STS	sodium thiosulfate
WR-2721	[*S*-2-(3-aminopropylamino)ethyl phosphorothioic acid]
WR-1065	dephosphorylated WR-2721
mesna	sodium 2-mercaptoethane sulfonate
d	deoxyribo
G	guanine
Guo	guanosine
GMP	guanosine monophosphate
p	phosphate group (irrespective of charge)
r_b	number of Pt atoms bound per nucleobase
CSA	chemical shift anisotropy

Acknowledgments

This research has been sponsored by the Netherlands Organisation for Chemical Research (SON), with financial aid of the Netherlands Organisation for the Advancement of Research (NWO). The authors acknowledge EEC support (Grant ST2J-0462-C) allowing regular scientific exchange with the group of Dr. J. C. Chottard (Paris). Dr. Annemarie J. Fichtinger-Schepman is thanked for continuous collaboration and many useful suggestions. The authors also wish to thank many colleagues and co-workers in Leiden (whose names are listed as co-authors) for many useful discussions. We are also indebted to Johnson & Matthey (Reading, UK) for their generous loan of $K_2[PtCl_4]$.

References

1. Rosenberg, B., Van Camp, L., and Krigas, T., *Nature (London)* **205,** 698 (1965).
2. Rosenberg, B., Renshaw, E., Van Camp, L., Hartwick, J., and Drobnik, J., *J. Bacteriol.* **93,** 716 (1967).
3. Rosenberg, B., Van Camp, L., Grimley, E. B., and Thomson, A. J., *J. Biol. Chem.* **242,** 1347 (1967).
4. Rosenberg, B., Van Camp, L., Trosko, J. E., and Mansour, V. H., *Nature (London)* **222,** 385 (1969).
5. Rosenberg, B., *in*"Nucleic Acid Metal Ion Interactions" (T. G. Spiro, ed.), p. 3. Wiley, New York 1980.
6. Roberts, J. J., and Thomson, A. J., *Prog. Nucleic Acid Res. Mol. Biol.* **22,** 71 (1979).
7. Higby, D. J., Wallace, H. J., and Holland, J. F., *Cancer Chemother. Rep.* **57,** 459 (1973).
8. Prestayko, A. W., Crooke, S. T., and Carter, S. K., eds., "Cisplatin: Current Status and New Developments." Academic Press, New York, 1980.
9. Loehrer, P. J., and Einhorn, L. H., *Ann. Intern. Med.* **100,** 704 (1984).
10. Barnard, C. F. J., *Platinum Met. Rev.* **33,** 162 (1989).
11. Bajetta, E., Rovej, R., Buzzoni, R., Vaglini, M., and Bonadonna, G., *Cancer Treat. Rep.* **66,** 1299 (1982).
12. Prestayko, A. W., D'Aoust, J. C., Issell, B. F., and Crooke, S. T., *Cancer Treat. Rev.* **6,** 17 (1979).
13. Douple, E. B., *Platinum Met. Rev.* **29,** 118 (1985).
14. Cleare, M. J., and Hoeschele, J. D., *Bioinorg. Chem.* **2,** 187 (1973).
15. Cleare, M. J., Hydes, P. C., Malerbi, B. W., and Watkins, D. M., *Biochimie* **60,** 835 (1978).
16. Wilkinson, R., Cox, P. J., Jones, M., and Harrap, K. R., *Biochimie* **60,** 851 (1978).
17. Cleare, M. J., *J. Clin. Hematol. Oncol.* **7,** 1 (1977).

18a. Pendyala, L., Cowens, J. W., Madajewicz, S., and Creaven, P. J., *in* "Platinum Coordination Complexes in Cancer Chemotherapy" (M. P. Hacker, E. B. Douple, and I. H. Krakoff, eds.), p.114. Martinus Nijhoff, Boston, Massachusetts, 1984.

18b. Chaney, S. G., Gibbons, G. R., Wyrick, S. D., and Podhasky, P., *Cancer Res.* **51,** 969 (1991).

19. Van der Veer, J. L., Peters, A. R., and Reedijk, J., *J. Inorg. Biochem.* **26,** 137 (1986).
20. Marcelis, A. T. M., and Reedijk, J., *Recl. Trav. Chim. Pays-Bas* **102,** 121 (1983).
21. Cleare, M. J., *Coord, Chem. Rev.* **12,** 349 (1974).
22. Barnard, C. F. J., Cleare, M. J., and Hydes, P. C., *Chem. Br.* **22,** 1001 (1986).
23. Farrell, N., Kiley, D. M., Schmidt, W., and Hacker, M. P., *Inorg. Chem.* **29,** 397 (1990).
24. Hollis, L. S., Amundsen, A. R., and Stern, E. W., *J. Med. Chem.* **32,** 128 (1989).
25. Schönenberger, H., Wappes, B., Jennerwein, M., and Berger, M., *Cancer Treat. Rev.* **11,** 125 (1984).
26. Bowler, B. E., Ahmed, K. J., Sundquist, W. I., Hollis, L. S., Whang, E. E., and Lippard, S. J., *J. Am. Chem. Soc.* **111,** 1299 (1989).
27. Sundquist, W. I., Bancroft, D. P., Chassot, L. D. P., and Lippard, S. J., *J. Am. Chem. Soc.* **110,** 8559 (1988); Sundquist, W. I., Bancroft, D. P., and Lippard, S. J., *ibid.* **112,** 1590 (1990).
28. Hollis, L. S., Miller, A. V., Amundsen, A. R., Schurig, J. E., and Stern, E. W., *J. Med. Chem.* **33,** 105 (1990).
29. Farrell, N. P., de Almeida, S. G., and Skov, K. A., *J. Am. Chem. Soc.* **110,** 5018 (1988); Farrell, N. P., 2nd and Qu, Y., *Inorg. Chem.* **28,** 3416 (1989).

30. Rochon, F. D., and Kong, P. C., *Can. J. Chem.* **64,** 1894 (1986).
31. Roberts, J. J., van der Vijgh, W. J. F., Vermorken, J. B., and Douple, E. B., *Cancer Chemother. Annu.* **6,** 118 (1984).
32. Farrell, N., and Skov, K. A., *J. Chem. Soc., Chem. Commun.* p. 1043 (1987).
33. Reishus, J. W., and Martin, D. S., Jr., *J. Am. Chem. Soc.* **83,** 2457 (1961).
34. Lee, K. W., and Martin, D. S., Jr., *Inorg. Chim. Acta* **17,** 105 (1976).
35. Lim, M. C., and Martin, R. B., *J. Inorg. Nucl. Chem.* **38,** 1911 (1976).
36. Miller, S. E., and House, D. A., *Inorg. Chim. Acta* **166,** 189 (1989).
36a. Faggianni, R., Lippert, B., Lock, C. J. L., and Rosenberg, B., *Inorg. Chem.* **17,** 1941 (1978).
36b. Appleton, T. G., Hall, J. R., Ralph, S. F., and Thompson, C. S. M., *Inorg. Chem.* **28,** 1989 (1989).
36c. Miller, S. A., and House, D. A., *Inorg. Chim. Acta* **173,** 53 (1990).
37. Martin, R. B., *ACS Symp. Ser.* **209,** 231 (1983).
38. Basolo, F., Gray, H. B., and Pearson, R. G., *J. Am. Chem. Soc.* **82,** 4200 (1960).
39. Gray, H. B., and Olcott, R. J., *Inorg. Chem.* **1,** 481 (1962).
40. Johnson, N. P., Hoeschele, J. D., and Rahn, R. O., *Chem.-Biol. Interact.* **30,** 151 (1980).
41. Bancroft, D. P., Lepre, C. A., and Lippard, S. J., *J. Am. Chem. Soc.* **112,** 6860 (1990).
41a. Know, R. J., Friedlos, F., Lydall, D. A., and Roberts, J. J., *Cancer Res.* **46,** 1972 (1986).
42. Reslova, S., *Chem.-Biol. Interact.* **4,** 66 (1971).
43. Howle, J. A., and Gale, G. R., *Biochem. Pharmacol.* **19,** 2757 (1970).
43a. Harder, H. C., and Rosenberg, B., *Int. J. Cancer* **6,** 207 (1970).
44. Montine, T. J., and Borch, R. F., *Cancer Res.* **48,** 6017 (1988); Tay, L. K., Bregman, C. L., Masters, B. A., and williams, P. D., *Cancer Res.* **48,** 2538 (1988).
45. Brouwer, J., Van de Putte, P., Fichtinger-Schepman, A. M. J., and Reedijk, J., *Proc. Natl. Acad. Sci. U.S.A.* **78,** 7010 (1981).
45a. Burnouf, D., Daune, M., and Fuchs, R. P. P., *Proc. Natl. Acad. Sci. U.S.A.* **84,** 3758 (1987); De Boer, J. G., and Glickman, B. W., *Carcinogenesis (London)* **10,** 1363 (1989); Mis, J. R. A., and Kunz, B. A., *Carcinogenesis (London)* **11,** 633 (1990).
46. Alazard, R., Germanier, M., and Johnson, N. P., *Mutat. Res.* **93,** 327 (1982).
47. Plooy, A. C. M., Van Dijk, M., Berends, F., and Lohman, P. H. M., *Cancer Res.* **45,** 4178 (1985).
48. Dijt, F. J., Fichtinger-Schepman, A. M. J., Berends, F., and Reedijk, J., *Cancer Res.* **48,** 6058 (1988).
49. Pinto, A. L., and Lippard, S. J., *Biochim. Biophys. Acta* **780,** 167 (1985); Martin, R. B., *Acc. Chem. Res.* **18,** 32 (1985); Köpf-Maier, P., and Köpf, H., *Naturwissenschaften* **73,** 239 (1986).
50. Reedijk, J., *Pure Appl. Chem.* **59,** 181 (1987).
50a. Pasini, A., and Zunino, F., *Angew. Chem.* **26,** 615 (1987).
50b. Reedijk, J., Fichtinger-Schepman, A. M. J., van Oosterom, A. T., and van de Putte, P., *Struct. Bonding (Berlin)* **67,** 53 (1987).
50c. Sherman, S. E., and Lippard, S. J., *Chem. Rev.* **87,** 1153 (1987); Eastman, A., *Pharmacol. Ther.* **34,** 155 (1987).
51. Johnson, N. P., *Prog. Clin. Biochem.* **10,** 1 (1989).
51a. Lippert, B., *Prog. Inorg. Chem.* **37,** 1 (1989).
52. Farrell, N., "Transition Metal Complexes as Drugs and Chemotherapeutic Agents." Kluwer Academic Publishers, Dordrecht, Netherlands, 1989.
53. Sundquist, W. I., and Lippard, S. J., *Coord. Chem. Rev.* **100,** 293 (1990).

54. Chottard, J. C., Girault, J. P., Chottard, G., Lallemand, J. Y., and Mansuy, D., *J. Am. Chem. Soc.* **102,** 5565 (1980).
54a. Lallemand, J. Y., Soulié, J., and Chottard J. C., *J. Chem. Soc., Chem. Commun.* p. 436 (1980).
54b. Van der Veer, J. L., Ligtvoet, G. J., and Reedijk, J., *J. Inorg. Biochem.* **29,** 217 (1987).
55. Dijt, F. J., Chottard, J. C., Girault, J. P., and Reedijk, J., *Eur. J. Biochem.* **179,** 333 (1989); Van Garderen, C. J. Altona, C., and Reedijk, J., *Inorg. Chem.* **29,** 1481 (1990); van Hemelryck, B., Girault, J.-P., Chottard, G., Valadon, P., Laoui, A., and Chottard, J.-C., *Inorg. Chem.* **26,** 787 (1987); Girault, J.-P., Chottard, J.-C., Guittet, E. R., Lallemand, J.-Y., Huynh-Dinh, T., and Igolen, J., *Biochem. Biophys. Res. Commun.* **109,** 1157 (1982); Girault, J.-P., Chottard, G., Lallemand, J.-Y., and Chottard, J.-C., *Biochemistry* **21,** 1352 (1982); Girault, J. P., Chottard, G., Lallemand, J. Y., Huguenin, F., and Chottard, J. C., *J. Am. Chem. Soc.* **106,** 7227 (1984); Laoui, A., Kozelka, J., and Chottard, J. C., *Inorg. Chem.* **27,** 2751 (1988); Lepre, C. A., Chassot, L., Costello, C. E., and Lippard, S. J., *Biochemistry* **29,** 811 (1990); Caradonna, J. P., and Lippard, S. J., *Inorg. Chem.* **27,** 1454 (1988); Rice, J. A., Crothers, D. M., Pinto, A. L., and Lippard, S. J., *Proc. Natl. Acad. Sci. U.S.A.* **85,** 4158 (1988); Reily, M. D., and Marzilli, L. G., *J. Am. Chem. Soc.* **108,** 6785 (1986); Spellmeyer-Fouts, C., Marzilli, L. G., Byrd, R. A., Summers, M. F., Zon, G., and Shinozuka, K., *Inorg. Chem.* **27,** 366 (1988); Marzilli, L. G., Kline, T. P., Live, D., and Zon, G., *ACS Symp. Ser.* **402,** 119 (1989).
56. Filipski, J., Kohn, K. W., Prather, R., and Bonner, W. M., *Science* **204,** 181 (1979).
57. Pinto, A. L., Naser, L. J., Essigman, J. M., and Lippard, S. J., *J. Am. Chem. Soc.* **108,** 7405 (1986).
58. Srivastava, R. C., Froehlich, J., and Eichhorn, G. L., *Biochimie* **60,** 879 (1978).
59. Hermann, D., Houssier, C., and Guschlbauer, W., *Biochim. Biophys. Acta* **564,** 456 (1979).
60. Cohen, G. L., Bauer, W. R., Barton, J. K., and Lippard, S. J., *Science* **203,** 1014 (1979).
61. Macquet, J. P., and Butour, J. L., *Biochimie* **60,** 901 (1978).
62. Scovell, W. M., and Capponi, V. J., *Biochem. Biophys, Res. Commun.* **107,** 1138 (1982).
63. Fichtinger-Schepman, A. M. J., van der Veer, J. L., den Hartog, J. H. J., Lohman, P. H. M., and Reedijk, J., *Biochemistry* **24,** 707 (1985).
64. Lippard, S. J., Ushay, H. M., Merkel, C. M., and Poirier, M. C., *Biochemistry* **22,** 5165 (1983); Terheggen, P., Ph. D. Dissertation, Leiden University (1990).
64a. Fichtinger-Schepman, A. M. J., Baan, R. A., Luiten-Schuite, A., van Dijk, M., and Lohman, P. H. M., *Chem.-Biol Interact.* **55,** 275 (1985); Fichtinger-Schepman, A. M. J., Van Oosterom. A. T., Lohman, P. H. M., and Berends, F., *Cancer Res.* **47,** 3000 (1987).
64b. den Hartog, J. H. J., Altona, C., van Boom, J. H., Marcelis, A. T. M., van der Marel, G. A., Rinkel, L. J., Wille-Hazeleger, G., and Reedijk, J., *Eur. J. Biochem.* **134,** 485 (1983).
64c. Kozelka, J., Petsko, G. A., Quigley, G. J., and Lippard, S. J., *Inorg. Chem.* **25,** 1075 (1986); Kozelka, J., Archer, S., Petsko, G. A., Lippard, S. J., and Quigley, G. J. *Biopolymers* **26,** 1245 (1987).
64d. Sherman, S., Gibson, D., Wang, A. H. J., and Lippard, S. J., *Science* **230,** 412 (1985); Sherman, S., Gibson, D., Wang, A. H. J., and Lippard, S. J., *J. Am. Chem. Soc.* **110,** 7368 (1988).

64e. Admiraal, G., van der Veer, J. L., de Graaff, R. A. G., den Hartog, J. H. J., and Reedijk, J., *J. Am. Chem. Soc.* **109,** 592 (1987).
64f. Inagaki, K., Dijt, F. J., Lempers E. L. M., and Reedijk, J. *Inorg. Chem.* **27,** 382 (1988).
65. Macquet, J. P., and Butour, J. L. *JNCI, J. Natl. Cancer Inst.* **70,** 899 (1983).
66. Macquet, J. P., Butour, J. L., and Johnson, N. P., *ACS Symp. Ser.* **209,** 75 (1983).
67. Raudaschl-Sieber, G., Schöllhorn, H., Thewalt, U., and Lippert, B., *J. Am. Chem. Soc.* **107,** 3591 (1985).
68. van Garderen, C. J., van Houte, L. P. A., van den Elst, H., van Boom, J. H., and Reedijk, J., *J. Am. Chem. Soc.* **111,** 4123 (1989).
69. Lempers, E. L. M., Bloemink, M. J., and Reedijk, J., *Inorg. Chem.* **30,** 201 (1991).
70. Bitha, P., Morton, G. O., Dunne, T. S., Delos Santos, E. F., Lin, Y., Boone, S. R., Haltiwanger, R. C., and Pierpont, C. G., *Inorg. Chem.* **29,** 645 (1990).
71. Lempers, E. L. M., Bloemink, M. J., Brouwer, J., Kidani, y., and Reedijk, J., *J. Inorg. Biochem.* **40,** 23 (1990).
72. Lippard, S. J., personal communication (1990); *Proc. Int. Symp. Platinum and Other Metal Coordination Compounds in Cancer Chemotherapy, 6th,* San Diego, 1991.
73. Basolo, F., and Pearson, R. G., "Mechanisms of Inorganic Reactions," pp. 351–453. Wiley, New York, 1967.
74. Eastman, A., *Chem.-Biol. Interact.* **61,** 241 (1987).
75. Andrews, P. A., Murphy, M. P., and Howell, S. B., *Mol. Pharmacol.* **30,** 643 (1986).
76. Dedon, P. C., and Borch, R. F., *Biochem. Pharmacol.* **36,** 1955 (1987).
77. Kulamowicz, I., Olinski, R., and Walter, Z., *Z. Naturforsch., C: Biosci.* **39C,** 180 (1984).
78. Burchenal, J. H., Kalaher, K., Dew, K., Lokys, L., and Gale, G., *Biochimie* **60,** 961 (1978).
79. Newman, A. D., Ridgeway, H., Speer, R. J., and Hill, J. M., *J. Clin. Hematol. Oncol.* **9,** 208 (1979).
80. Sternson, L. A., Repta, A. J., Shih, H., Himmelstein, K. J., and Patton, T. F., *in* "Platinum Coordination Complexes in Cancer Chemotherapy" (M. P. Hacker, E. B. Douple, and I. H. Krakoff, eds.), p. 126. Martinus Nijhoff, Boston, Massachusetts, 1984.
81. Berners-Price, S. J., and Kuchel, P. W., *J. Inorg. Biochem.* **38,** 327 (1990).
82. Zwelling L. A., Filipski, J., and Kohn, K. W., *Cancer Res.* **39,** 4989 (1979).
83. Micetich, K., Zwelling, L. A., and Kohn, K. W., *Cancer Res.* **43,** 3609 (1983).
84. Eastman, A., and Barry, M. A., *Biochemistry* **26,** 3303 (1987).
85. Lippard, S. J., and Hoeschele, J. D., *Proc. Natl. Acad. Sci. U.S.A.* **76,** 6091 (1979).
86. Olinski, R., Wedrychowski, A., Schmidt, W. N., Briggs, R. C., and Hnilica, L. S., *Cancer Res.* **47,** 201 (1987).
87. Kraker, A., Schmidt, J., Krezoski, S., and Petering, D. H., *Biochem. Biophys. Res. Commun.* **130,** 786 (1985).
88. Bakka, A., Endresen, L., Johnsen, A. B. S., Edminson, P. D., and Rugstad, H. E., *Toxicol. Appl. Pharmacol.* **61,** 215 (1981).
89. Sharma, R. P., and Edwards, I. R., *Biochem. Pharmacol.* **32,** 2665 (1983).
90. Zelazowski, A. J., Garvey, J. S., and Hoeschele, J. D., *Arch. Biochem. Biophys.* **229,** 246 (1984).
91. Andrews, P. A., Murphy, M. P., and Howell, S. B., *Cancer Res.* **45,** 6250 (1985).
92. Louie, K. G., Behrens, B. C., Kinsella, T. J., Hamilton, T. C., Grotzinger, K. R., McKoy, W. M., Winker, M. A., and Ozols, R. F., *Cancer Res.* **45,** 2110 (1985).
93. Hamilton, T. C., Winker, M. A., Louie, K. G., Batist, G., Behrens, B. C., Tsuruo, T.,

Grotzinger, K. R., McKoy, W. M., Young, R. C., and Ozols, R. F., *Biochem. Pharmacol.* **34,** 2583 (1985).
94. Kraker, A. J., and Moore, C. W., *Cancer Res.* **48,** 9 (1988).
95. Richon. V. M., Schulte, N., and Eastman, A., *Cancer Res.* **47,** 2056 (1987).
96. Andrews, P. A., Murphy, M. P., and Howell, S. B., *Eur. J. Cancer Clin. Oncol.* **25,** 619 (1989).
97. Hospers, G. A. P., Mulder, N. H., de Jong, B., de Ley, L., Uges, D. R. A., Fichtinger-Schepman, A. M. J., Scheper, R. J., and de Vries, E. G. E., *Cancer Res.* **48,** 6803 (1988); Meijer, C., Mulder, N. H., Hospers, G. A. P., Uges, D. R. A., and de Vries, E. G. E., *Br. J. Cancer* **62,** 72 (1990).
98. Chen, G., Frei, E., and Zeller, W. J., *Cancer Lett.* **46,** 207 (1989).
99. Sekiya, S., Oosaki, T., Andoh, S., Suzuki, N., Akaboshi, M., and Takamizawa, H., *Eur. J. Cancer Clin. Oncol.* **25,** 429 (1989).
100. Mansouri, A., Henle, K. J., Benson, A. M., Moss, A. J., and Nagle, W. A., *Cancer Res.* **49,** 2674 (1989).
101. Fram, R. J., Woda, B. A., Wilson, J. M., and Robichaud, N., *Cancer Res.* **50,** 72 (1990).
102. Andrews, P. A., Schiefer, M. A., Murphy, M. P., and Howell, S. B., *Chem.-Biol. Interact.* **65,** 51 (1988).
103. Kelley, S. L., Basu, A., Teicher, B. A., Hacker, M. P., Hamer, D. H., and Lazo, J. S., *Science* **241,** 1813 (1988).
104. Schilder, R. J., Hall, C., Monks, A., Handel, L. M., Fornace, A. J., Jr., Ozols, R. F., Fojo, A. T., and Hamilton, T. C., *Int. J. Cancer* **45,** 416 (1990).
105. Farnworth, P. G., Hillcoat, B. L., and Roos, I. A. G., *Chem.-Biol. Interact.* **69,** 319 (1989).
106. Krakoff, I. H., *Cancer Treat. Rep.* **63,** 1523 (1979).
107. Dentino, M., Luft, F. C., Yum, M. N., Williams, S. D., and Einhorn, L. H., *Cancer (Philadelphia)* **41,** 1274 (1978).
108. Von Hoff, D. D., Schilsky, R., Reichert, C. M., Reddick, R. L., Rozencwerg, M., Young, R. C., and Muggie, F. M., *Cancer Treat. Rep.* **63,** 1527 (1979).
109. Hayes, D. M., Cvitkovic, E., Golbey, R. B., Scheiner, E., Helson, L., and Krakoff, I. H., *Cancer (Philadelphia)* **39,** 1372 (1979).
110. Ozols, R. F., Corden, B. J., Collins, J., and Young, R. C., *in* "Platinum Coordination Complexes in Cancer Chemotherapy" (M. P. Hacker, E. B. Douple, and I. H. Krakoff, eds.), p. 321. Martinus Nijhoff, Boston, Massachusetts, 1984.
111. Borch, R. F., and Pleasants, M. E., *Proc. Natl. Acad. Sci. U.S.A.* **76,** 6611 (1979).
112. Levi, J., Jacobs, C., Kalman, S. M., McTigue, M., and Weiner, M. W., *J. Pharmacol. Exp. Ther.* **213,** 545 (1980).
113. Weiner, M. W., and Jacobs, C., *Fed. Proc., Fed. Am. Soc. Exp. Biol.* **42,** 2974 (1983).
114. Daley-Yates, P. T., and McBrien, D. C. A., *Chem.-Biol. Interact.* **40,** 325 (1982).
115. Sugiyama, S., Hayakawa, M., Kato, T., Hanaki, Y., Shimizu, K., and Ozawa, T., *Biochem. Biophys. Res. Commun.* **159,** 1121 (1989).
116. Singh, G., *Toxicology* **58,** 71 (1989).
117. Bodenner, D. L., Dedon, P. C., Keng, P. C., Katz, J. C., and Borch, R. F., *Cancer Res.* **46,** 2751 (1986).
118. Evans, R. G., Wheatley, C., Engel, C., Nielsen, J., and Ciborowski, L. J., *Cancer Res.* **44,** 3686 (1984).
119. Gringeri, A., Keng, P. C., and Borch, R. F., *Cancer Res.* **48,** 5708 (1988).
120. Borch, R. F., Katz, J. C., Lieder, P. H., and Pleasants, M. E., *Proc. Natl. Acad. Sci. U.S.A.* **77,** 5441 (1980).

121. Gale, G. R., Atkins, L. M., and Walker, E. M., *Ann. Clin. Lab. Sci.* **12,** 345 (1982).
122. Khandekar, J. D., *Res. Commun. chem. Pathol. Pharmacol.* **40,** 55 (1983).
123. Borch, R. F., Bodenner, D. L., and Katz, J. C., *in* "Platinum Coordination Complexes in Cancer Chemotherapy" (M. P. Hacker, E. B. Douple, and I. H. Krakoff, eds.), p. 154. Martinus Nijhoff, Boston, Massachusetts, 1984.
124. Bodenner, D. L., Dedon, P. C., Keng, P. C., and Borch, R. F., *Cancer Res.* **46,** 2745 (1986).
125. Jones, M. M., and Basinger, M. A., *J. Appl. Toxicol.* **9,** 229 (1989).
126. Sunderman, F. W., Jr., and Fraser, C. B., *Ann. Clin. Lab. Sci.* **13,** 489 (1983).
127. Gonias, S. L., Oakley, A. C., Walther, P. J., and Pizzo, S. V., *Cancer Res.* **44,** 5764 (1984).
128. Dedon, P. C., Qazi, R., and Borch, R. F., *in* "Biochemical Mechanisms of Platinum Antitumor Drugs" (D. C. H. McBrien and T. F. Slater, eds.), p. 199. IRL Press, Oxford, 1986.
129. Boelrijk, A. E. M., Boogaard, P. J., Lempers, E. L. M., and Reedijk, J., *J. Inorg. Biochem.* **41,** 17 (1991).
130. Lempers, E. L. M., Inagaki, K., and Reedijk, J., *Inorg. Chim. Acta* **152,** 201 (1988).
131. Lempers, E. L. M., and Reedijk, J., *Inorg. Chem.* **29,** 217 (1990).
132. Andrews, P. A., Wung, W. E., and Howell, S. B., *Anal. biochem.* **143,** 46 (1984).
133. Hegedus, L., van der Vijgh, W. J. F., Klein, I., Kerpel-Fronius, S., and Pinedo, H. M., *Cancer Chemother. Pharmacol.* **20,** 211 (1987).
134. Hacker, M. P., and Roberts, J. D., *in* "Platinum and Other Metal Coordination Compounds in Cancer Chemotherapy" (M. Nicolini, ed.), p. 163. Martinus Nijhoff, Boston, Massachusetts, 1987.
135. Jones, M. M., Basinger, M. A., Mitchell, W. M., and Bradley, C. A., *Cancer Chemother. Pharmacol.* **17,** 38 (1986).
136. Jones, M. M., Basinger, M. A., Craft, W. D., Domingo, J. L., and Llobet, J. M., *Arch. Toxicol.* **59,** 167 (1986).
137. Shea, M., Koziol, J. A., and Howell, S. B., *Clin. Pharmacol, Ther.* **35,** 419 (1984).
138. Howell, S. B., and Taetle, R., *Cancer Treat. Rep.* **64,** 611 (1980).
139. Howell, S. B., Pfeifle, C. L., Wung, W. E., Olshen, R. A., Lucas, W. E., Yon, J. L., and Green, M., *Ann. Intern. Med.* **97,** 845 (1982).
140. Howell, S. B., Pfeifle, C. L., Wung, W. E., and Olshen, R. A., *Cancer Res.* **43,** 1426 (1983).
141. Goel, R., Cleary, S. M., Horton, C., Kirmani, S., Abramson, I., Kelly, C., and Howell, S. B., *JNCI, J. Natl. Cancer Inst.* **81,** 1552 (1989).
142. Aamdal, S., Fodstad, O., and Pihl, A., *Cancer Chemother. Pharmacol.* **21,** 129 (1988).
143. Elferink, F., van der Vijgh, W. J. F., and Klein, I., *Clin. Chem. (Winston-Salem, N.C.)* **32,** 641 (1986).
144. Glover, D., Glick, J. H., Weiler, C., Yuhas, J., and Kligerman, M. M., *Int. J. Radiat. Oncol. Biol. Phys.* **10,** 1781 (1984).
145. Glover, D., Fox, K. R., Weiler, C., Kligerman, M. M., Turrisi, A., and Glick, J. H., *Pharmacol. Ther.* **39,** 3 (1988).
146. Yuhas, J. M., *Cancer Res.* **40,** 1519 (1980).
147. Calabro-Jones, P. M., Aguilera, J. A., Ward, J. F., Smoluk, G. D., and Fahey, R. C., *Cancer Res.* **48,** 3634 (1988).
148. McComb, R. B., Bowers, G. N., Jr., and Posen, S., "Alkaline Phosphatase," Chapters 3 and 12. Plenum, New York, 1979.
149. Kennedy, K. A., Teicher, B. A., Rockwell, S., and Sartorelli, J. R., *in* "Molecular

Targets and Actions for Cancer Chemotherapeutic Agents" (A. C. Sartorelli, J. R. Bertino, and J. S. Lazo, eds.), p. 85. Academic Press, New York, 1981.
150. Zunino, F., Tofanetti, O., Besati, A., Cavalletti, E., and Savi, G., *Tumori* **69,** 105 (1983).
151. Oriana, S., Böhm, S., Spatti, G., Zunino, F., and Di Re, F., *Tumori* **73,** 337 (1987).
152. Böhm, S., Oriana, S., Spatti, G. B., Tognella, S., Tedeschi, M., Zunino, F., and Di Re, F., *in* "Platinum and Other Metal Coordination Compounds in Cancer Chemotherapy" (M. Nicolini, ed.), p. 456. Martinus Nijhoff, Boston, Massachusetts, 1987.
153. Zunino, F., Pratesi, G., Micheloni, A., Cavalletti, E., Sala, F., and Tofanetti, O., *Chem.-Biol. Interact.* **70,** 89 (1989).
154. Hahn, R., Wendel, A., and Flohé, L., *Biochim. Biophys. Acta* **539,** 324 (1978).
155. Litterst, C. L., Bertolero, F., and Uozumi, J., *in* "Biochemical Mechanisms of Platinum Antitumor Drugs" (D. C. H. McBrien and T. F. Slater, eds.), p. 227. IRL Press, Oxford, 1986.
156. Ishikawa, M., Takayanagi, Y., and Sasaki, K., *Res. Commun. Chem. Pathol. Pharmacol.* **67,** 131 (1990).
157. Mayer, R. D., Lee, K., and Cockett, A. T. K., *Cancer Chemother. Pharmacol.* **20,** 207 (1987).
158. Naganuma, A., Satoh, M., and Imura, N., *Cancer Res.* **47,** 983 (1987).
159. Satoh, M., Naganuma, A., and Imura, N., *Cancer Chemother. Pharmacol.* **21,** 176 (1988).
160. LeRoy, A. F., and Thompson, W. C., *JNCI, J. Natl. Cancer Inst.* **81,** 427 (1989).
161. Repta, A. J., and Long, D. F., *in* "Cisplatin: Current Status and New Developments" (A. W. Prestayko, S. T. Crooke, and S. K. Carter, eds.), p. 285. Academic Press, New York, 1980.
162. Corden, B. J., *Inorg. Chim. Acta* **137,** 125 (1987).
163. Otvos, J. D., Petering, D. H., and Shaw, C. F., *Comments Inorg. Chem.* **9,** 1 (1989).
164. Djuran, M., Lempers, E. L. M., and Reedijk, J., *Inorg. Chem.* in press (1991).
165. Cole, W. C., and Wolf, W., *Chem.-Biol. Interact.* **30,** 223 (1980).
166. Daley-Yates, P. T., and McBrien, D. C. H., *Biochem. Pharmacol.* **33,** 3063 (1984).
167. Eastman, A., *Biochem. Pharmacol.* **36,** 4177 (1987).
168. Friedman, M. E., and Teggins, J. E., *Biochim. Biophys. Acta* **350,** 263 (1974).
169. Aull, J. L., Rice, A. C., and Tebbetts, L. A., *Biochemistry* **16,** 672 (1977).
170. Aull, J. L., Allen, R. L., Bapat, A. R., Daron, H. H., Friedman, M. E., and Wilson, J. F., *Biochim. Biophys. Acta* **571,** 352 (1979).
171. Melius, P., Andersson, L. A., and Lee, Y. Y., *J. Med. Chem.* **23,** 685 (1980).
172. Gonias, S. L., and Pizzo, S. V., *J. Biol. Chem.* **256,** 12478 (1981).
173. Gonias, S. L., and Pizzo, S. V., *J. Biol. Chem.* **258,** 5764 (1983).
174. McQuire, J. P., Friedman, M. E., and McAuliffe, C. A., *Inorg. Chim. Acta* **91,** 161 (1984).
175. Pizzo, S. V., Roche, P. A., Feldman, S. R., and Gonias, S. L., Biochem. *J.* **238,** 217 (1986).
176. Gonias, S. L., Swaim, M. W., Massey, M. F., and Pizzo, S. V., *J. Biol. Chem.* **263,** 393 (1988).
177. Pizzo, S. V., Swaim, M. W., Roche, P. A., and Gonias, S. L., *J. Inorg. Biochem.* **33,** 67 (1988).
178. Hannemann, J., and Baumann, K., *Res. Commun. Chem. Pathol. Pharmacol.* **60,** 371 (1988).
179. Smith, S. L., and Douglas, K. T., *Biochem. Biophys. Res. Commun.* **162,** 715 (1989).

180. DeConti, R. C., Toftness, B. R., Lange, R. C., and Creassey, W. A., *Cancer Res.* **33,** 1310 (1973).
181. Robbins, A. B., *Chem.-Biol. Interact.* **38,** 349 (1982).
182. Bongers, J., Bell, J. U., and Richardson, D. E., *J. Inorg. Biochem.* **34,** 55 (1988).
183. Volshtein, L. M., Krylova, L. F., and Mogilevkina, M. F., *Russ. J. Inorg. Chem. (Engl. Transl.)* **11,** 333 (1966).
184. Bell, J. D., Norman, R. E., and Sadler, P. J., *J. Inorg. Biochem.* **31,** 241 (1987).
185. Appleton, T. G., Connor, J. W., and Hall, J. R., *Inorg. Chem.* **27,** 130 (1988).
186. Norman, R. E., and Sadler, P. J., *Inorg. Chem.* **27,** 3583 (1988).
187. Odenheimer, B., and Wolf, W., *Inorg. Chim. Acta* **66,** L41 (1982).
188. Appleton, T. G., Connor, J. W., Hall, J. R., and Prenzler, P. D., *Inorg. Chem.* **28,** 2030 (1989).
189. Berners-Price, S. J., and Kuchel, P. W., *J. Inorg. Biochem.* **38,** 305 (1990).
190. Ismail, I. M., and Sadler, P. J., *in* "Platinum, Gold and other Metal Chemotherapeutic Agents" (S. J. Lippard, ed.), p. 171. Am. Chem. Soc., Washington, D.C. 1983.
191. Shanjin, W., Peiyan, D., Li, Y., and Kui, W., *Fenzi Kexue Yu Huaxue Yanjiu* **4,** 537 (1984); *Chem. Abstr.* **102,** 88998q (1985).
192. Volshtein, L. M., Krylova, L. F., and Mogilevkina, M. F., *Russ. J. Inorg. Chem. (Engl. Transl.)* **10,** 1077 (1965).
193. Lempers, E. L. M., and Reedijk, J., *Inorg. Chem.* **29,** 1880 (1990).
194. Toney, J. H., Donahue, B. A., Kellett, P. J., Bruhn, S. L., Essigmann, J. M., and Lippard, S. J., *Proc. Natl. Acad. Sci. U.S.A.* **86,** 8328 (1989).
195. Donahue, B. A., Augot, M., Bellon, S. F., Treiber, D. K., Toney, J. H., Essigmann, J. M., and Lippard, S. J., *Biochemistry* **29,** 5872 (1990).
196. Chu, G., and Chang, E., *Proc. Natl. Acad. Sci. U.S.A.* **87,** 3324 (1990).
197. Canon, J. L., Humblet, Y., and Symann, M., *Eur. J. Cancer* **26,** 1 (1990).
198. Hospers, G. A. P., *Pharmacol. Weekbl. (Sci. Ed.)* **11,** 183 (1989).
199. Kelley, S. L., and Roxencweig, M., *Eur. J. Cancer* **25,** 1135 (1989).
200. Daugaard, G., and Abildgaard, U., *Cancer Chemother. Pharmacol.* **25,** 1 (1989).
201. Fillastre, J. P., and Raguenez-Viotte, G., *Toxicol. Lett.* **46,** 163 (1989).
202. Safirstein, R., Winston, J., Moel, D., Dikman, S., and Guttenplan, J., *Int. J. Androlo.* **10,** 325 (1987).
203. Borch, R. F., and Markman, M., *Pharmacol. Ther.* **41,** 371 (1989).
204. Harrap, K. R., Murrer, B. A., Giandomenico, C., Morgan, S. E., Kelland, L. R., Jones, M., Goddard, P. M., and Schurig, J., *Proc. Int. Symp. Platinum and Other Metal Coordination Compounds in Cancer Chemotherapy, 6th.* San Diego, 1991.
205. Visse, R., de Ruijter, M., Brouwer, J., Brandsma, J. A., and van de Putte, P., *J. Biol. Chem.* in press (1991).

RECENT ADVANCES IN OSMIUM CHEMISTRY

PETER A. LAY* and W. DEAN HARMAN†

* Department of Inorganic Chemistry, University of Sydney, New South Wales, 2006 Australia and † Department of Chemistry, University of Virginia, Charlottesville, Virginia 22901

I. Introduction
  A. History
  B. Scope of This Review
II. Survey of Coordination Complexes
  A. Oxidation States
  B. General Synthetic Methods
  C. Mononuclear Complexes
  D. Dinuclear and Polynuclear Complexes
  E. Polymers
III. Electrochemistry of Coordination Complexes
  A. General
  B. Monomers
  C. Dinuclear and Polynuclear Complexes
  D. Polymers
IV. Spectroscopic and Magnetic Properties of Coordination Complexes
  A. Monomers
  B. Dimers
  C. Polymers
V. Reactivity of Coordination Complexes
  A. Addition Reactions of Osmium(VIII)
  B. Substitution Reactions
  C. Electron Transfer Reactions
  D. Linkage Isomerizations
  E. Reactions of Ligands on Osmium
Abbreviations and Trivial Nomenclature
References

I. Introduction

A. History

Osmium was first discovered in 1803 by Tennant (*1*), who also isolated the first coordination complex [OsO_4]. Although the discovery of Os preceded that of Ru by 41 years (*2*), the known chemistry of Ru is more extensive than that of Os. The reasons for this are that, compared with Ru, Os chemistry is considerably more expensive, and second, most Os complexes are prepared via more difficult routes starting from the hazardous (*3*) [OsO_4]. In spite of this, there are now literally thousands of Os coordination complexes, not to mention a quite extensive organometallic chemistry (*4*). In recent years, the number of complexes has expanded rapidly because of the development of more convenient synthetic methods. This review is aimed at describing these recent developments.

B. Scope of This Review

1. *Reviews on Osmium Chemistry*

A summary of the reviews published on Os chemistry up until 1986 is given in Griffith's 1987 article (*2*). More extensive reviews of the literature up until 1980 are found in Gmelin (*5, 6*), and brief annual surveys of publications during 1986–1988 have appeared (*7*).

In the area of Os organometallic chemistry, a comprehensive review of the literature up until 1982 has been produced (*4*). More recent reviews include the annual surveys on the organometallic chemistry of Ru and Os (*8–10*), organometallic arene chemistry of Ru and Os (*11*), and terminal methylene complexes (*12*).

A 1987 review surveys the methods available for the determination of Os and other platinum group metals (*13*), with a more recent review examining the use of thiourea complexes in the determination of Os (*14*). In the following text, reviews on more specific topics will be referred to where appropriate.

2. *Scope*

This review on the synthesis, properties, and chemistry of coordination complexes will concentrate on the literature since the end of 1985, referring to earlier literature only where relevant. It includes a general survey of recent publications, with more detailed discussions of interesting developments in the latter sections. Analogous Ru chemistry will also be included in particular sections, to show the similarities and differences in the chemistry of these elements. Appropriate literature dealing with organometallic chemistry will be referred to where relevant to the discussion of the coordination chemistry, because Os chemistry often transcends these traditional boundaries. However, no attempt has been made to review recent developments in Os organometallic chemistry; therefore, this review is confined largely to the chemistry of Os in oxidation states II or higher.

II. Survey of Coordination Complexes

A. Oxidation States

1. *General Comments*

Osmium complexes exist in every oxidation state from II− to VIII, but generally the coordination chemistry is restricted to oxidation states II through VIII (*2, 5, 6*). The low-oxidation-state chemistry is dominated by ligands that are good σ donors (e.g., amine ligands) and π acceptors (e.g., N heterocycles). By contrast, higher oxidation states are stabilized by the addition of ligands that are both strong σ and π donors, e.g., O^{2-} and N^{3-}. The strengths of π bonding and π backbonding are so great with Os that changing one ligand in the coordination sphere can have a dramatic effect on the stability of an oxidation state. This has an important influence on both the physical properties and the reactivities of Os complexes, and as such, will be discussed in detail where appropriate.

2. *Osmium(0)*

This oxidation state is rare for nonorganometallic complexes, being found only in the trigonal-bipyramidal $[Os(PR_3)_5]$ complexes and in reactive intermediates such as $[Os(PR_3)_4]$ and $[Os(PR_3)_3]$ (*15*).

3. *Osmium(I)*

Os(I) is confined to complexes containing ligands that are very good at stabilizing low oxidation states, such as those with the cyano ligand (*2, 5, 6, 16*).

4. *Osmium(II)*

An extensive pentaammine chemistry has emerged in recent years, but most Os(II) complexes are stronger reductants than are their Ru(II) analogs and are sensitive to aerial oxidation. Many air-stable Os(II) complexes with N-heterocyclic, phosphine, cyano, and CO ligands are known. In general, Os(II) is octahedral and low spin (*2, 5, 6*). However,

$[Os(PPh_3)_3Cl_2]$ (*17*) and $[Os(PPr^i_3)_2(CO)(H)(Cl)]$ (*18*) are five coordinate, and $[Os(bpy)(PPh_3)_2(CO)(H)_2]$ is seven coordinate (*19*).

5. Osmium(III)

This oxidation state is dominated by amine complexes, and a large variety of complexes with N, O, S, and P donors. Complexes with N-heterocyclic ligands tend to be strong oxidants. Invariably, Os(III) complexes are low spin and octahedral (*2, 5, 6*).

6. Osmium(IV)

In order to stabilize Os(IV), it is necessary to have either several ligands that are good π bases (e.g., Cl^- or Br^-), or one ligand that is a strong π base (e.g., O^{2-}) (*2, 5, 6*). Many Os(IV) complexes contain one or more halo ligands. There is also a growing number of oxo- and nitrido-bridged complexes.

Most Os(IV) complexes are low spin and octahedral. Although they have two unpaired electrons, they often have anomalous magnetic properties at room temperature. This is due to quenching of the electron spin by the orbital spin, as a consequence of the large spin-orbit coupling constant (*20*). Recently, low-spin, diamagnetic, tetrahedral complexes have been characterized that contain four sterically hindered alkyl or aryl groups (*21*), and the square-planar $[Os(NAr)_2(PMe_2Ph)_2]$ (Ar = 2,3-$Pr^i_2C_6H_3$) is also known (*22*).

7. Osmium(V)

Os(V) complexes are relatively rare and generally unstable toward disproportionation or other reactions. Authentic complexes that have been isolated include $[OsF_5]_4$, $[Os_2Cl_{10}]$, salts of $[OsX_6]^-$ (X = F^- or Cl^-), and mixed halo/aqua or halo/oxo complexes, all of which are octahedral (*2, 5, 6*). Recently, the tetrahedral $[Os(2\text{-}MeC_6H_4)_4]^+$ and five-coordinate $[Os(O)(ehba)_2]^-$ complexes have been isolated and characterized (*23*).

8. Osmium(VI)

Generally, two strong π bases are required to stabilize Os(VI), hence its chemistry is dominated by octahedral complexes with the *trans*-dioxo moiety. However, a complex containing three doubly deprotonated 2-aminobenzenethiol ligands, $[Os(abt)_3]$, has been characterized (*24*). In addition, a growing number of five-coordinate complexes that have either an oxo, nitrido, or imido ligand, together with four good σ and π-donor ligands, have been reported. Examples include [Os-O(Ctmen-2H)(Ctmen-H)]ClO_4 (*25*), similar complexes with four alkyl groups (*26*) or two diolato(2−) ligands (*27*), and $[Os(N)(R)_4]^-$ and [Os-(NMe)$(R)_4$] (*28*). Five-coordinate complexes with two oxo ligands are also known (*29, 30*) and the diamagnetic tetrahedral complexes [Os-$(O)_2(R)_2$] (R = 2,6-xylyl or 2,4,6-mesityl), $[Os(O)(NBu^t)(mes)_2]$, and $[Os(O)_2(SSO_3)_2]^{2-}$ have been characterized recently (*31–33*). The isolation and characterization of the air-stable trigonal-planar complex, $[Os(NAr)_3]$ (Ar = 2,3-$Pr^i{}_2C_6H_3$) (*22*), is even more remarkable given that most complexes are octahedral (*2, 5, 6*).

9. Osmium(VII)

The few isolated complexes of Os(VII) contain at least four ligands that are both strong π and σ donors, e.g., the tetrahedral $[OsO_4]^-$ ion, the octahedral $[OsOF_5]$ complex, and the pentagonal-bipyramidal $[OsF_7]$ (*2, 5, 6*).

10. Osmium(VIII)

Tetrahedral $[OsO_4]$ is quite stable and forms five-coordinate adducts with a number of ligands. It also adds two OH^- ligands to form *trans*-$[Os(O)_4(OH)_2]^{2-}$, or substitutes one or more oxo groups for other strong ligands, such as N^{3-} or NR^{2-} (*2, 5, 6*). The unusual six-coordinate Os(VIII) complexes, *trans*-$[Os(O)_2(OSiMe_3)_2(NP(Ph)_2CH_2P(Ph)_2N)]$ (*34*) and *trans*-$[Os(O)_2(pda)_2]$ (*24*), have been synthesized recently.

B. General Synthetic Methods

1. Overview

With the exception of the neutral halo complexes, which are prepared by direct reaction of the halogen with the metal, and a number of complexes generated by the direct reaction of the metal with a strong acid, coordination complexes of Os are prepared directly or indirectly from $[OsO_4]$ (*2*). The latter is a toxic, volatile solid, normally purchased in 1-g ampuls and opened within the reaction mixture contained in a fume hood to prevent escape of the toxic vapor (*3*) and loss of the reactant. It is also available as an aqueous solution of "osmic acid," which is suitable for some reactions.

General routes for the syntheses of complexes with oxo and N-donor ligands are presented. More extensive compilations of synthetic methods are given elsewhere (*2, 5, 6*).

2. *Complexes with Oxo Ligands*

Five-coordinate Os(VIII) complexes of the type $[Os(O)_4L]$ contain a weak Os—L bond and are prepared by the addition of the ligand to $[OsO_4]$ (*2*). The *trans*-dioxoOs(VI) complexes are prepared in a number of ways, including the reduction of $[OsO_4]$ by an alkene or alkyne to form an osmyl diester with the oxidized ligand (*2*), the reduction of $[OsO_4]$ in the presence of excess ligand (*2, 35, 36*), and the ligand substitution reactions of *trans*-$[Os^{VI}O_2X_4]^{2-}$ complexes (*2, 5, 6*). The *cis*- and *trans*-dioxoOs(VI) complexes are also prepared from the oxidations of Os(II) or Os(III) complexes or geometric isomerizations of cis complexes (*2, 37–40*). This chemistry is summarized in Scheme 1.

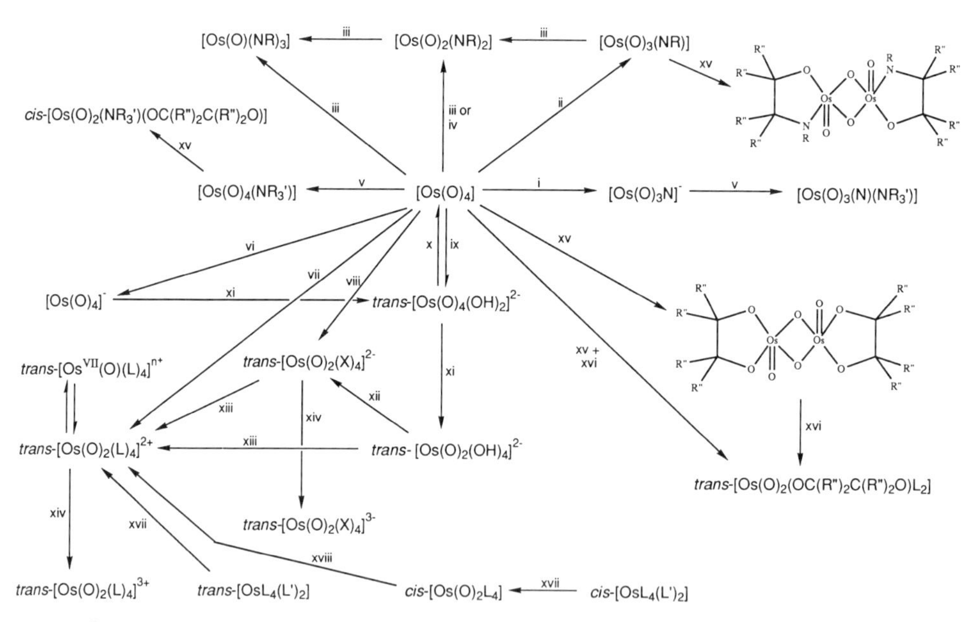

SCHEME 1. Preparation of oxo complexes. Reagents: (i) NH_3, (ii) RNH_2 (R = tertiary alkyl group), (iii) $RN{=}PPh_3$, (iv) $RNH{-}SiMe_3$, (v) NR_3', (vi) reductant, nonaqueous solvent (vii) excess L, (viii) excess HX, (ix) OH^-, (x) nonreducing acid, (xi) reductant, (xii) excess HX or X^-, (xiii) L, (xiv) $h\nu$ plus reductant or electrochemical reduction, (xv) $R_2''C{=}CR_2''$, (xvi) L = py, etc., (xvii) oxidation of Os(II), Os(III), or Os(IV) complexes, and (xviii) Δ

3. *Nitrido and Nitrido-Bridged Complexes*

a. Mononuclear Complexes. $[Os^{VIII}(O)_3(N)]^-$ is best prepared from the reaction of NH_3 with $[OsO_4]$ in strongly basic media (*2, 5, 6*). It is the only known Os(VIII) complex with the N^{3-} ligand, and reacts with HX to form the Os(VI) complexes $[Os(N)X_5]^{2-}$, $[Os(N)X_4]^-$, and *trans*-$[Os(N)X_4(OH_2)]^-$. These, in turn, undergo ligand exchange reactions to produce a large range of other Os(VI) nitrido complexes (*2, 5, 6*).

Alternative methods for the synthesis of mononuclear Os(VI) nitrido complexes are the chemical or electrochemical oxidation of Os(II) or Os(III) ammine complexes (*41–43*) and the oxidation of Os(IV) complexes by organic azides (*44*).

b. Dinuclear Complexes. The original syntheses of dinuclear Os(IV)$_2$ μ-nitrido complexes involved heating $(NH_4)_2[OsX_6]$ in the presence of HX or NH_3 to yield *trans,trans*-$[(H_2O)X_4OsNOsX_4(OH_2)]^{3-}$ and $[X(NH_3)_4OsNOs(NH_3)_4X]^{3+}$, respectively. Heating *trans,trans*-$[(H_2O)X_4OsNOsX_4(OH_2)]^{3-}$ in excess ammonia also produces *trans,trans*-$[X(NH_3)_4OsNOs(NH_3)_4X]^{3+}$ complexes. The X^- ligands can be substituted to prepare a range of complexes (*45*), whereas the octaammine complex ($X = Cl^-$) and $[Os_2N(NH_3)_7Cl_3]Cl_2$ are also obtained from heating $Na_2[OsCl_6]$ in aqueous NH_3 (*46*). Recently, the reaction of $NH_2NH_2 \cdot H_2O$ with solid $(NH_4)_2[OsCl_6]$ was found to yield the decaammine complex, $[(NH_3)_5OsNOs(NH_3)_5]^{5+}$ (*47*). This complex undergoes trans substitution reactions, resulting in a high-yielding and simple entry into the aforementioned series (*47*). The μ-N^{3-} chemistry is summarized in Scheme 2.

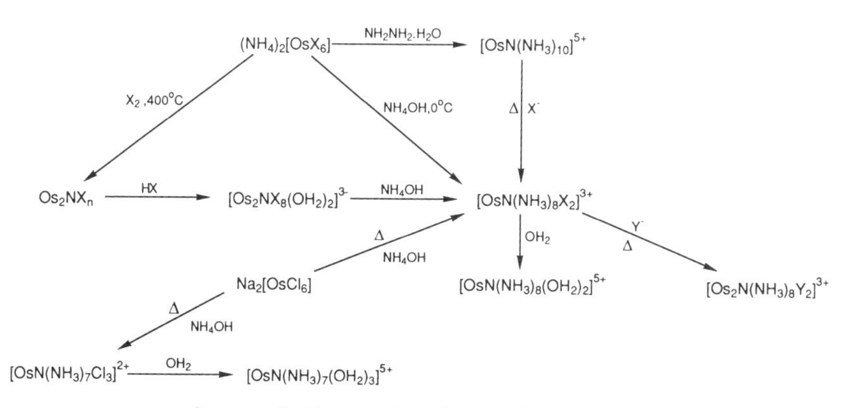

SCHEME 2. Preparation of μ-nitrido complexes.

4. Osmium Ammine Complexes

a. Pentaammineosmium(III) and -osmium(II) Complexes. Though original methods for preparing these complexes involved heating $[OsX_6]^{2-}$ (X = Cl or Br) with excess NH_3 (*48–53*), all modern synthetic methods use $[Os(NH_3)_5(N_2)]^{2+}$ as an intermediate. This complex is generated from the reaction of $[OsCl_6]^{2-}$ with excess $NH_2NH_2{\cdot}H_2O$, and because of its synthetic importance, the reaction conditions have been improved to give overall yields of $\geq$90% (*47, 54–59*). The lability of the N_2 ligand in the Os(III) oxidation state, following oxidation of the Os(II) precursor, is utilized in preparative procedures that rely on this intermediate (*56, 58, 60, 61*). However, such *in situ* reactions are limited in their applications, and prior to the early 1980s this prevented the development of $[Os(NH_3)_5L]^{n+}$ chemistry that paralleled the extensive $[Ru(NH_3)_5L]^{n+}$ chemistry (*62, 63*). More recently, pentaammineosmium chemistry has been opened up by the synthesis of $[Os(NH_3)_5(OSO_2CF_3)](CF_3SO_3)_2$. Unlike triflato complexes of the other inert metal ions, which are normally prepared from reactions of the neat solvent with chloro complexes (*64, 65*), the Os(III) complex is prepared in quantitative yield by the oxidation of $[Os(NH_3)_5(N_2)]Cl_2$ in neat CF_3SO_3H (*59, 66, 67*). The triflato complex has the synthetic advantages of being readily substituted by many ligands and being soluble in most polar organic solvents. However, the use of this complex as a synthetic intermediate is still limited to ligands that are better σ donors to Os(III) than $CF_3SO_3^-$. Moreover, the ligands must not be sufficiently basic so as to bring about competition between substitution and base-catalyzed disproportionation of Os(III) amine complexes to Os(II) and Os(IV). The latter leads to multiple substitution and is a particular problem with the preparation of N-heterocyclic complexes (*68*). Unlike pentaammineruthenium chemistry, wherein these problems are overcome by the use of labile $[Ru(NH_3)_5(OH_2)]^{2+}$ or other Ru(II) solvent complexes (*62, 63*), the analogous complexes in Os chemistry are not of general synthetic utility. This is because of the propensity of $[Os(NH_3)_5(OH_2)]^{2+}$ to reduce water (*68, 69*) and of the other solvent complexes to be either substitutionally inert or to undergo chemical reactions to produce inert complexes. Nevertheless, the $[Os(NH_3)_5(OH_2)]^{2+}$ ion has been used as a synthetic intermediate on a number of occasions (*70–72*). The claim, however, that the Os(II) aqua complex is in equilibrium with η^2-heterocyclic complexes (*72*) has been shown to be incorrect (*73*).

In order to overcome the difficulties of coordinating weak σ donors, several methods have been developed by Harman and Taube for the preparations of labile Os(II) synthetic intermediates (*74, 75*). Together with the triflato complex, such methods have resulted in an explosion of our knowledge of pentaammineosmium chemistry. These synthetic methods involve the *in situ* reduction of $[Os(NH_3)_5(OSO_2CF_3)]^{2+}$ in dma/dme (see the list of abbreviations at the end of this article) to form a labile solvent complex (*75*), or the use of $[Os^{II}(NH_3)_5(1,2,3,4\text{-tetramethylbenzene})]^{2+}$ as an intermediate (*74*). An array of dinuclear decaamminediosmium and heterodinuclear complexes have been prepared using such intermediates (see Section II,D). Current synthetic methods for pentaammine and decaammine complexes are summarized in Scheme 3.

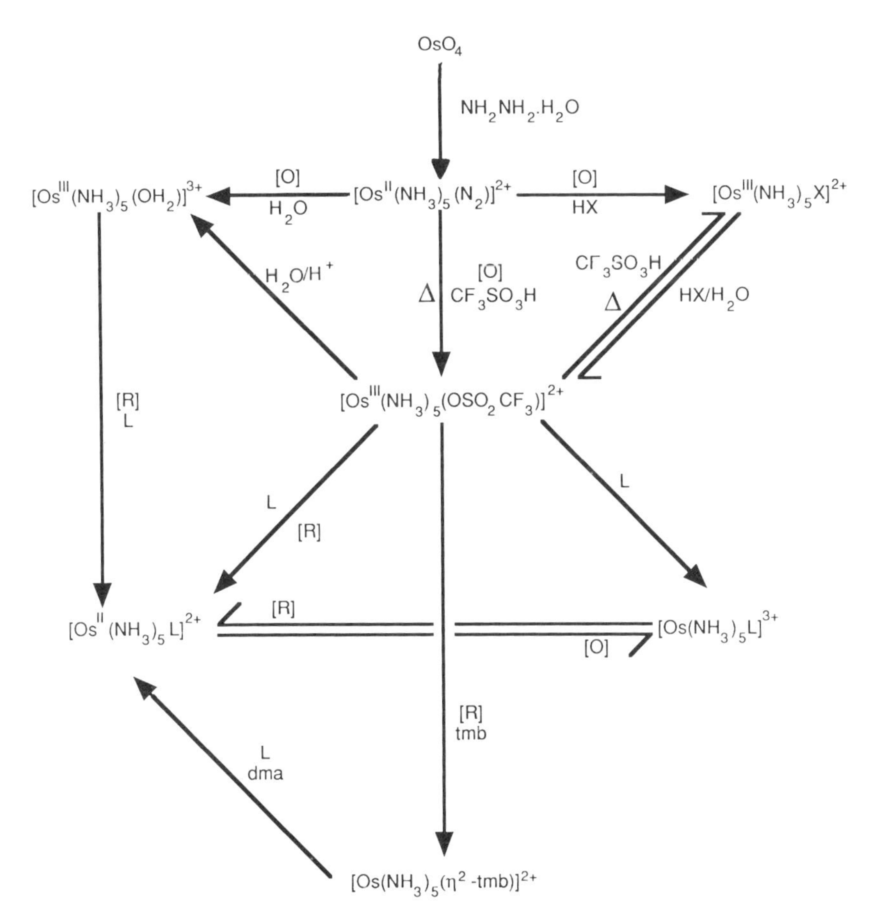

SCHEME 3. Preparation of pentaammine and decaammine complexes.

b. Tetraammineosmium(II), -osmium(III), and -osmium(IV). The conventional routes to the synthesis of *cis*-tetraammine complexes involve diazotization of $[Os(NH_3)_5(N_2)]^{2+}$ to form *cis*-$[Os(NH_3)_4(N_2)_2]^{2+}$ (*76, 77*). One of the N_2 ligands is readily replaced with other ligands by gentle warming (*77–80*). Heating in aqueous HX solutions result in the loss of both N_2 ligands with concomitant oxidation to form *cis*-$[Os(NH_3)_4X_2]^+$ (*61*). *cis*-$[Os(NH_3)_4(OSO_2CF_3)_2](CF_3SO_3)$ is prepared most conveniently from oxidation of *cis*-$[Os(NH_3)_4(N_2)_2]Cl_2$ in neat CF_3SO_3H (*81*) and undergoes substitution chemistry similar to that of the pentaammine analog. Mixed-ligand complexes are prepared by treating *cis*-$[Os(NH_3)_4(N_2)_2]^+$ with an excess of the appropriate ligand to give *cis*-$[Os(NH_3)_4(N_2)L]^{2+}$ (L = PR_3, N heterocycle, etc.), which are used to generate the corresponding Os(III) monotriflato complex. Although such chemistry is in its infancy, it is likely to lead to the synthesis of a large number of *cis*-tetraammine complexes. An alternative method is to use complexes such as *cis*-$[Os^{III}(NH_3)_4(pz)Cl]^{2+}$, generated as previously described (*77, 79*), to synthesize triflato intermediates. *cis*-$[Os(NH_3)_4(heterocycle)_2]^{3+/2+}$ also result from reactions of $[Os(NH_3)_5(OSO_2CF_3)]^{2+}$ with excess ligand (*68*). In addition, dinuclear nonaammine- and octaamminediosmium complexes are prepared using triflato or dinitrogen intermediates in the manner just described. The chemistry of the *cis*-tetraammine complexes is summarized in Scheme 4.

Apart from the nitrosyl complexes, which are discussed in detail elsewhere (*2, 82*), the *trans*-tetraammineOs(III) complexes and related complexes with chelating amine ligands are prepared by geometric isomerization of the cis complexes. This requires refluxing for several days in the presence of HX (*61*). An alternative route is the reduction of the *trans*-$[Os^{VI}(NH_3)_4(O)_2]^{2+}$ precursor in HX (*83*). The substitution chemistry of the triflato group is used to synthesize a variety of complexes starting from the halo complexes (*81, 84*) and requires further exploration. As is the case for the *trans*-$[MA_4Cl_2]^+$ complexes of Rh(III), Ir(III), and Ru(III) (*65, 81*), only one of the chloro ligands is replaced to produce *trans*-$[Os(NH_3)_4Cl(OSO_2CF_3)]^+$, even after prolonged heating in neat CF_3SO_3H (*81*). This is synthetically useful because it enables the sequential displacement of the chloro ligands. Other synthetic routes into the *trans*-tetraammine series include the activation of the *trans*-NH_3 group toward substitution by the formation of either the Os(II)–ylide complex of 2,6-lutidine (and other N heterocycles in which the ligand is bound to Os via the para carbon) (*85*), or the formation of an Os(III)—SO_3^{2-} bond (*86*). *cis*- and *trans*-$[Os^{IV}(NH_3)_4X_2]^{2+}$ are prepared from the Fe(III) oxidation of Os(III) complexes (*83*). The synthesis of the *trans*-tetraammine complexes is summarized in Scheme 5.

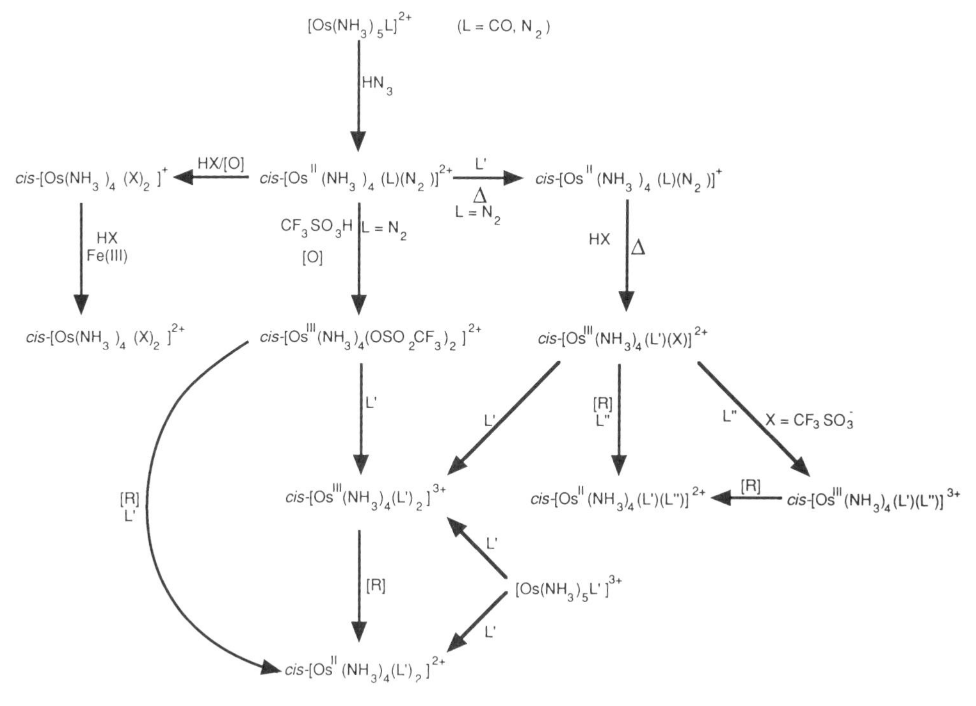

SCHEME 4. Preparation of *cis*-tetraammine complexes.

c. Triammineosmium(II), -osmium(III), and -osmium(IV). The mer complexes are prepared by the diazotization of the appropriate *trans*-tetraammine complex to form *mer*-$[Os^{II}(NH_3)_3(N_2)X_2]$, which is oxidized in HX to *mer*-$[Os(NH_3)_3X_3]$ (*87*). These complexes are readily oxidized further to their Os(IV) analogs, *mer*-$[Os(NH_3)_3X_3]^+$ (*83*).

The fac isomers are probably among the products formed from the diazotization of the *cis*-tetraammine complexes, but have yet to be obtained pure (*87*). When *cis*-$[Os(NH_3)_4(OSO_2CF_3)_2]^+$ is reduced in the presence of an arene ligand, an ammine and two triflato ligands are displaced to form $[Os^{II}(NH_3)_3(\eta^6\text{-arene})]^{2+}$. The benzene complex is also a product of the prolonged heating of $[(NH_3)_5Os(\mu\text{-}\eta^2\text{:}\eta^2\text{-benzene})Os(NH_{35}]^{4+}$ along with $[(NH_3)_5Os(\mu\text{-}\eta^2\text{:}\eta^6\text{-benzene})Os(NH_3)_3]^{4+}$, which has a triammine unit (*88*). The naphthalene ligand is readily substituted for other ligands to form *fac*-$[Os^{II}(NH_3)_3L_3]^{2+}$ complexes (*81, 89*), which provides a convenient synthetic route for the preparation of the fac isomers. Imidazole activates ligands to substitution and the reaction of $[Os(NH_3)_5(OSO_2CF_3)]^{2+}$ with excess ligand results in *fac*-$[Os(NH_3)_3(im)_3]^{2+}$ (*90*). The chemistry of the triammine

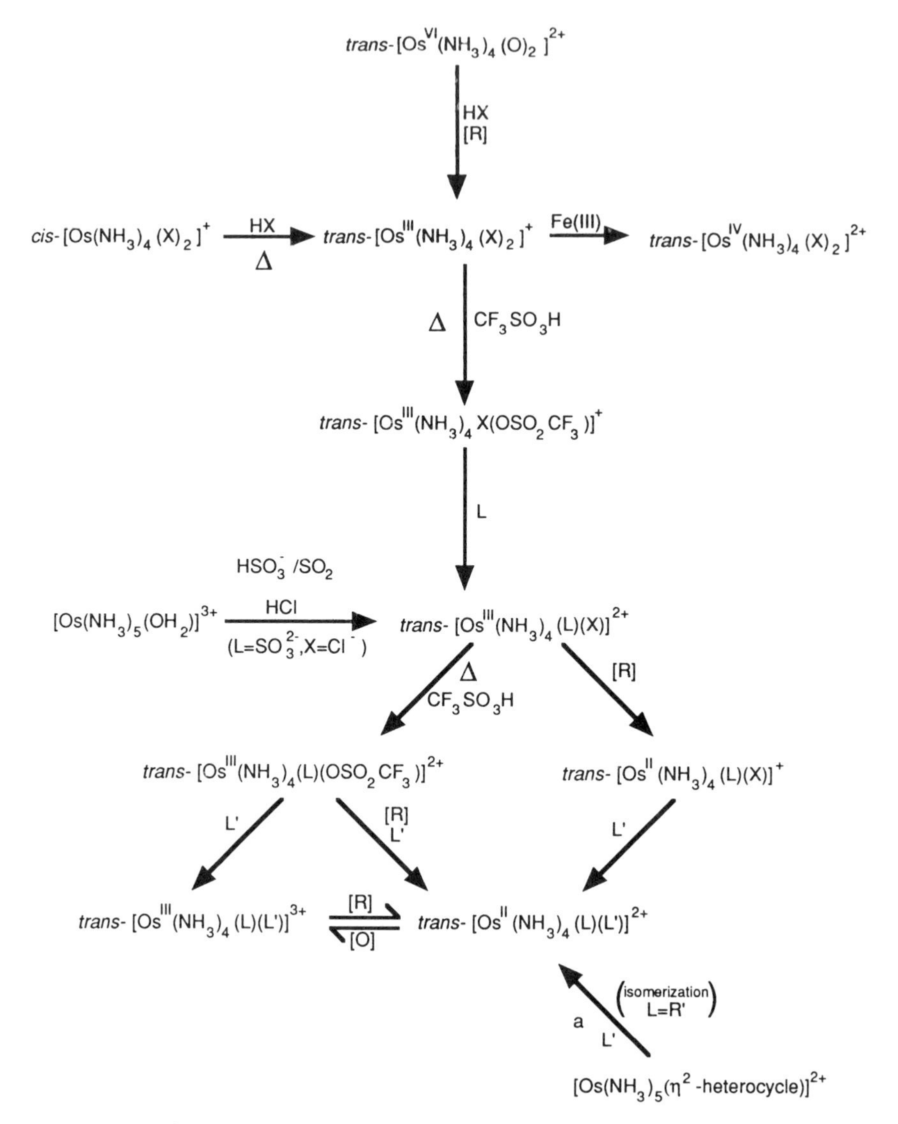

SCHEME 5. Preparation of *trans*-tetraammine complexes.

complexes, apart from the nitrosyl complexes that are discussed in detail elsewhere (*2, 82*), is summarized in Scheme 6.

5. *N-Macrocyclic Complexes*

At the time of Griffith's review (*2*), no macrocyclic complexes of Os had been reported except for the porphyrin and phthalocyanine com-

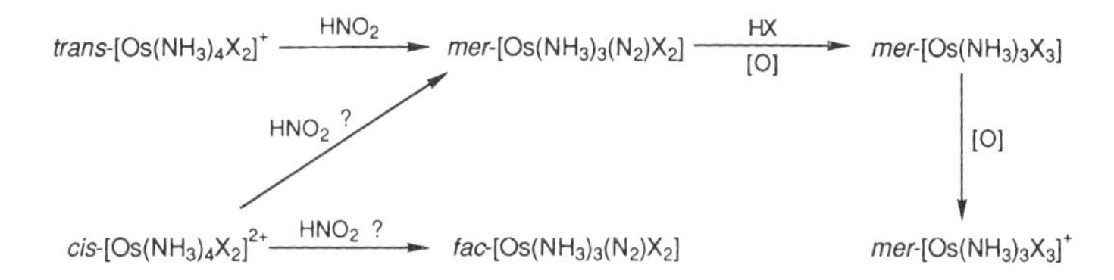

SCHEME 6. Preparation of triammine complexes.

plexes, which are reviewed in Section II,C,4,d and elsewhere (*2, 91, 92*). In recent times, methods have been devised for the synthesis of the macrocyclic complexes of Os. This chemistry has been reviewed (*39, 93*) and is summarized in Scheme 7.

6. *Complexes With bpy, trpy, phen, and Related Ligands*

General methods for the preparation of N-heterocyclic complexes have been well summarized previously (*2*). One of the new synthetic procedures involves the use of triflato complexes, e.g., *cis*-

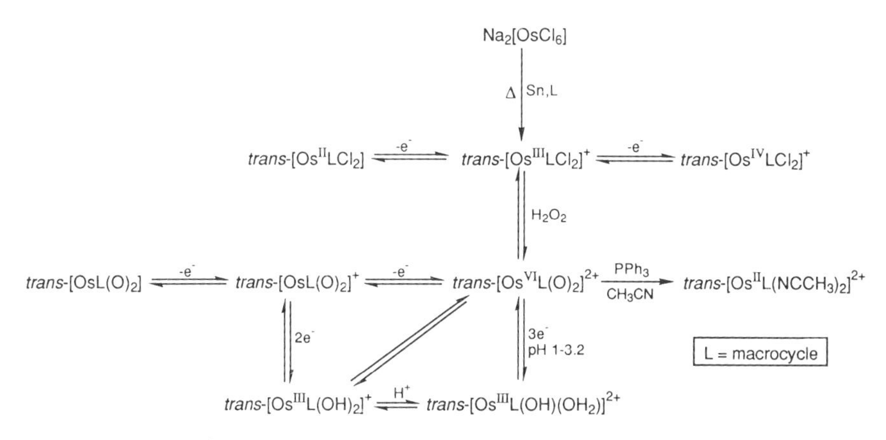

SCHEME 7. Preparation of macrocyclic complexes.

$[Os(bpy)_2(OSO_2CF_3)_2]^+$ and $[Os(trpy)(bpy)(OSO_2CF_3)]^{2+/+}$, as intermediates (*94–96*).[1] Although such complexes provide convenient entries into the aqua/hydroxo/oxo chemistry and they are also useful intermediates for the preparation of other complexes (*42*), their synthetic utility has only just begun to be explored. Other recently developed synthetic routes utilize intermediates that contain bidentate carbonato ligands, for synthesizing *cis*-$[Os(bpy)_2L_2]^{n+}$ and *cis*-$[Os(bpy)(dppe)L_2]^{n+}$ complexes (*97, 98*). This new chemistry is summarized in Scheme 8.

7. *Osmium(III) Complexes*

Most Os(III) complexes are prepared by either the direct reduction of $[OsO_4]$ or $[OsX_6]^{2-}$ ($X = Cl^-$ or Br^-) with an excess of the ligand (*2, 5, 6*). The latter is made from the reduction of $[OsO_4]$ in the presence of HCl or HBr (*99*). Other reactions include those of $OsCl_3$ with the ligands and the reduction of Os(VI) or Os(IV) complexes (*2, 5, 6*).

C. Mononuclear Complexes

Because of the sheer volume of work published over the last 5 years, only certain specialist topics will be discussed in any depth (Sections III–V). Whereas this survey covers many areas in a cursory fashion, it amply illustrates the rapid growth of knowledge in Os chemistry. It also serves to indicate where future advances are likely to occur, especially in the "organometallic" chemistry of $[Os(NH_3)_nL_m]^{x+}$ complexes, which often mimics or surpasses the extensive chemistry normally associated with phosphine ligands. Examples wherein novel organometallic chemistry has occurred are the η^2-ketone and η^2-arene complexes of pentaammineosmium.

1. *Boron Ligands*

Osmaboranes and Osmacarboranes. The first osmaboranes were reported in 1983 from the reactions of $[Os(CO)(Cl)(H)(PPh_3)_3]$ with *arachno*-$[B_3H_8]^-$ or *nido*-$[B_5H_8]^-$, to yield *arachno*-$[(HOsB_3H_8)(CO)(PPh_3)_2]$ (**I**) and *nido*-$[(OsB_5H_9)(CO)(PPh_3)_2]$ (**II**), respectively. Both complexes have two hydride and one or more boron atoms bound to

[1] Although some aspects of the preparation and reactivity of $[Os(trpy)(bpy)(OSO_2CF_3)]^{2+}$ appeared in the literature in 1984 (*96*), this paper was actually submitted after our paper dealing with $[Os(trpy)(bpy)(OSO_2CF_3)]^{2+/+}$ (*95*). The preparation of these complexes was communicated to the authors of Ref. *96* in 1982.

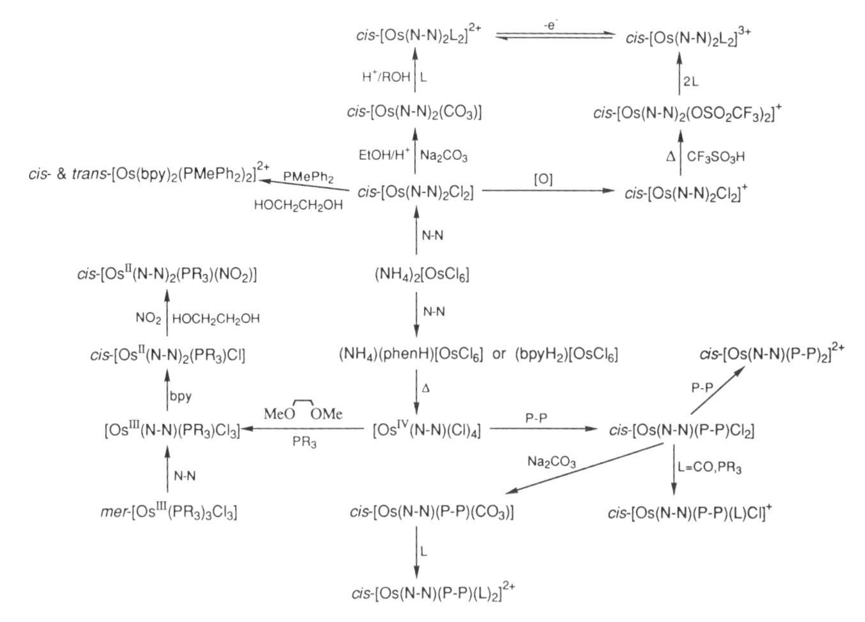

SCHEME 8. New methods for the synthesis of complexes with bidentate N-heterocyclic ligands.

Os (*100*). Treatment of the nido complex with NaH yields Na-$[(Ph_3P)_3(CO)Os(B_5H_8)]$ (**III**), which reacts with $[PtCl_2(PMe_2Ph)_2]$ to give $[(Ph_3P)_2(CO)Os(PhMe_2P)(Cl)(H)Pt(B_5H_7)]$ (**IV**). X-Ray crystallography shows that the borane and hydride ligands act as bridges (*101*). Similarly, *mer*-$[OsCl_3(PMe_2Ph)_3]$ reacts with $[NBu_4][B_9H_{14}]$ (*102*) or *closo*-$[B_{10}H_{10}]^{2-}$ in EtOH (*103*) to yield [6,6,6-$(PMe_2Ph)_3$-*nido*-6-OsB_9H_{13}] (**V**) and $[(PMe_2Ph)_2OsB_{10}H_8(OEt)_2]$ (**VI**), respectively. The structure of the former has been determined by X-ray crystallography (*102*). $[OsCl_2(PPh_3)_3]$ and *closo*-$[B_{10}H_{10}]^{2-}$ yield $[(PPh_3)(Ph_2PC_6H_4)OsB_{10}H_7(OEt)_2]$ (**VII**), for which the X-ray structure shows that one of the PPh_3 ligands has orthometalated to a boron atom (*103*). Similarly, $[(\eta^6\text{-}C_6Me_6)OsCl_2]_2$ reacts with *closo*-$[B_{10}H_{10}]^{2-}$ to form [1-$(\eta^6\text{-}C_6Me_6)OsB_{10}H_{10}$] (**VIII**), with Tl[*arachno*-B_3H_8] to form [2-$(\eta^6\text{-}C_6Me_6)ClOsB_3H_8$] (**IX**), or with *arachno*-$[B_9H_{14}]^-$ to form [6-$(\eta^6\text{-}C_6Me_6)OsB_9H_{13}$] (**X**) (*104*).

closo-[1-$Os(CO)_3$-2,3-$\{(CH_3)_3Si\}_2$-2,3-$C_2B_4H_4$] (**XI**) is made from $[Os_3(CO)_{12}]$ and *closo*-$[Sn\{(CH_3)_3Si\}_2C_2B_4H_4]$ or *nido*-$[\{(CH_3)_3Si\}_2C_2B_4H_6]$ and has been characterized by ^1H, ^{11}B, ^{13}C, and ^{29}Si NMR and mass spectroscopies (*105*). More recently, *nido*-[9,9,9-$(CO)(PPh_3)_2$-9,6-

$OsCB_8H_{10}$-5-PPh_3] (**XII**) was prepared from [Os(Cl)(CO)(H)$(PPh_3)_3$] and *arachno*-$[CB_8H_{13}]^-$, and its X-ray structure was determined (*106*). The structures of the osmaboranes and osmacarboranes are given in Fig. 1.

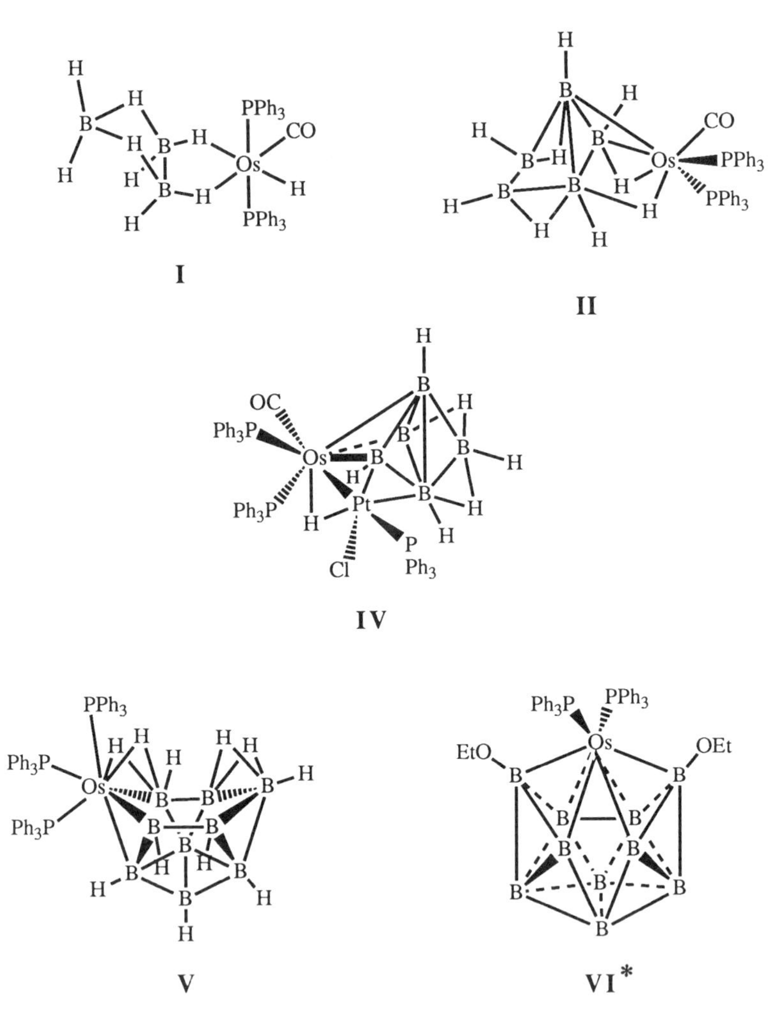

FIG. 1. Structures of the osmaboranes and osmacarboranes. Structure **III** (not shown) is the same as **II,** except a hydride is removed; only borons of the borohydrides are shown in structures **VI–X** and **XII** (asterisks).

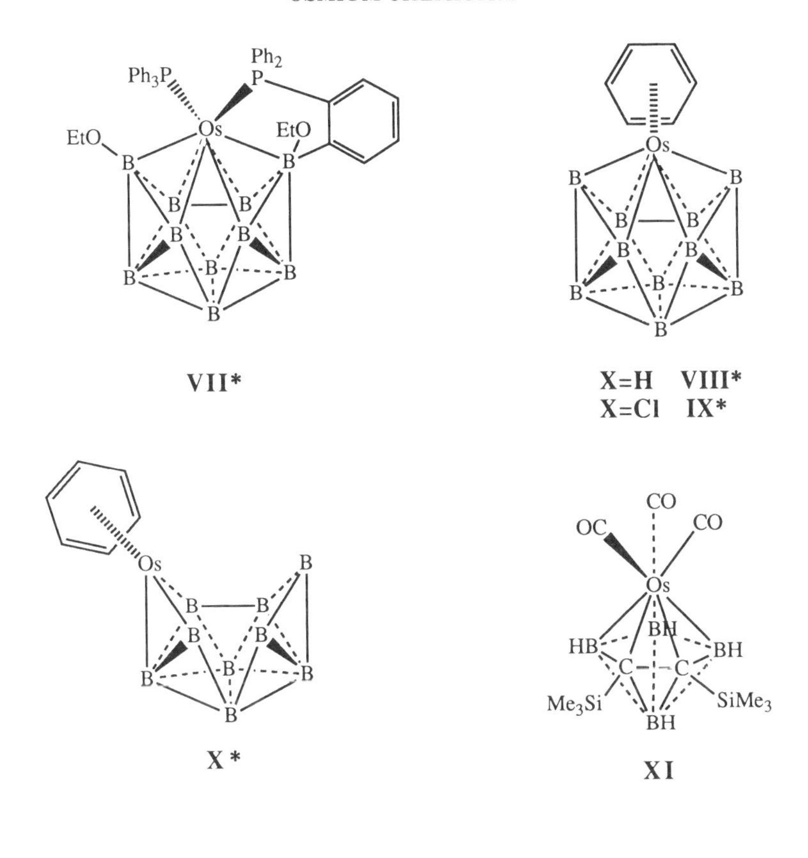

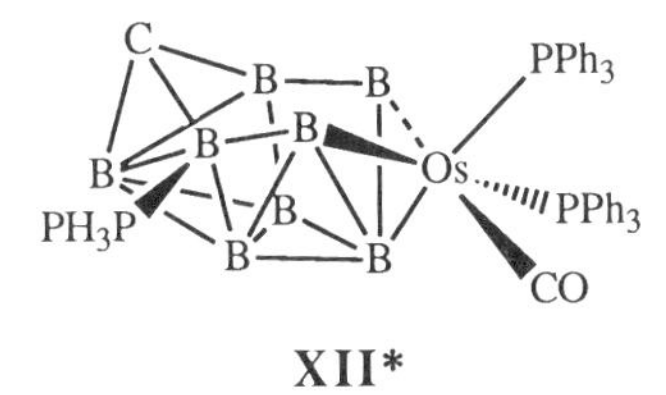

XII*

FIG. 1. (*Continued*)

2. *Carbon Ligands*

a. Cyano Complexes. $[Os^{I}(CN)_5Cl]^{5-}$ and *trans*-$[Os^{I}(CN)_4Cl_2]^{5-}$ are produced by X-radiation of $K_4[Os(CN)_6]$ in a KCl matrix. These complexes are stable at room temperature and have been characterized by the hyperfine and superhyperfine coupling observed with EPR spectroscopy (*16*).

trans-$[Os(CN)_4(NCCH_3)_2]^{2-}$ is prepared by irradiating *trans*-$[Os^{VI}(O)_2(CN)_4]^{2-}$ in an acetonitrile solution containing an alkene or other substrate (*107*) and $[Os(CN)_2\{CNCH(Ph)CH_2OH\}_4]$ is prepared from the reaction of $[Os(CN)_2(CNH)_4]$ with an epoxide (*108*). $[Os(N)(CN)_5]^{2-}$, *trans*-$[Os(N)(CN)_4(OH)]^-$, *trans*-$[Os(NH)(CN)_4(OCOCF_3)]$, and *trans*-$[Os\{N(COCF_3)\}(CN)_4(OCOCF_3)]^-$ are discussed in Sections II,C,4,q and II,C,6,o. The well-known complexes *cis*-$[Os(LL)_2(CN)_2]$ (LL = bpy or phen) have been studied by X-ray photoelectron spectroscopy (XPS) and cyclic voltammetry (*109*). Lattice dynamics of $K_4[Os(CN)_6] \cdot 3H_2O$ have been evaluated by NMR spectroscopy at temperatures near its phase transition and the dynamics have been compared with the isostructural Fe(II) and Ru(II) complexes (*110*). The nonmagnetic $Th[Os(CN)_6] \cdot 5H_2O$ has been prepared and its structure characterized by X-ray powder diffraction and thermogravimetric analysis (*110a*). X-ray crystal structures of the isomorphous series $Na_4[M(CN)_6] \cdot 10\ H_2O$ (M = Fe, Ru, Os) have also been reported and the IR spectra of the three complexes compared (*110b*). The heterogeneous electron transfer rate of the $[Os(CN)_6]^{3-/4-}$ couple has been measured at a carbon fiber electrode using very fast cyclic voltammetry (*111*). An outer-sphere charge-transfer complex, $\{[Pt(NH_3)_5Cl]^{3+} \cdot [Os(CN)_6]^{4-}\}^-$, has been prepared and its spectroscopy studied (*111a*).

The *trans*-dioxoOs(VI) complexes that contain two trans or cis cyano ligands are obtained from the reactions of *trans*-$[Os(O)_2(OH)_4]^{2-}$ and CN^- in a 1:2 molar ratio in the presence of a carboxylic acid. If oxalic acid is used, the product is *trans-O,O-cis*-$[Os(O)_2(CN)_2(ox)]^{2-}$; however, if acetic acid is used, the produce is *trans,trans,trans*-$[Os(O)_2(CN)_2(OH)_2]^{2-}$. A brief report on the latter complex appeared some time ago (*2, 112*), but its substitution reactions were not explored. The OH^- ligands are substituted by other ligands to produce *trans,-trans,trans*-$[Os(O)_2(CN)_2X_2]^{2-}$ (X^- = NCO^-, NCS^-, MeO^-, and $SeCN^-$), *trans,trans,trans*-$[Os(O)_2(CN)_2(py)_2]$, and *trans-O,O*-$[Os(O)_2(CN)_2(bpy)]$. All complexes were characterized by infrared (IR), Raman, and UV/Vis spectroscopies, the latter being strongly vibronically coupled (*113*). *trans*-$[Os(O)_2(CN)_4]^{2-}$ undergoes photochemical or electrochemical reduction to the Os(V) analog (*107, 114*), and the ^{13}C and ^{18}O derivatives of the Os(VI) complex have been prepared to study the vibrational fine structure that occurs within the charge-transfer transitions in the UV/Vis spectrum (*115*).

b. Carbonyl Complexes. A very large number of carbonyl complexes of Os have been prepared and characterized, most of which have been reviewed elsewhere (*6, 8–12, 116*). The only CO complexes discussed here are those with ligands that are normally associated with classical coordination chemistry, e.g., $[Os(NH_3)_5(CO)]^{2+}$. Most carbonyl complexes in oxidation states of Os(II) or higher contain halide and/or hydride ligands, and complexes with halides have been reviewed recently (*116*). The unusual Os(IV) CO complex, $[OsCl_5(CO)]^-$, is prepared from Cl_2 oxidation of *trans*-$[OsCl_4(CO)_2]^-$. The former is reduced reversibly to the Os(III) complex and both have been studied by near-infrared (NIR), IR, Raman, and UV/Vis spectroscopies and electrochemistry (*117*).

The synthesis of carbonyl complexes is achieved in several ways. Although, the direct reaction of CO with an Os complex is a general route, preparation is also via oxidative dehydrogenation of coordinated methanol (Section V,E,2,b), dehydration of coordinated formate (Section V,E,4,g), and elimination of RH or $(CH_3)_2NH$ from coordinated aldehydes (RCH = O) or dmf, respectively (Section V,E,4,h). The direct reactions of CO with $[Os(NH_3)_5L]^{2+}$ complexes have not as yet yielded the CO complex, but all of the above reactions of coordinated ligands can be used in the synthesis of $[Os(NH_3)_5(CO)]^{2+}$ (*55, 118–120*). $[Os(NH_3)_5(CO)]^{2+}$ is oxidized to the Os(III) complex by $[IrCl_6]^{2-}$, isolated as $[Os^{III}(NH_3)_5CO][Ir^{III}Cl_6]$ (*58*). The Os(III) complex is quite acidic, with the first pK_a value of the ammine ligands being ~2.5 (*87*). It undergoes base-catalyzed disproportionation to give $[Os^{II}(NH_3)_5CO]^{2+}$ and Os(IV) and/or Os(V) nitride carbonyl complexes. The latter are intermediates in the coupling reaction to form *cis,cis*-$[(CO)(NH_3)_4Os(N_2)Os(NH_3)_4(CO)]^{4+}$ (Section V,E,1,c) (*58*). $[Os(NH_3)_5(CO)]^{2+}$ is diazotized to form *cis*-$[Os(NH_3)_4(N_2)(CO)]^{2+}$ (*55, 87*).

Reactions of coordinated ligands are also standard preparative methods for the synthesis of Os hydride/carbonyl/phosphine or arsine complexes, which in turn are the starting materials for a large range of organometallic or coordination complexes (*116*). The general synthetic methods are summarized in Scheme 9 (*18, 19, 106, 121–137*). The five-coordinate $[Os(H)(X)(CO)(PR_3)_2]$ complexes are coordinatively unsaturated and very effective hydrogenation catalysts (*125, 135, 137*). $[Os(H)_2(CO)(PMe_2Ph)_3]$ is also prepared by the reduction of CO_2 by the hydride ligands in $[(cod)Rh(H)_3Os(PMe_2Ph)_3]$ (*138*).

Oxidation of 2-methoxyethanol or 1,2-ethanediol by $[OsO_4]$, in the presence of a porphyrin, is used to prepare $[Os^{II}(P)(CO)(X)]$ complexes [P is a porphyrinato(2−) ligand] (*139*). Recently, more convenient preparations of porphyrinato complexes were developed using the reactions of $[Os_3(CO)_{12}]$ with the appropriate porphyrin (Section II,C,4,d) (*38, 40, 140, 141*). Electrochemical studies show that both the Os(III) and Os(III)/porphyrin–cation–radical complexes are moderately stable

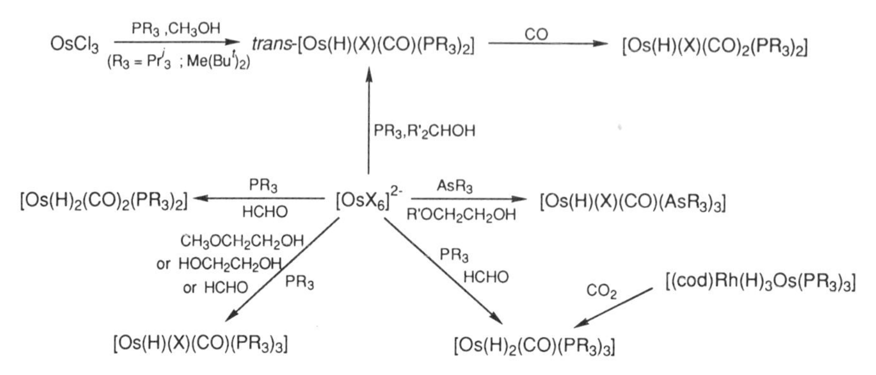

SCHEME 9. Preparation of osmium carbonyl/hydride/phosphine or arsine synthetic intermediates via the reactions of coordinated ligands.

with respect to CO dissociation (*142*). Despite this, the CO ligand is sufficiently labile on Os(III) for oxidation of the CO complexes to be useful routes in the synthesis of Os porphyrin complexes of oxidation state III and higher. Recently, the use of [Os(oep)(PBu_3)(CO)] (*143*), [Os(tpp)(PBu_3)(CO)] (*143*), [Os(*p*-Xtpp)(CO)(EtOH)] (X = Cl, H, OMe, or Me) (*40*), and [Os(tmp)(CO)(EtOH)] (*40*) in such reactions has been reported. [Os(mix)(CO)(EtOH)] has also been prepared and reconstituted into Ru-modified and native sperm whale myoglobins to form carbonyl osmoglobin ([Os^{II}(CO)][Mb]) (*144*).

Complexes prepared by the direct reaction of CO with a complex include *trans*-[$OsCl_2$(CO)(PMe_2Ph)$_3$] from *trans*-[$OsCl_2$(PMe_2Ph)$_3$] (*145*), *trans-P,P-cis-C,C*-[Os(H)(Cl)(CO)$_2$($PMeBu^t_2$)$_2$] from [Os(H)-(Cl)(CO)($PMeBu^t_2$)$_2$] (*124*), and *cis*-[OsL(CO)(PPh_3)] from *trans*-[OsL(PPh_3)$_2$] (L = hba-b or chba-dcb) (*146*). The latter is a remarkable reaction because it involves the direct coordination of CO to an Os(IV) center to form a stable complex. *trans*-[$OsCl_2$(CO)(PMe_2Ph)$_3$] undergoes isomerization to form *cis*-[$OsCl_2$(CO)(PMe_2Ph)$_3$] (*145*).

Two series of complexes, [Os(bpy)$_2$(CO)(R)] (R = H^- or alkyl) and [Os(bpy)(phen)(CO)X], have been characterized and their excited states studied (*147, 148*). The former are also used in the autocatalytic reduction of CO_2 (*149*).

c. Complexes with Thiocarbonyl, Selenocarbonyl, and Tellurocarbonyl Ligands. The syntheses of [$OsCl_2$(CS)(PPh_3)$_3$], [OsH(H_2O)(C-S)(PPh_3)$_3$]$^+$, [OsH(CO)(CS)(PPh_3)$_3$]$^+$, and [Os(CO)(CS)(PPh_3)$_3$] have been described (*17*); [$OsCl_2$(CS)(PPh_3)$_3$] reacts with bsd to give [Os-Cl_2(CS)(PPh_3)$_2$(bsd)] (Section II,C,7,h) (*150*). *trans-P,P-cis-Cl,-*

Cl-$[OsCl_2(CO)(CE)(PPh_3)_2]$ (E = O, S, Se, or Te) have been prepared from reactions of HE^- or H_2E with $[OsCl_2(CCl_2)(CO)(PPh_3)_2]$. X-Ray crystallography has shown that the trans influences of the chalcocarbonyl ligands increase in the order CO < CS ≤ CSe < CTe. Six-coordinate $[OsCl_2(CCl_2)(CS)(PPh_3)_2]$, $[OsCl_2(CS)_2(PPh_3)_2]$, $[OsCl_2(CS)(CSe)(PPh_3)_2]$, and $[OsCl(NCCH_3)(CO)(CTe)(PPh_3)_2]^+$ (*151*) and five-coordinate *trans*-$[OsX(NO)(CS)(PPh_3)_2]$ (X^- = Cl^- or I^-), *trans*-$[Os(NO)(CS)(PPh_3)_2L]^+$ (L = CO or PPh_3), *trans-P,P-cis-I,I*-$[Os(I)_2(NO)(CS)(PPh_3)_2]I_3$, and *trans*-$[Os(HSe)(NO)(CS)(PPh_3)_2]$ have also been prepared (*152*). $[Os(CS)(Cl)(4\text{-}MeC_6H_4)(PPh_3)_2]$ reacts with $SNNMe_2$ to give $[Os(CS)(Cl)(4\text{-}MeC_6H_4)(S\text{-}SNNMe_2)(PPh_3)_2]$ (*153*).

d. Carbon Dioxide Complexes. No stable mononuclear carbon dioxide complexes have been reported, but a CO_2 complex is likely to be an intermediate in the formation of $[Os(H)_2(CO)(PMe_2Ph)_3]$ from $[(cod)Rh(H)_3Os(PMe_2Ph)_3]$ and CO_2 (Section II,C,2,b). When $[Os(NH_3)_5(OSO_2CF_3)]^{2+}$ is reduced in dme in the presence of CO_2, $[Os(NH_3)_5(CO)]^{2+}$ is recovered in ~ 40% yield. Presumably, this arises from the formation of an $[Os(NH_3)_5(CO_2)]^{2+}$ intermediate, although such an intermediate has not been characterized (*120*).

e. Carbon Disulfide and Related Complexes. $[OsX(NO)(CS_2)(PPh_3)_2]$ ($X = Cl^-$ or I^-), which have an η^2-(*C,S*) binding mode for the CS_2 ligand, are methylated to form $[OsX(NO)\{\eta^2\text{-}(C,S)\text{-}S{=}CSMe\}(PPh_3)_2]$. The CS_2 complex ($X = Cl^-$) is also converted into $[OsCl(NO)(CSSe)(PPh_3)_2]$, which exists as both the η^2-*C,S* and η^2-*C,Se* linkage isomers (Section V,D,2), and evidence has been obtained for the existence of a transient $[OsCl(NO)\{\eta^2\text{-}(C,S)\text{-}CSTe\}(PPh_3)_2]$ intermediate. The $\eta^2\text{-}(C,S)\text{-}S{=}CSMe^-$ ligand undergoes a linkage isomerization reaction on the addition of donor ligands to form $\eta^1\text{-}(C)\text{-}C(S)SMe^-$ complexes, *trans*-$[OsI(X)\{\eta^1\text{-}C(S)SMe\}(NO)(PPh_3)_2]$, ($X = Cl^-$ or $SCNMe_2^-$), $[OsI(CNBu^t)\{\eta^1\text{-}C(S)Me\}(NO)(PPh_3)_2]^+$, and *trans-P,P-cis-H,H*-$[Os(H)_2(NO)\{\eta^1\text{-}C(S)SMe\}(PPh_3)_2]$. These complexes were characterized by NMR (^1H and ^{31}P) and IR spectroscopies (*152*).

f. Alkyl and Aryl Complexes. Though complexes with alkyl and aryl ligands are organometallic complexes, by definition, there are many examples of alkyl and aryl complexes that fit comfortably within the classes of complexes discussed in this review. Others contain only alkyl or aryl groups but exhibit many of the properties traditionally associated with coordination complexes. Where appropriate, these two types of alkyl and aryl complexes are discussed, but for more comprehensive treatments of this chemistry other reviews should be consulted (*4, 8–10*).

Alkyl ligands are isoelectronic with amines, but because they carry a negative charge they are much stronger σ donors than are amines, and hence stabilize higher oxidation states. Os(VI) chemistry of the air and thermally stable alkyl complexes $[OsO(CH_2SiMe_3)_4]$, $[OsN(R)_4]^-$ (R = CH_3, CH_2Ph, or CH_2SiMe_3), $[Os(NR')(R)_4]$ (R = Me, CH_2Ph, or CH_2SiMe_3, R′ = Me; R = CH_2SiMe_3, R′ = Et or $SiMe_3$), *cis*-$[OsN(CH_2SiMe_3)_2X_2]^-$ [$X^- = Cl^-$, NCS^-, ReO_4^-, $\frac{1}{2}CO_3^{2-}$, $\frac{1}{2}CrO_4^{2-}$, $\frac{1}{2}SO_4^{2-}$, or $\frac{1}{2}(CH_2S^-)_2$], *cis*-$[OsN(CH_2SiMe_3)_2(NCS)(SCN)]^-$, and $[Os(Y)(COR)_n(R)_{4-n}]$ (Y = CH_3N^{2-} and O^{2-}; R = CH_2SiMe_3, R = CH_3; $n = 1$ or 2) is growing rapidly (*26, 28, 154–159*). The much greater σ-donor capacity of an alkyl group as compared to an amine group is illustrated by the fact that with amine ligands a *trans*-dioxo arrangement is required to stabilize Os(VI), whereas four alkyl groups will stabilize Os(VI) with one oxo ligand.

Recently, the tetrahedral oxo/aryl complexes $[Os(O)_2(R)_2]$ [R = 2,6-xylyl (*31*) or 2,4,6-mesityl (*32*)] were prepared via the reaction of $[OsO_4]$ with RMgBr and subsequent oxidation of the red complexes [believed to be the Os(V) complexes $[Os(O)_2(R)_2]^-$] to give the deep-green Os(VI) complexes. A similar reaction of $[Os(O)_3(NBu^t)]$ with mesMgBr results in a mixture of $[Os(O)_2(mes)_2]$ and $[Os(O)(NBu^t)(mes)_2]$ (*30*). These new complexes were characterized by mass, IR, ^1H, and ^{13}C NMR spectroscopies, conductivity, cyclic voltammetry, and X-ray crystallography. The Os — C bond lengths in $[Os(O)_2(xylyl)_2]$ [2.058(6) and 2.060(6) Å (*31*)] and $[Os(O)_2(mes)_2]$ [2.053(8) and 2.047(8) Å (*32*)] are significantly shorter than those in $[Os(O)(NBu^t)(mes)_2]$ [2.119(13) and 2.161(14) Å (*30*)], which is probably an indication of the better donor properties of Bu^tN^{2-} compared with O^{2-}. This is consistent with the electrochemistry, because it is more difficult to reduce $[Os(O)(NBu^t)(mes)_2]$ than the dioxo analog, and the Os(V) complex that is produced is less stable. By contrast, $[Os(O)(NBu^t)(mes)_2]$ is oxidized reversibly to its Os(VII) counterpart, although its dioxo analog undergoes an irreversible multielectron oxidation in acetonitrile at more positive potentials. $[Os(O)_2(mes)_2]$ reacts with pyridine ligands to form octahedral, *trans-O,O-cis*-$[Os(O)_2(mes)_2(L)_2]$ (L = py, 4-Bu^tpy, or $\frac{1}{2}$bpy), with 2,6-$Me_2C_6H_3NC$ to give the trigonal-bipyramidal $[Os(O)_2(xylylNC)(mes)_2]$ (in which the mes ligands occupy the axial positions), and with tertiary phosphines to form *trans-C,C-cis*-$[Os(O)_2(PR_3)_2(mes)_2]$ (PR_3 = PMe_3, PMe_2Ph, or $PMePh_2$), which are rare examples of *cis*-dioxoOs(VI) complexes. All of these complexes were characterized by mass, IR, and ^1H NMR spectroscopies. The tetrahedral bis-dioxo starting material undergoes a complex series of redox reactions with N_2O_3 or NO_2 to give [mesN_2]-[$Os^{VI}(O)_2(ONO_2)_2$(mes)], which was characterized by IR and ^{1}H NMR spectroscopies and an X-ray structural analysis. It is a distorted trigonal bipyramid in which the nitrato ligands occupy axial positions and the Os — C bond length of 2.053(13) Å is comparable to that of the parent complex (*30*). This chemistry is summarized in Scheme 10.

Again, the ability of the alkyl and aryl groups to stabilize oxidation states higher than those of their isoelectronic amine and N-heterocyclic counterparts is evidenced by the isolation and characterization of remarkable stable peralkylated and perarylated Os(IV) and Os(V) complexes, $[Os(R)_4]^{0/1+}$ [R = cyclohexyl, phenyl, or 2-tolyl) (*21, 23, 160–162*)]. Both oxidation states have distorted tetrahedral geometries as shown by single-crystal X-ray diffraction. The Os(IV) complexes decompose in the presence of strong π acids such as RNC, CO, and PMe_3 to give a variety of products with alkyl ligands, viz. [Os{η^6-2-(2-MeC_6H_4)}2-$MeC_6H_4)_2$L] (L = CO or PMe_3), *cis*-[Os(2-$MeC_6H_4)_2(CNR)_4$] (R = Me_3C or 2,6-$Me_2C_6H_3$), *fac*-[Os{*C,N*-3-Me[2-C(2-$MeC_6H_4NCMe_3]C_6H_3$}(2-MeC_6H_4)(CNR)] (R = CH_3 or 2,6-$Me_2C_6H_3$), and *cis,fac*-[Os(2-$MeC_6H_4)_2(CNCMe_3)_3$(CO)] (*160*). In addition to the reversible oxidation of [Os(2-tolyl)$_4$] to the Os(V) complex, two reversible reductions to the Os(III) and Os(II) complexes are observed in the CV. The Os(V) complex has been characterized by IR and electron paramagnetic resonance (EPR) spectroscopies (*162*).

There are a large number of alkyl Os(II) complexes. These are not discussed in detail, but recent examples are given below. The structure has been determined of the osmacycle, [Os{$CH_2CH_2S(NSO_2C_6H_4$Me-4)O}Cl(NO)$(PPh_3)_2$] (Fig. 2), which is prepared from the reaction of $OSNSO_2C_6H_4$Me-4 with [OsCl(NO)$(C_2H_4)(PPh_3)_2$] (*163, 164*). The

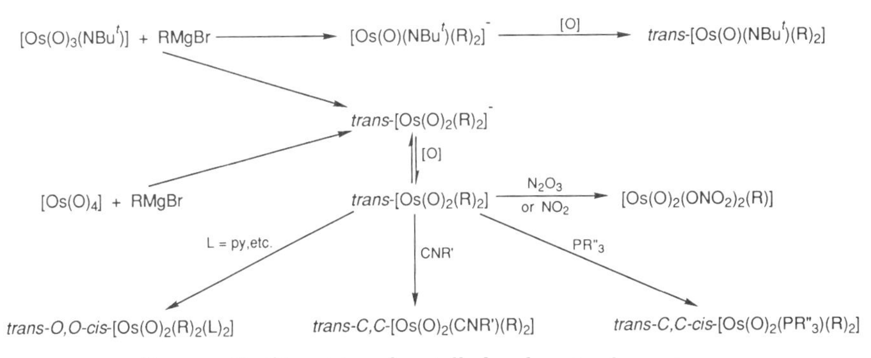

SCHEME 10. Chemistry of oxo/alkyl and oxo/aryl complexes.

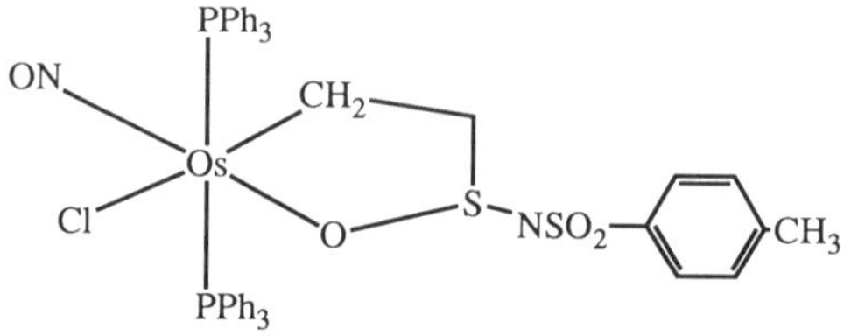

FIG. 2. Structure of [Os{$CH_2CH_2S(NSO_2C_6H_4Me$-4)O}Cl(NO)$(PPh_3)_2$].

structure of *cis*-[Os$(PMe_3)_4$(H)(neopentyl)] has also been determined and its reactions with N_3^-, CO, CH_4, 2-xylyl isocyanide, and Me_3CNC have been studied (*165*). With CH_4, it yields a cyclometalated product and *cis*-[Os$(PMe_3)_4$(H)(CH_3)] (*165*). The orthometalated complex [Os-(PPh_3)($Ph_2PC_6H_4$)($B_{10}H_7(OEt)_2$)] (Fig. 1) is mentioned in Section II,C,1 (*103*). *trans,cis-X,Y*-[Os(NO)(X)(Y)(CF_3)$(PPh_3)_2$] (X = Y = Cl^-; X = Cl^-, Y = I^-; X = Y = I^-) have been obtained by X_2 oxidation of [Os(NO)(Cl)($=CF_2$)$(PPh_3)_2$] in the presence of F^- and the complexes were characterized by X-ray diffraction and IR spectroscopy (*166*). The alkenyl complexes [Os{(*E*)-CH$=$CHPh}(Cl)(CO)$(PMeBu^t{}_2)_2$] and [Os(CO){(*E*)-CH$=$CHPh}{η^2-(*N,O*)-N(O)$=$CRR'}$(PMeBu^t{}_2)_2$)] have also been reported (*124*). The *cis*- and *trans*-[Os$(bpy)_2$(CO)R]$^+$ complexes are discussed in Section II,C,2,b.

Os(II) aryl complexes, [$Os^{II}(NH_3)_5$(aryl)]$^{n+}$ and *trans*-[$Os^{II}(NH_3)_4$ L(aryl)]$^{n+}$, have been prepared with pyridinium, *N*-methylpyridinium, *N*-methyl-4-picolinium, 2,6-lutidinium, and 2,6-lutidine bound through the para carbon (Section V,D,3) (*85, 90, 167*). Similarly, a complex containing a 2-C-bound *N,N'*-dimethylimidazolium ligand, [$Os^{III}(NH_3)_5$(2-*C*-diMeim)]$^{2+}$, is formed via Mg^0 reduction of [Os-$(NH_3)_5(OSO_2CF_3)$]$^{2+}$ in the presence of an excess of the iodide salt of the ligand in neat nmp or dma. Surprisingly, a similar reduction in dme using the triflate salt of the ligand resulted in the η^2-C,C linkage isomer. The two linkage isomers have not been interconverted, as yet (*90*). All of the C-bound N heterocycles reported above undergo reversible oxidations to their Os(III) counterparts (*90*).

g. Carbene Complexes. *trans*-[Os(NO)(Cl)($=CF_2$)$(PPh_3)_2$] is prepared by the reaction of [OsCl(NO)$(PPh_3)_3$] with $Cd(CF_3)_2$ (*166*). Carbene ether and alcohol complexes have also been reported (Section II,C,2,o).

h. Alkene Complexes. Like Ru(II), Os(II) has a high affinity for olefins and stable [$Os(NH_3)_5$(alkene)]$^{2+}$ complexes that have been prepared from either reactions of olefins with labile $[Os^{II}(NH_3)_5(solvent)]^{2+}$ complexes (*70, 71, 120, 168*), the partial hydrogenation of η^2-bound arenes (Section V,E,4,c) (*120, 169*), or other reactions of coordinated ligands (Section V,E,4) (*70, 120, 168*), including those with ethene, propene, *iso*-butylene, 1,3-butadiene, 1,5-hexadiene, *cis*-2-methoxy-2-butene, *trans*-2-methoxy-2-butene, *cis*-2-hydroxy-2-butene, *trans*-2-hydroxy-2-butene, 3,4-dibromo-1-butene, styrene, cyclohexene, 1,2-dihydronaphthalene, 1,4-dihydronaphthalene, 3-methoxycyclohexene, 3,6-dimethoxycyclohexene, 2-cyclohexen-1-one, and 1,3-cyclohexadiene; $[Os(NH_3)_5(\eta^2\text{-}(C,C)\text{-3-methoxycyclohexene})]^{2+}$ has been characterized by X-ray crystallography (*169*). All complexes show reversible Os(III/II) redox couples, and the Os(III) complexes are much more kinetically stable than are their Ru(III) congenors. Generally, the olefin bound to Os(III) is released over a matter of hours under ambient conditions (*120, 169*).

An Os(II) complex with a chelating diene, norbornadiene, is prepared from $[Os(bpy)_2(CO_3)]$ and the ligand in the presence of acid (*97*), whereas $[Os(tpp)(\eta^2\text{-}C_2H_4)]$ is formed from the reductive elimination reaction of 1,2-dibromoethane with $[Os(tpp)]^{2-}$ (*170*). Similar complexes are formed from the reactions between alkenes and Os(II) phosphines, e.g., *trans*-$[OsCl_2(PR_3)_3(\eta^2\text{-}CH_2{=}CHR')]$ [R_3 = Me_2Ph; R′ = H (*145*)], $[Os(H)(Cl)(CO)(PPr^i_3)_2(\eta^2\text{-}CH_2CHR)]$ [R = H, CO_2Me, CN, or COMe (*18*), in which the alkene is trans to the hydrido ligand], and *trans*-$[OsCl(NO)(\eta^2\text{-}C_2H_4)(PPh_3)_2]$ (*163*).

i. Alkyne Ligands. $[Os(NH_3)_5(alkyne)]^{3+/2+}$ (alkyne = ethyne, 2-butyne, and diphenylacetylene) have been isolated and characterized (*71, 120, 168, 171*). The thermodynamically stable linkage isomers of both the Os(III) and Os(II) oxidation states of $[Os(NH_3)_5(diphenylacetylene)]^{3+/2+}$ are the η^2-alkyne complexes, which have been characterized by NMR spectroscopy and electrochemistry (*171*). The alkyne complexes exhibit reversible electrochemical oxidations to Os(III) analogs.

j. η^2-Bound Arene and Heterocyclic Ligands. A recent interesting development has been the synthesis and characterization of a variety of η^2-arene complexes of the type $[Os(NH_3)_5(\eta^2\text{-arene})]^{2+}$. They have been characterized by NMR spectroscopy and electrochemistry. These Os(II) complexes are much more stable than their Ru(II) analogs, none of which has been isolated, although it is possible that an η^2 Ru(II)/benzene complex is the reactive intermediate in the hydrogenation of benzene by $[Ru(NH_3)_5]^{2+}$ (*172*). $[Ru(NH_3)_5(\eta^2\text{-arene})]^{2+}$ coordination has only to date been observed in heterodinuclear Os(II)–Ru(II) dimers, where the coordination of osmium has enhanced the donor abilities of the arene (Section II,D,2,c) (*172*). The ligands that bind in this fashion to Os include benzene (*172, 173*), toluene (*75*), *p*-xylene (*75*), *tert*-butylbenzene (*75*), cumene (*75*), α,α,α-trifluorotoluene [(trifluoromethyl)benzene] (*75*), 1,2,3,4-tetramethylbenzene (*74*), biphenyl (*75*), diphenylacetylene (*171*), naphthalene (*172*), anthracene (*174*), phenanthrene (*174*), pyrene (*174*), aniline (*175*), *N,N*-dimethylaniline (*175*), benzonitrile (*176*), benzophenone (*177*), 2,2-dimethylpropiophenone (*177*), phenol (*178*), methoxybenzene (*169*), and 1,4-dimethoxybenzene (*169*). There is also evidence for the existence of small amounts of the η^2-arene linkage isomer of $[Os(NH_3)_5(styrene)]^{2+}$ from CV data reported for the oxidation of the Os(II) complex (Section III,B), although no mention of this possibility was made in the original paper (*71*).

N-, O-, and S-heterocyclic ligands also form $[Os(NH_3)_5\{\eta^2\text{-}(C,C)\text{-}L\}]^{2+}$ complexes [L = 2,6-lutidine, 2,6-lutidinium, pyridinium, *N*-methylpyridinium, and *N*-methyl-4-picolinium (*85, 167*), *N,N'*-dimethylimidazolium (*90*), pyrrole (*90, 179*), *N*-methylpyrrole (*90, 179*), thiophene (*90, 179*), furan (*90, 179*), and 1,3-dimethyluracil (*72, 73*)]. On oxidation to Os(III), arene ligands are rapidly lost from the coordination sphere, or in the case of the substituted arene ligands with good σ donors, rapid linkage isomerization reactions occur (Section V,D).

k. η^6-Arene Ligands. $[Os(NH_3)_3(\eta^6\text{-arene})]^{2+}$ (arene = benzene or naphthalene) are prepared by either the thermal degradation of $[(NH_3)_5Os(\eta^2{:}\eta^2\text{-arene})Os(NH_3)_5]^{4+}$ or the reaction of the appropriate arene with *cis*-$[Os(NH_3)_4(solvent)_2]^{2+}$ (*88, 89*) (see Section II,B,4,c).

l. Allyl Complexes. The Os(IV) η^3-aryl complexes, $[Os(NH_3)_5(\eta^3\text{-}CH_2CRCH)]^{3+}$, have been reported (*180*) (Section V,E,2,a) and show reactivity typical of complexes with phosphine ligands, e.g., $[Os(H)(CO)(\eta^3\text{-}CH_2CRCH_2)(PPh_3)_2]$ (R = Me or H) (*133*).

m. η^2-(C,O)-Ketones and -Aldehydes. $[Os(NH_3)_5]^{2+}$ has a remarkable ability to stabilize η^2-aldehyde and ketone complexes, which are prepared by the reduction of $[Os(NH_3)_5(OSO_2CF_3)]^{2+}$ in a solution of the appropriate ligand. Not only are these complexes very inert toward ligand substitution, but they are also moderately air stable in the solid state (*120, 168, 169, 177, 181*). Complexes that bind in this η^2 fashion are acetaldehyde, acetone, 2-butanone, cyclopentanone, cyclobutanone, 2-cyclohexen-1-one, 2,2-dimethylpropiophenone, and benzophenone. The Os(III) analogs are very unstable toward linkage isomerization reactions (Sections V,D,1). The X-ray structure of $[Os(NH_3)_5\{\eta^2\text{-}(C,O)\text{-acetone}\}]^{2+}$ indicates a considerable contribution from the Os(IV) metallocycle resonance structure (Fig. 3). Thus, the C—O bond length is intermediate between a C—O single bond and double bond, and the coordinated C has an sp^3 tetrahedral geometry (*181*).

Recently, $[Os(H)Cl(CO)(PPr^i_3)_2\{\eta^2\text{-}(CD_3)_2CO\}]$ has been postulated to be in rapid equilibrium with $[Os(H)Cl(CO)(PPr^i_3)_2]$ and $[OsCl\{OCH(CD_3)_2\}(CO)(PPr^i_3)_2]$ in d_6-acetone (*181a*), but there does not appear to be any examples of stable η^2-ketone or aldehyde complexes of Os with phosphine ligands.

As is the case for the η^2-arene complexes, the chemistry of Ru and Os differs markedly. Thus, whereas $[Ru(NH_3)_5\{\eta^2\text{-}(C,O)\text{-}O{=}CR_2\}]^{2+}$ complexes have been isolated, they are sensitive to moisture, oxygen, and nitrogen and are thermally unstable, unlike their substitutionally inert Os(II) counterparts (*182–184*).

n. Isonitrile Complexes. *cis*-$[Os(2\text{-}MeC_6H_4)_2(CNR)_4]$ (R = Me_3C or 2,6-$Me_2C_6H_3$) and *cis,fac*-$[Os(2\text{-}MeC_6H_4)_2(Me_3CNC)_3(CO)]$ are prepared from $[Os(2\text{-}MeC_6H_4)_4]$ (*160*). $[Os(CNR)(C_6H_4Me\text{-}4)(CO)(PPh_3)_2(SNNMe_2)]$ (R = 2-xylyl and Me_3C) (*185*), $[Os(hba\text{-}b)(CO)(CNBu^t)]$ and $[Os(chba\text{-}dcb)(CNBu^t)_2]$ (*146,*) $[Os(pc)(CNBu^t)_2]$ (*187*), $[Os(H)(CNR)L_4]^+$ [L = $P(OMe)_3$, $P(OEt)_3$, or $PPh(OEt)_2$; R = 4-$MeOC_6H_4$ or 4-MeC_6H_4] (*187*), $[OsL_4(4\text{-}MeC_6H_4NC)(4\text{-}MeC_6H_4N{=}NH)]^+$ [L = $P(OEt)_3$ or $PPh(OEt)_2$] (*188*), $[Os(O)_2(CNxylyl)(mes)_2]$ (*30*), and $[OsI\{\eta^1\text{-}C(S)SMe\}(NO)(CNBu^t)(PPh_3)_2]^+$ (*152*) are discussed in Sections II,C,7,j, II,C,4,x, II,C,4,d, II,C,5,b, II,C,4,m, II,C,2,f, and II,C,2,e, respectively.

cis-$[Os(bpy)_2(CNR)_2]^{2+}$ (R = Me or Ph; Section V,E,4,b) undergo reversible oxidations to the Os(III) analogs and two reversible one-electron reductions that are probably bpy-ligand centered. The electronic absorption and emission spectra of the Os(II) complexes have also been reported (*97*).

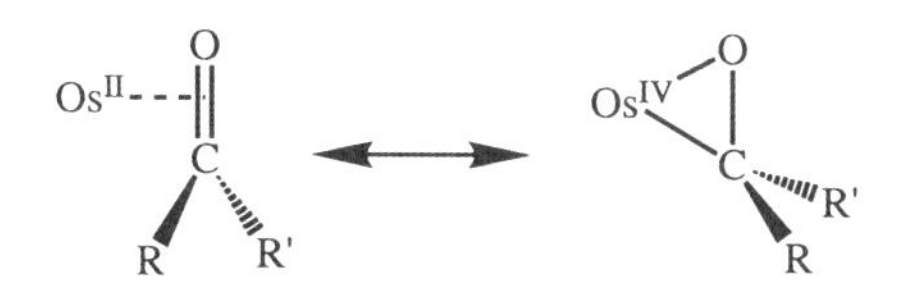

FIG. 3. Resonance forms of $Os^{II}\text{-}\eta^2\text{-}O{=}CR_2$ complexes.

o. C-Bound Aldehyde and Ketone Complexes. The reactivity of the C-bound formyl complexes *trans*-[$Os(CHO)(CO)(P\text{-}P)_2$]($SbF_6$) [P-P = $Ph_2PCH_2CH_2PPh_2$ or *o*-$(Ph_2P)_2C_6H_4$] with electrophiles is discussed. CF_3SO_3R (R = Me or H) attack the oxygen atom at −30°C to form the carbene ether and alcohol complexes, *trans*-[M(CHOR)(CO)(P-P)$_2$]$SbF_6 \cdot CF_3SO_3$, which were characterized by NMR (^1H and ^{31}P) and IR spectroscopies and extended Hückel calculations (*189*). [Os(Y)$(COR)_n(R)_{4-n}$] (Y = CH_3N^{2-} or O^{2-}; R = CH_2SiMe_3, R = CH_3; n = 1 or 2) were prepared from CO insertion reactions with [Os(Y)$(R)_4$] and characterized by NMR and IR spectroscopies (*159*).

3. Silyl Complexes

The reaction between [Os(Ph)(Cl)(CO)$(PPh_3)_2$] and R_3SiH, (R = Et or Cl) leads to the elimination of C_6H_6 to form [Os(SiR_3)(Cl)(CO)$(PPh_3)_2$]. The $SiCl_3^-$ complex undergoes substitution reactions at Si to produce other silyl complexes, e.g., the complex with R = Me. The reaction of [Os(H)Cl(CO)$(PPh_3)_2$] with $Hg(SiMe_3)_2$ results in the same complex as a minor product, but the major product isolated was [Os(CO) {η^2-(C^2,*P*)-$Ph_2PC_6H_4$} {η^2-(*P*,*Si*)-*o*-$Ph_2PC_6H_4SiMe_2$} (PPh_3)] (Fig. 4). These complexes were characterized by NMR (^1H and ^{31}P) and IR spectroscopies, and the latter two, by X-ray crystallography (*190*). [Os$(PPh_2CH_2CH_2SiR^1R^2)_2(CO)_2$] ($R^1 = R^2$ = Me or Ph; R^1 = Me, R^2 = Ph) are prepared from [$Os_3(CO)_{12}$] and $Ph_2PCH_2CH_2SiR^1R^2H$ at 140°C. The crystal structure of the complex, where $R^1 = R^2 = CH_3$, has been determined (*191*). *mer*-[$Os(CO)_3(PR_3)(SiMeCl_2)_2$] [$PR_3 = PPh_3$ or $P(OCH_2)_3CEt$] has also been prepared and characterized by IR, ^1H NMR, and mass spectroscopies (*192*).

The reaction of [Os(H)Cl(CO)$(PPr^i_3)_2$] with triethylsilane produces

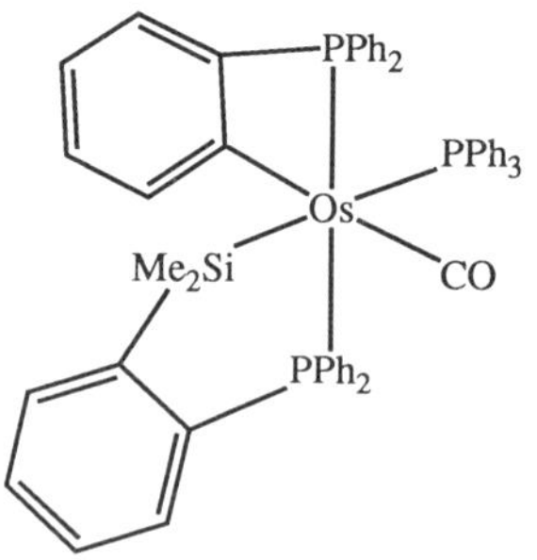

FIG. 4. Structure of [Os(CO){η^2-(C^2,*P*)-$C_6H_4PPh_2$}{η^2-(*P*,*Si*)-*o*-$Ph_2PC_6H_4SiMe_2$}-(PPh_3)].

the dihydrogen complex $[Os(SiEt_3)Cl(\eta^2\text{-}H_2)(CO)(PPr^i_3)_2]$ which acts as a catalyst for the addition of triethylsilane to phenylacetylene. The Os(II) dihydrogen complex is in equilibrium with a small amount of the Os(IV) hydrido complex $[Os(H)_2(SiEt_3)Cl(CO)(PPr^i_3)_2]$, as shown by ^1H NMR spectroscopy (*192a*).

4. *Nitrogen Ligands*

a. Ammine Complexes. An improved synthesis of $[Os(NH_3)_6]^{3+}$ involves the reaction of $[Os(NH_3)_5(OSO_2CF_3)](CF_3SO_3)_2$ in anhydrous liquid ammonia (*59, 67*). This complex has been recrystallized as the mixed salt $[Os(NH_3)_6](ClO_4)_2Cl\cdot KCl$ (*193*), which has had its structure determined by X-ray crystallography.

A large variety of pentaammine, tetraammine, and triammine complexes have been prepared and characterized in recent years using the synthetic procedures outlined in Section II,B,4. The various classes of pentaammine complexes that have been prepared are summarized in the appropriate sections (Table I) (*2, 47, 54–60, 66–71, 74, 75, 80, 85, 90, 118–120, 167–169, 171–181, 194–201*). Dimeric and polymeric complexes are described in Section II,D.

b. Amine Complexes. Many amine adducts of $[OsO_4]$ have been studied and are discussed in Sections II,C,6,b, V,A, and V,E,1,b. The aniline complexes $[Os(NH_3)_5(NH_2Ph)]^{3+/2+}$ and $[Os(NH_3)_5(NMe_2\text{-}Ph)]^{3+/2+}$ were characterized by spectroscopic and electrochemical techniques (Section V,D,3) (*175*). Other $[Os(NH_3)_5(NH_2R)]^{3+}$ complexes are intermediates in the oxidative dehydrogenation reactions of coordinated amines (Section V,E,2,b).

The doubly deprotonated Os(IV) complex $[Os(en)(en\text{-}H)_2]^{2+}$, produced from the reaction of $[OsBr_6]^{2-}$ with neat 1,2-ethanediamine, was partially characterized in the 1950s (*202, 203*) and recently its structure was confirmed by X-ray crystallography (*204, 205*). It is reduced to $[Os(en)_3]^{3+}$ by a number of reductants, and the structure of *rac*-$[Os(en)_3](CF_3SO_3)_3$ has been published (*205*). The Os(III) complex is reversibly reduced to its Os(II) analog, but this complex has not been structurally characterized (*204, 205*). All complexes are oxidized by air to form ethanediimine complexes, i.e., $[Os^{II}(en)_2(diim)]^{2+}$ and $[Os^{II}(en)(diim)_2]^{2+}$, and have been characterized by NMR (^1H and ^{13}C) and UV/Vis spectroscopies (Scheme 11). Recently, $[Os^{IV}(Ctmen)(Ctmen\text{-}H)_2]^{2+}$ has been prepared and found to be very stable because the Ctmen ligand cannot undergo oxidation without breaking a C—C bond (*206*). It has been characterized by X-ray crystallography (*206*) and has

TABLE I

PENTAAMMINEOSMIUM COMPLEXES $[Os(NH_3)_5L]^{n+}$

Ligand	Section	Ref.
Os(II) and Os(III)		
NH_3	II,C,4,a	*59, 67, 193*
CO	II,C,2,b	*55, 58, 118–120*
C-imidazolium	II,C,2,f	*90*
C-(*N*-heterocyclic)	II,C,2,f	*85, 90, 167*
η^2-alkene	II,C,2,h	*70, 71, 120, 168, 169*
η^2-alkynes	II,C,2,i	*71, 120, 168, 171*
η^2-(*C,C*)-arenes	II,B,4,a and II,C,2,j	*71, 74, 75, 169, 172–178*
η^2-(*C,C*)-*N*-heterocycles	II,C,2,j	*90*
η^2-(*C,C*)-*O*-heterocycles	II,C,2,j	*90*
η^2-(*C,C*)-*S*-heterocycles	II,C,2,j	*90*
η^2-(*C,O*)-aldehydes	II,C,2,m	*177*
η^2-(*C,O*)-ketones	II,C,2,m	*120, 168, 169, 177, 181*
Alkyl- and arylamines	II,C,4,b, V,D,3, and V,E,2,b	*118, 175*
N-imidazoles	II,C,4,e	*90*
N-pyridines	II,C,4,e	*60, 66, 68*
N-pyrazine	II,C,4,e	*60, 66, 68*
N-pyrimidine	II,C,4,e	*60, 66, 68*
N-pyridazine	II,C,4,e	*60, 66, 68*
N-4,4′-bipyridine	II,C,4,e	*68*
NO	II,C,4,f	*2*
N_2	II,B,4,a and II,C,4,j	*47, 54–59*
N-imines	II,C,4,k	*194*
N-oximes	II,C,4,l and V,E,5,g	*195*
N-nitriles	II,C,4,n	*67, 176, 196*
N-NCS^-	II,C,4,t	*197, 198*
N-$NCSe^-$	II,C,4,u	*198*
OH_2	II,B,4,a and II,C,6,a	*69, 199*
OH^-	II,C,6,a	*69*
Alcohols	II,C,6,c	*118*
O-dmso	II,C,4,k and V,D, 4	*67, 120, 200*
Ethers	II,C,6,l	*120*
O-amides	II,C,6,g and V,E,5,h	*67, 120*
O-aldehydes	II,C,6,j	*177*
O-ketones	II,C,6,j, V,D,1, and V,E,3	*67, 120, 168, 169, 177, 181*
Trialkyl phosphates	II,C,6,e and V,E,3	*67, 68, 80*
Carboxylates	II,C,6,o and V,E,5,g	*119*
Triflate	II,B,4,a and II,C,6,q	*59, 66, 67*
S-SO_2	II,C,7,j	*60*
S-SO_3^{2-}	II,C,7,s	*60*
S-dmso	II,C,7,l and V,D,4	*67, 120, 200*
S-thiophene	II,C,7,c	*90, 179*
Cl^-, Br^-, and I^-	II,C,8,c	*68*
η^2-H_2	II,C,9,a	*201*
H^-	II,C,9,c	*201*
Os(IV)		
Allyl	II,C,2,1	*180*

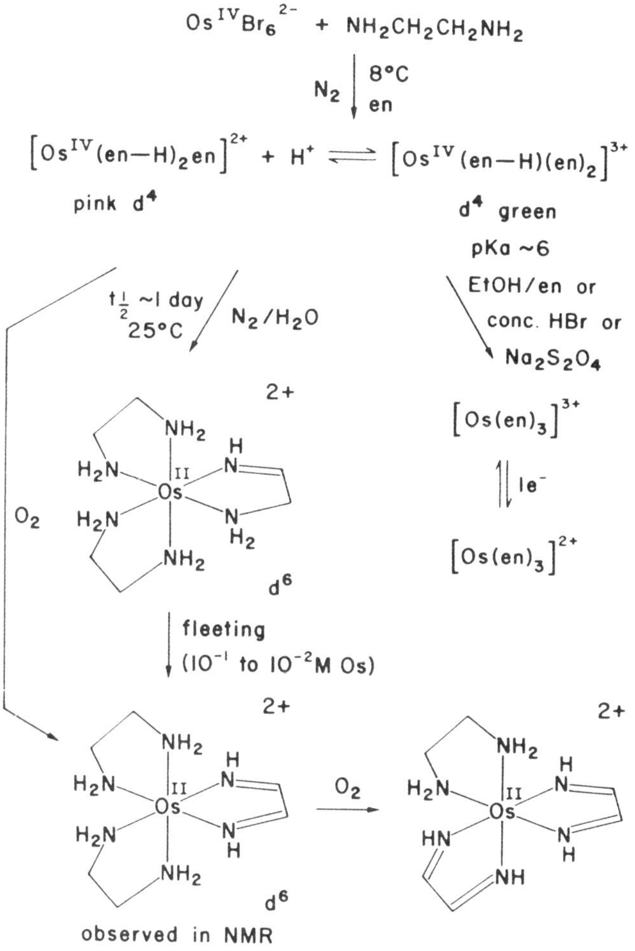

SCHEME 11. Chemistry of the Os/en system. Reprinted with permission from the *Journal of the American Chemical Society*, Ref. *204*. Copyright 1982, American Chemical Society.

Os—N bond lengths similar to those observed in the en analog (*204*). $[Os(Ctmen)_3]^{3+/2+}$ have been isolated and characterized (*206*) and *trans*$[Os(Ctmen)_2Cl_2]Cl$ has been prepared by the Sn reduction of $Na_2[OsCl_6]$ in an ethanolic solution of the ligand. It is oxidized to *trans*-$[Os^{VI}(O)_2(Ctmen)_2]^{2+}$ by aqueous H_2O_2 and the dioxo complex has been characterized by IR and UV/Vis spectroscopies. The five-coordinate $[Os^{VI}(O)(Ctmen\text{-}2H)(Ctmen\text{-}H)]^+$ is prepared by either the reaction of the six-coordinate complex with collidine in acetonitrile, or the reaction of K[Os(O)$_3$(NBu^t)] in methanol. It was characterized by X-ray crystallography and IR spectroscopy and adopts a square-pyramidal geometry (*24*). The equilibrium between [Os(O)(Ctmen-2H)(Ctmen-H)]$^+$ and *trans*-[Os(O)$_2$(Ctmen)$_2$]$^{2+}$ in basic media (*24*) provides a mechanism for the previously observed (*207*) rapid ^{18}O exchange of *trans*-[Os^{VI}(en)$_2$(O)$_2$]$^{2+}$.

1,2-Phenylenediamine (pdaH$_2$) reacts with [OsO$_4$] to form *trans*-[Os^{VIII}(O)$_2$(pda)$_2$], which is an unusual example of an Os(VIII) complex. It has been characterized by IR and NMR spectroscopies (*24*). Four-coordinate diaminato–Os(VI) complexes of the type [Os(O)-(NAr)(ArNCHRCHRNAr)] (Ar = 2,6-Pr$^i_2$C$_6$H$_3$; R = H; R + R = cyclopentane or norbornane) have been prepared from the reactions of [Os(O)(NAr)$_3$] with alkenes and characterized by NMR (^{1}H and ^{13}C) and IR spectroscopies, and an X-ray structure was determined for R = H) (*22*). Similarly, [Os(O)$_2$(NBu^t)$_2$] and fumaronitrile give [Os-(O)$_2${Bu^{t}N(CHCN)$_2$NBu^t}], in which the ligand is bound via the deprotonated amines, as deduced from IR, Raman, and ^{1}H NMR spectroscopies (*208*). The reactions of [OsO$_4$] with Ntmen and HX yield *trans*-[Os(O)$_2$X$_2$(Ntmen)] (X = Cl$^-$ or Br$^-$), which have been characterized by UV/Vis, IR, and ^{1}H NMR spectroscopies (*209*).

c. Macrocyclic Complexes. Since the first report of an Os macrocyclic complex in 1986 (*210*), many complexes have been prepared. *trans*-[Os^{VI}(14-tmc)(O)$_2$]$^{2+}$, which is obtained from *trans*-[Os^{III}(14-tmc)Cl$_2$]$^+$ (Scheme 7), undergoes a reversible three-electron reduction to form *trans*-[Os^{III}(14-tmc)(OH)(OH$_2$)]$^{2+}$ at pH values of 1–3.2. In the pH range 3.2–6.5, the reduction product is [Os^{III}(14-tmc)(OH)$_2$]$^+$, whereas at pH values >7, the reversible reduction to [Os^V(14-tmc)(O)$_2$]$^+$ and the subsequent two-electron reduction to *trans*-[Os(14-tmc)(OH)$_2$]$^+$ are observed (*210, 211*). This contrasts with analogous Ru chemistry, wherein the Ru(IV) complex is also a stable intermediate. Like most other *trans*-dioxo Os(VI) complexes, [Os(14-tmc)(O)$_2$]$^{2+}$ exhibits considerable vibrational structure in its UV/Vis spectrum (*210*). Electrochemistry in acetonitrile exhibits two one-electron reversible reductions to form [Os^V(14-tmc)(O)$_2$]$^+$ and [Os^{IV}(14-tmc)(O)$_2$], respectively. The Os(V) complex was isolated by this method and has been characterized by IR and electronic absorption spectroscopies. It also exhibits vibronic coupling in its UV/Vis bands (*212*).

The preparation of the macrocyclic complexes has been extended to include *trans*-[Os(L)Cl$_2$]$^+$ complexes (L = 15-tmc, 16-tmc, [14]aneN$_4$, [15]aneN$_4$, [16]aneN$_4$, crMe$_3$, and tet*a*) (*39, 114, 213*), and the crystal structure of *trans*-[Os(16-tmc)Cl$_2$)]ClO$_4$ has been determined (*213*).

The complexes exhibit reversible reductions to their Os(II) analogs, but oxidations to their Os(IV) analogs are irreversible for the secondary amine macrocycles. This is probably due to oxidative dehydrogenation of the ligand (Section V,E,2,b). By contrast, the macrocycle 16-tmc cannot undergo oxidative dehydrogenation reactions and a reversible Os(IV/III) redox couple is observed to form *trans*-$[Os^{IV}(16\text{-tmc})(Cl)_2]^{2+}$. This complex is moderately stable and its UV/Vis spectrum and reduction back to Os(III) by ascorbate have been reported (*213*). All of the complexes, $[Os(n\text{-tmc})(O)_2]^{2+}$ and $[Os(crMe_3)(O)_2]^{2+}$, have been prepared by the method described above (*114, 213*) and are weak oxidants, with no reaction occurring with benzyl alcohol or styrene even at 70°C. Refluxing an acetonitrile solution of *trans*-$[Os(14\text{-tmc})(O)_2]^{2+}$ with PPh_3 results in the formation of *trans*-$[Os(14\text{-tmc})(NCCH_3)_2]^{2+}$ and $OPPh_3$ (*39*). The redox chemistry is summarized in two recent reviews (*39, 93*), and the electrochemistry of $[Os(14\text{-tmc})(O)_2]^{2+}$ has been studied in detail (*211*). *trans*-$[Os(n\text{-tmc})(O)_2]^{2+}$ and $[Os(crMe_3)(O)_2]^{2+}$ are also powerful photooxidants and the rate constants for their excited-state quenching with arenes in acetonitrile to form the corresponding cation radicals and dioxo Os(V) complexes, are reported (*107, 211a*). The excited states also undergo oxo transfer reactions with trialkylphosphines, dialkylsulfides, and alkenes (*114*).

d. Porphyrin and Phthalocyanine Complexes. Osmium porphyrin complexes were first prepared from the reaction of $[OsO_4]$ with the porphyrin at 200°C in $Me(OCH_2CH_2)_2OH$ to give $[Os^{II}(P)(CO)X]$ complexes (P = porphyrin) (*139*). More recently, a less hazardous method which involves refluxing the porphyrin with $[Os_3(CO)_{12}]$ was developed (*140*). This is used to prepare [Os(P)(CO)] [P = oep (*38, 140*), mix-dme (*140*), tpp (*38, 40*), mix (*144*), and *p*-Xtpp; X = Me, Cl, or OMe (*40*)], which react readily with nucleophiles to form six-coordinate complexes such as [Os(P)(CO)(ROH)] (P = oep, mix-dme, tpp, or *p*-Xtpp; X = Cl, OMe, or Me; R = Me, Et, Pr^i, or Ph) (*38, 40, 140, 144, 214*), $[Os^{II}(P)(CO)(PBu^n_3)]$ (P = oep or tpp), and [Os(P)(CO)(py)] (*140*). The latter are oxidized by Br_2 to form $[Os^{III}(P)(PBu^n_3)(Br)]$, which undergo reversible oxidations to $[Os^{IV}(P)(PBu^n_3)(Br)^+$ (*143*). $[Os^{II}(P)(CO)]$ (P = oep or tpp) are also oxidized by air in CH_2Cl_2 to give $[Os^{III}(P)(CO)(Cl)]$, which are reversibly reduced by $SnCl_2$ to reform the starting material (*38*).

Oxidations of [Os(P)(CO)] with *tert*-butyl hydroperoxide or 3-chloroperoxybenzoic acid give *trans*-$[Os(P)(O)_2]$ (*38, 40, 140, 214*) (P = mix-dme, oep, tpp, or *p*-Xtpp; X = Cl, Me, or MeO), and the crystal structures of $[Os(p\text{-Metpp})(O)_2]\cdot$thf (*40*) and $[Os(oep)(O)_2]$ (*215*) have been reported. Reductions of [Os(P)$(O)_2$] with ascorbate produce [Os^{IV}-(P)$(OH)_2$] (P = tpp or *p*-Xtpp; X = Cl, Me, or MeO), which are unstable toward aerial oxidation back to the *trans*-dioxoOs(VI) complexes (*40*). Os(IV) complexes containing alcoholate ligands [Os^{IV}(P)$(OR)_2$] are also prepared via the reduction of [Os(P)$(O)_2$] in the presence of the appropriate alcohol. The reductants used include N_2H_4 (*216*), PPh_3 (*140*), $SnCl_2$ (*38*), and ascorbate (*214*). These Os(IV) complexes are also intermediates in the oxidation of [Os^{II}(P)(CO)(EtOH)] complexes to [Os^{VI}(P)$(O)_2$] in ethanol (*40*). The complexes prepared by these methods include those in which P = mix-dme, oep, tpp, or *p*-Xtpp (X = Cl, Me, or MeO), and in which R = Me, Et, Pr^i, and Ph. Most of these Os(IV) complexes exhibit reversible one-electron oxidations to form [Os^V-(P)$(OR)_2]^+$ and reductions to form [Os^{III}(P)$(OR)_2]^-$. The Os(V) complexes are unstable on the slower bulk electrolysis time scale and disproportionate to form [Os(P)$(O)_2$] and other products (*214*). [Os^{IV}-(P)(OP)$r^i)_2$] is also prepared by the γ-radiolysis of [Os(P)(CO)] in CCl_4/Pr^iOH/H_2O (*38*). The 2-propanol complex undergoes reversible protonation to form [Os^{IV}(P)$(HOR)_2]^{2+}$ and reduction to form [Os-(P)$(HOR)_2$] (*38*). The crystal structures of [Os^{IV}(tpp)$(OPr^i)_2$], [Os^{IV}-(tpp)$(OPh)_2$], and [Os(oep)$(OEt)_2$] have been mentioned (*214, 217*), but no details were given.

[Os(P)$(O)_2$] is also reduced by PhSH to give [Os^{IV}(P)$(SPh)_2$] (P = oep or tpp) (*214*); by Br_2, to give [Os^{IV}(oep)Br_2] (*214*); and by PPh_3, to give [Os^{IV}(P)(O)$(OPPh_3)$] intermediates en route to [Os^{II}(P)$(OPPh_3)_2$] (*218*). All of the Os(IV) complexes are paramagnetic except [Os-(P)$(SPh)_2$], which is diamagnetic (*214*). Unlike the alcoholate complexes, [Os(oep)$(Br)_2$] is oxidized at the porphyrin ring. Both [Os-(oep)$(Br)_2$] and [Os(oep)$(SPh)_2$] are reversibly reduced to their Os(III) analogs (*214*).

In the presence of excess PPh_3, [Os^{VI}(oep)$(O)_2$] and [Os^{VI}(tpp)$(O)_2$] are converted to [Os^{II}(oep)$(OPPh_3)_2$] and [Os^{II}(tpp)$(PPh_3)_2$], respectively. The structures of the Os(II) complexes were determined by X-ray crystallography. The $OPPh_3$ complex is converted to [Os^{II}(oep)(P-$Ph_3)_2$] by refluxing in benzene with excess PPh_3 (*218*). Dioxo complexes are also reported to be useful epoxidation catalysts (*219*). [Os^{III}(P)(P-$Bu^n{}_3$)Br] (P = oep or tpp) reacts with cyclohexene in the presence of PhIO to form the corresponding epoxide and enone. The active catalysts are not [Os(P)$(O)_2$] complexes, because they do not oxidize styrene to styrene oxide, whereas this catalytic system does. Turnovers for the oxidation of cyclohexene ($\geq$120), are much higher than for the Ru analogs (*143*).

[Os(oep)$(pz)_2$] is prepared from the reaction of $[Os(oep)]_2$ with excess pz (*220*), and resonance Raman spectroscopy has been used as a probe for π backbonding in the series [Os(oep)(L)(L′)] (L = CO, L′ = py; L = L′ = py; L = L′ = NH_3) (*221*). K_2[Os(tpp)] reacts with dibromoethane to produce [Os(tpp)(η^2-C_2H_4)] and with water to yield K[O-s^{II}(tpp)(H)] (*170*). The latter reacts with benzoic acid or benzoic acid-d_1 to produce the η^2-dihydrogen complexes [Os^{II}(tpp)(H_2)] and [Os-^{II}(tpp)(HD)], in which the H—H and H—D bonds are very weak (*140*).

Mononuclear complexes of the dimeric dpb (Fig. 5) ligand are prepared from [Zn(H_2dpb)] and [$Os_3(CO)_{12}$] to form [(CO)(OH_2)Os(dpb)Zn], and then [Os(H_2dpb)(CO)($HOCH_3$)]. Pyridine displaces CH_3OH to give [Os(H_2dpb)(CO)(py)], or, under irradiation [Os-(H_2dpb)$(py)_2$]. [Os(H_2dpb)(CO)($HOCH_3$)] and [Os(H_2dpb)(CO)(py)] exist as two geometric isomers in which the CO ligand points toward the second porphyrin ring or is on the opposite side (*141*).

[Os^{II}(mix)(CO)(EtOH)] and [Os^{II}(mix)$(dmf)_2$] have been inserted into native and ruthenated myoglobin to form carbonyl myoglobin [Os^{II}(CO)(Mb)] and the oxidized myoglobins [Os^{III}][Mb] and [Os^{III}][Ru_3Mb]. The ascorbate reduction of O_2 to water is catalyzed by [Os^{III}][Ru_3Mb] (*144*).

Although some phthalocyanato(2−) complexes have been reported (*2*), pure [Os(Pc)] has been prepared for the first time by reacting 2-cyanobenzamide with $OsCl_3$ in molten naphthalene with subsequent heating of the crude [Os(pc)L_x] in a stream of N_2 at less than 400°C. The pure complex was characterized by its IR and FD mass spectra. It reacts with excess ligand to produce [Os(pc)L_2] (L = py, dmso, pz, and Me_3CNC), which were characterized by electronic, IR, mass, and NMR (^{1}H and ^{13}C) spectroscopies (*186*).

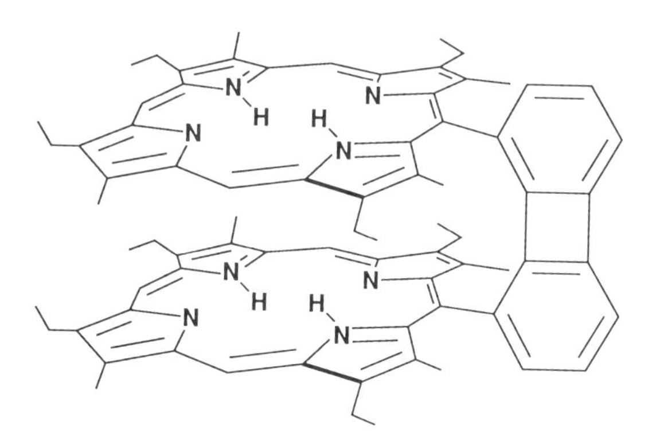

FIG. 5. Structure of the dinuclear porphyrin ligand $dpbH_4$.

e. Complexes with N Heterocycles. Much of the extensive literature on N-heterocyclic complexes has been reviewed elsewhere, particularly those involving bpy, phen, trpy, and related ligands (*2, 222–225*). Table II (*41, 42, 60, 66, 68, 90, 226–236*) summarizes some of the more recent work performed on N-heterocyclic complexes and several of these complexes are discussed in other parts of this review (*94–98*). Their spectroscopy is discussed in Section IV,A,2.

f. Nitrosyl and Nitrosonium Complexes. Os and Ru nitrosyl chemistries have been discussed in recent reviews (*2, 82*). Though Os chemistry is not as extensive as that of Ru, it is expected that with further research, many new Os–NO compounds will be found. As with nitrosyl complexes of other elements, there is considerable debate about whether the ligand is best considered as an NO^-, NO^0, or NO^+ for a particular complex. This will not be discussed here, as it is well covered in other reviews (*2, 82*). Recent publications on nitrosyl complexes are collected in Table III (*2, 43, 153, 163, 166, 237–248a*). They are arranged according to the formalism that the ligand is in the NO^+ oxidation state, with the superscript denoting the number of d electrons (*2*), i.e., $[Os(NO)]^6$ represents an Os(II) nitrosyl complex under this formalism.

g. Nitrosoarene and Nitrosoamine Complexes. $[Os(Cl)(NO(PPh_3)_2(L)]$ ($L = C_2H_4$ or PPh_3) reacts with nitrosobenzene to form $[Os(Cl)(NO)\{\eta^2\text{-}(N,O)\text{-}ONPh\}(PPh_3)_2]$, which was characterized by IR and ^{31}P NMR spectroscopies. Unlike its Ru analog, this complex is quite air and thermally stable (*248*).

The six-electron electrochemical oxidation of $[Os^{II}(trpy)(bpy)(NH_3)](PF_6)_2$ in the presence of Et_2NH or morpholine gives $[Os^{II}(trpy)(bpy)\{N(O)NR_2\}]^{2+}$. The same complexes are produced in low yields from the reaction between $[Os^{II}(trpy)(bpy)(NO)]^{3+}$ and excess amine. This appears to be the only example of coordination of nitrosoamines through the central (nitroso) N atom. The complexes have been characterized by IR (including ^{15}N labeling) and ^1H NMR spectroscopies. The latter shows that though there is limited rotation about the N=N double bond, the barriers to rotation are reduced to 57 kJ mol^{-1} (R = Et) and 63–67 kJ mol^{-1} (R_2 = morpholine) from the free ligand values of ~96 kJ mol^{-1} (*249*).

h. Hydroxylamine and Hydrazine Complexes. $[Os(Cl)(NO)\{\eta^2\text{-}(N,O)\text{-}ONPh\}(PPh_3)_2]$ reacts reversibly with HCl to give the deprotonated hydroxylamine complex, $[Os(Cl)_2\{N(OH)Ph\}(NO)(PPh_3)_2]$, which has been characterized by IR and NMR (^1H and ^{31}P) spectroscopies (*248*).

TABLE II

RECENT EXAMPLES OF COMPLEXES WITH *N*-HETEROCYCLIC LIGANDS

Complex	Studies and references
$[Os(NH_3)_5L]^{n+}$, L = py, pz, pd, pyr, 4,4′-bpy, isn, Mepz Etpz, pzH, Phpy, im, or pyrrole	Preparation, electrochem. UV/Vis/NIR, IR (*60, 66, 68*)
cis-$[Os(NH_3)_4L_2]^{n+}$, L = pz or Phpy	Preparation, electrochem., UV/Vis/NIR, IR (*68*)
trans-$[Os(NH_3)_4(L)X]^{n+}$	Section II,B,4,b
fac-$[Os(NH_3)_3(im)_3]^{3+/2+}$	Preparation, IR, UV/Vis (*90*)
$[Os(trpy)(NH_3)Cl_2]^+$	Preparation, UV/Vis, IR, electrochem. (*41, 42, 226*)
$[Os(L\text{-}L)(NH_3)Cl_3]$, L-L = 2py or 4,4′-$Me_2$bpy	Preparation, UV/Vis, IR, electrochem. (*42, 226*)
mer-$[OsCl_3(pic)_2(NCCH_3)]$	Preparation, X-ray, UV/Vis, ^{1}H NMR, IR (*42, 226*)
$[Os(H)(CO)(Hbim)(PPh_3)_2]$	Preparation, IR, ^{1}H and ^{31}P NMR (*227*)
$[(H)(CO)(PPh_3)_2Os(bim)Ir(cod)]$	Preparation, IR, ^{1}H and ^{31}P NMR (*227*)
$[Os(trpy)(bpy)(4,4'\text{-}bpy)]^{2+}$	Preparation, (*228*)
$[Os(tterpy)(X\text{-}tterpy)]^{n+}$, X = H, mv^{2+}, or ptz	CV, emission, intramolecular e^- transfer (*229*)
$[Os(Me\text{-}tterpy)(X\text{-}tterpy)]^{n+}$, X = Me, MeO, Br, mv^{2+}, ptz, or diaa	CV, UV/Vis, ^{1}H NMR, mass spectrosc. (*230*)
cis-$[Os(4,4'\text{-}X_2\text{-}5,5'\text{-}Y_2bpy)_2(CO)Cl]^+$, X = H, Me, Cl, NEt_2, or OMe, Y = H; X = Me, Y = H or Me	^{1}H NMR, UV/Vis absorption and emission, IR (*231*)
cis-$[Os(4,4'\text{-}X_2\text{-}5,5'\text{-}Y_2bpy)_2Cl_2]$, X = H, Me, Cl, NEt_2, or OMe, Y = H; X = Me, Y = H or Me)	^{1}H NMR, UV/Vis, IR (*231*)
$[Os(bpy)_2(dpq)]^{2+}$	Preparation, CV (*232*)
$[Os(dpp)_3]^{2+}$	Preparation, emission (*233*)
$[Os(4,4'\text{-}dcbpy)_3]^{4-}$	Preparation, emission (*233*)
fac-$[Os(py)_3X_3]^{0/1+}$, X = Cl^-, or Br^-	Preparation, CV, UV/Vis, IR, mag., cond. (*234*)
mer-$[Os(py)_3X_3]^{0/1+}$, X = Cl^- or Br^-	Preparation, CV, UV/Vis, IR, mag., cond. (*234*)
mer-$[Os(py)_3X_3]^{0/1+}$, X = Cl^-	X-ray (*234*)
trans-$[Os(O)_2(py)_3(OH_2)]^{2+}$	Preparation, CV, ^{1}H NMR, UV/Vis, IR, Raman (*235*)
trans-$[Os(O)_2(py)_2X_2]^{2+}$, X = Cl^- or Br^-	Preparation, CV, ^{1}H NMR, UV/Vis, IR, Raman (*235*)
$[Os(Me\text{-}tterpy)(O)_2(OH)]^+$	Preparation, ^{1}H NMR, mass spectrosc., CV (*230*)
trans-$[Os(O)_2(5\text{-}SO_3^-\text{-}bpy)(chd)]$	Preparation, ^{1}H NMR, IR (*236*)
trans-$[Os(N)(trpy)X_2]^+$, X = Cl^- or Br^-	CV, ^{1}H NMR, UV/Vis (*41, 42, 226*)
trans-$[Os(N)(trpy)X_2]^+$, X = Cl^-	X-Ray (*41*)
trans-$[Os(N)(L)_2Cl_3]$, L = py, pic, or Bu^tpy	Preparation ^{1}H NMR, UV/Vis, IR (*42, 226*)
mer-$[Os(N)(L\text{-}L)Cl_3]$, L-L = bpy, 4,4′-Me_2bpy, η^2-trpy, or η^2-HC(py)$_3$	Preparation, UV/Vis, ^{1}H NMR, IR (*42, 226*)

TABLE III

NITROSYL COMPLEXES

Complex	Studies and references
$[Os(NO)]^4$	
cis-$[Os(NO)(NSCl)Cl_4]^-$	Preparation, IR, X-ray (indicates an $NSCl^{2-}$ ligand) (*237*)
$[Os(NO)]^5$	
$[Os(NO)F_5]^-$	X-Ray, XPS, EPR, electrochem. (*238*)
$[Os(NO)]^6$	
$[Os(NH_3)_5(NO)]^{3+}$	(*2*)
$[Os(NO)Cl_3]$	Reaction with trithiazyl chloride (*237*)
$[Os(NO)F_5]^{2-}$	Variable-temp. IR and Raman, including assignments and normal coordinate analyses; X-ray (*239*)
$[Os(NO)Cl_5]^{2-}$	CNDO/2 calculations of atomic charges and orbital populations (*240, 241*); variable-temp. IR and Raman, including assignments and normal coordinate analyses; X-ray (*239*); synthesis (*237*); reaction with pycaH (*242*)
$[Os(NO)Br_5]^{2-}$	Variable-temp. IR and Raman, including assignments and normal coordinate analyses; X-ray (*239*)
$[Os(NO)I_5]^{2-}$	Variable-temp. IR and Raman, including assignments and normal coordinate analyses; X-ray (*239*)
trans-$[Os(NO)Cl_4(OH_2)]^-$	Calculation of atomic charges and core binding energies (*241*)
mer,trans-NO,O-$[Os(NO)Cl_3(\eta^2\text{-}N,O\text{-}pyca)]^-$	Preparation, IR (*242*)
trans-$[Os(NO)(NH_3)_4X]^{2+}$, X = F^-, Cl^-, Br^-, I^-, OH^-, or NO_3^-	Variable-temp. IR and Raman, including assignments and normal coordinate analyses (*243–245*)
trans-$[Os(NO)(NH_3)_4L]^{3+}$	Variable temp. IR and Raman, including assignments, normal coordinate analyses (L = OH_2, NH_3), and hydrogen bonding between OH_2 ligand and Cl^- anions (*243, 244*)
trans-NO,OH$_2$-$[Os(NO)(NH)_3X(OH_2)]Y_2$, X = Cl^-, Y = Cl^-, Br^-, or I^-, X = Y = Br^- or I^-	Hydrogen bonding between the OH_2 ligand and Y^- counterions as studied by variable-temp. IR and Raman (*244*)
trans,trans-Cl,C-$[OsCl(NO)(L\text{-}L)(PPh_3)_2]$, L = η^2-*C,O*-	Preparation from *trans*-$[OsCl(NO)(C_2H_4)(PPh_3)_2]$ and

TABLE III (*Continued*)

Complex	Studies and references
$CH_2CH_2S(NSO_2C_6H_4Me\text{-}4)O^{2-}$	O=S=$NSO_2C_6H_4Me$-4 and characterization by ^1H and ^{31}P NMR, IR, and single-crystal X-ray diffraction (*163*)
trans-$[Os(NO)X_3L_2]$, L = PPh_3 or $AsPh_3$; X_3 = Cl_3, $Cl_2(Br)$, $Cl(Br)_2$, or Br_3	Preparation and characterization by IR and m.p. (*246*) (L = PPh_3; X = Cl^-); byproduct (*166*); atomic charges and core binding energies (*241*)
$[Os(NO)Cl_2(S_2CNR_2)(PPh_3)]$	Synthesis and characterization by IR, UV/Vis, and m.p. (R = Me or Et) (*246*)
trans,trans-NO,X-$[Os(NO)Cl(X)(CF_2)(PPh_3)_2]X$	Intermediates in the reactions of *trans*-$[Os(NO)Cl(CF_2)(PPh_3)_2]$ with X_2 (X = Cl or I) (*166*)
trans,trans-NO,X-$[Os(NO)Cl(X)(CO)(PPh_3)_2]X$	Intermediates in the reactions of *trans,trans-NO,X*-$[Os(NO)Cl(X)(CF_2)(PPh_3)_2]X$ with OH_2 (X = Cl or I) (*166*)
trans,cis-X,X-$[Os(NO)Cl(X)_2(PPh_3)_2]$	Decomposition product of *trans,trans-NO,X*-$[Os(NO)Cl(X)(CO)(PPh_3)_2]X$ (X = Cl^- or I^-) (*166*)
trans,trans-NO,X-$[Os(NO)Cl(X)(CF_3)(PPh_3)_2]$	Preparation, IR spectra (X = Cl^- or I^-) and X-ray structure (X = Cl^-) (*166*)
{*trans,trans-NO,I*-$[Os(NO)ClI(CF_3)(PPh_3)_2]\}_2$·*trans-P,P-cis*-$[Os(NO)I_2(CF_3)(PPh_3)_2]$	X-Ray structure (*166*)
trans,trans-NO,I-$[OsCl(I)(\eta^1\text{-}C\text{-}CS_2Me)(NO)(PPh_3)_2]$	Preparation, IR, ^1H NMR, and m.p. (*153*)
trans,trans-NO,CS₂Me-$[Os(NO)I(\eta^1\text{-}C\text{-}CS_2Me)(CNBu^t)(PPh_3)_2]^+$	Preparation, IR, ^1H NMR, and m.p. (*153*)
trans,trans-NO,C-$[Os(NO)I(\eta^2\text{-}C,S\text{-}S{=}C(SMe))(PPh_3)_2]^+$	Preparation, IR, ^1H NMR, and m.p. (*153*)
trans,trans-NO,C-$[Os(NO)Cl(\eta^2\text{-}C,S\text{-}C{=}S(SMe))(PPh_3)_2]^+$	Preparation, IR, ^1H NMR, and m.p. (*153*)
trans,trans-NO,C-$[Os(NO)I(\eta^1\text{-}C\text{-}CS_2Me)(\eta^1\text{-}S_2CNMe_2)(PPh_3)_2]$	Preparation, IR, ^1H NMR, and m.p. (*153*)
trans,cis-$[Os(NO)(H)_2(\eta^1\text{-}C\text{-}CS_2Me)(PPh_3)_2]$	Preparation, IR, ^1H and ^{31}P NMR, and m.p. (*153*)
trans-P,P-cis-$[Os(NO)(CS)I_2(PPh_3)_2]^+$	Preparation, IR, and m.p. (*153*)
$[Os(trpy)(bpy)(NO)]^{2+}$	CV (*43*)
cis-$[Os(N\text{-}N)_2(NO)(X)]^{2+}$, N-N = bpy or phen; X = Cl^-, Br^-, OH^-, or NO_2^-	Synthesis, IR, CV (*247*)
cis-$[Os(N\text{-}N)_2(NO)(OH_2)]^{3+}$, N-N = bpy or phen	Synthesis, IR, CV (*247*)
$[OsCl(NO)(ONPh)(PPh_3)_2]$	Section II,C,4,g (*248*)

(continued)

TABLE III (*Continued*)

Complex	Studies and references
[Os(NO)]^7	
cis-$[Os(N\text{-}N)_2(NO)(X)]^{2+}$, N-N = bpy or phen; X = Cl^-, Br^-, OH^-, or NO_2^-	CV (*247*)
cis-$[Os(N\text{-}N)_2(NO)(OH_2)]^{3+}$, N-N = bpy or phen	CV (*247*)
[Os(NO)]^8	
trans-$[OsCl(NO)(PPh_3)_2]$	Reactive intermediate in equilibrium with the six-coordinate alkene complexes (*163*)
$[Os(NO)Cl(PPh_3)_3]$	Reaction with $Cd(CF_3)_2$ (*166*)
trans-$[Os(NO)Cl(AsPh_3)_3]$	Preparation and characterization by IR and m.p. (*246*)
trans-$[Os(NO)(CS)(PPh_3)_3]^+$	Preparation, IR, and m.p. (*153*)
trans-$[Os(NO)Cl(CF_2)(PPh_3)_2]$	Preparation and reaction with X_2/F^- (X = Cl or Br) (*166*)
trans-$[Os(NO)I(CS)(PPh_3)_2]$	Preparation, IR, and m.p. (*153*)
trans-$[Os(NO)Cl(CS)(PPh_3)_2]$	Preparation, IR, and m.p. (*153*)
trans-$[Os(NO)(CO)(CS)(PPh_3)_2]^+$	Preparation, IR, and m.p. (*153*)
trans-$[Os(NO)(SeH)(CS)(PPh_3)_2]$	Preparation, IR, ^1H NMR, and m.p. (*153*)
trans-$[OsCl(NO)(C_2H_4)(PPh_3)_2]$	Reaction with $O{=}S{=}NSO_2C_6H_4Me\text{-}4$ (*163*)
trans-$[OsCl(NO)\{\eta^2\text{-}(C,S)\text{-}CS_2\}(PPh_3)_2]$	Synthesis and methylation (*153*)
trans,trans-N,N-$[OsCl(NO)(NO)(L\text{-}L)(PPh_3)_2]$	(L-L = η^2-*N,S*-$S(O)N(SO_2C_6H_4Me\text{-}4)$). Preparation and characterization by IR and ^1H and ^{31}P NMR (*163*)
$[OsCl(NO)_2(PPh_3)_2]$	X-Ray, ^{15}N NMR (*248a*)

Hydrazine ligands such as *N*′-(2-hydroxybenzoly)-*N*-(4-toluenesulfonyl)hydrazine, *N*-(4-hydroxy-3-methoxylbenzylidene)hydrazine carbothioamide, and *N*′-benzoyl-*N*-(4-toluenesulfonyl)hydrazine react with $K_2[Os(O)_2(OH)_4]$ to form $[Os^{VI}(O)_2L_2]$. These are used as the basis of spectrophotometric determinations of Os (*250–253*). The value of the formation constant of the complex between the second ligand and $[Os(O)_2(OH)_4]^{2-}$ in acetic acid solutions is reported to be 6.8 at 28°C (*251*). It is uncertain whether any chemical transformations of the ligands have occurred during binding, because the products are not well characterized.

i. Thionitrosyl Complexes. The atomic charges and core binding energies of $[Os(NS)Cl_5]^{2-}$, $[Os(NS)Cl_3(PPh_3)_2]$, and $[Os(NS)Cl_4(OH_2)]^-$ have been calculated and compared with nitrosyl and Ru analogs. The conclusion is drawn that the NS ligand is a better electron acceptor than the NO ligand and that, consistent with experimental results, it should be better at labilizing a *trans* ligand (*240, 241*).

j. Dinitrogen Complexes. The preparation of $[Os(NH_3)_5(N_2)]Cl_2$ from $(NH_4)_2[OsCl_6]$ and $NH_2NH_2 \cdot H_2O$ is discussed in Section II, B,4,a (*47, 59*)

The reactions of *mer*-$[Os(X)_2(N_2)(PMe_2Ph)_3]$ ($X = Cl^-$ or Br^-) with RS^- in acetone result in the displacement of X^- to give *mer*-$[Os(X)(SR)(N_2)(PMe_2Ph)_3]$ (R = Ph, C_6F_5, Me, or CF_3). The structure of one complex ($X = Cl^-$, $R = C_6F_5$) was determined by X-ray crystallography. If the starting material is reacted with a dithiocarbamate ligand or RS^- at elevated temperatures, the dinitrogen ligand is lost (Section II,C,7,q) (*254*).

Reduction of *mer*-$[Os(Cl)_3(PMe_2Ph)_3]$ to *mer*-$[Os(Cl)_3(PMe_2Ph)_3]^-$ in the presence of N_2 yields *trans,mer*-$[OsCl_2(PMe_2Ph)_3(N_2)]$ (^{31}P NMR spectroscopy and CV), which reverts to the thermodynamically stable cis isomer. Both cis and trans isomers exhibit reversible Os(III/II) couples in thf, demonstrating that the corresponding Os(III) complexes are stable on the CV time scale. It appears that a range of other *trans,mer*-$[OsCl_2L_3(N_2)]$ complexes have been prepared inadvertantly following the reduction of *mer*-$[OsCl_3L_3]$ in CH_2Cl_2 (L = $PMePh_2$, $PEtPh_2$, PEt_2Ph, PEt_3, $PPr^n{}_3$, or $PBu^n{}_3$) under an N_2 atmosphere (*145*). The original authors did not recognize the reason for the irreversibility of the reduction process (*255*).

k. Imine Complexes. The preparations of $[Os(en)_2(enim)]^{2+}$, $[Os(en)_2(diim)]^{2+}$, and $[Os(en)(diim)_2]^{2+}$ are discussed in Sections II,C,4,b and V,E,2,b. The diim ligand imparts similar properties on both the Os(III/II) redox potentials, and the Os(II) charge-transfer spectra, as the N-heterocyclic ligands, bpy, phen, etc. Cyclic voltammetry shows that the Os(III) complexes are stable on the CV time scale even at low scan rates (*204, 205*). $[Os(bpy)_2(impy)]^{2+}$ is also prepared by an oxidative dehydrogenation reaction (*256*).

$[Os^{II}(NH_3)_5\{NH{=}C(Ch_3)R\}]^{2+}$ (R = H or CH_3) are prepared from the condensations of acetone or acetaldehyde with traces of $[Os(NH_3)_6]^{3+}$ in the presence of a large excess of $[Os(NH_3)_6]^{2+}$. The latter reduces the Os(III) condensation products with concomitant regeneration of the $[Os(NH_3)_6]^{3+}$ catalyst and in so doing results in the quantitative conversion of $[Os(NH_3)_6]^{2+}$ to the corresponding Os(II) imine complex. The imine complexes are reversibly oxidized to their corresponding Os(III) complexes on the CV time scale (*194*).

l. η^1-(N)- and η^1-(N,O)-Oxime Complexes. $[Os(H)(Cl)(CO)(PR_3)_2]$ add aldoximes and ketoximes to give $[Os(H)(Cl)(CO)\{(N)N(OH){=}CR'R''\}(PR_3)_2]$ ($R_3 = Pr^i_3$, R′ = R″ = Me or R′ + R″ = cyclohexyl; $R_3 = MeBu^t_2$, R′ = H, R″ = Me, R′ = R″ = Me or R′ + R″ = cyclohexyl). Similarly, $[Os(H)(Cl)(CO)(PMeBu^t_2)_2]$ reacts with Na[ON=CR′R″], or $[Os(H)(CO)\{\eta^1\text{-}(N)\text{-}N(OH){=}CR'R''\}(PMeBu^t_2)_2]$ reacts with NaH, to yield complexes containing deprotonated aldoxime or ketoxime ligands, $[Os(H)(CO)\{\eta^2\text{-}(N,O)\text{-}ON{=}CR'R''\}(PMeBu^t_2)_2]$. The latter (R′ = R″ = Me) reacts with CO to give the O-bound ketoxime complex, *trans-P,P-cis-C,C*-$[Os(H)(CO)_2\{ON{=}C(CH_3)_2\}(PMeBu^t_2)_2]$. $[Os\{(E)\text{-}CH{=}CHPh\}(CO)\{\eta^2\text{-}(N,O)\text{-}N(O){=}C(Me)_2\}(PMeBu^t_2)]$ is also prepared from the reaction of $[Os\{(E)\text{-}CH{=}CHPh\}(Cl)(CO)(PMeBu^t_2)_2]$ with $Na[N(O){=}C(CH_3)_2]$ (*124*).

$[Os(NH_3)_5(OSO_2CF_3)]^{2+}$ reacts with HONRR′ to give Os(III) oxime complexes, which are reduced to Os(II) to yield nitrile complexes (Section V,E,4,g) (*195*).

m. Diazene Complexes. Reactions of $[Os(H)_2L_4]$ [L = $P(OEt)_3$ or $PPh(OEt)_2$] with arenediazonium cations result in the preparations of $[Os(RN{=}NH)_2L_4](BPh_4)_2$ and $[Os(H)(RN{=}NH)L_4](BPh_4)$ (R = Ph, 4-MeC_6H_4, or 4-$MeOC_6H_4$). They were characterized by NMR (^1H and ^{31}P) spectroscopy and ^{15}N isotopic substitution. $[Os(4\text{-}MeC_6H_4\text{-}N{=}NH)(4\text{-}MeC_6H_4NC)L_4](BPh_4)_2$ was prepared in a similar manner. Deprotonation of the bis(aryldiazene) complexes with Et_3N results in the formation of the five-coordinate Os(II) complexes, $[Os(RN_2)L_4](BPh_4)$ (R = Ph or 4-MeC_6H_4). Protonation of these complexes with CF_3CO_2H or HBF_4 gives the six-coordinate $[Os(CF_3CO_2)(RN{=}NH)L_4]^+$ and the five-coordinate $[Os(RN{=}NH)L_4]BPh_4$ (R = 4-MeC_6H_4), respectively (*188*).

n. Nitrile Complexes. Pentaammine complexes with acetonitrile (*59, 67, 176*), propionitrile (*176*), *tert*-butylnitrile (*176*), acrylonitrile (*176*), benzonitrile (*67, 176*), pentafluorobenzonitrile (*176*), 9-cyanoanthracene (*176*), 1,2-, 1,3-, and 1,4-dicyanobenzene (*196*), and glutaronitrile (*196*) are prepared from $[Os(NH_3)_5(OSO_2CF_3)](CF_3SO_3)_2$ and exhibit reversible Os(III/II) couples. There is a progressive decrease in the C≡N stretching frequency for both the Os(II) and Os(III) complexes along the series R = Me, $CH_2{=}CH$, Ph, or C_5F_5 (*176*). The intensities of the coordinated and free C≡N stretching modes in the IR spectra have also been the subject of detailed analysis for the complexes with dinitrile ligands, $[Os(NH_3)_5L]^{3+/2+}$ (L = 1,2-dcb, 1,3-dcb, 1,4-dcb, and gn). These intensities were compared with analogous Co(III), Rh(III), Ir(III), Ru(III), and Ru(II) complexes and interpreted in terms of π backbonding and other factors (*196*). $[Os(NH_3)_5(NCCH_3)]^{3+}$ is less susceptible to nucleophilic attack at the nitrile than is its Ru analog (*67*). The IR spectra and reactivity patterns of nitrile complexes provide strong evidence for the importance of π-backbonding stabilization of the Os(III) complexes, in addition to the Os(II) complexes, albeit to a lesser extent for the former. The UV/Vis spectra of the Os(III) and Os(II) complexes have also been studied in detail (*176*).

trans-$[Os^{II}(NH_3)_4(\text{4-lutdn-H})(NCR)]^{2+}$ (R = Me or Ph) are prepared by dissolving $[Os(NH_3)_5(\text{4-lutdn-H})]^{2+}$ in the nitrile (*85*). *trans*-$[Os^{II}(CN)_4(NCMe)_2]^{2-}$, $[Os(CN)_2\{CNCH(Ph)CH_2OH\}_4]$ (Section II,C, 2,a), and *trans*-$[Os^{II}(\text{14-tmc})(NCCH_3)_2]^{2+}$ have also been reported (Section II,C,4,c) (*39, 107, 108*). The Os(IV) nitrile complex *cis*-$[OsCl_4(NCCH_3)_2]\cdot\frac{1}{2}CH_3CN$ is prepared from the reaction of $[OsCl_5]$ with CH_3CN and has been characterized by X-ray crystallography and IR and Raman spectroscopies (*257*).

Luminescent spectra of *cis*-$[OsL_2(CO)\{NC(CH_2)_nMe\}]^{2+}$ (L = bpy or phen, n = 0–19) and *cis*-$[Os(bpy)_2(CO)(NCPh)]^{2+}$ (*258, 259*) have shown long-lived excited states at room temperature. The nitrile ligands undergo addition reactions with nucleophiles such as alcohols and are also photolabile. By contrast, the MLCT (metal–ligand charge transfer) excited states of the aminobenzonitrile complexes are short-lived and decay rapidly to the LLCT (ligand–ligand charge transfer) states, *cis*-$[Os(bpy)(bpy^{\cdot-})(CO)(abn^{\cdot+})]^{2+}$ (*259*). Heating *cis*-$[Os(bpy)(dppe)Cl_2]$ in CH_3CN results in the displacement of one chloride to produce the *cis*-$[Os(bpy)(dppe)(NCCH_3)Cl]^+$ isomer in which the nitrile is trans to an Os—P bond. This isomer is thermally unstable and isomerizes to the cis isomer in which the nitrile is trans to an Os—N bond. The rate constants for the substitution and isomerization reactions were studied (*98*). CVs of these complexes reveal reversible Os(III/II) couples (*98, 260*). *mer*-$[OsCl_3(\text{4-pic})_2(NCCH_3)]$ is one of the products of the coupling of nitrido complexes to give dinitrogen. It has been characterized by UV/Vis, IR, and ^1H NMR spectroscopies and by an X-ray structure (*226*).

$[Os(H)(Cl)(PHRR')(CO)(Cl)(PPh_3)_2]$ (R = R′ = H; R = H, R′ = Ph; R = R′ = Ph) and $HClO_4/CH_3CN$ give an isomeric mixture of $[Os(Cl)(PHRR')(NCCH_3)(CO)(PPh_3)_2]^+$ in which the nitrile is a labile ligand (*131, 261*). The reactions of *trans*-$[OsCl_2(PMe_2Ph)_3]$ with RCN give $[OsCl_2(PMe_2Ph)_3(NCR)]$ (R = CH_3 or Ph) (*145*) and $[OsCl(NCCH_3)(CO)(CTe)(PPh_3)_2]ClO_4$ have been reported (Section II, C,2,c) (*151*). The first η^1-N-bound complexes of tetracyanoethylene, *cis*-$[Os(S_2PR_2)_2(PPh_3)(tcne)]$, have also been reported (see Section II,C,7,r).

$[OsNX_5]^{2-}$ (X = Cl^- or Br^-) react with acetonitrile to give *trans*-$[OsNX_4(NCCH_3)]^-$, which were characterized by ^{35}Cl and 8^1Br NQR spectroscopy and other techniques (*262, 263*). $[Os^{IV}(cp)_2(NCCH_3)]^{2+}$ is obtained from the disproportionation of $[(cp)_2OsOs(cp)_2]^{2+}$ in acetonitrile and its reactions with nucleophiles have been studied (*264*).

o. Chlorothionitrene Complex. Molten $(NSCl)_3$ reacts with $[OsCl_3(NO)]$ to yield a mixture of products, including $(PPh_4)\{$*cis*-$[OsCl_4(NO)(NSCl)]\}$, which has been characterized by IR spectroscopy and X-ray crystallography. The NO and NSCl ligands are disordered in the structure (*237*).

p. Imido Complexes. $[Os^{VIII}(NBu^t)_4]$ was prepared recently from the reaction of $[OsO_4]$ with excess $Bu^tNH(SiMe_3)$ and was characterized by mass, IR, and ^1H and ^{13}C NMR spectroscopies, and electrochemistry (*265*). The complex undergoes a reversible reduction to its Os(VII) analogue in CH_2Cl_2. A remarkable trigonal-planar Os(VI) complex $[Os(NAr)_3]$ (Ar = 2,6-$Pr^i{}_2C_6H_3$), has been prepared from the reaction of $[OsO_4]$ with ArNCO. The air-stable complex has been characterized by X-ray crystallography and IR and NMR (^1H and ^{13}C) spectroscopies. It is believed to be stabilized by a combination of steric hindrance and the strong σ- and π-donor properties of the ligand. It reacts with Me_3NO to give $[Os^{VIII}(O)(NAr)_3]$ and with PR_3 [R_3 = Me_2Ph, Me_3, $(OMe)_3$, or $(OPh)_3$] to give the square-planar $[Os^{IV}(NAr)_2(PR_3)_2]$. All complexes have been characterized by IR and NMR (^1H and ^{13}C) spectroscopies, and for the complex with PR_3 = PMe_2Ph, by X-ray crystallography (*22*). $[Os^{VI}(NR)_3]$ (R = 2,6-$Me_2C_6H_3$, 2,6-$Pr^i{}_2C_6H_3$) have also been prepared by heating $[OsO_4]$ in neat $NHR(SiMe_3)$ for several hours at 100–120°C. Both show two irreversible one-electron reductions (*265a*). Reactions of these Os(VIII) complexes with alkenes are discussed in Sections II,C,4,b and V,E,1,c.

The bis(imido) Os(VIII) complex $[Os(\eta^4$-hba-b$)(NPh)_2]$ is believed to be an intermediate in the formation of the amido complex, *cis*-β-$[Os\{\eta^2$-(N,N')-$PhNC_6H_4NH\}(\eta^4$-hba-b$)]$ from *trans*-$[Os^{IV}(\eta^4$-hba-b$)(PPh_3)_2]$ and phenyl azide (*44*). Complexes of the type $[Os(NR')(R)_4]$ (R′ = CH_3, CH_2Ph, or CH_2SiMe_3) and $[Os(O)(NBu^t)(mes)_2]$ are discussed in Section II,C,2,f, and *trans*-$[Os(NH)(CN)_4(OCOCF_3)]$ is discussed in Section II,C,6,o.

TABLE IV

MONONUCLEAR NITRIDO COMPLEXES

Complex	Studies and references
Os(VIII)	
$[Os(O)_3(N)]^-$ and $[Os(O)_3(N)L]^-$	Sections II,B,2 and II,C,6,b
Os(VI)	
$[Os(N)(CN)_5]^-$	Preparation, IR, UV/Vis, X-ray *(266)*
$[OsNR_4]^-$	Section II,C,2,f
cis-$[OsNR_2X_2]^-$	Section II,C,2,f
$[OsNX_4]^-$	IR, Raman, emission, Xtal Vis *(267)*; ^{35}Cl, ^{81}Br, and ^{127}I NQR *(262, 263)*; XPS *(268)*
$[Os(N)(NH_3)_4]^{3+}$	Preparation, react., UV/Vis, photochem., CV *(269, 270)*
$[Os(N)(\eta^4\text{-hba-b})]^-$	Section II,C,4,x
$[Os(N)X_5]^{2-}$	^{35}Cl, ^{81}Br, and ^{127}I NQR *(263)*; XPS *(268)*
trans-$[Os(N)X_4L]^-$	^{35}Cl, ^{81}Br, and ^{127}I NQR (L = $NCCH_3$, H_2O) *(262, 263)*, XPS (L = $NCCH_3$, H_2O, py) *(268)*, preparation, cond., IR, photochem., reactions (X = CN^-, L = OH^-) *(266)*
$[Os(N)(trpy)X_2]^+$	Section II,C,4,e
$[Os(N)(L)_2Cl_3]$, L = N heterocycle	Section II,C,4,e
Os(V)	
trans-$[Os(N)L_2(OH_2)]$, L = 1-amidino-2-thioureas	Mag., therm. anal., IR *(271)*
trans$[Os(N)(CN)_4(OH)]^{3-}$	Preparation and reactions *(266)*
$[Os(trpy)(bpy)(N)]^{2+}$	Reactions *(43)*

q. Nitrido Complexes. Apart from the μ-nitrido complexes that are discussed in Sections II,B,3,b and II,D, a large number of mononuclear complexes have been prepared. General synthetic routes have been discussed in Section II,B,3,a and in Griffith's review *(2)*. A summary of publications since Griffith's review is given in Table IV *(43, 262, 263, 266–271)*.

r. Azido Complexes. $[Os(N_3)(C_6H_4Me\text{-}4)(CO)(PPh_3)_2(SNNMe_2)]$ has been reported *(185)*.

s. Cyanato Complexes. *trans*-$[Os(O)_2(NCO)_4]^{2-}$ is prepared by the reaction of *trans*-$[Os(O)_2(OR)_4]^{2-}$ (R = H or CH_3) with KCNO in methanol. The binding mode has been assigned on the basis of IR and Raman spectroscopies and its UV/Vis spectrum has been reported (*272*). *trans-O,O-cis-C,C*-$[Os(O)_2(CN)_2(NCO)_2]^{2-}$ has been characterized by IR and Raman spectroscopies (Section II,C,2,a) (*113*). On the basis of IR spectroscopy, the reaction product of $[OsF_6]$ with $(CH_3)_3Si(NCO)$ is believed to be either $[OsF_4(NCO)]$ or $[OsF_5(NCO)]$, but the product was not well characterized (*273*).

t. Thiocyanato Complexes. $[OsX_6]^{2-}$ (X = Cl, Br, or I) or $[Os(O)_2(OH)_4]^{2-}$ react with aqueous or methanolic solutions of KSCN to give $[Os^{III}(NCS)_n(SCN)_{6-n}]^{3-}$. The isomeric mixture is separated by chromatography to yield all possible isomers, i.e., $[Os(NCS)_6]^{3-}$, $[Os(NCS)_5(SCN)]^{3-}$, *cis*- and *trans*-$[Os(NCS)_4(SCN)_2]^{3-}$, *fac*- and *mer*-$[Os(NCS)_3(SCN)_3]^{3-}$, *cis*- and *trans*-$[Os(NCS)_2(SCN)_4]^{3-}$, $[Os(NCS)(SCN)_5]^{3-}$, and $[Os(SCN)_6]^{3-}$. Most of these complexes have been isolated previously (*2*), but it is the first time that the complete series has been characterized. The complexes are reversibly oxidized to the Os(IV) complexes and generally irreversibly reduced to Os(II). Chemical oxidation with Ce(IV) occurs without linkage isomerization to give the Os(IV) series, except for $[Os(SCN)_6]^{3-}$, which is obtained pure only by repeated crystallization. A complete IR, Raman, and electronic spectroscopic analysis of all of the Os(III) and Os(IV) complexes, including a study of the near-infrared electronic transitions, has been completed (*274, 275*).

$[Os(NH_3)_5(NCS)]^{2+}$ was first prepared by the reaction of $[Os(NH_3)_5(OH_2)]^{3+}$ with KSCN in water (*197*), and, more recently, by the reaction of $[Os(NH_3)_5(OSO_2CF_3)]^{2+}$ with NH_4SCN in acetone (*198*). It undergoes a reversible reduction to the Os(II) complex and an irreversible oxidation to Os(IV). A complete analysis of the Raman and IR spectra (including deuteration experiments) has been reported, and the electronic spectra, including near-infrared bands, are discussed. $[Os(NH_3)_4(NCS)_2]^+$ and $[Os(NH_3)_3(NCS)_3]$ are by-products of the reactions, but have not been characterized (*198*).

trans-$[Os(O)_2(OR)_4]^{2-}$ (R = H or CH_3) react with KSCN in methanol to give *trans*-$[Os(O)_2(NCS)_4]^{2-}$, which on the basis of its Raman and IR spectra has all the NCS^- ligands N bound. The UV/Vis spectrum is strongly vibronically coupled at 10 K (*272*). $[OsO_4]$ reacts with either $NO_2^-/H^+/NCS^-$ or $NO_2^-/H^+/NCS^-$/phen to produce $[Os(NO)(NCS)_5]^{2+}$ and $[Os(NO)(NCS)_3(phen)]$, previously synthesized using NH_2OH instead of NO_2^- (*276*). *trans,trans,trans*-$[Os(O)_2(CN)_2(NCS)_2]^{2-}$ has been prepared and characterized by IR and Raman spectroscopies (Section II,C,2,a) (*113*). *cis*-$[Os(N)(CH_2SiMe_3)_2(NCS)(X)]^-$ ($X = SCN^-$ or NCS^-) are obtained by the addition of two equivalents of KSCN to *cis*-$[Os(N)(CH_2SiMe_3)_2Cl_2]^-$ (*156*).

u. Selenocyanato Complexes. $[Os(NCSe)_n(SeCN)_{6-n}]^{3-}$ [n = 0–3, including cis and trans (n = 2) and fac and mer (n = 3) isomers] and $[Os^{III}(NH_3)_5(NCSe)]^{2+}$ have been prepared, purified, and characterized by IR, Raman, and UV/Vis spectroscopies in the same manner as their NCS^- analogs (*198, 277*). $[Os(NH_3)_4(NCSe)_2]^+$ and $[Os(NH_3)_3(NCSe)_3]$ are by-products of the preparation of $[Os(NH_3)_5(NCSe)]^{2+}$, but have not as yet been characterized (*198*).

v. Nitro and Thionitro Complexes. Treatment of *cis*-$[Os(L\text{-}L)(NO)(OH_2)]^{3+}$ with NO_2^-/NH_3 results in the formation of *cis*-$[Os(L\text{-}L)_2(NO_2)_2]$, which reacts with acid to form *cis*-$[Os(L\text{-}L)_2(NO)(NO_2)]^{2+}$ (L-L = bpy or phen). The complexes were characterized by IR spectroscopy (*247*). $[Os(trpy)(bpy)(NO_2)]^{2+}$ is prepared by the reaction of $[Os(trpy)(bpy)(NO)]^{3+}$ with OH^-. It undergoes reversible oxidation to the Os(III) analog (*43*). The reaction of $[Os(bpy)_2(CO)(Cl)]BF_4$ with Me_3NO and NO_2^- results in the synthesis of *cis*-$[Os(bpy)(NO_2)Cl]$ (*278*).

$[M(NO)(NH_3)_4(OH)][M'(NO)(NO_2)_4(OH)]$ (M = Os, M′ = Os or Ru; M = Ru, M′ = Os) have been prepared and characterized by electrical conductivity, X-ray phase and thermal analysis, and IR spectroscopy (*279*).

$[Os(NSO)(X)_2(PPh_3)_2]$ (X = Cl^- or Br^-) are obtained from reactions of $[Os(NO)(Cl)_3(PPh_3)_2]$ with elemental sulfur in benzene. They were characterized by magnetic measurements and IR and electronic absorption spectroscopies. No indication of the binding mode of the NOS^- ligand has been given, but it is presumably N bound. It is the first example of the binding of such a ligand to Os (*246*).

w. Schiff Base Complexes. Complexes of the type *trans*-$[Os(O)_2(salen)]$ have been prepared and characterized by CV and by IR, UV/Vis, and ^1H NMR spectroscopies, and the X-ray structure of *trans*-$[Os(O)_2(3\text{-}Bu^t\text{-}saltmen)]$ has been reported. All complexes exhibit irreversible oxidations and reductions in acetonitrile (*280*). *trans*-$[Os^{VI}(salen)(O)_2]$ is converted to *trans*-$[Os(salen)(ER)_2]$ (ER = OMe^-, OEt^-, or SPh^-) by methods similar to those of their porphyrinato analogs (Section II,C,4,d). The complexes were characterized by IR, UV/Vis, and ^1H NMR spectroscopies, and in the case of the phenylthiolato complex, by X-ray diffraction (*281*). More recently, a series of Schiff base complexes of the type *trans*-$[Os^{VI}(O)_2L]$ (L = $(ba)_2en$, $(aa)_2en$, $(Bu^t)_2en$)) and *trans*-$[Os^{IV}(SR)_2\{(ba)_2en\}]$ (R = C_6H_5, $CH_2\text{-}C_6H_5$, 3-MeC_6H_4, 4-FC_6H_4, 2-naphthyl, Et, Bu, cyclohexyl) have been prepared by similar methods and characterized by ^1H NMR, UV/Vis, and IR spectroscopies and electrochemistry. The X-ray structures of *trans*-[$Os^{VI}(O)_2\{(ba)_2en\}$] and *trans*-[$Os^{IV}(SCH_2C_6H_5)_2\{(ba)_2en\}$] · 0.5 H_2O were also reported (*281a*). [OsL_3] and [OsL_2Cl_2] [HL = 2-$HOC_6H_4CR{=}NR'$; R = Me, R′ = 4-XC_6H_4 (X = H, Me, OMe, CO_2Et, or Cl)] have been prepared and characterized by CV, EPR, and UV/Vis/NIR spectroscopies. The complexes in the Os(II), Os(III), and Os(IV) oxidation states have been characterized from the electrochemistry (*282*). [$OsL(H)(CO)(PPh_3)_2$] and [$OsL(Cl)(CO)(PPh_3)_2$] [HL = 2-$HOC_6H_4CH{=}NR$; R = Ph, $C_6H_5CH_2$, or 2- or 4-XC_6H_4 (X = CH_3, Cl, Br, or NO_2)] have also been prepared and characterized by NMR spectroscopy (*129*).

x. Chelating Amide Ligands. Although, this area of Os chemistry has only been developed recently, it has resulted in a considerable amount of new and interesting chemistry. The structures of the chelating amide ligands are given in Fig. 6, along with the structures of the cis-α and cis-β isomers referred to in this section.

The reaction of $bpbH_2$ with $K_2[Os(O)_2(OH)_4]$ yields *trans*-[Os^{III}-($bpbH_2$)$(O)_2$]Cl_2. This is deprotonated by Et_3N to give *trans*-[Os-(bpb)$(O)_2$] or reacts with PPh_3 in MeOH to give *trans*-[Os-(bpb)(PPh_3)(Cl)]. The structure of the latter has been determined by X-ray crystallography, and its CV exhibits a reversible Os(IV/III) redox couple and a quasi-reversible Os(III/II) redox couple. [Os(bpb)(PPh_3)Cl] reacts with PhIO to form a complex that is believed to be an Os(V) oxo complex. This species is a useful catalytic intermediate in the epoxidation of cyclohexane by PhIO (*283*).

Recently, complexes with tetradentate amide/phenol ligands have been studied extensively because of the abilities of the ligands to stabilize high oxidation states (*44, 146, 283–290*). $[Os(O)_2(OH)_4]^{2-}$ reacts with H_4chba-Et to yield *trans*-$[Os(\eta^4\text{-chba-Et})(O)_2]^{2-}$, which, in the presence of O_2/PPh_3, gives K_2[{*trans*-Os(η^4-chba-Et)($OPPh_3$)}$_2$O] (*290*) or, in the presence of py or 4-Bu^{t}py, gives *trans*-$[Os^{III}(\eta^4\text{-chba-Et})(Xpy)_2]^-$ (*289*). The X-ray structure of the oxo dimer has been determined (*290*). The bis(pyridine) complexes exhibit rich redox chemistry that is both metal and ligand centered; they are reversibly reduced to Os(II) and are oxidized to Os(IV). An X-ray structure of *trans*-[$Os^{IV}(\eta^4$-chba-Et)$(py)_2$] has been determined and the Os(IV) complexes are irreversibly oxidized at the ligands. This oxidation occurs in stages, giving *trans*-[Os(η^4-chba-ethylene)L_2], *trans*-[Os(η^4-chba-*t*-1,2-diRO-Et)L_2] (R = H, Me, or Et), *trans*-[Os(η^4-chba-*t*-1-OH-2-OMe-Et)L_2], and *cis*- and *trans*-[Os(η^2-fo-chba)$_2L_2$] (L = py or 4-Bu^{t}py) (*289*). Heating the latter in acid results in the decarbonylation of the complexes to form *cis*- and *trans*-[Os(η^2-chba)$_2L_2$]. These complexes are also prepared via

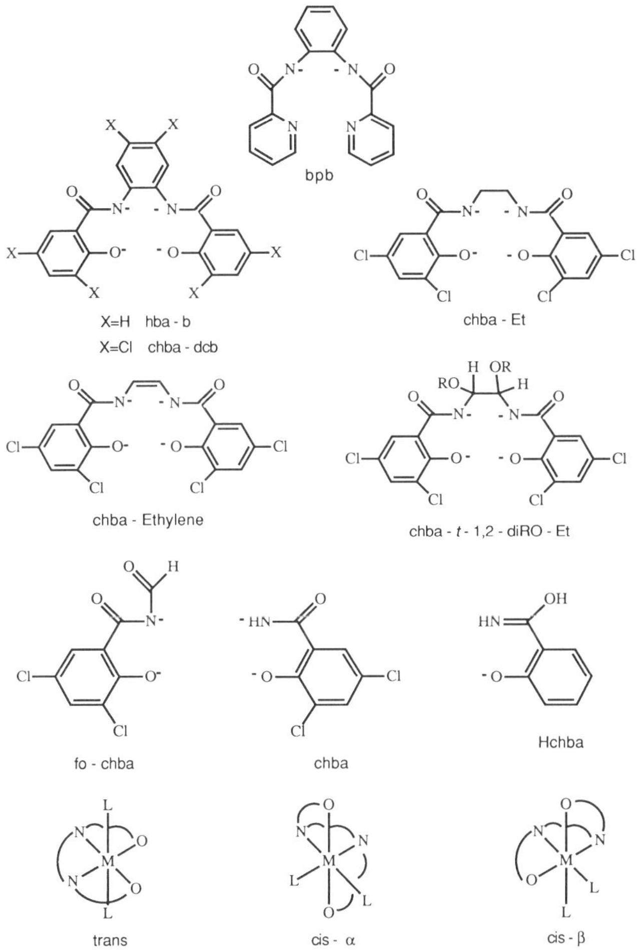

FIG. 6. Structures of the chelating amide ligands and the trans, cis-α, and cis-β geometric isomers of their complexes.

the $[Os(\eta^2\text{-chba})_2(O)_2]^{2-}$ intermediate (*287*). If $K_2[Os(\eta^2\text{-chba})_2(O)_2]$ is allowed to react with two equivalents of PPh_3, in the presence of 4-Bu^tpy, *cis*- and *trans*-$[Os(\eta^2\text{-chba})_2(4\text{-}Bu^t py)(OPPh_3)]$ are obtained. *cis*-α- or *trans*-$[Os(\eta^2\text{-chba})_2(4\text{-}Bu^t py)_2]$ is protonated without isomerization to give *cis*-α- or *trans*-$[Os(\eta^2\text{-Hchba})(4\text{-}Bu^t py)_2]^{2+}$, respectively (*287*). The structures of *trans*-$[Os(\eta^4\text{-chba-ethylene})(4\text{-}Bu^t py)_2]$ (*287*), *trans*-$[Os(\eta^4\text{-chba-}t\text{-1-OH-2-OMe-Et})(py)_2]$ (*287*), *cis*- and *trans*-$[Os(\eta^2\text{-fo-chba})_2(4\text{-}Bu^t py)_2]$ (*289*), and *trans*-$[Os(\eta^2\text{-chba})_2(4\text{-}Bu^t py)_2]$ (*287*) have been determined by X-ray crystallography and the complexes have been extensively characterized by NMR spectroscopy and other techniques. *trans*-[Os(η^4-chba-ethylene)L_2] exhibit two reversible one-electron reductions and reversible one-electron oxidations to the nominal Os(V) complexes in the absence of ROH (*289*). The electrochemistry of *trans*-[Os(η^4-chba-Et)$(py)_2$], *trans*-[Os(η^4-chba-ethylene)$(py)_2$], *trans*-[Os(η^4-chba-*t*-1,2-diEtO-Et)$(py)_2$], and *cis*-[Os(η^2-fo-chba)$_2$-$(py)_2$] have also been studied in liquid SO_2 at 40°C. All complexes show a series of reversible oxidations in addition to a reversible Os(IV/III) couple (*284*). The chemistry of this system is summarized in Scheme 12.

The complexes in which the ethylene bridge has been replaced by a phenylene bridge are considerably more stable to oxidation, and a

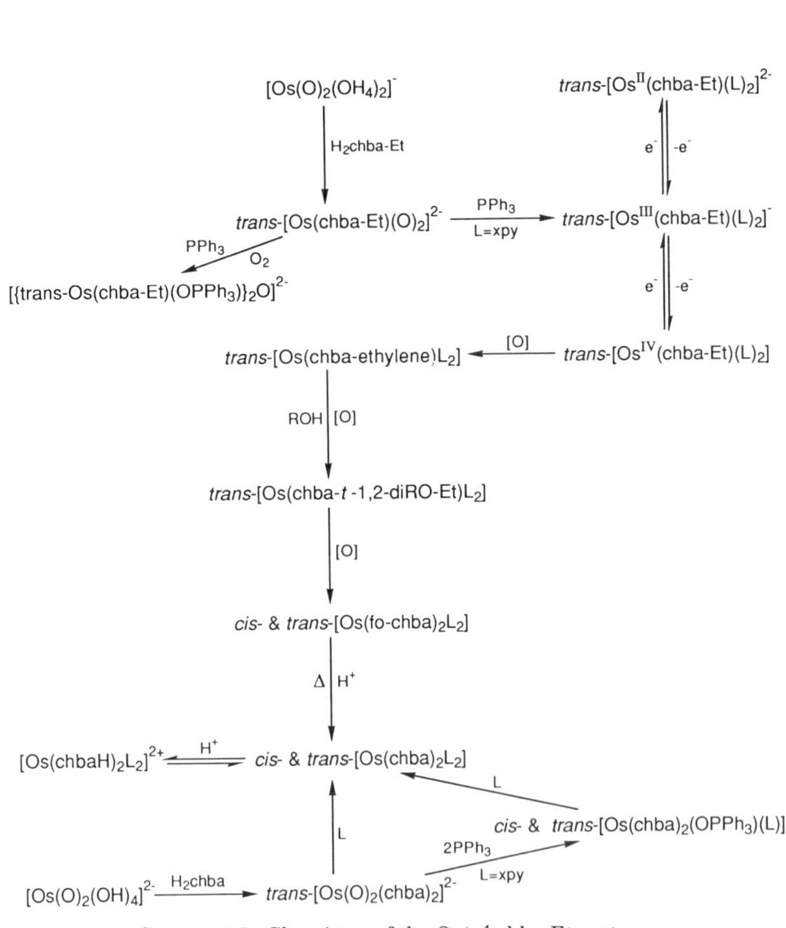

SCHEME 12. Chemistry of the Os/η^4-chba-Et system.

different type of chemistry ensues. $K_2[Os(O)_2(L)]$ (L = chba-dcb or hba-b) are prepared as indicated above (*288, 289*). Treatment with PPh_3 and CF_3CO_2H results in the synthesis of *trans*-$[Os^{IV}L(PPh_3)_2]$, or treatment with PPh_3 and 4-Xpy results in *trans*-$[Os^{IV}L(4\text{-}Xpy)_2]$ (L = chba-dcb, X = H, MeO, Et, Me, Cl, Br, Ac, or Bu^t) (*288, 289*). *trans*-$[Os(\eta^4\text{-hba-b})(PPh_3)_2]$ reacts with one equivalent of CO to form *cis*-α-$[Os(\eta^4\text{-hba-b})(PPh_3)(CO)]$ (*146*); it reacts with phen to give *mer*-$[Os^{III}(\eta^3\text{-Hhba-b})(PPh_3)(phen)]$ (*286*) and with Bu^tNC or dppe to give *cis*-α-$[Os(\eta^4\text{-hba-b})(PPh_3)(CNBu^t)]$ and *cis*-β-$[Os(\eta^4\text{-hba-b})(dppe)]$, respectively (*288*). The structures of the first two complexes have been determined by X-ray crystallography (*146, 286*). *mer*-$[Os^{III}(\eta^3\text{-Hhba-b})(PPh_3)(phen)]$ is reversibly reduced to the corresponding Os(II) complex and oxidized to the Os(IV) complex. The ligand is also acetylated using Ac_2O to form *mer*-$[Os\{\eta^3\text{-}(CH_3CO)\text{hba-b}\}(PPh_3)(phen)]$ (*286*). $[Os(\eta^4\text{-hba-b})(PPh_3)_2]$ also reacts with azides to give $[Ph_3N\text{-}NH_2][Os^{VI}(\eta^4\text{-hba-b})(N)]$, in the case of Me_3SiN_3 and *cis*-β-$[Os^{VI}(\eta^4\text{-hba-b})(\eta^2\text{-}PhNC_6H_4NH)]$, in the case of PhN_3. Both products have been characterized by X-ray crystallography (*44*).

trans-$[Os^{IV}(\eta^4\text{-chba-dcb})(PPh_3)_2]$ reacts with one equivalent, or excess 4-Bu^tpy, to give *trans*-$[Os(\eta^4\text{-chba-dcb})(PPh_3)(4\text{-}Bu^tpy)]$ and *trans*-$[Os(\eta^4\text{-chba-dcb})(4\text{-}Bu^tpy)_2]$, respectively; it reacts with bpy, Bu^tNC, or dppe to give *cis*-α- and *cis*-β-$[Os(\eta^4\text{-chba-dcb})(bpy)]$, *cis*-$\alpha$-$[Os(\eta^4\text{-chba-dcb})(PPh_3)(CNBu^t)]$, or *cis*-$\beta$-$[Os(\eta^4\text{-chba-dcb})(dppe)]$, respectively. It is also oxidized by iodosylbenzoic acid to give *trans*-$[Os(\eta^4\text{-chba-dcb})(OPPh_3)_2]$ (*146, 288, 289*). The structures of *cis*-$[Os(\eta^4\text{-chba-dcb})(bpy)]$ and *cis*-α-$[Os(\eta^4\text{-chba-dcb})(PPh_3)(CNBu^t)]$ have been determined by X-ray crystallography. Treatment of *trans*-$[Os(\eta^4\text{-chba-dcb})(O)_2]$ with PPh_3 and $CNBu^t$ followed by oxidation was initially reported to yield *trans*-$[Os(\eta^4\text{-chba-dcb})(CNBu^t)_2]$ (*289*), but the isomer was shown subsequently to be *cis*-α-$[Os(\eta^4\text{-chba-dcb})(CNBu^t)_2]$ (*288*). These complexes undergo reversible oxidations to their corresponding Os(V) complexes and two reversible one-electron reductions to their Os(III) and Os(II) analogs. On either oxidation or reduction, the complexes undergo geometric isomerization reactions. In the case of *cis*-α-$[Os(\eta^4\text{-chba-dcb})(CNBu^t)_2]$, a two-electron reduction followed by a two-electron oxidation results in the preparation of *cis*-β-$[Os(\eta^4\text{-chba-dcb})(CNBu^t)_2]$. Oxidation of *trans*-$[Os(\eta^4\text{-chba-dcb})(L)_2]$ (L = 4-*t*-Bupy or $OPPh_3$) followed by reduction at low temperatures with Fc results in the isolation of the *cis*-α-$[Os(\eta^4\text{-chba-dcb})(L)_2]$ isomers together with their trans analogs. Generally, the equilibrium constants for the geometric isomerizations are oxidation-state dependent, with *cis*-α-$[Os(\eta^4\text{-chba-dcb})(4\text{-}Xpy)_2]$ isomers becoming more stable relative to the trans isomers as the higher oxidation states are obtained. The cis complexes are rare examples of complexes with nonplanar amides. Their formation can be rationalized by consideration of the electronic demands of the higher oxidation states, because nonplanar amides maximize π donation. The equilibrium constants for the cis-α to trans isomerizations for X = Ac, Cl, Br, H, Me, Et, Bu^t, and MeO have been determined in the formal oxidation states Os(V), Os(IV), Os(III), and Os(II). They fall in the range of 0.1–1, 10–10^2, 10^{11}–10^{12}, and 10^{13}–10^{15}, respectively, for the four oxidation states at ambient temperatures, consistent with the electronic arguments discussed above. The equilibrium constants correlate with the Hammett substitutent constant of *X* (*288, 289*), and the kinetics of the isomerization reactions of *cis*-α- and *trans*-[Os(η^4-chba-dcb)($OPPh_3$)$_2$]$^{0/+}$ have been followed using UV/Vis and ^{1}H NMR spectroscopies. There is no ligand exchange during the isomerization reaction, and together with other evidence, this suggests a twist mechanism (*285*). The electrochemistry of *trans*-[Os(η^4-chba-dcb)(L)$_2$] (L = py or PPh_3) has also been followed in liquid SO_2 at − 60°C. In addition to the reversible Os(IV/III) couple, three other reversible oxidations are observed, together with a less reversible fourth oxidation. It appears that these processes are ligand centered. The electrochemistry of *cis*-α-[Os(η^4-chba-dcb)($CNBu^t$)$_2$] has also been studied in liquid SO_2, but the oxidations are less reversible in this case (*284*).

The related complex containing a chelating amido/aldehyde ligand, H_2phenba, reacts with $[Os(O)_2(OH)_4]^{2-}$ to give *trans*-[Os(O)$_2$(phenba)], which reacts with PPh_3 to give *trans*-[Os^{II}(phenba)(PPh_3)$_2$]. All complexes have been characterized by UV/Vis, IR, and ^{1}H NMR spectroscopies and cyclic voltammetry (*291*).

y. Quinolol Complexes. $(NH_4)_2[OsX_6]$ react with Hquin to give [Os^{IV}(quin)$_2$X$_2$], which, on further reaction with Hquin, yield [Os^{III}-(quin)$_3$], (X = Cl^- or Br^-). [Os(quin)$_2$X$_2$] is reduced by $NH_2NH_2 \cdot H_2O$ to yield [Et_4N][Os^{III}(quin)$_2$X$_2$], and [Os(quin)$_3$] is oxidized by Ce(IV) to give [Os^{IV}(quin)$_3$]ClO_4. The tris bidentates are assigned as having a mer structure and the bis complexes are assigned a trans,trans,trans structure on the basis of IR spectra and other considerations. All complexes exhibit reversible Os(III/II) and Os(IV/III) redox couples and a further oxidation is observed. The near-infrared and EPR spectra have also been recorded and interpreted (*282*).

z. Benzotriazole, 1,3-Diaryltriazene, 2-(Phenylazo)pyridine, and (Phenylazo)acetaldoxime Complexes. [Os(H)$_2$(CO)(PPh_3)$_3$] reacts with benzotriazole to give *trans*-[Os(bta···H···bta)(H)(CO)(PPh_3)$_2$], which

X-ray crystallography has shown to have two hydrogen-bonded bta ligands to form a chelate. Similar reactions with $[OsCl_2(PPh_3)_3]$ result in the linkage isomers, $[Os(bta\cdots H\cdots bta)_2\{\eta^1\text{-}(N^2)\text{-btaH}\}(PPh_3)]$ and $[Os(bta\cdots H\cdots bta)_2\{\eta^1\text{-}(N^3)\text{-btaH}\}(PPh_3)]$. All complexes have been characterized by IR and NMR (^1H, ^{13}C, and ^{31}P) spectroscopies (*292*).

$[OsCl_2(PPh_3)_3]$ reacts with ArNNNAr (Ar = C_6H_5, 4-MeC_6H_4, or 4-ClC_6H_4) in aerobic benzene solutions to give $[OsCl_2\{\eta^2\text{-}(N,N'')\text{-ArNNNAr}\}(PPh_3)_2]$, which were characterized by IR spectroscopy and magnetic measurements (*293*). The complexes are identical to those prepared by the reactions of Li[PhNNNPh] with $[OsX_4(PPh_3)_2]$ or $[OsX_2(O)_2(PPh_3)_2]$ (*255*). $[Os(ArNNNAr)_3]$ are prepared from $[OsCl_6]^{2-}$ and the ligand and have been characterized by magnetic measurements and IR, Raman, and ^1H NMR spectroscopies, and $[OsCl_2(ArNNNAr)_2]$ have also been reported (*293*).

Geometric isomers of $[OsX_2(L\text{-}L)_2]$ [L-L = 2-(R-phenylazo)pyridine (R = H, 3-Me, or 4-Me_2N); X = Cl^- or Br^-], in which the ligand is bound via both the azo and pyridine nitrogens, have been characterized by ^1H NMR spectroscopy (*294*). The X-ray structure of the cis,cis,cis complex with R = 3-Me and X = Cl^- was determined (*295*). (Phenylazo)acetaldoxime (HL) reacts with $[OsBr_6]^{2-}$ to give *fac*-$[Os^{II}L_3]^-$. The uncoordinated oxygens of the oxime groups act as tridentate ligands for other metal ions to form trimers of the type $[OsL_3ML_3Os]$ (M = Mg, Mn, Co, or Ni) and $[OsL_3FeL_3Os]^+$. These complexes have been characterized by electrochemistry, ^1H NMR, and EPR spectroscopies (*296*).

5. *Phosphorus, Arsenic, and Antimony*

a. Phosphine Complexes. Complexes containing only phosphine ligands will be discussed. Table V lists relevant sections for phosphine complexes with other ligands, except for halide and hydride complexes, which are listed in Tables VI (*15, 97, 145, 234, 293, 297–305*) and VII (*128, 138, 297, 299, 301, 306–309*), respectively.

$[Os^0(PMe_3)_5]$ has been prepared from the Na reduction of *trans*-$[OsCl_2(PMe_3)_4]$ in thf containing an excess of PMe_3 and a catalytic amount of naphthalene. At room temperature, a singlet is observed in the $^{31}P\{^1H\}$, $^{13}C\{^1H\}$, and ^1H NMR spectra, characteristic of fluxional behavior, whereas the spectra at − 100°C in thf are consistent with a trigonal-bipyramidal structure. The complex reacts with traces of moisture or triflic acid to produce the hydride complex, $[Os(PMe_3)_5H]^+$. In the absence of excess ligand, $[Os(PMe_3)_5]$ is in equilibrium with $[Os(PMe_3)_4]$, which undergoes an intramolecular oxidative addition reaction

TABLE V

PHOSPHINE COMPLEXES APART FROM HYDRIDES AND HALIDES

Complex or coligand	Section	Complex or coligand	Section
Osmaboranes	II,C,1	Osmacarboranes	II,C,1
CO	II,C,2,b	CS, CSe, CTe	II,C,2,c
CO_2	II,C,2,d	CS_2, CSSe, $SCSMe^-$	II,C,2,e
Alkyl and aryl	II,C,2,f	Carbene	II,C,2,g
Alkene	II,C,2,h	Allyl	II,C,2,l
Isonitrile	II,C,2,n	Formyl and C esters	II,C,2,o
SiR_3	II,C,3	Porphyrins	II,C,4,d
N heterocycles	II,C,4,e	NO, NO^+	II,C,4,f
Nitrosoarene	II,C,4,g	Hydroxylamine	II,C,4,h
NS, NS^+	II,C,4,i	N_2	II,C,4,j
Oximes	II,C,4,l	NCR	II,C,4,n
Imido	II,C,4,p	N_3^-	II,C,4,r
NOS^-	II,C,4,v	Schiff bases	II,C,4,w
Chelating amides	II,C,4,x	Diaryltriazenes	II,C,4,z
Benzotriazole	II,C,4,z	η^2-CH_2PR_2	II,C,5,c
Stibines	II,C,4,f	H_2O and OH^-	II,c,6,a
ROH, RO^-	II,C,4,c	Maltolate and tropolonate	II,C,6,d
Amide	II,C,6,g	Diketonates	II,C,6,k
O_2	II,C,6,m	CO_3^{2-}	II,C,6,n
RCO_2^-	II,C,6,o	RSO_3^-	II,C,6,q
O-sulfinate	II,C,6,r	RS^-	II,C,7,c
S_2	II,C,7,d	HSe^-	II,C,7,e
btd and bsd	II,C,7,h	*S*-dmso	II,C,7,i
$SNNMe_2$	II,C,7,j	Iminooxosulfane	II,C,7,k
SO_2, S_2O	II,C,7,l	*S*-sulfinato	II,C,7,m
$R_2NCS_2^-$ and xanthate	II,C,7,q	$R_2PS_2^-$	II,C,7,r
PyS, pymS	II,C,7,v	BH_4^-	II,C,9,b
H_2	II,C,9,c		

to form [$Os^{II}(PMe_3)_3(\eta^2$-$CH_2PMe_2)(H)$]. The latter is also prepared from the Na reduction of *trans*-[$OsCl_2(PMe_3)_4$] in the absence of excess PMe_3. [$Os(PMe_3)_5$] undergoes slow ligand exchange reactions with $P(CD_3)_3$, and the activation energy for the process, which is believed to occur via the [$Os(PMe_3)_4$] intermediate, is ~116 kJ mol^{-1} (*15*). [$Os(PMe_3)_4$] and [$Os(PMe_3)_3$] are also believed to be intermediates in the thermolysis of *cis*-[$Os(PMe_3)_4(R)(H)$] (*15, 165, 306, 310, 310a*).

b. Phosphite Complexes. Although less numerous, these complexes exhibit chemistry similar to that of their phosphine analogs, e.g., [Os(4-MeC_6H_4N=NH)(4-$MeC_6H_4NC)L_4]^+$ [L = $P(OEt)_3$ or $PPh(OEt)_2$; see

TABLE VI

MONONUCLEAR PHOSPHINE/HALIDE COMPLEXES

Complex	Studies and references
Os(II)	
mer-$[OsCl_3(PMe_2Ph)_3]^-$	CV, chloride loss (*145*)
$[OsCl_2(PMe_2Ph)_3]$	CV, addition of π-acid ligands (*145*)
$[OsCl_2(PPh_3)_3]$	Reactions with diarylazides (*293*)
cis-$[Os(PMe_3)_4(H)Cl]$	Prep. (*297*)
cis-$[OsX_2(PMe_3)_4]$, X = Cl^-, Br^-, or I^-	Preparation (*298*)
cis-$[OsCl_2(PR_2CH_2CH_2PR_2)_2]$	Preparation and reaction with H_2 (*299*)
cis-$[OsX_2(L\text{-}L)_2]$, L-L = dcpe, X = Cl^- or Br^-	Preparation, IR, UV/Vis, cond., ^{31}P NMR (*300*)
cis-β-$[OsCl_2($*meso*-tetraphos$)]$	Reaction with H_2 (*301*)
trans-$[OsCl_2(PMe_3)_4]$	Na reduction (*15*)
trans-$[OsX_2(PR_3)_4]$, X = Cl^- or Br^-	Prep. (*302*)
trans-$[OsX_2(L\text{-}L)_2]$, L-L = dcpe, X = Cl^- or Br^-	Preparation, IR, UV/Vis, cond., ^{31}P NMR (*300*)
trans-$[OsX_2(L\text{-}L_2]$, L-L = dmpe, dppe, or dppm; X = Cl^- or Br^-	Prep. (*302*)
trans-$[OsX_2(L\text{-}L)_2]$, L-L = $(Ph_2P)]_2C{=}CH_2$	Preparations, X-ray (*303*)
trans-$[Os(H)Cl($*meso*-tetraphos$)]$	Reaction with H_2 (*301*)
trans-*P*,*P*-$[OsBr_2(CO)_2(PPh_3)_2]$	X-Ray (*304*)
$[OsCl(L\text{-}L)_2]^+$, L-L = $Ph_2P(CH_2)_3PPh_2$ or $Ph_2P(CH_2)_2(2\text{-py})$	Epoxidation catalysts (*305*)
$[OsX(L\text{-}L)_2]^+$, L-L = dcpe, X = Cl^- or Br^-	Preparation, IR, UV/Vis, cond., ^{31}P NMR, addition reactions with PhCN, MeCN, and CO (*300*)
Os(III)	
trans-$[OsX_4(PR_3)_2]^-$, L = PEt_3, PPh_3, PEt_2Ph, or $PEtPh_2$; X = Cl^- or Br^-	Preparation, IR, UV/Vis, mag., NMR, CV (L = PEt_3, X = Cl^-) X-ray (*302a*)
mer-$[OsL_3X_3]$, X = Cl^- or Br^-; L = PMe_3, PEt_3, $PEtPh_2$, PEt_2Ph, or PMe_2Ph	Preparation, UV/Vis, IR, CV (*97, 145, 234, 302*)
fac-$[Os(PEt_2Ph)_3Cl_3]$	Preparation, UV/Vis, IR, CV, X-ray (*234*)
trans-$[OsX_2(PR_3)_4]^+$	Preparation (*302*)
Os(IV)	
trans-$[OsX_4(PR_3)_2]$, L = PEt_3, PPh_3, PEt_2Ph, or $PEtPh_2$; X = Cl^- or Br^-	Preparation, IR, UV/Vis, mag., NMR, CV (*302*)
mer-$[OsL_3X_3]^+$, X = Cl^- or Br^-; L = PMe_3, PEt_3, $PEtPh_2$, or PEt_2Ph	Preparation, UV/Vis, IR, CV (*234*)

TABLE VII

MONONUCLEAR PHOSPHINE/HYDRIDE COMPLEXES

Complex	Studies and references
Os(II)	
cis-$[Os(H)_2(PMe_3)_4]$	Preparation (*297*)
cis-$[Os(H)_2(PR_2CH_2CH_2PR_2)_2]$	Preparation (*299*)
cis-α-$[Os(H)_2$(*rac*-tetraphos)]	Reaction with HBF_4 (*301*)
cis,mer-$[Os(H)_2(PMe_2Ph)_3(CO)]$	Preparation (*138*)
$[Os(H)(PMe_3)_4]^+$	Intermediate (*297, 306*)
$[Os(H)(PMe_3)_5]^+$	Preparation (*297, 306*)
Os(IV)	
$[Os(H)_4(PR_3)_3]$	Protonation (Section II,C,9,c), photochemistry (*307*)
$[Os(H)_4(CO)(PPr^i{}_3)_2]$	Preparation (*128*)
$[Os(H)_3(PR_3)_4]^+$	Preparation, R = Me (*297*); preparation, X-ray, ^{31}P NMR, fluxionality, R = Ph (*308*)
Os(VI)	
"$[Os(H)_5(PR_3)_3]^+$"	Reassignment as $[Os^{IV}(\eta^2\text{-}H_2)(H)_3(PR_3)_3]^+$ (Section II,C,9,c)
$[Os(H)_6(PPhPr^i{}_2)_2]$	Neutron diffraction (*309*)

Section II,C,4,m] and $[Os(H)(CNR)(L)_4]^+$ [L = $PPh(OEt)_2$, $P(OMe)_3$, or $P(OEt)_3$; R = 4-$MeOC_6H_4$ or 4-MeC_6H_4], which were characterized by NMR (^1H and ^{31}P) and IR spectroscopies (*187*), and *trans*-$[OsCl_2\{P(OPh)_3\}_4]$ (*18*) and $[Os(CO)_3\{P(OCH_2)_3CEt\}(SiMeCl_2)_2]$ (Section II,C,3). The hydride complexes $[Os(H)_2L_4]$ (L = $P(OEt)_3$ or $PPh(OEt)_2$) have also been prepared and characterized (*188*), and complexes with H_2 ligands have been reported (Section II,C,9,c).

c. η^2-CH_2PR_2 Complexes. The intramolecular oxidative addition of $[Os(PMe_3)_4]$—produced from the thermolysis of *cis*-$[Os(PMe_3)_4(H)(R)]$ or reduction of *trans*-$[OsCl_2(PMe_3)_4]$—gives $[Os(H)(\eta^2\text{-}CH_2PMe_2)(PMe_3)_3]$ (see Section II,C,5,a) (*15, 310*), and its hydrogenolysis has been studied (*297*).

d. Arsine Complexes. The new arsine complexes, *mer*-$[Os(AsEt_3)_3X_3]$ (X – Cl^- or Br^-) and *fac*-$[Os(AsMe_2Ph)_3Cl_3]$, are prepared (*234*) using standard methods (*2*). Oxidation of the mer isomers by aqueous nitric acid gives the Os(IV) complexes, *mer*-$[Os(AsEt_3)_3X_3]^+$, but the fac isomers are too unstable to be isolated (*234*). *trans*-[Os^{IV}-$Br_4(AsPh_3)_2$] is prepared from the reaction of [$(Bu^n)_4N]_2[OsBr_6$] with $AsPh_3$ and NaOAc in acetic acid/acetic anhydride. Its structure has been determined by X-ray crystallography and it has an Os—As bond length of 2.569(1) Å (*311*). The syntheses, the IR, UV/Vis, and NMR spectroscopies, and the electrochemistry of *trans*-[$OsX_4(AsR_3)_2]^{0/-}$ (X = Cl^- or Br^-; R = Et or Ph) have been studied (*302a*).

Reaction of [Os(H)(CO)Cl$(AsPh_3)_3$], [Os(CO)$Cl_2(AsPh_3)_3$], or [Os-$(H)_2$(CO)$(AsPh_3)_3$] with NOCl in CH_2Cl_2 under reflux results in [Os-(NO)$Cl_3(AsPh_3)_2$] and in *trans-As,As-cis*-[Os$(CO)_2(Cl)_2(AsPh_3)_2$] as a by-product. [Os(H)(CO)Cl$(AsPh_3)_3$], [Os(CO)$Cl_2(AsPh_3)_3$], and [Os-$(H)_2$(CO)$(AsPh_3)_3$] react with NOBr under the same conditions to produce [Os(NO)Cl$(Br)_2(AsPh_3)_2$], [Os(NO)Cl_2Br$(AsPh_3)_2$], and [Os-(NO)$Br_3(AsPh_3)_3$], respectively. In each instance, *trans-As,As-cis, cis*-[Os$(CO)_2Br_2(AsPh_3)_2$] was the by-product. All the complexes were characterized by IR spectroscopy. [Os(NO)$X_3(AsPh_3)_2$] react with $AsPh_3$ in benzene under reflux to produce [Os(NO)X$(AsPh_3)_3$], which are believed to be distorted trigonal bipyramidal in which the arsines occupy equatorial positions. These complexes were characterized by IR and electronic absorption spectroscopy (*246*).

[OsO_4] reacts with AsR_3 in concentrated HCl/EtOH at -50°C to produce *trans,trans,trans*-[Os$(O)_2Cl_2(AsR_3)_2$] (AsR_3 = $AsEt_3$, As-Me_2Ph, or $AsPh_3$), which have been characterized by IR and ^{1}H NMR spectroscopies. If the same reaction is performed at 0°C, a mixture of *trans*-[$OsCl_4(AsR_3)_2$] and *trans,trans,trans*-[Os$(O)_2Cl_2(AsR_3)_2$] is obtained (*209*).

cis-[Os$(bpy)_2(AsPh_3)Cl]^+$ is prepared by heating *cis*-[Os$(bpy)_2Cl_2$] with $AsPh_3$ in 1:1 EtOH/water. It has been characterized from its UV/Vis spectrum and electrochemistry in CH_3CN, wherin it exhibits two reversible oxidations to the Os(III) and Os(IV) analogs and a reversible reduction. A second reduction is accompanied by loss of the Cl^- ligand (*97*).

e. Diarsine Ligands. Complexes of the type *trans*-[Os$(O)_2X_2$(L-L)] [X = Cl^- or Br^-; L-L = 1,2-$C_6H_4(AsR_2)_2$ (R = Me, Ph), $Ph_2As(CH_2)_2$-$AsPh_2$, $Me_2As(CH_2)_3AsMe_2$, or *cis*-$Ph_2AsCH{=}CHAsPh_2$] are prepared by the reactions of HX and the ligand with [OsO_4] at low temperatures in EtOH. These complexes are readily converted to [OsX_4(L-L)] by heating with HX in aqueous ethanol. Although no details of the preparation or properties of the tetrahalo complexes were given, they are reported to be prepared more conveniently from the reaction of [Os-

$X_6]^{2-}$ with L-L. The *trans*-dioxo complexes were characterized by IR, ^{1}H NMR, and electronic absorption spectroscopies (*209*).

The tris-bidentate complexes, $[Os(phen)_2(das)]^{2+}$, $[Os(bpy)_2(das)]^{2+}$, $[Os(phen)_2(dpae)]^{2+}$, $[Os(phen)(das)_2]^{2+}$, and $[Os(bpy)(das)_2]^{2+}$, are prepared by refluxing *cis*-$[Os(N\text{-}N)_2Cl_2]$ or $[Os(N\text{-}N)Cl_4]$ with the appropriate diarsine in ethylene glycol. These complexes exhibit reversible Os(III/II) oxidations and a reversible reduction that is centred at the phen or bpy ligand. In the case of $[Os(bpy)_2(das)]^{2+}$, a second ligand-centered reduction is observed in acetonitrile. The electronic absorption and emission spectra of these complexes have also been reported (*97*). $[Os(bpy)(das)_2]^{2+}$ has been used as an effective sensitizer in a photoelectrochemical half-cell (*312*).

f. Stibine Complexes. The Os(III) complexes (*2*) *mer*-$[Os(SbPh_3)_3X_3]$ ($X = Cl^-$ or Br^-) exhibit reversible reductions to *mer*-$[Os(SbPh_3)_3X_3]^-$ and oxidations to *mer*-$[Os(SbPh_3)_3X_3]^+$, but the Os(II) and Os(IV) complexes are too unstable to isolate (*234*). The structure of *mer*-$[Os(SbPh_3)_3Br_3]$ has been reported; it is the first structural determination of a complex with an Os—Sb bond. The mutually trans Os—Sb bond lengths [2.640(2) Å and 2.654(2) Å] and that trans to an Os—Br bond [2.644(2) Å] do not differ significantly (*313*). The complex containing the mixed P–Sb chelate $[Os(O)_2Cl_2\{o\text{-}C_6H_4(PMe_2)(SbMe_2)\}]$ has been prepared by the reaction of the ligand with $[OsO_4]$ and HCl in EtOH at 0°C (*209*). Similar reactions with distibine ligands result in chlorination of the ligands, rather than formation of Os complexes (*209*). The syntheses, the IR, UV/Vis, and NMR spectroscopies, and the electrochemistry of *trans*-$[OsX_4(SbPh_3)_2]^{0/-}$ have been studied (*302a*).

6. *Oxygen Donor Ligands*

a. Aqua and Hydroxo Complexes. $[Os(OH_2)_6]^{3+}$ has not been reported, but recent evidence suggests it may be obtained from reduction of $[OsO_4]$ or *trans*-$[Os(O)_2(OH)_4]^{2-}$ in aqueous acid. The complex has not been characterized fully as yet, so it is possible that it is a mixed hydride/aqua complex (*314*).

The species previously assigned as the hydroxo complex, $[Hpy]_2$-$[Os(O)_2(OH)_2Cl_2]$ (*315*), and the aqua complex, $[Hpy]_2[OsO_3Cl_2(H_2O)]$ (*316*), have had their structures reassigned as $[Os(O)_2(py)_2Cl_2]$ and $[Os(O)_2(py)_2Cl_2]$, respectively (*235*).

A summary of the aqua and hydroxo complexes studied recently is given in Table VIII (*2, 17, 37, 46, 59, 70, 72, 94, 95, 113, 126, 133, 210, 211, 240, 241, 243–245, 247, 263, 266, 279, 298, 317–326*) and μ-hydroxo complexes are discussed in Section II,D,5,c.

TABLE VIII

AQUA AND HYDROXO COMPLEXES

Complex	Section	Ref.
Os(II)		
$[Os(NH_3)_5(OH_2)]^{2+}$	II,B,4,a	*70, 72*
trans-$[Os(NO)(NH_3)_4(OH_2)]Cl_3 \cdot H_2O$	II,C,4,f	*243, 244*
$[Os(NO)(NH_3)_3X(OH_2)]Y_2$, $X = Cl^-$, $Y = Cl^-$, Br^-, or I^-; $X = Y = Br^-$ or I^-	II,C,4,f	*244*
cis,cis-$[Os(NO)(NH_3)_2X_2(OH)]$, $X = Cl^-$ or Br^-	II,C,4,f	*317*
trans,trans-$[Os(NO)(NH_3)_2X_2(OH)]$, $X = Cl^-$ or Br^-	II,C,4,f	*317*
trans-$[M(NO)(NH_3)_4(OH)]$*trans*-$[M'(NO)(NO_2)_4(OH)] \cdot nH_2O$ (M = Os, M′ = Os, Ru; M = Ru, M′ = Os)	II,C,4,f	*279*
trans-$[Os(NO)(NH_3)_4(OH)][Os(NO)Cl_5]$	II,C,4,f	*279*
trans-$[Os(NO)(NH_3)_4(OH)]Cl_2$	II,C,4,f	*243, 279*
trans-$[Os(NO)(NH_3)_4(OH)]^{2+}$	II,C,4,f	*245*
$Na_2\{$*trans*-$[Os(NO)(NO_2)_4(OH)]\}$	II,C,4,f	*279*
trans-$[Os(NX)Cl_4(OH_2)]^-$, X = O, S	II,C,4,f and II,C,4,i	*240, 241*
cis-$[Os(bpy)_2(NO)(OH_2)]^{3+}$	II,C,4,f	*247*
cis-$[Os(phen)_2(NO)(OH_2)]^{3+}$	II,C,4,f	*247*
cis-$[Os(PMe_3)_4(H)(OH)]$	—	*298*
$[Os(OSO_2R)_2(OH_2)(CO)(PPh_3)_2]$, R = Me, CF_3, or 4-MeC_6H_4	II,C,6,q	*126*
$[Os(H)(CO)(OH_2)(L)(PPh_3)_2]$, L = PPh_3 or CO	—	*133*
$[Os(H)(CS)(OH_2)(PPh_3)_3]^+$	II,C,2,c	*17*
cis-$[Os(bpy)_2(OH_2)_2]^{2+}$	—	*318*
$[Os(trpy)(bpy)(OH_2)]^{2+}$	—	*95*
Os_2^{5+}		
$[Os_2Cl_4(chp)_2(OH_2)]^-$	II,D,1,d	*319*
Os(III)		
$[Os(NH_3)_5(OH_2)]^{3+}$	II,B,4,a	*59, 70, 72*
$[Os(OH_2)Cl_5]^{2-}$	—	*320*
$[Os(bpy)_3(OH_2)]^{3+}$	—	*321*
$[Os(bpy)_3(OH)]^{2+}$	—	*321*
cis-$[Os(bpy)_2(OH_2)_2]^{3+}$	—	*94*
$[Os(trpy)(bpy)(OH_2)]^{3+}$	—	*95*
trans-$[Os(14\text{-}tmc)(OH)(OH_2)]^{2+}$	II,C,4,c	*210, 211*
trans-$[Os(14\text{-}tmc)(OH)_2]^+$	II,C,4,c	*210, 211*
$[Os_2Cl_4(chp)_2(OH_2)]$	II,D,1,d	*319, 322*

(continued)

TABLE VIII (*Continued*)

Complex	Section	Ref.
Os(IV)		
$[Os(OH_2)Cl_5]^-$	—	*320, 323*
$[Os(OH_2)_2Cl_4]$	—	*323*
$[Os_2(\mu\text{-}N)(NH_3)_8(OH_2)_2]^{5+}$	II,B,3,b and II,D,3,o	*46*
$[Os_2(\mu\text{-}N)(NH_3)_7(OH_2)_3]^{5+}$	II,B,3,b and II,D,3,o	*46*
$[Os(tpp)(OH)_2]$	II,B,4,d	*37*
Os(V)		
trans-$[Os(N)(CN)_4(OH)]^{3-}$	II,C,4,q	*266*
Os(VI)		
trans-$[Os(O)_2(OH)_4]^{2-}$	—	*2, 113, 324*
trans-$[Os(N)(CN)_4(OH)]^{2-}$	II,C,4,q	*266*
trans,trans,trans-$[Os(O)_2(CN)_2(OH)_2]^{2-}$	II,C,2,a	*113*
trans-$[Os(trpy)(O)_2(OH)]^+$	—	*325*
trans-$[Os(phen)(O)_2(OH)_2]$	—	*325*
$K[Os(N)X(OH_2)]$, X = Cl^- or Br^-	II,C,4,q	*263*
Os(VIII)		
$[Os(O)_4(OH)(H_2AsO_3)]^{2+}$	II,C,6,u	*326*

b. Oxo Complexes. Over the past 5 years literally hundreds of publications have appeared in which $[OsO_4]$ and related species have been used as oxidants in organic chemistry, for staining and fixing biological samples, as selective agents for DNA structure, and many other fundamental and applied studies. Even a cursory review of this work is far beyond the scope of this article. Therefore, only a few recent key references that have considerably advanced the understanding of the chemistry in each of these areas will be cited. An excellent account of the chemistry and applications of $[OsO_4]$ and related complexes, up to the mid-1980s, is given in the review by Griffith (*2*). Recently, a number of brief reviews and Ph.D. theses have been presented on the organic oxidation chemistry of Os(VIII) oxo complexes (*35, 36, 327–332*) and $[OsO_4]$ and its regeneration (*333*).

Recent developments in the understanding of the mechanisms of catalytic and asymmetric dihydroxylation reactions are discussed in Section V,E,1,b. An important aspect of this work is the kinetics and thermodynamics of the formation of adducts with N heterocycles, which have an important role in promoting many reactions. The crystal structure of the $[OsO_4]$ adduct with the cinchona alkaloid ligand (dimethylcarbamoyl)dihydroquinidine, which is a useful catalyst for asymmetric cis hydroxylation of olefins, shows the ligand bound via the apical aza group of the bicyclo[2.2.2]octane group. A similar structure has been assigned in solution on the basis of NMR experiments (*334*). The formation and thermodynamic stability of a range of $[OsO_4L]$ complexes in water and CCl_4 have been studied, wherein L is an amine ligand (Ctmen, hmt, and dabco) or an N-heterocyclic ligand (im, py, bpds, 4-pic, phen, and bpy) (*335, 336*).

Though $[OsO_4]$ and its adducts continue to find many applications as biological stains and fixatives for both optical and electron microscopy (e.g., *337–347*), $[OsO_4]$ is finding increased applications in selective staining of polymers with alkene functionalities (*348–350*), in the determination of DNA structure (e.g., *351, 352*), in the detection of point mutations in DNA due to mismatched T (and to a lesser extent C) groups (*353*), and in the chemiluminescent determination of proteins (*354*). There have also been many studies on the kinetics and mechanisms of the stoichiometric and the $[OsO_4]$-catalyzed oxidations of both organic and inorganic substrates. These are summarized in Table IX (*26, 28, 267, 324, 326, 334, 335, 355–397*).

$[OsO_4]^-$ has been prepared recently and studied by IR spectroscopy and cyclic voltammetry. It has also been used as a stoichiometric oxidant for benzylic and allylic alcohols to ketones and aldehydes (*398*). The ion pairs $\{[arene]^+[OsO_4]^-\}$ are prepared from irradiating the arene · $[OsO_4]$ charge-transfer bands, and such intermediates are involved in both the photochemical and thermal cis hydroxylation of arenes (*392*).

Numerous studies have been performed on dioxoOs(VI) and other oxo complexes; many of these are discussed in the relevant sections of this review. Those complexes that have been characterized in recent years by X-ray crystallography are summarized in Table X (*22, 24–26, 28, 30–33, 40, 215, 280, 281a, 290, 334, 392, 399–405*).

c. ROH, RO^-, and R_3SiO^- Complexes. For examples of those complexes prepared from the reaction of an alkene with Os(VIII), see Sections II,B,2 and II,C,6,b. There are many hundreds of examples of such complexes, but few have been characterized, as they are only synthetic intermediates in the cis dihydroxylation reactions. Alcohol complexes are also likely to be intermediates in the oxidation of ROH with $[OsO_4]$ (Table IX).

Porphyrinato complexes containing alcoholate and alcohol ligands are discussed in Section II,C,4,d; those containing cyano ligands are discussed in Section II,C,2,a, and those containing Schiff bases are

TABLE IX

KINETIC AND MECHANISTIC STUDIES ON OXIDATION REACTIONS OF $[OsO_4]$ AND $[OsO_4L]$ COMPLEXES

Reaction	Ref.
$[OsO_4]$ and/or *trans*-$[Os(O)_4(OH)_2]^{2-}$	
Reduction by Fc	*355*
Oxidation of arsenite	*326*
Catalysis of H_2O_2 decomposition	*356,357*
Catalysis of the $[Fe(CN)_6]^{3-}$ oxidation of selenium(IV)	*358*
Catalysis of the chlorate oxidation of the hydrazinium cation	*358a*
Catalysis of the diperiodatoargentate(III) oxidation of phosphite	*359*
Catalysis of the periodate oxidation of Na_2HAsO_3	*360*
Oxidation of alkenes; theoretical study	*361*
Oxidation of 1-octene	*362*
Oxidation of unsaturated organic molecules in H_2SO_4/AcOH	*363*
Oxidation of alcohols, diols, and 2-hydroxyacids in alkaline solution	*364*
Oxidation of 3-cresol	*365*
Oxidation of 2-methylpropan-1-ol and 2-butanol	*366*
Oxidation of butanol	*367*
Oxidation of methylglycol and diacetone alcohol	*368*
Oxidation of bis(2-hydroxyethyl) ether	*369*
Oxidation of ethyl digol	*370*
Oxidation of ethyl methyl ketone	*324*
Oxidation of triethylamine	*371*
Catalysis of the diperiodatoargentate(III) oxidations of cycloalkenones and acetophenone	*372,373*
Catalysis of the periodate oxidation of cyclic ketones	*374*
Catalysis of the oxidation of allyl and crotyl alcohols with chloramine-T, chloramine-B, bromamine-T, and bromamine-B	*375*
Catalysis of the chloramine-T oxidation of glycolic and lactic acids	*376*
Catalysis of the chloramine-T oxidation of cinnamaldehyde	*377*
Catalysis of the chloramine-T oxidation of hypophosphite	*377a*
Catalysis of the bromamine-B oxidation of MeSPh to MeS(O)Ph	*378*
Catalysis of the Fe(VI) oxidation of dmso	*379*
Catalysis of the diperiodatocuprate(III) oxidation of Me_2SO to Me_2SO_2	*380*
Catalysis of the ditelluratocuprate(III) oxidation of bis(2-aminoethyl) disulfide	*381*
Catalysis of the $[Fe(CN)_6]^{3-}$ oxidations of alkanals	*382*
Catalysis of the $[Fe(CN)_6]^{3-}$ oxidation of β-bromopropionic acid	*383*
Catalysis of the $[Fe(CN)_6]^{3-}$ oxidations of benzylphenylglycolic acids	*384*
Catalysis of the $[Fe(CN)_6]^{3-}$ oxidations of benzoin and its derivatives	*385*
Catalysis of the $[Fe(CN)_6]^{3-}$ oxidations of benzyl alcohol and benzylamine	*267*
Catalysis of the iodate oxidations of styrene and stilbene	*386*
Catalysis of the *N*-bromosuccinimide oxidation of 2-propanol	*387*
Catalysis of the Tl(III) acetate oxidation of benzilic acid	*388*
Metathesis polymerization of norbornene	*389*

TABLE IX (*Continued*)

Reaction	Ref.
Catalysis of the *N*-morpholine oxide cis hydroxylation of alkenes	*26*
Catalysis of the trimethylamine *N*-oxide cis hydoxylation of cyclohexene and α-pinene	*390*
Catalysis of the cis dihydroxylation of olefins	*391*
Light-catalyzed cis dihydroxylation of arenes	*392*
$[Os(O)_4L]$	
L = dihydroquinine and dihydroquinidine 4-chlorobenzoate; enantioselective cis hydroxylation of alkenes	*28*
L = dimethylcarbamoyl dihydroquinidine; preparation and solid-state and solution structures	*334*
L = (*S*,*S*)-*N*,*N*′-bis(2,4,6-trimethylbenzylidene)-1,2-diphenyl-1,2-diamine; catalytic asymmetric cis dihydoxylation of alkenes	*393*
L = dihydroquinidine or dihydroquinine 4-nitrobenzoate; catalytic asymmetric cis dihydoxylation of *trans*-stilbene	*394*
L = Xdhqd (X = Ac, dmc, MeO, or ClBz) or ClBzdhq; conformational changes of cinchona alkaloids on complexation to $[Os(O)_4]$	*395*
L = pyridine; cis dihydroxylation of arenes	*392*
L = RNH_2 (R = Et, Bu^n, or Pr^i), Et_2NH, Et_3N, or amino alcohols); catalysis of the chloroamine-T oxidation of amines and amino alcohols	*396,397*
L = imidazoles; oxidations of imidazoles	*335*

discussed in Section II,C,4,w. *trans*-$[Os^{IV}Br_3(OMe)(PPh_3)_2]$ is prepared by refluxing a mixture of $(Bu_4N)_2[OsBr_6]$, PPh_3, phthalic acid, and phthalic anhydride in CH_2Cl_2, and its structure has been determined by X-ray crystallography (*406*). $[Os^{III}Cl_3(PPh_3)_2(MeOH)]$ is prepared from $[OsCl_6]^{2-}$ in PPh_3/MeOH solutions and has been characterized using IR, UV/Vis, and EPR spectroscopies. It is a useful intermediate in the synthesis of *trans-P,P*-$[Os^{III}Cl_2(PPh_3)_2L]$ complexes, where L is an anionic bidentate oxygen ligand (*407*).

A number of *trans*-$[Os(O)_2(py)_2L]$ complexes (H_2L = glycolic, 2-hydroxyisobutyric, (*S*)-(+)-mandelic, or 2-salicylic acid) have been prepared from $[Os(O)_2(OMe)_4]^{2-}$ and py with the ligand. They have been characterized by IR, mass, and NMR (^1H and ^{13}C) spectroscopies (408) and X-ray structures for L = glycolate(2−) and salicylate (*401, 402*). The Os(V) complex, $(PPh_4)[Os(O)(ehba)_2]$ has been prepared from the reaction of the ligand with $(PPh_4)[OsO_4]$ and has been characterized by IR spectroscopy (*23a*).

TABLE X

STRUCTURAL STUDIES OF OXO COMPLEXES

Complex	Section	Ref.
Os(IV)		
$K_2[\{trans\text{-}Os(\eta^4\text{-chba-Et})(OPPh_3)_2\}_2O]$	II,C,4,x	*290*
Os(VI)		
$Ba[OsO_4]$	—	*399*
$[Os(O)(NBu^t)(mes)_2]$	II,C,2,f	*30*
$[Os(O)(NAr)(ArNCH_2CH_2NAr)]$[a]	II,C,4,b	*22*
$[Os(O)_2(mes)_2]$	II,C,2,f	*32*
$[Os(O)_2(xylyl)_2]$	II,C,2,f	*31*
$[Os(O)_2(SSO_3)_2]^{2-}$	II,C,7,n	*33*
$[Os(O)(Ctmen\text{-}2H)(Ctmen\text{-}H)]^+$	II,C,4,b	*24*
$[Os(O)Cl_4]$	—	*400*
$[Os(O)(CH_2SiMe_3)_4]$	—	*25*
$[Os(O)(L)_2]$[b]	—	*26*
cis-$[Os(O)_2L(L\text{-}L)]$[c]	—	*28*
$[Os(O)_2(ONO_2)_2(mes)]^-$	II,C,2,f	*30*
$[Os(O)_2(4\text{-Metpp})] \cdot thf$	II,C,4,d	*40*
$[Os(O)_2(oep)]$	II,C,4,d	*215*
trans-$[Os(O)_2(3\text{-}Bu^t\text{-saltmen})]$	II,C,4,w	*280*
trans-$[Os(O)_2\{(ba)_2en\}] \cdot 0.5H_2O$	II,C,4,w	*281a*
trans-$[Os(O)_2(glycolate)_2(py)_2] \cdot MeOH$	II,C,6,c	*401*
trans-$[Os(O)_2(2\text{-oxobenzoato})(py)_2]$	II,C,6,c	*402*
$[CtmenH_2] \cdot$ *trans*-$[Os(O)_2Cl_4]$	—	*403*
cis-$[Os(O)_2(L\text{-}L)(L'\text{-}L')]$[d]	—	*404*
$[Me_4N]_2[\{Os(O)_2(OCOMe)_2(\mu\text{-}OMe)\}_2]$	II,D,5,d	*405*
anti-$\{trans\text{-}O,O\text{-}[Os(O)_2L_2]_2(\mu\text{-}L')\}$[e]	—	*392*
Os(VIII)		
$[Os(O)_4L]$[f]	—	*334*

[a] $Ar = 2,6\text{-}Pr^i{}_2C_6H_3$.

[b] L = (3*aS*,5*S*,5*aS*,9*aR*,9*bR*)-octahydro-5,5*a*-dihydroxy-3*a*,6,6,-9*a*-tetramethyl-1*H*-benz[*e*]indene-3,7(2*H*,3*aH*)-dionato(2−)-O^5, O^{5a}.

[c] L = dihydroquinidine 4-chlorobenzoate, L-L = (3*S*,4*S*)-2,2,5,-5-tetramethyl-3,4-hexanediolato(2−); L = dihydroquinine 4-chlorobenzoate, L-L = (3*R*,4*R*)-2,5-dimethyl-3,4-hexanediolato-(2−).

[d] L-L = {*N*,*N*′-bis(neohexyl)-2,2′-bipyrrolidine}, L′-L′ = {(*S*,*S*)-1,2-diphenyl-1,2-ethanediolato(2−)}.

[e] L = py; L′ = η^2-O^3,O^4-η^2-O^5,O^6,-thch; η^2-O^1,O^2-η^2-O^3,O^4-ththa.

[f] L = (dimethylcarbamoyl)dihydroquinidine.

trans-$[Os(O)_2(OSiMe_3)_4]^{2-}$ is prepared from the reaction of *trans*-$[Os(O)_2Cl_4]^{2-}$ with $NaOSiMe_3$, and its reactions with $Mg(CH_2SiMe_3)_2$ or $ClMg(CH_2SiMe_3)$ have been studied (*26*). *trans*-$[Os(O)_2(OSiMe_3)_2\{NP(Ph)_2CH_2P(Ph)_2N\}]$ has also been reported (*33*).

d. Catechol and Quinone and Related Complexes. The preparation, characterization (X-ray structure, UV/Vis/NIR, and NMR), electrochemistry, and fluxional behavior of *mer*-$[Os(dpq)_3]$ have been studied. The complex exhibits a reversible oxidation and reduction, with a further one-electron quasi-reversible oxidation and reduction (*409*). $[Os(trop)_3]$ has been prepared from the reaction between $[OsCl_6]^{3-}$ and excess tropH, and has been characterized by mass, IR, and Raman spectroscopies (*410*).

$[Os(bpy)_2LL]^{2+/1+/0}$ (L-L are quinone, semiquinone, or catecholato ligands derived from catechol, 3,5-di-*tert*-butylcatechol, or tetrachlorocatechol) have been characterized by UV/Vis/NIR and EPR spectroscopies. These spectroscopic properties and the crystal structure of $[Os(bpy)_2(dbcat)]ClO_4$ confirm an Os(III)–catecholate ground state for the +1 ion. This contrasts with the ground state of the +1 ions of Ru analogs, which are best described as Ru(II)–semiquinone complexes (*411*).

Catecholate complexes of the type *trans*-$[Os(O)_2L_2]^{2-}$ (H_2L = dopa, dopamine, adrenaline, noradrenaline, or isoproterenol) are prepared from the reactions between *trans*-$[Os(O)_2(OH)_4]^{2-}$ and H_2L. The complexes have been characterized using Raman, IR, and NMR (^1H and ^{13}C) spectroscopies, which indicate that the ligands are bound via the catecholate oxygens. These types of complexes are believed to be models for the staining of catecholamine rich sites in biological tissues (*412*). Similar reactions occur with maltol (maltH) (*413*), tropolone (tropH) (*410*), and 2,3-naphthalenediol (ndH_2) (*414*), to give *trans*-$[Os(O)_2L_2]^{n-}$ (n = 0, L = malt or trop; n = 2, L = nd). The complexes have been characterized by mass, IR, Raman, and NMR (^1H and ^{13}C) spectroscopies.

$[Os^{III}Cl_3(PPh_3)_2(MeOH)]$ reacts with maltH and tropH to give $[Os^{III}Cl_2(PPh_3)_2L]$ (L = malt or trop). IR spectroscopy indicates that the phosphines are mutually trans. The complexes exhibit reversible oxidation and reductions to their Os(IV) and Os(II) analogs. The Os(III) complexes have been characterized by UV/Vis and EPR spectroscopies (*407*). Similarly, the methanol complex reacts with a variety of catechols (catechol, 4-methylcatechol, 3,5-di(*tert*-butyl)catechol, tetrachlorocatechol, tetrabromocatechol, and 2,3-dihydroxynaphthalene) to form the corresponding Os(III)-semiquinonate complexes $[OsCl_2(PPh_3)_2(SQ)]$. The complexes were characterized by electrochemistry, and UV/Vis and IR spectroscopies. They are good catalysts for promoting the *N*-morpholine oxide oxidation of alcohols (*414a*).

e. Triethylphosphate and Phosphine Oxide Complexes. Dissolution of $[Os(NH_3)_5(OSO_2CF_3)]^{2+}$ in $(EtO)_3PO$ yields $[Os(NH_3)_5\{O\text{-}P(OEt)_3\}]^{3+}$, which has been characterized by IR spectroscopy. It is a useful intermediate for the synthesis of other pentaammine complexes (*67*).

$OPPh_3$ complexes with porphyrinato ligands and chelating amides are discussed in Sections II,C,4,d and II,C,4,x, respectively, and *cis*- and *trans*-$[Os(bpy)_2(OPR_3)_2]^{2+}$ complexes are discussed in Section V,E,1,b. The latter have been characterized by UV/Vis spectroscopy and CV. They are reversibly oxidized to their Os(III) analogs (*415*).

f. Pyridine Oxide Complexes. $[OsO_4(pyO)]$ is prepared by the reaction of $[OsO_4]$ with pyO in acetone/CCl_4. It has been characterized by IR and Raman spectroscopies and electrochemistry. A variety of *trans*-dioxo Os(VI) complexes containing diolato(2−) ligands are prepared by the reaction of this complex with an alkene in the presence of excess pyO. These include $[Os(O)_2(pyO)_2(O_2R)]$ ($R = CH_2CH_2$ or cyclohexane), which have been characterized by electrochemistry and UV/Vis, IR, and Raman spectroscopies (*235*).

g. Amide Complexes. $[Os(NH_3)_5(OSO_2CF_3)]^{2+}$ reacts with neat amides to form $[Os(NH_3)_5(O\text{-}L)]^{3+}$ (L = dmf or dma) (*67, 120, 177*). These complexes exhibit reversible reductions, but the Os(II) dmf complex undergoes elimination reactions (Section V,E,4,h). The porphyrin complex $[Os^{II}(mix)(dmf)_2]$ (*144*), *cis*- and *trans*-$[OsCl_2(PMe_2Ph)_3(dmf)]$ (*145*), and *cis*-$[Os(bpy)(dppe)(dmf)_2]^{2+}$ (*98*) have been reported.

h. η^1-(O)-Oxime Complexes. See Section II,C,4,l.

i. O-Dimethyl Sulfoxide Complexes. $[Os(NH_3)_5(O\text{-dmso})]^{3+}$ is prepared from the reaction of $[Os(NH_3)_5(OSO_2CF_3)]^{2+}$ with dmso (*67, 120, 200*). At sufficiently fast scan rates, it is reversibly reduced to $[Os(NH_3)_5(O\text{-dmso})]^{2+}$, which is unstable with respect to its S-bound linkage isomer (Section V,D,4). *trans*-$[Os(O\text{-dmso})(S\text{-dmso})_3Br_2]$ has also been reported (Section II,C,7,i).

j. η^1-(O)-Aldehyde and -Ketone Complexes. The reaction of aldehydes and ketones with $[Os(NH_3)_5(OSO_2CF_3)]^{2+}$, or oxidation of $[Os^{II}(NH_3)_5(\eta^2\text{-}O{=}CR_2)]^{2+}$, results in the formation of $[Os^{III}\text{-}$

$(NH_3)_5(\eta^1\text{-}O{=}CR_2)]^{2+}$ (*67, 120, 168, 169, 177, 181*). The complexes prepared in these ways include acetaldehyde, acetone, 2-butanone, cyclopentanone, cyclobutanone, 2-cyclohexen-1-one, 2,2-dimethylpropiophenone, and benzophenone. These complexes have been characterized by CV and IR spectroscopy.

Acetone complexes are also prepared by the trans activation of the ammine ligands in $[Os^{II}(NH_3)_5(C\text{-Mepy})]^{2+}$ to form *trans*-$[Os(NH_3)_4(C\text{-Mepy})(O\text{-acetone})]^{2+}$ (*167*).

k. Diketonate Complexes. $[Os^{II}Cl_2(PPh_3)_3]$ or $[Os^{III}Cl_3(PPh_3)_2(MeOH)]$ reacts with 2,4-pentanedione ligands to give $[Os^{III}Cl_2\{RC(O)CHC(O)R'\}(PPh_3)_2]$ ($R = R' = CH_3$; $R = CH_3$, $R' = CF_3$; $R = R' = CF_3$; $R = Me$, $R' = Ph$; $R = R' = Ph$), in which the phosphines are believed to be mutually trans from IR spectroscopy. The complexes exhibit reversible oxidation and reductions to their Os(IV) and Os(II) analogs and the Os(III) complexes have been characterized by UV/Vis and EPR spectroscopies (*407, 416*). *trans*-$[Os(H)(acac)(CO)(PPr^i_3)_2]$ is prepared from *trans*-$[Os(H)(Cl)(CO)(PPr^i_3)_2]$ (*18*) in a fashion similar to that described for the PPh_3 analogs (*417*), and has been characterized by IR, and ^1H and $^{31}P\{^1H\}$ NMR spectroscopies (*18*).

l. Ether Complexes. $[Os(NH_3)_5L]^{3+/2+}$ (L = thf, dme, furan, and anisole) are prepared by standard methods (*75, 120*). A dinuclear carboxylato containing thf has also been reported (*418*) (Section II,D,1,c).

m. Dioxygen Complexes. $[Os(H)(Cl)(CO)(PR_3)_2]$ ($PR_3 = PMeBu^t_2$ or PPr^i_3) (*419*) or $[Os(H)(Cl)(CO)(PPr^i_3)_2(\eta^2\text{-}H_2)]$ (*420*) bind O_2 to form $[Os(CO)(Cl)(H)(PR_3)_2(\eta^2\text{-}O_2)]$. These complexes have been characterized by IR and NMR spectroscopies.

n. Carbonato Complexes. $[Os(bpy)_2(CO_3)]$ and $[Os(bpy)(dppe)(CO_3)]$ are useful synthetic intermediates for the preparation of a large number of complexes. They are prepared from the reactions of the corresponding halo complexes with CO_3^{2-} and have been characterized by IR and electronic absorption spectroscopies (*97, 98*). Carbonate also displaces the Cl^- ligands in $[Os(N)(CH_2SiMe_3)_2Cl_2]^-$ to form $[Os(N)(CH_2SiMe_3)_2(CO_3)]^-$ (*157*). All of the complexes contain the η^2-O,O′ coordination mode.

o. Carboxylato Complexes. The complexes $[Os(NH_3)_5(OCOR)]^{2+}$ ($R = H$ or CH_3) have been prepared and characterized by electrochemistry and IR spectroscopy. Reduction of the formato complex leads to the carbonyl complex (Section V,E,4,g) (*119*). *cis*-$[Os(PPh_3)_2(CO)X\{\eta^2$-$(O,O')$-$O_2CCH_3\}]$ ($X = Cl^-$ or Br^-) have been prepared and characterized by IR and NMR (^1H and ^{31}P) spectroscopies (*134*). The complexes are catalysts for the hydrogenation of aldehydes and ketones (*135*). The X-ray structure of the complex with $X = Br^-$ has a cis,cis geometry in which Br^- is trans to an O atom of the acetato ligand (*134, 421*). *trans*-$[OsBr_2\{\eta^2$-(O,O')-$O_2CCH_3\}(PPh_3)_2]$ is prepared from refluxing $[OsBr_6]^{2-}$ in $MeCO_2H/(MeCO)_2O$ and its structure was determined by X-ray crystallography (*422*). The syntheses of $[Os(H)\{\eta^2$-(O,O')-$O_2CCH_3\}(CO)(PPr^i_3)_2]$ (*18*), $[Os(\eta^2$-O,O-$OAc)(PPh_3)_3(H_2)]^+$ (*423*), and *cis*-$[Os(O_2CCF_3)_2(PMe_3)_4]$ (*298*) were reported. $[Os(H)\{\eta^2$-(O,O')-$O_2CCF_3\}(CO)(PPh_3)_2]$ and $[Os(O_2CCF_3)_2(CO)(PPh_3)_2]$ are oligomerization catalysts for $C_6H_5C{\equiv}CH$ (*424*). The reactions of *trans*-$[Os(N)(CN)_4(OH)]^{2-}$ with CF_3CO_2H or $(CF_3CO_2)O$ yield *trans*-$[Os(NR)(CN)_4(OCOCF_3)]^-$ (R = H or $OCOCF_3$, respectively) (*266*).

Complexes with 2-hydroxycarboxylic acids, pyridine-2-carboxylate, and dicarboxylates are described in Sections II,C,6,c, II,C,4,f, and II,C,6,p, respectively.

p. Oxalate and Malonate Complexes. *trans*-$[Os(O)_2(CN)_2(C_2O_4)]^{2-}$ is prepared from the reaction of *trans*-$[Os(O)_2(OH)_4]^{2-}$ with stoichiometric amounts of CN^- and $H_2C_2O_4$. It has been characterized by IR, Raman, and UV/Vis spectroscopies (*113*). *trans*-$[^{191}Os(O)_2(\text{malonate})_2]^{2-}$ is a useful parent complex for a $^{191}Os \rightarrow {}^{191m}Ir$ generator for medical applications (*425*).

q. Alkyl- and Arylsulfonato Complexes. The trifluoromethanesulfonato (triflato) ligand is a very good leaving group and the synthesis of complexes containing this ligand has opened up many new areas of Os ammine chemistry. The synthetic procedures for preparation of these complexes are outlined in Sections II,B,4 and II,B,6 and the kinetics of their substitution reactions are discussed in Section V,B. The complexes prepared include $[Os(NH_3)_5(OSO_2CF_3)]^{2+}$ (*59, 66, 67*), *cis*-$[Os(NH_3)_4(OSO_2CF_3)_2]^+$ and *trans*-$[Os(NH_3)_4(OSO_2CF_3)Cl]^+$ (*81*), *cis*-$[Os(bpy)_2(OSO_2CF_3)_2]^+$ (*94*), and $[Os(trpy)(bpy)(OSO_2CF_3)]^{2+/+}$ (*95, 96*). The ammine complexes have been used extensively in the synthesis of other pentaammine and tetraammine complexes (*26, 55, 59, 66–68, 70, 71, 74, 75, 80, 81, 85, 87–90, 118–120, 167–169, 171–181, 193–195, 198, 200, 201, 426–432*).

cis-$[Os^{II}(PMe_3)_4(H)(OSO_2CF_3)]$ is prepared from either the reaction of *fac*-$[Os(PMe_3)_3(\eta^2$-C,P-$CH_2PMe_2)(H)]$ with CF_3SO_3H (*306*), or *cis*-

$[Os(PMe_3)_4(H)(CH_3)]$ with CF_3SO_3H or $[Os(PMe_3)_4(H)_3]CF_3SO_3$ (*297*). The triflato complex has been characterized by ^1H and $^{31}P\{^1H\}$ NMR spectroscopies. In thf solution, the triflato complex is in dynamic equilibrium with a second species that is postulated to be $[Os(PMe_3)_4H]^+$ on the basis of NMR experiments (*306*), although the possibility of a six-coordinate *cis*-$[Os(PMe_3)_4(thf)(H)]^+$ intermediate cannot be discounted. The triflato complex is a useful intermediate for a number of substitution reactions (*297, 306*).

$[Os(OSO_2R)_2(OH_2)(CO)(PPh_3)_2]$ and $[Os(OSO_2R)_2(CO)_2(PPh_3)_2]$ (R = Me, CF_3, or 4-$CH_3C_6H_4$) and Ru analogs are obtained when RSO_3H reacts with $[M(H)_2(CO)(PPh_3)_3]$ or $[M(H)_2(CO)_2(PPh_3)_2]$, respectively (*126, 433*). In $[Os(OSO_2R)_2(OH_2)(CO)(PPh_3)_2]$, the sulfonato ligands are trans to PPh_3 and CO, with the OH_2 ligand being trans to the second PPh_3 ligand. $MeSO_3H$ also reacts with $[Os(H)(Cl)(CO)(PPh_3)_3]$ in $MeOCH_2CH_2OH$ to give $[OsCl(O_3SMe)(CO)(PPh_3)_3]$ (*433*).

r. O-Sulfinato Complex. $[OsCl(OS(O)C_6H_4Me\text{-}4)(CO)_2(PPh_3)_2]$ is prepared by a linkage isomerization reaction (Section V,D,5) (*434*).

s. Benzeneseleninato Complexes. $[Os(XC_6H_4SeO_2)_3]$, $[Os(XC_6H_4SeO_2)_2Y]$, $[Os(XC_6H_4SeO_2)Y_2]$, and $[Os_2(XC_6H_4SeO_2)_3Y_3]$ (X = H, 4-Cl, 3-Cl, 4-Br, 3-Br, or 4-Me; Y = Cl^- or Br^-) have been prepared from the reactions of the appropriate quantity of the ligand with $OsCl_3$ in aqueous methanolic solutions. Based on IR and UV/Vis spectroscopic data, the benzeneseleninato ligands are believed to act as O,O′ chelates. The complexes were also characterized using magnetic susceptibility measurements (*435*).

u. Nitrato Complexes. $[Os(O)_2(ONO_2)_2(mes)]$ (Section II,C,2,f) (*30*) and $[Os(cp)_2(ONO_2)]^+$ (*264*) have been reported recently. Detailed IR and Raman spectroscopic studies have been performed on mixed NO/NH_3/NO_3^- complexes, including *trans*-$[Os(NO)(NH_3)_4(ONO_2)]^{2+}$ (*243, 436*).

v. Sulfato, Chromato, and Perrhenato Complexes. Sulfate displaces the Cl^- ligands in $[Os(N)(CH_2SiMe_3)_2Cl_2]^-$ to form $[Os(N)(CH_2SiMe_3)_2\{\eta^2\text{-}(O,O')\text{-}SO_4\}]^-$ (*157*). Similarly, Ag_2CrO_4 reacts with *cis*-$[Os(N)R_2Cl_2]^-$ (R = CH_2SiMe_3 or Me) in light to produce $[Os(N)R_2\{\eta^2\text{-}(O,O')\text{-}CrO_4\}]^-$. These are selective oxidants for the conversion of primary alcohols to aldehydes (*154*). By contrast, perrhenate reacts with $[Os(N)(CH_2SiMe_3)_2Cl_2]^-$ to form *cis*-$[Os(N)(CH_2SiMe_3)_2(OReO_3)_2]^-$, in which the perrhenato ligands are monodentate (*157*).

u. Arsenito and Hydrogenarsenito Complexes. The details of the kinetics of the oxidation of arsenite by Os(VIII) have been studied and the initial adduct is believed to be *trans*-$[Os^{VIII}(O)_4(OH)(H_2AsO_3)]^{2-}$, which decomposes to *trans*-$[Os(O)_2(OH)_4]^{2-}$ and NaH_2AsO_4. *trans*-$[Os(O)_4(H_2AsO_3)_2]^{2-}$ is also thought to be a reactive intermediate at high concentrations of $H_2AsO_3^-$ (*326*).

7. Sulfur, Selenium, and Tellurium

a. Complexes with Thioethers and Selenoethers. *mer*-$[Os^{III}LX_3]$ [X = Cl^- or Br^-; L = bis(3-methylthiopropyl) sulfide] has been reported in Griffith's review (*2, 437*), but more recently the triselenoether analogs have been prepared from the reaction of $[OsX_6]^{2-}$ with the appropriate ligand (*438*). The Os(IV) complex [L = $S(CH_2CH_2CH_2SMe)_2$, X = Br^-] has been isolated and characterized by UV/Vis and IR spectroscopies, electrochemistry, magnetic measurements, and solution conductivities. The other Os(IV) complexes of the series have been observed in cyclic voltammograms, because the oxidation of the Os(III) analogs is reversible, but the complexes have been too unstable to isolate. The Os(II) analogs *mer*-$[OsLX_3]^-$ (X = Cl^- or Br^-) have also been observed in the cyclic voltammograms, but have not been isolated (*234*).

$[Os^{IV}(L\text{-}L)X_4]$ [X = Cl^-, L-L = $RSe(CH_2)_2SeR$ (R = Me or Ph), $MeSe(CH_2)_3SeMe$, *cis*-MeSeCHCHSeMe, or $2Me_2Se$; X = Br^-, L-L = $MeSe(CH_2)_2SeMe$] are prepared by heating the ligand with $[OsX_6]^{2-}$ in 2-methoxyethanol and were characterized by IR and UV/Vis spectroscopies and by magnetic measurements (*439*). The syntheses, the IR, UV/Vis, and NMR spectroscopies, and the electrochemistry of *trans*-$[OsX_4(SeMe_2)_2]^{0/-}$ have also been studied (*302a*).

b. Complexes with Thioether Macrocycles. In a recent review by Blake and Schröder (*440*) on complexes with thioether macrocycles, a number of Os complexes were reported. Those that have been characterized crystallographically are $[Os([9]aneS_3)_2]^{2+}$ (*440*) and $[Os(4\text{-}MeC_6H_4Pr^i)([9]aneS_3)]^{2+}$ (*440, 441*). The only other complexes containing thioether macrocycles are $[OsCl_4([14]aneS_4)]$ (*437*) and $[Os_2Cl_2(arene)([18]aneS_6)]^{2+}$; both have the macrocycle bound in a bidentate fashion, the latter having the macrocycle as a bis(bidentate) bridging ligand (*441*).

c. Thiolato Complexes. $[Os^{IV}(oep)(SPh)_2]$ and $[Os^{IV}(tpp)(SPh)_2]$ (*214*) and *trans*-$[Os^{IV}(salen)(SPh)_2]$ and related Schiff base complexes

(*281*) have been discussed in Sections II,C,4,d and II,C,4,w, respectively. Dinuclear complexes with MeS^- ligands are discussed in Section II,D,6,a.

The reaction of *mer*-$[OsX_3(PR_3)_3]$ (PR_3 = PMe_2Ph, PEt_2Ph, $PMePh_2$, or $PEtPh_2$) with 1.5 mol equivalents of $Pb(SC_6F_5)_2$ yields *mer*-$[Os(SC_6F_5)_3(PR_3)_2]$. These paramagnetic Os(III) complexes have been characterized by IR and NMR (^1H, ^{31}P, and ^{19}F) spectroscopies. They tend to dimerize in solution, but in the solid state they are believed to be octahedral, with the sixth coordination site being occupied by an ortho F of a $C_6F_5S^-$ ligand. This assignment is based on the crystal structure of the analogous Ru complex, $[Ru\{\eta^2\text{-}(S,F)\text{-}SC_6F_5\}(\eta^1\text{-}(S)\text{-}SC_6F_5)_2(PMe_2Ph)_2]$ (*442*). In the presence of Zn/CO, they react to form $[Os(SC_6F_5)_2(CO)_2(PR_3)_2]$, with the preferred geometric isomer being dependent on the solvent and the reaction time. If the reaction is carried out in acetone, the trans,trans,trans isomer is obtained as the main product. However, the trans-P,P-cis,cis isomer is obtained if the reaction is performed in thf instead of acetone. *trans-C,C-cis,cis*-$[Os(SC_6F_5)_2(CO)_2(PR_3)_2]$ (PR_3 = PMe_2Ph or $PEtPh_2$) is obtained as the predominant isomer at longer reaction times and hence is the most thermodynamically stable. A fourth geometric isomer, trans-S,S-cis,cis, has been obtained from the reaction of *mer*-$[OsBr_2(N_2)(PEt_2Ph)_3]$ with $Pb(SC_6F_5)_2$ under a CO atmosphere. Therefore, only the cis,cis,cis isomer is yet to be isolated. All of the complexes have been characterized by ^1H, ^{19}F, and ^{31}P NMR and IR spectroscopies, and in the case of *trans,trans,trans*-$[Os(SC_6F_5)_2(CO)_2(PEt_2Ph)_2]$, by an X-ray structural analysis (*443*). The Os — S bond length [2.447(1) Å] (*443*) is shorter than that in *mer*-$[OsCl(SC_6F_5)(N_2)(PMe_2Ph)_3]$ [2.507(1) Å] (*254, 444*), where the Os — S bond is trans to an Os — P bond.

mer-$[OsX_2(N_2)(PMe_2Ph)_3]$ react with either $Pb(SR)_2$ (R = Ph or C_6F_5), NaSMe, or $AgSCF_3$ to give *mer*-$[OsX(SR')(N_2)(PMe_2Ph)_3]$ ($X = Cl^-$, R′ = Me, CF_3, Ph, or C_6F_5; $X = Br^-$, R′ = C_6F_5). The complexes were characterized by NMR (^1H, ^{19}F, and ^{31}P) and IR spectroscopies. The $SC_6F_5^-$ ligand is trans to a phosphine ligand with the X^- and N_2 ligands being trans (*254, 444*).

Either *cis*- or *trans*-$[Os(N)(CH_2SiMe_3)_2Cl_2]^-$ react with 1,2-ethanedithiol in the presence of base to give $[Os(N)(CH_2SiMe_3)_2(SCH_2CH_2S)]^-$. This reacts with MeI to give the thioether complex, $[Os(N)(CH_2SiMe_3)_2(SCH_2CH_2SCH_3)]$. All complexes have been characterized by NMR (^1H and ^{13}C) and IR spectroscopies (*156*). Spectrophotometric determinations of the complexation of Os(VI) with 2,3-dimercaptopropanesulfonate indicate the formation of $[Os(O)_2(dmps)_2]^{4-}$ in which the ligand is anticipated to act as a dithiolato chelate (*445*).

d. Sulfide and Polysulfide Complexes. $[Os(CO)Cl(NO)(PPh_3)_2]$ reacts with HS^- and $[OsCl(NO)(PPh_3)_3]$ reacts with S_8 to give $[OsCl(NO)(PPh_3)_2(S_2)]$ (*446*). $[OsO_4]$ reacts with S_2Cl_2, S_nCl_2, or S_2Br_2 to yield $[OsS_2O_2Cl_3]$, $[OsS_4Cl_2]$, and $[OsS_2Br_4]$, respectively. All of these complexes are poorly characterized and probably contain polysulfide groups (*447*).

e. Hydrogen Selenido Complexes. *trans*-$[Os(SeH)(NO)(CS)(PPh_3)_2]$ has been prepared and characterized by ^1H NMR and IR spectroscopies (*152*).

f. Thiophene Complexes. Until recently, relatively few transition metal thiophene complexes have been reported, but the ligand is now known to coordinate in at least six different modes (*448*). $[Os(NH_3)_5\{\eta^2$-(*C,C*)-thiophene$\}]^{2+}$ (Section II,C,2,j) apparently rearranges to form $[Os(NH_3)_5\{(S)$-thiophene$\}]^{3+}$ on oxidation (*90, 179*).

g. Thiourea and 1-Amidino-2-thiourea Complexes. The preparation of various salts of the well-known (*2*) $[Os(thio)_6]^{3+}$ ion has been reported by the reactions of Os(IV) chloro complexes with HCl and the ligand, or $[OsO_4]$ and H_2SO_4 with the ligand. These salts include $[Os(thio)_6]Cl_3 \cdot H_2O$, $[Os(thio)_6](HSO_4thio)_3 \cdot 3H_2O$, $[Os(thio)_6][Os^{IV}Cl_6]Cl$, and $[Os^{III}(thio)_6][Os^{III}Cl_6]$. In addition, the new Os(III) complexes, $[Os(thio)_5Cl]Cl_2$ and $[Os(thio)_5Cl][OsCl_6]$, have been reported. All complexes were studied by EPR, IR, and UV/Vis spectroscopies, X-ray powder diffraction, electrical conductivity, and thermal gravimetric analysis (*449*). They have also been studied by XPS spectroscopy, together with *trans*-$[Os(O)_2(thio)_4]SO_4$ (*450*), and detailed IR spectroscopic studies have revealed that all the thio ligands are S bound (*451*).

$[OsO_4]$ forms a 1:1 complex with allylthiourea, which is intensely absorbing and is the basis of a spectrophotometric method for the determination of Os(VIII) (*452, 453*). $[Os(cp)_2(thio)]^{2+}$ has also been reported (*264*).

$K[Os(O)_3N]$ reacts with the amidinothiourea ligands to give *trans*-$[Os^VNL_2(H_2O)]$ (HL = 1-amidino-2-thiourea, *N*-methyl-1-amidino-2-thiourea, or *N*-ethyl-1-amidino-2-thiourea) in which the ligands are believed to act as N,S donors with the sulfurs of the two ligands being mutually trans. The complexes were characterized by IR spectroscopy, thermogravimetric analysis, and magnetic measurements (*271*).

h. 2,1,3-Benzothiadiazole and 2,1,3-Benzoselenadiazole Complexes. The first Os complexes of these ligands, $[OsCl(C_6H_4Me-$

4)(CO)(PPh_3)$_2$L] (L = bsd or btd), have been prepared from the reactions of [OsCl(C_6H_4Me-4)(CO)(PPh_3)$_2$] with btd or bsd. The PPh_3 ligands are mutually trans, with the btd and bsd ligands being trans to the alkyl group. These S-donor or Se-donor ligands are readily displaced from the coordination sphere by CO or $SNNMe_2$. [Os-Cl_2(CS)(PPh_3)$_3$] also reacts with bsd to form *trans-P,P-cis*-[OsCl_2(CS)(PPh_3)$_2$(bsd)]. All of the complexes have been characterized by IR and NMR (^{1}H and ^{31}P) spectroscopies (*150*).

i. S-Dimethyl Sulfoxide Complexes. $[Os(NH_3)_5(S\text{-dmso})]^{2+}$ is prepared by either the linkage isomerization of the *O*-dmso linkage isomer, or $[Os^{II}(NH_3)_5(\text{solvent})]^{2+}$ with dmso (*67, 120, 200*). The Os(III) complex is observed in the CV at fast scan rates, but is unstable with respect to $[Os(NH_3)_5(O\text{-dmso})]^{3+}$ (Section V,D,4). $[Os^{II}(NH_3)_5(\text{4-lutdm})]^{2+}$ reacts with dmso to yield *trans*-$[Os^{II}(NH_3)_4(\text{4-lutdm})(\text{dmso})]^{2+}$. Although, the donor atom of the dmso ligand was not specified (*85*) it is expected to be S bound by analogy with $[Os^{II}(NH_3)_5(S\text{-dmso})]^{2+}$ and other Os(II) dmso complexes. [Os(pc)(dmso)$_2$] has also been reported (Section II,C,4,d) (*186*), and, from the X-ray structure of its Fe(II) analog (*454*), it is expected to have both dmso ligands bound via S.

The thermal decompositions of *trans*-[Os(*S*-dmso)$_4$$Cl_2$] and *trans*-[Os(*O*-dmso)(*S*-dmso)$_3$$Cl_2$] [first prepared in 1980 (*445*)] to give μ-MeS^- dinuclear complexes have been described recently (*456*) (Section II,D,6,a). *trans*-[Os(*S*-dmso)$_4$$Br_2$] is prepared from the prolonged heating of $[Bu^n{}_4N]_2[OsBr_6]$ in dmso (*457*) and its crystal structure is isomorphous with its Ru analog. The Os — S bond lengths [2.351(2) Å] are similar to that found in [Os(η^6-4-cymene)(*S*-dmso)Cl_2] [2.324 Å (*458*)], which is prepared from the reaction of [{Os(η^6-4-cymene)Cl}$_2$(μ-Cl)$_2$] with dmso (*459*). The dmso product reacts with Al_2Me_6 in the molar ratios of 1:0.6 or 1:1.6 to give [Os(η^6-4-cymene)(*S*-dmso)(Me)Cl] and [Os(η^6-4-cymene)(*S*-dmso)(Me)$_2$], respectively. All of the dmso products have been characterized by ^{1}H and ^{13}C{^{1}H} NMR spectroscopies (*459*). [Os(η^6-4-cymene)(*S*-dmso)(Me)Cl] undergoes an orthometalation reaction with $PhCO_2Ag$ to give [Os(η^6-4-cymene){η^2-(*O*,C^2)-$O_2CC_6H_4$}(*S*-dmso)] (*460*).

Electrochemical reduction of *mer*-[Os(PMe_2Ph)$_3$$Cl_3$] in dmso results in the formation of *mer,trans*-[Os(PMe_2Ph)$_3$(dmso)Cl_2] in which the dmso ligand is believed to be S bound. This complex has been characterized by ^{31}P{^{1}H} NMR spectroscopy and exhibits a reversible oxidation to the Os(III) analog (*145*).

j. Thionitrosoamine Complexes. $SNNMe_2$ reacts with [Os(CO)-(Cl)(H)(PPh_3)$_3$] to give [Os(CO)(Cl)(H)($SNNMe_2$)(PPh_3)$_2$], in which the hydrido ligand is trans to $SNNMe_2$, whereas *cis,trans*-$[Os(CO)(H)(SNNMe_2)_2(PPh_3)_2]^+$ is obtained when the reactant is $[Os(H)(OH_2)(CO)(PPh_3)_3]^+$. $[Os(L)(Cl)(4\text{-}MeC_6H_4)(PPh_3)_3]$ (L = CO or CS) react with the ligand to give $[Os(L)(Cl)(4\text{-}MeC_6H_4)(SNNMe_2)(PPh_3)_2]$ (*153*). In the X-ray crystal structure of $[Os(Cl)(C_6H_4Me\text{-}4)(CO)(PPh_3)_2(SNNMe_2)]$, the thionitrosoamine ligand is essentially planar and is bound through its sulfur atom. The Os — S bond is short [2.411(2) Å] and the *trans*-Os — Cl bond is long [2.476(3) Å]. The long Os — Cl bond facilitates the substitution reactions of this complex to give $[Os(X)(C_6H_4Me\text{-}4)(CO)(PPh_3)_2(SNNMe_2)]$ (X = N_3^-, CO, 2-xylyl isocyanide, or Me_3CCN) (*185*). The complex with X = Cl^- was also prepared from the reaction of the ligand with $[OsCl(C_6H_4Me\text{-}4)(CO)(PPh_3)_2(bsd)]$ (*26*).

k. Iminooxosulfane Ligands. The reaction of [(4-tolylsulfonyl)imino]oxo-λ^4-sulfane with the square-planar $[Os(NO)(Cl)(PPh_3)_2]$ yields the six-coordinate complex, *trans*-$[Os(NO)(Cl)(\eta^2\text{-}OSNSO_2C_6H_4Me\text{-}4)(PPh_3)_2]$, in which the ligand is bound via the N and S atoms of the sulfane group. However, if the same ligand is allowed to react with the octahedral $[Os(NO)(Cl)(\eta^2\text{-}CH_2{=}CH_2)(PPh_3)_2]$, then either electrophilic attack at the ethene occurs to give the metallocycle, *trans*-$[Os(NO)(Cl)\{\eta^2\text{-}CH_2CH_2S(O){=}NSO_2C_6H_4Me\text{-}4\}(PPh_3)_2]$ (Section II, C,2,f), or ethylene substitution occurs to give the iminooxosulfane products. Similar reactions occur with other alkene or alkyne complexes. The products were characterized by IR and NMR (^1H and $^{31}P\{^1H\}$) spectroscopies (*163, 164, 446*).

l. Disulfur Oxide and Sulfur Dioxide Complexes. $[Os(NH_3)_5(SO_2)]^{2+}$ has been prepared and characterized by electrochemistry and by UV/Vis and IR spectroscopies (*60*). $[OsCl(NO)(PPh_3)_2(S_2)]$ is oxidized to $[OsCl(NO)(PPh_3)_2(S_2O)]$ by $3\text{-}ClPhCO_3H$ (*446*) and $[Os(NO)Cl(PPh_3)_2(SO_2)]$ is prepared from the hydrolysis of $[Os(NO)Cl(PPh_3)_2(\eta^2\text{-}OSNSO_2C_6H_4Me\text{-}4)]$ (*446*). $[OsCl(C_6H_4Me\text{-}4)(CO)(PPh_3)_2]$ reacts with SO_2 to form $[OsCl(C_6H_4Me\text{-}4)(CO)(PPh_3)_2(SO_2)]$, which is unstable toward rearrangement to the *S*-sulfinato complex, $[OsCl\{S(O)_2C_6H_4Me\text{-}4\}(CO)(PPh_3)_2]$ (*434*).

m. S-Sulfinato Complexes. $[OsCl\{S(O)_2C_6H_4Me\text{-}4\}(CO)(PPh_3)_2]$ is discussed above (*434*).

n. S-Thiosulfato Complexes. Thiosulfate reacts with $[OsO_4]$ to form the tetrahedral $[Os(O)_2(SSO_3)_2]^{2-}$ complex, which has had its structure determined by X-ray crystallography. This complex is only the third example of a tetrahedral Os(VI) complex, and is the first example without aryl groups. The thiosulfate ligands are bound via S, with Os — S bond lengths of 2.218(1) Å. It is diamagnetic and Raman, IR, and ^{17}O NMR spectroscopies indicate that the tetrahedral structure is maintained in solution. The complex exhibits a reversible reduction to the Os(V) complex, $[Os(O)_2(SSO_3)_2]^{3-}$ (*33*).

o. Sulfito Complexes. $[Os(NH_3)_5(SO_3)]^+$ has been reported and studied by electrochemistry and UV/Vis spectroscopy (*60*). Sulfite reacts with *cis*-$[Os(NO)(bpy)_2Cl]^{2+}$ at pH 6.9–9.2 to give *cis*-$[Os(bpy)_2(NO)(SO_3)]$, which has been characterized by IR and electronic absorption spectroscopies. This complex undergoes a reversible one-electron reduction and an irreversible one-electron oxidation in dmf (*461*).

p. Isothiocyanato and Isoselenocyanato Complexes. For the synthesis and properties of $[Os(NCS)_n(SCN)_{6-n}]^{2-/3-}$ (*274, 275*) and $[Os(N)(CH_2SiMe_3)(NCS)(SCN)]^-$ (*156*), see Section II,C,4,t. $[Os(cp)_2(SCN)]SCN$ is prepared by the reaction of $[(cp)_2Os(SS)Os(cp)_2]$ with CN^-, and the X-ray structure has been determined (*462*).

$[Os(NCSe)_n(SeCN)_{6-n}]^{3-}$ ($n = 0–3$) have been discussed in Section II,C,4,u. *trans,trans,trans*-$[Os(O)_2(CN)_2(SeCN)_2]^{2-}$ (*113*) and *trans*-$[Os(O)_2(SeCN)_4]^{2-}$ (*272*) are prepared from the reactions of *trans,trans,trans*-$[Os(O)_2(CN)_2(OH)_2]$ and *trans*-$[Os(O)_2(OH)_4]^{2-}$, respectively, with the ligand. The structures were assigned in each case on the basis of IR and Raman spectroscopies (*113, 272*). The UV/Vis spectrum of *trans*-$[Os(O)_2(SeCN)_4]^{2-}$ is highly vibrationally coupled (*272*).

q. Dithiocarbamate and Xanthate Complexes. The chemistry and electrochemistry of Os dithiocarbamates were reviewed in 1984 and 1986 (*2, 463*). Recently, the reactions of $[Os(NO)Cl_3(PPh_3)_2]$ with $Na(S_2CNR_2)$ (R = Me or Et) have been studied (*246*). If the reactants are stirred together in benzene at room temperature, $[Os(NO)Cl_2(S_2CNR_2)(PPh_3)]$ results, but if the solution containing excess ligand is refluxed, the product is $[Os(S_2CNR_2)_3]$. The complexes were characterized by IR and electronic absorption spectroscopies and magnetic measurements. *trans*-$[OsI(NO)\{\eta^1\text{-}(C)\text{-}CS_2Me\}(S_2CNMe_2)(PPh_3)_2]$, in which the NO and CS_2Me ligands are mutually trans, has also been reported (Section II,C,2,e) (*152*). *mer*-$[OsCl_2(N_2)(PMe_2Ph)_3]$ reacts with $Na[S_2CNMe_2]$ to give *cis*-$[Os(S_2CNMe_2)_2(PMe_2Ph)_2]$ and with $Tl[S_2CNMe_2]$ to give $[OsCl(S_2CNMe_2)(PMe_2Ph)_3]$ as a 1:1.35 mixture of the mer:fac isomers. The complexes were characterized by ^1H

and ^{31}P NMR spectroscopies (*254*) and are identical to the products obtained if *mer*-[$OsCl_3(PMe_2Ph)_3$] is used as a starting material (*464*).

The complexes [$Os^{III}L_3$] (L = anilinodithiocarbamate, 2-, 3-, or 4-fluoroanilinodithiocarbamate, *N*-methylpiperizinodithiocarbamate, dicyclohexyldithiocarbamate, methylcyclohexyldithiocarbamate, benzylxanthate, and butylxanthate) have been prepared by the reaction of the K^+ salt of the ligands with $K_2[OsCl_6]$ or $Na_2[OsCl_6]$ (*465*). These complexes and their Ru analogs, were screened for antitumor activity in rats bearing Ehrlich ascites, P_{388}, and ADJ/PC6 tumors, and for antitrypanosomal activity in rats infected with *Trypanosoma bruceii bruceii, Trypanosoma congolense,* and *Trypanosoma cruzi*. The Os complexes were more active than their Ru analogs and dithiocarbamate complexes were more active than xanthate complexes in all assays. The most interesting complex was that in which L = anilinodithiocarbamate, which was of low nephrotoxicity and toxicity and is effective in all assays (*465*).

Recently, the synthesis and properties of *cis*- and *trans*-[Os^{II}-(Rxanthate)$_2$(PPh_3)$_2$] and *cis*- and *trans*-[Os^{III}(Rxanthate)$_2$(PPh_3)$_2$]$^+$ (R = Me, Et, Pr^i, CH_2Ph) have been reported (*465a*). These complexes have been studied by UV/Vis/NIR and EPR spectroscopies, magnetic measurements, and electrochemistry. The X-ray crystal structures of *cis*-[Os(Mexanthate)$_2$(PPh_3)$_2$], *trans*-[Os(Mexanthate)$_2$(PPh_3)$_2$] and *trans*-[Os(Mexanthate)$_2$(PPh_3)$_2$]$PF_6 \cdot 2H_2O$ were also reported. Using electrochemical and UV/Vis spectroscopic techniques, the kinetics and thermodynamics of the geometric isomerizations in both oxidation states have been studied. The *cis* isomers are favored thermodynamically for Os(II) and the *trans* isomers for Os(III) (*465a*). Similar results are found in related complexes with thioxanthate ligands and the crystal structures of *cis*-[$Os(EtSCS_2)_2(PPh_3)_2$], *trans*-[$Os(EtSCS_2)_2(PPh_3)_2$], and *trans*-[$Os(EtSCS_2)_2(PPh_3)_2$]PF_6 have been reported (*465b*).

r. Complexes and Dithiophosphinates. [Os(2,4,5-$Me_3C_6H_2PS_2$)$_3$] is prepared from the reaction of the ligand with $(NH_4)_2[OsCl_6]$ and have been characterized by EPR and UV/Vis spectroscopies and magnetic measurements (*466*). *cis*-[$Os(S_2PR_2)_2(PPh_3)_2$] (R = Me, Ph, or OEt) react with tetracyanoethylene to give *cis*-[$Os(S_2PR_2)_2(PPh_3)$(tcne)]. These complexes have been characterized by IR and UV/Vis spectroscopies and the X-ray structures of two complexes (R = Me or Ph) have been determined. They exhibit two reversible one-electron reductions that are centered on the tcne ligand and reversible oxidations to the Os(III) analogs (*467*).

s. Cysteine Complexes. Cysteine is reported to form a 1 : 1 complex with Os(VIII) in aqueous solution, but this complex is poorly characterized (*468*).

t. 2-(Tolylthio)picolinamide. $[OsL_2(OH)\cdot H_2O]$ and $[Os(HL)_2(OH)_2Cl]H_2O$ [HL = 2-(tolythio)picolinamide] are prepared by the reactions of HL with $[OsO_4]$ or $[OsCl_6]^{2-}$. In weakly acidic solutions, L^- is bidentate and coordinates via the pyridine N and the N or S atom of the thioamide group. In strongly acidic solutions, HL is monodentate and binds via the thioketone S atom (*469*).

u. 2-Aminobenzenethiol Complexes. 2-Aminobenzenethiol reacts with $[OsO_4]$ to give *fac*-$[Os^{VI}(abt)_3]$, which has been characterized by X-ray crystallography and mass, IR, and NMR spectroscopies. It is diamagnetic and contains three doubly deprotonated ligands. Electrochemical studies in acetonitrile show two one-electron reversible reductions to the Os(V) and Os(IV) analogs and two reversible one-electron oxidations to the formal oxidation states of Os(VII) and Os(VIII) (*24*). Os(VI) reacts with 4-sulfo-2-aminobenzenethiol (H_2L) to form 1 : 2 complexes that presumably contain the N,S chelates, *trans*-$[Os(O)_2(L)_2]$. The formation of this complex has been used as the basis of a spectrophotometric determination of Os(VI) (*470*).

v. 2-Pyridinethiolato, 2-Pyrimidinethiolato, and Thiopyrine Complexes. When $[Os(H)_2(CO)(PPh_3)_3]$ is refluxed with pySH in toluene, a mixture of *trans,trans-S,H-* and *trans,trans,-S,C*-$[OsH(pyS)(CO)(PPh_3)_2]$ in the ratio of ~4 : 1 was obtained. If the starting material was refluxed with pySSpy for longer periods of time, *cis,trans-S,S*-$[Os(pyS)_2(CO)(PPh_3)]$ results. Refluxing $[Os(H)_4(PPh_3)_3]$ with pySH in benzene leads to *trans,trans-S,C*-$[Os(H)(pyS)(CO)(PPh_3)_2]$, whereas the reaction between $[Os(H)(Cl)(CO)(PPh_3)_3]$ and pySH in cold benzene results in a mixture of *trans,trans-S,C*-$[Os(H)(pyS)(CO)(PPh_3)_2]$ and *trans,trans-S,H-* and *trans,trans-S,C*-$[Os(H)(Cl)(HpyS)(CO)(PPh_3)_2]$. The pyS ligand is bidentate, but the HpyS ligand is bound via S and protonated at N. If the same reaction is carried out in boiling toluene (18 hours), *trans,trans-S,H-* and *trans,trans-S,C*-$[Os(H)(pyS)(CO)(PPh_3)_2]$ and *trans*-$[OsCl(pyS)(CO)(PPh_3)_2]$ result, in the ratio 4 : 1 : 15. The geometric isomer of the latter is uncertain with respect to the ligand that is trans to the S atom. When pySSpy is reacted instead of the ligand, the product is solely $[OsCl(pyS)(CO)(PPh_3)_2]$. Similar reactions yield *trans,trans-S,H-* and *trans,trans-S,C*-$[Os(H)(Br)(HpyS)(CO)(P-$

$Ph_3)_2$] and [OsBr(pyS)(CO)(PPh_3)$_2$]. All of the complexes were characterized by IR and NMR (^1H and ^{31}P) spectroscopies (*136*). Complexes in which pyS reacts as a bridging ligand are described in Section II,D,6,a.

Heating [Os(H)$_2$(CO)(PPh_3)$_3$] in boiling toluene or heating [Os(H)-(Cl)(CO)(PPh_3)$_3$] in boiling benzene with Me_2pymSH yields two geometic isomers of *trans*-[Os(H)(Me_2pymS)(CO)(PPh_3)$_2$], in which the RS^- group of the chelate is trans to either the H^- or CO ligand. [Os-Cl_2(PPh_3)$_3$] reacts with an excess of the same ligand to form *trans*-[Os(Me_2pymS)$_2$(PPh_3)$_2$]. The complexes have been characterized by NMR (^1H, ^{13}C, and ^{31}P) and IR spectroscopies (*471*). 1-Pyridyl- and 1-(4′-phenylthiazolyl)-4,4,6-trimethyl-1*H*,4*H*-pyrimidine-2-thiol have been used as complexing agents for the spectrophotometric determination of Os, but the details of the complexes formed have not been given (*472*).

Os(VIII) reacts with thiopyrine to form [Os(O)$_2$($C_{11}H_{12}N_2S$)$_2$]-(ClO_4)$_2$, which is used as a method of extraction into $CHCl_3$ for the spectrophotometric determination of Os(VIII) (*473*).

w. Pyrimidinethione and Thiobarbituric Acid Complexes. Os(VIII) is reported to react with 3,4-dihydro-4,4,6-trimethyl-2-(1*H*)-pyrimidinethione (*474*) or 2-thiobarbituric acid (*475*) to form highly colored [Os(L)$_4$] complexes as a basis of spectrophotometric determinations of Os, but these complexes have not been well characterized.

8. *Halo Ligands*

Most halo complexes have been discussed in previous sections (Table XI). Table XII (*239, 268, 323, 400, 403, 476–499*) summarizes studies on complexes that contain only halo ligands or halo/oxo complexes.

a. Fluoro Complexes. Matrix isolation methods have been used to study the IR spectra of the molecular species [Os(O)F_4], [Os(O)$_2F_3$], and [Os(O)$_3F_2$] that have been obtained by the vaporization of the solids. The species have also been identified by mass spectrometry via the molecular ions [Os(O)$_3$F]$^+$, [Os(O)$_2F_2$]$^+$, [Os(O)F_3]$^+$, [Os(O)$_3F_2$]$^+$, and [Os(O)$_2F_3$]$^+$. Other products of the vaporization include [OsF_6] and another fluoride believed to be [OsF_5]. These products are consistent with the generation of the molecular ions, [OsF_5]$^+$ and [OsF_4]$^+$, in the mass spectra (*494, 500*).

b. Organofluorine Complexes. *mer*-[Os(SC_6F_5)$_3$(PR_3)$_2$] complexes (PR_3 = PMe_2Ph, PEt_2Ph, $PMePh_2$, or $PEtPh_2$) are believed to be octa-

TABLE XI

HALO COMPLEXES WITH OTHER LIGANDS

Complex or coligand	Section	Complex of coligand	Section
Osmaboranes	II,C,1	CN^-	II,C,2,a
CO	II,C,2,b	CS, CSe, CTe	II,C,2,c
CS_2, CSSe, $SCSMe^-$	II,C,2,e	Alkyl and aryl	II,C,2,f
Carbene	II,C,2,g	Alkene	II,C,2,h
Isonitrile	II,C,2,n	SiR_3	II,C,3
Ammine	II,C,4,a	Amines	II,C,4,b
Macrocycles	II,C,4,c	Porphyrins	II,C,4,d
N-heterocycles	II,C,4,e	NO, NO^+	II,C,4,f
Nitrosoarene	II,C,4,g	Hydroxylamine	II,C,4,h
NS, NS^+	II,C,4,i	N_2	II,C,4,j
Oxime	II,C,4,l	NCR	II,C,4,n
NSCl	II,C,4,o	N^{3-}	II,C,4,q
NCO^-	II,C,4,s	NO_2^-, NOS^-	II,C,4,v
Schiff bases	II,C,4,w	Quinololates	II,C,4,y
Diaryltriazenes	II,C,4,z	Penylazopyridines	II,C,4,z
Phosphines	II,C,5,a	Phosphites	II,C,5,b
η^2-CH_2PR_2	II,C,5,c	Arsines	II,C,5,d
Diarsines	II,C,5,e	Stibines	II,C,4,f
H_2O, OH^-	II,C,6,a	Oxo	II,C,6,b
ROH, RO^-	II,C,4,c	Maltolate and tropolonate	II,C,6,d
O-dmso	II,C,6,i	Diketonates	II,C,6,k
O_2	II,C,6,m	RCO_2^-	II,C,6,o
RSO_3^-	II,C,6,q	*O*-sulfinate	II,C,6,r
Benzeneseleninato	II,C,6,s	Thioether and selenoether	II,C,7,a
Thioether macrocycles	II,C,7,b	RS^-	II,C,7,c
S_2	II,C,7,d	Thiourea	II,C,7,g
btd and bsd	II,C,7,h	*S*-dmso	II,C,7,i
$SNNMe_2$	II,C,7,j	Iminooxosulfane	II,C,7,k
SO_2, S_2O	II,C,7,l	*S*-sulfinato	II,C,7,m
$R_2NCS_2^-$ and xanthate	II,C,7,q	2-(tolylthio)picolinamide	II,C,7,t
Pys, pymS	II,C,7,v	ECl_4 (E = S, Se, or Te)	II,C,8,d
BH_4^-	II,C,9,b	H_2	II,C,9,c

hedral, with one of the $SC_6F_5^-$ ligands acting as an η^2-S,F chelate via a fluoro substituent in the 2-position (Section II,C,7,c).

c. Chloro, Bromo, and Iodo Complexes. $[Os(NH_3)_5X]^{2+}$ ($X = Cl^-$, Br^-, or I^-) has been obtained from the triflato complex by heating in aqueous HX. The crystal structure of $[Os(NH_3)_5Cl]Cl_2$ has been determined. By comparison with isomorphous complexes, it is deduced that Os—Cl π bonding is comparable to that of Ru in these complexes (*501*).

TABLE XII

HALO AND OXO/HALO COMPLEXES

Complex	Studies and references
Os(II)	
$[OsCl_6]^{4-}$	CV (*476*)
Os(III)	
$[OsCl_6]^{3-}$	CV (*476,477*)
Os(IV)	
$[OsCl_4]$	Reaction with ECl_4 (E = Se or Te) (*478*)
$[OsBr_xCl_{4-x}]$, $0 < x < 2.3$	X-ray (*479*)
$[OsBr_4]$	X-ray (*479*)
$[OsF_6]^{2-}$	IR (*239*), Raman (*239*), normal coordinate analysis (*239*), XPS (*268*), CV (*476*)
$[OsF_nCl_{6-n}]^{2-}$	^{19}F NMR (*480*), oxidation (*481*)
$[OsCl_6]^{2-}$	IR (*239,482*) Raman (*239*), normal coordinate analysis (*239*), polarized absorption spectrum (*483*), X-ray (*482,484,485*), XPS (*268*), CV (*476,477*), electron density distribution (*485*), NQR (*486–488*), high-pressure NQR (*489*), DTA (*482*), aquation (*323*), ^{191m}Ir generator (*490*), doped into AgBr, photochem. (*491*)
$[OsCl_nBr_{6-n}]^{2-}$	IR, Raman, normal coordinate analysis (*492,493*)
$[OsCl_5I]^{2-}$	Far IR, normal coordinate analysis (*493*)
cis-$[OsCl_4I_2]^{2-}$	Far IR, normal coordinate analysis (*493*)
fac-$[OsCl_3I_3]^{2-}$	Far IR, normal coordinate analysis (*493*)
$[OsBr_6]^{2-}$	IR (*239*), Raman (*239*), normal coordinate analysis (*239*), X-ray (*484*), XPS (*268*)
$[OsI_6]^{2-}$	IR (*239*), Raman (*239*), normal coordinate analysis (*239*), XPS (*268*)
Os(V)	
$[OsF_6]^-$	CV (*476*)
$[OsF_nCl_{6-n}]^-$	Preparation, IR, Raman (*481*)
$[OsCl_6]^-$	CV (*476*)
Os(VI)	
trans-$[Os(O)_2Cl_4]^{2-}$	X-Ray (*403*)
$[OsOF_4]$	Matrix-isolation IR and UV/Vis (*494*), mass spectrum (*494*)
$[OsOCl_4]$	Electron diffraction (*400*)
$[OsF_6]$	Matrix-isolation IR and UV/Vis (*494*), mass spectrum (*494*), calc. of optical and magnetic properties (*495*), graphite interchelate, neutron scattering, EPR, Raman, mag. (*496–498*)
$[OsOF_3]^+$	Mass spectrum (*494*)

TABLE XII (*Continued*)

Complex	Studies and references
Os(VII)	
$[OsOF_5]$	Matrix-isolation IR and UV/Vis (*494*), mass spectrum (*494*)
$[OsOF_4]^+$	Mass spectrum (*494*)
Os(VIII)	
$[Os(O)_3F_2]$	Matrix-isolation IR and UV/Vis (*494*), mass spectrum (*494*), force constants (499)

d. Sulfur, Selenium, and Tellurium Tetrahalide Complexes. The reaction of $[OsO_4]$ with SCl_2 yields *trans*-$[Os(SCl_4)_2Cl_4]$, which reacts with ECl_4 (E = Se or Te), to yield *trans*-$[Os(ECl_4)_2Cl_4]$. The structures of all three complexes have been determined by IR and ^{35}Cl NQR spectroscopies and X-ray crystallography. They contain long E—Cl bonds between the ECl_3^+ and $[OsCl_6]^{2-}$ groups (*447, 478, 502*). *trans*-$[Os(SeBr_4)_2Br_4]$ and $[Os(SF_4)F_5]$ have also been reported (*447, 503*).

e. Dichlorobis(triphenylphosphine)argentate(I) Complexes. $[OsCl_6]^{2-}$ reacts with Ag(I) and PPh_3 to form *trans*-$[Os\{\mu\text{-}\eta^2\text{-}(Cl,Cl)\text{-}AgCl_2(PPh_3)_2\}_2Cl_2]$, which has been characterized by X-ray crystallography (*504*).

9. *Hydride and Dihydrogen Complexes*

a. Hydride Complexes. The Os(II) hydride complexes, $Mg_2[OsH_6]$ and $Mg_2[OsD_6]$, are prepared from Os powder and MgH_2 and MgD_2, respectively. Powder X-ray and neutron diffraction of the hydride and deuteride, respectively, show that the complexes possess a $K_2[PtCl_6]$-type structure, with Os—D bond lengths of 1.68(1) Å (*505*). Other hydride complexes are discussed elsewhere in this review (Table XIII).

b. Borohydride Complexes. The reactions of $[Os(H)(Cl)(CO)(PR_3)_2]$ (R_3 = $Pr^i{}_3$ or $MeBu^t{}_2$) with $NaBH_4$ in MeOH yield the octahedral complexes, *trans*-$[Os(H)(CO)(\eta^2\text{-}BH_4)(PR_3)_2]$, which contain an η^2-H,H′ borohydride ligand. The structure is only rigid below −30°C; at higher temperatures, the bridging and terminal hydrides of the BH_4^- ligand undergo exchange on the NMR time scale. There is no exchange

TABLE XIII

HYDRIDE COMPLEXES WITH OTHER LIGANDS

Complex or coligand	Section	Complex or coligand	Section
Osmaboranes	II,C,1	H_2	II,C,9,c
CS, CSe, CTe	II,C,2,c	CO	II,C,2,b
CS_2, CSSe, $SCSMe^-$	II,C,2,e	CO_2	II,C,2,d
Alkene	II,C,2,h	Alkyl and aryl	II,C,2,f
Isonitrile	II,C,2,n	Allyl	II,C,2,l
Silyl	II,C,3	Ammine	II,C,4,a
Porphyrins	II,C,4,d	N heterocycles	II,C,4,e
NO, NO^+	II,C,4,f	Oxime	II,C,4,l
Diazenes	II,C,4,m	Schiff bases	II,C,4,w
Benzotriazole	II,C,4,z	Phosphines	II,C,5,a
Phosphites	II,C,5,b	η^2-CH_2PR_2	II,C,5,c
Arsines	II,C,5,d	Diketonates	II,C,6,k
O_2	II,C,6,m	RCO_2^-	II,C,6,o
RSO_3^-	II,C,6,q	$SNNMe_2$	II,C,7,j
PyS, pymS	II,C,7,v	BH_4^-	II,C,9,b

between the H^- ligand and the BH_4^- hydrides (*128*). This contrasts with the pentagonal-bipyramidal complex, $[Os(\eta^2\text{-}BH_4)(H)_3\{P(cyclo\text{-}C_5H_9)_3\}_2]$, in which the hydrides and the two sites occupied by the BH_4^- hydrides exchange rapidly on the NMR time scale at 90°C (*506*). The former BH_4^- complex, with $R_3 = MeBu^t{}_2$, is also prepared from the reaction of $[Os(H)(Cl)(CO)\{N\text{-}N(OH){=}(CH_3)_2\}(PMeBu^t{}_2)_2]$ with $NaBH_4$ (*124*).

c. Dihydrogen Complexes. The reduction of $[Os(NH_3)_5(O\text{-}SO_2CF_3)]^{2+}$ in methanolic or aqueous solutions yields the η^2-dihydrogen complex, $[Os^{II}(NH_3)_5(H_2)]^{2+}$, which is reversibly oxidized to the Os(III) complex, $[Os^{III}(NH_3)_5(H_2)]^{3+}$. The Os(III) complex is only stable in strong acid and readily deprotonates to give the hydride complex, $[Os^{III}(NH_3)_5H]^{2+}$. The presence of the dihydrogen ligand has been established by NMR and IR spectroscopies in conjunction with deuterium labeling experiments (*201*). The complexes *cis*-$[Os(NH_3)_4(H_2)(\pi\ acid)]^{2+}$ and *cis*-$[Os^{III}(NH_3)_4(H_2)(X)]^{2+}$ (X = halide) are prepared from the reactions of a π acid or X^- with the Os(IV) complex, *cis*-$[Os(NH_3)_4(H)_2]^{2+}$ (*89*). The porphyrin dihydrogen complexes, $[Os(oep)(H_2)]^{2+}$ and $[Os(oep)(HD)]^{2+}$, have also been reported (Section II,C,4,d) (*507*).

Recently, *trans*-$[Os(\eta^2\text{-}H_2)(H)(depe)_2]^+$ and its Ru and Fe analogs have been prepared and characterized by NMR spectroscopy. The barrier for exchange involving the dihydrogen and hydride ligands, at 300 K, is lower for Os (53 kJ mol^{-1}) than for Ru (>63 kJ mol^{-1}) and is approximately equal to that for the Fe (54 kJ mol^{-1}) analogs. The Os complex has a weak H—H bond and a strong Os—H_2 bond. When small changes to the bidentate phosphine ligand are made, i.e., replacing depe by dppe, the dihydrogen analogs only exist for the Fe and Ru complexes. The Os complex exists as the classical trihydride complex. The depe dihydrogen complex is made from the reaction of *trans*-$[Os(H)(Cl)(depe)_2]$ with H_2 and the H—H bond length is estimated to be 1.12 ± 0.03 Å from NMR relaxation techniques. This is much longer than the typical H—H bond lengths of ~0.9 Å that are estimated for Ru and Fe analogs (*508, 509*). The crystal structure of the dihydrogen complex has been determined, but disorder has precluded an analysis of the H—H bond length (*510*). An even longer bond has been estimated from ^1H NMR experiments on *trans*-$[Os(\eta^2\text{-}H_2)(H)(Et_2PCH_2CH_2\text{-}PPh_2)]^+$ (*511*). Reactions of either *cis*-β-$[Os(Cl)_2$(*meso*-tetraphos)] or *trans*-[Os(H)(Cl)(*meso*-tetraphos)] with H_2 in thf yield *trans*-$[Os(\eta^2\text{-}H_2)(H)$(*meso*-tetraphos)$]^+$. *cis*-α-$[Os(\eta^2\text{-}H_2)(H)$(*rac*-tetraphos)$]^+$ is formed by protonation of *cis*-α-$[Os(H)_2$(*rac*-tetraphos)] with HBF_4. There is no exchange on the NMR time scale at 293 K between $\eta^2\text{-}H_2$ and H^- for the trans complex, but the cis complex undergoes extremely rapid exchange (*301*). Related *trans*-$[Os(\eta^2\text{-}H_2)(H)(PR_3)_4]$ and *trans*-$[Os(\eta^2\text{-}H_2)(Cl)(PR_3)_4]$ complexes are prepared from protonations of the hydride analogs (*299*). Similarly, protonation of $[Os(H)_2(L)_4]$ (L = $PhP(OEt)_2$ or $P(OEt)_3$] with HBF_4 results in the complexes $[Os(H)(\eta^2\text{-}H_2)L_4]$, which are assigned a trans structure on the basis of NMR experiments. The H_2 ligands in these complexes are readily displaced by other ligands, making them useful synthetic intermediates (*187, 512*). *trans*-$(\eta^2\text{-}H_2),H^-$ complexes are formed via the reversible binding of H_2 to *trans*-$[Os(H)(Cl)(CO)(PPr^i{}_3)_2]$ (*419, 420*) or *trans*-$[Os(H)(Cl)(CO)(PMeBu^t{}_2)]$ (*419*) to give *trans*-$[Os(\eta^2\text{-}H_2)(H)(Cl)(CO)L_2]$. $[Os^{II}(\eta^2\text{-}O,O'\text{-}OAc)(PPh_3)_3(H_2)]^+$ has also been reported recently (*423*). $[Os(SiEt_3)Cl(\eta^2\text{-}H_2)(CO)(PPr^i{}_3)_2]$ (Section II,C,3) and $[Os^{II}\{(\eta^2\text{-}O,O')\text{-}OAc)(PPh_3)_3(\eta^2\text{-}H_2)]^+$ have also been reported recently (*192a, 423*).

Although most of the stable hydrogen complexes have a formal oxidation state of Os(II), there is recent evidence for the formation of $[Os^{IV}(H)_3(\eta^2\text{-}H_2)L_3]^+$ complexes. These complexes were first assigned to be Os(VI) pentahydrides (*513*) and are obtained by protonations of the tetrahydrides, $[Os^{IV}(H)_4L_3]$ [L = PMe_2Ph, PPh_3, or P(4-tolyl)$_3$], with HBF_4 at low temperatures (*423, 513–516*). The criteria for distinguishing between classical polyhydride and η^2-H_2 complexes have also been reviewed (*423, 514*).

D. Dinuclear and Polynuclear Complexes

This section excludes the quite extensive dinuclear and polynuclear Os/CO cluster chemistry, which will only be referred to where it overlaps with classical coordination chemistry. The general classes of Os dimers that are often encountered are Os_2^{n+} (n = 4–8) complexes with Os—Os bonds, with or without bridging ligands; Os(II) dimers, $[(L)_6Os_2(\mu\text{-}L')]^+$; $[A_{10}Os_2(\mu\text{-}L)]^{n+}$; $[(bpy)_4Os_2(\mu\text{-}L)]^{n+}$; Os(VI) dimers, $[L_8Os_2(\mu\text{-}O)_2]$; and Os(VIII) dimers $[\{Os(O)_n(NR)_{4-n}\}_2(\mu\text{-}L)]$. Because the nature of the bridging group has a dominant influence on chemical and physical properties of dimers, these dimers are arranged according to the nature of these bridging ligands.

1. *Dimers with Os—Os Bonds*

The variations in the Os—Os bond lengths and variations in other M—M bond lengths have been the subject of theoretical modeling, and these variations in bond lengths have been rationalized (*517*).

a. No Bridging Ligands. The electrochemistry of dinuclear Os prophyrinato(2−) complexes containing an Os=Os double bond has been studied. $[(oep)Os^{II}{=}Os^{II}(oep)]$ undergoes a series of one-electron reversible oxidation and reduction processes. The complexes $[(oep^+)Os^{III}{\equiv}Os^{III}(oep^+)]^{4+}$, $[(oep^+)Os^{III}{\equiv}Os^{III}(oep)]^{3+}$, $[(oep)Os^{III}{\equiv}Os^{III}(oep)]^{2+}$, $[(oep)Os^{III}{\doteq}Os^{II}(oep)]^+$, and $[(oep)Os^{II}{=}Os^{II}(oep^-)]^-$ have been identified in the CV. $[Os^{II}(oep)]_2$ is oxidized by one equivalent of Fc^+ or Ag^+ to form $[Os(oep)]_2^+$ and by two equivalents of Ag^+ to form $[Os^{III}(oep)_2]^{2+}$. The Os_2^{II} porphyrin complex has the unusual property of having an Os=Os double bond and triplet ground state, whereas the Os_2^{III} dimer has an Os≡Os triple bond and a singlet ground state. The properties have been rationalized in terms of an MO scheme of metal–metal bonding. Both the Os_2^{II} and the mixed-valence ions are EPR silent down to 77 K (*518*). Detailed IR and Raman spectroscopic studies of the 0, +1, and +2 ions have been performed (*519*).

The Os(III) dimers, $[Os_2X_8]^{2-}$ (X = Cl^-, Br^-, or I^-) each have an Os≡Os triple bond and terminal halogens. They have been characterized by X-ray diffraction, IR and UV/Vis spectroscopies, and electrochemistry (*232, 520–523*). The complexes have also been subjected to SCF-X_α-SW calculations of the energies of their electronic states (*521*).

The complexes exhibit reversible oxidations to the $[Os_2X_8]^-$ analogs and irreversible reductions (*522, 523*). Decomposition and substitution chemistry of the complexes has also been studied (*302, 523*). The tetrameric complex $[I_4OsOsI_2(\mu\text{-}I)_2I_2OsOsI_4]^{2-}$ contains two $[Os_2I_8]$ units fused together via two bridging iodides, as shown by X-ray diffraction (*524*). $[(cp)_2Os{\equiv}Os(cp)_2]^{2+}$ has been prepared and characterized by X-ray diffraction (*264*).

b. μ-Porphyrinato Complexes. The dimeric complex $[(py)_2Os(dpb)\text{-}Os(py)_2]$ undergoes pyrolysis to form the Os dimer with an Os=Os double bond, [Os(dpb)Os]. [Ru(dpb)Os] is prepared in a similar manner from $[(py)_2Ru(dpb)Os(py)_2]$ (*141*).

c. μ-Carboxylato Complexes. $[Os_2(\mu\text{-}O_2CCH_3)_4Cl_2]$ is the standard starting material for the synthesis of a variety of diosmium complexes with metal–metal bonds (*303, 319, 322, 525–529*). Detailed IR and Raman spectroscopic studies, including excitation profiles, have been performed with $[Os_2(\mu\text{-}O_2CR)_4Cl_2]$ R = CH_3, CD_3, Et, or Pr^n). All resonance Raman spectra exhibit progressions of $\nu_1(Os{\equiv}Os)$ (*530, 531*). $[Os_2(\mu\text{-}O_2CR)_4Br_2]$ (R = Et or Pr^n) are prepared from their chloro analogs and have been characterized by electrochemistry and ^1H NMR and UV/Vis spectroscopies. They exhibit reversible oxidations to the mixed-valence Os(IV)–Os(III) complexes but the reductions are irreversible (*525*).

Treatment of $[Os_2(\mu\text{-}O_2CCH_3)_4Cl_2]$ with appropriate Grignard reagents leads to the formation of $[Os_2(\mu\text{-}O_2CCH_3)_2(R)_4]$ (R = CH_2CMe_3 or CH_2SiMe_3), but unlike the Ru analogs, further treatment with Grignard reagents does not remove the remaining acetate bridges. These complexes have been characterized by ^1H and ^{13}C NMR spectroscopies (*527*). $[Os_2(CO)_6(\mu\text{-}O_2CMe)_2]$ reacts with dppm or dppmS to form $[Os_2(CO)_4(\mu\text{-}O_2CMe)_2(\eta^1\text{-}L)_2]$, and the dppm complex is oxidized by H_2O_2 to give the analogous complex containing dppmO ligands. All complexes were characterized by IR and ^1H and ^{31}P NMR spectroscopies (*532*). The complexes in which L = PR_3, NCMe, or py are also known, but with thf only one CO ligand is lost to form $[Os_2(CO)_5(\mu\text{-}O_2CMe)_2(thf)]$. Reaction of the latter with PH_3 or PMe_2Ph gives $[Os_2(CO)_5(\mu\text{-}O_2CMe)_2(PR_3)]$, and reaction with Cl^- gives $[Os_2(CO)_5(\mu\text{-}O_2CMe)_2Cl]^-$, which was characterized by X-ray crystallography (*418*).

d. μ-2-Hydroxypyridine Complexes. The mixed-valence ions $[Os_2\text{-}(\mu\text{-}L)_4Cl]$ (L = chp or fhp), have been prepared from $[Os_2(\mu\text{-}O_2CCH_3)_4Cl_2]$ and were characterized by X-ray diffraction, UV/Vis spectroscopy, and electrochemistry. The Os(II) center is five coordinate, with four pyridine groups and the Os completing a square pyramid, whereas the Os(III) center is octahedral with the four alcoholates, Cl^-, and Os completing the coordination sphere (*319, 322, 528*). The fhp complex is reversibly oxidized but is irreversibly reduced, whereas both the reduction and oxidation of the chp complex are reversible (*322, 528*). The chp complex has also been characterized by magnetic measurements, IR spectroscopy, and low-temperature EPR spectroscopy (*322*). The other product of the reaction with chp is $[Os_2(\mu\text{-chp})_2Cl_4]$, which reacts with L′ to give $[Os_2(\mu\text{-chp})_2Cl_4(L')]$ (L′ = py or OH_2), both of which have been characterized by UV/Vis spectroscopy, electrochemistry, and X-ray diffraction. Both complexes exhibit reversible one-electron reductions (*319*).

$[Os_2(hp)_4Br_2]$ has been prepared from its chloro analog and has been studied by electrochemistry and UV/Vis and ^1H NMR spectroscopies (*525*).

e. μ-2-(Diphenylphosphine)pyridine Complexes. Two crystal structures of $[Os_2(\mu\text{-}O_2CMe)(\mu\text{-}Ph_2Ppy)_2Cl_4]$ have been reported in which each Os is octahedrally bound to P, N, O, Os, and two Cl donors. The complex undergoes reversible oxidation and reduction (*526, 529*).

f. μ-Orthometalated Triphenylphosphine Complexes. $[Os_2(\mu\text{-}O_2CMe)_2\{\mu\text{-}(C,P)\text{-}Ph_2PC_6H_4\}_2Cl_2]$ reacts with Me_3SiCl in refluxing thf to form $[Os_2\{\mu\text{-}(C,P)\text{-}Ph_2PC_6H_4\}_2Cl_4]$, in which the bridging ligands adopt a cis geometry, with each Os being bonded to two Cl and P, C, and Os donors. The complex has been characterized by IR spectroscopy, X-ray diffraction, and cyclic voltammetry. It reveals a reversible reduction to its $Os^{III}\text{–}Os^{II}$ analog (*533*).

g. μ-η^5,η^1-Cyclopentadienyl(2−) Complex $[(cp)_2Os{\equiv}Os(cp)_2]^{2+}$ is oxidized to the Os(IV) dimer $[(cp)Os(\mu\text{-}\eta^5,\eta^1\text{-}C_5H_4)_2Os(cp)]^{2+}$, which has been characterized by X-ray crystallography (*264*).

2. *Carbon Ligands*

a. μ-CO_2 Complexes. The dimeric complex $[(cod)Rh(H)_3Os(PMe_2Ph)_3]$ reacts with CO_2 to give *cis, mer*-$[Os(H)_2(CO)(PMe_2Ph)_3]$, $[(cod)_2Rh_2(CO_2)Os(PMe_2Ph)_3(H)_2]$, and H_2O. Both the CO in the first complex and the CO_2 in the second are derived from the gaseous CO_2, as shown by labeling experiments. The trimer was analyzed by X-ray crystallography, which shows the CO_2 bound to Os via the carbon and to each of the Rh atoms via an oxygen atom (*138*).

b. Isopropylcarbamoyl as a Bridging Ligand. The complex μ-bromo-μ-(isopropylcarbamoyl-*O,C*)bis(bromotricarbonylosmium) has been prepared and characterized by X-ray crystallography (*534*).

c. η^2-Arene-Bridged Complexes. Benzene, naphthalene, pyrene, and diphenylacetylene act as bridging ligands to form dinuclear $[(NH_3)_5Os(arene)M(NH_3)_5]^{4+}$ complexes [M = Os (*88, 171, 173, 174*), Ru (*172*)]. The former complexes are made from the self-condensation of $[Os(NH_3)_5(arene)]^{2+}$ with concomitant release of a mole of arene per mole of dimer (*173*), whereas the latter complexes are prepared from the reaction of $[Os(NH_3)_5(arene)]^{2+}$ with $[Ru(NH_3)_5(solvent)]^{2+}$ (*172*). These complexes are believed to have the ligand coordinated in a trans fashion to the 2,3 and 4,5 positions of one ring and this has been confirmed in the X-ray structure of $[(NH_3)_5Os(benzene)Os(NH_3)_5]^{4+}$ (*88*). In the case of the dinuclear pyrene complex, the osmium centers bind to the two exterior double bonds of the central rings in a trans geometry, as shown in the X-ray structure of $[(NH_3)_5Os(\mu\text{-pyrene})Os(NH_3)_5](CF_3SO_3)_4 \cdot 4(CH_3)_2CO$ (*174*). The mixed-valence +5 ions have been observed but lose an η^2-linkage to form the M(III) solvent complex and $[Os^{II}(NH_3)_5(arene)]^{2+}$. The timescales of these decomposition reactions are fast (seconds) for M = Ru (*172*) and the dinuclear Os pyrene complex (*174*); however, $[(NH_3)_5Os(\mu\text{-benzene})Os(NH_3)_5]^{5+}$ is stable for 0.5 hr or more at 20°C (*173*). Further oxidations to the +6 ions are irreversible on the conventional CV time scales and result in the loss of the arene ligands, and in the case of diphenylacetylene, formation of $[Os(NH_3)_5\{\eta^2\text{-}(alkyne)\text{-}PhC{\equiv}CPh\}]^{3+}$. Pyrolysis of $[(NH_3)_5Os(benzene)Os(NH_3)_5]^{4+}$ results in the formation of $[(NH_3)_5Os(\mu\text{-}\eta^2,\eta^6\text{-benzene})Os(NH_3)_3]^{4+}$ (*88*).

d. μ-1,3-Butadiene Complexes. $[(NH_3)_5M(1,3\text{-butadiene})M'(NH_3)_5]^{4+}$(M = M′ = Ru; M = Ru, M′ = Os; M = M′ = Os) have been prepared and characterized by ^{1}H NMR spectroscopy and electrochemistry (*70*).

e. μ-Cyano Complexes. $[(NH_3)_5Os^{III}NCM^{II}(CN)_5]^-$ (M = Fe, Ru, or Os) and $[(NC)_5OsCNCo(CN_5)]^{6-}$ have been prepared and their UV/Vis spectroscopy, electrochemistry, and photochemistry have been studied (*535*).

3. *Nitrogen Donor Ligands*

a. μ-Dinitrogen Complexes. Much of this chemistry has been covered in the review by Griffith (*2*). However, a full account of the chemistry is presented here to place it in context with recent developments. $[(NH_3)_5Os(N_2)Os(NH_3)_5]^{5+/4+}$ complexes were the first of the decaammine dinuclear complexes of Os to be reported. They were prepared from the reduction of $[Os(NH_3)_5(OH_2)]^{3+}$ in the presence of $[Os(NH_3)_5(N_2)]^{2+}$, but the yields were poor (*536*). Improved yields have been obtained from the partial oxidation of $[Os(NH_3)_5(N_2)]^{2+}$ by I_2 in neat CF_3SO_3H, or the reduction of $[Os(NH_3)_5(OSO_2CF_3)]^{2+}$ in dma/dme in the presence of $[Os(NH_3)_5(N_2)]^{2+}$ (*537*). The analogous +6 ion is unstable at room temperature, but it can be generated quantitatively in solution at 5°C by the oxidation of the +5 ion by Ce(IV). It is also obtained analytically pure from the solid-state oxidation of the chloride salt of the +5 ion by Cl_2. Provided that it is kept dry, it is stable for weeks at −4°C (*426, 537*).

Unlike the remarkably stable mixed-valence diosmium complex, the much less stable heterodinuclear $[(NH_3)_5Os(N_2)Ru(NH_3)_5]^{5+}$ decomposes to $[Os(NH_3)_5(N_2)]^{2+}$ and $[Ru(NH_3)_5(OH_2)]^{3+}$ in water over a period of hours (*537*). This is to be contrasted with the Ru_2 mixed-valence analog that decomposes over a matter of minutes at room temperature (*536*). The heterodinuclear +5 and +4 ions have been reported by a number of workers (*538–540*), though not completely characterized until recently (*537*). Similarly, the oxidation of the +5 ion to the +6 ion has been known for some time (*539*). However, its properties have only recently been studied in solution by Ce(IV) oxidation, or in the solid state, by the Cl_2 oxidation. It is much less stable than its diosmium analogs. There is also evidence for the formation of the unstable $[(NH_3)_5Os(N_2)Rh(NH_3)_5]^{5+}$ heterodinuclear species from the reaction of $[Os(NH_3)_5(N_2)]^{2+}$ with $[Rh(NH_3)_5(OSO_2CF_3)]^{2+}$ in neat CF_3SO_3H (*537*).

A large variety of nonaammine and octaammine complexes of the form *cis*-$[(NH_3)_5Os(N_2)Os(NH_3)_4X]^{5+/4+/3+}$ ($X = Cl^-$, Br^-, I^-, or $CF_3SO_3^-$), *cis*-$[(NH_3)_5Os(N_2)Os(NH_3)_4L]^{6+/5+/4+}$ ($L = OH_2$ or N_2), *cis,cis*-$[X(NH_3)_4Os(N_2)Os(NH_3)_4X]^{4+/3+/2+}$ ($X = Cl^-$, Br^-, I^-, or $CF_3SO_3^-$), *cis,cis*-$[L(NH_3)_4Os(N_2)Os(NH_3)_4X]^{5+/4+/3+}$ ($L = N_2$, $X = Cl^-$ or $CF_3SO_3^-$; $L = OH_2$, $X = Cl^-$), and *cis*-$[L(NH_3)_5Os(N_2)Os(NH_3)_4L]^{6+/5+/4+}$ ($L = N_2$) were prepared and characterized by methods similar to those described above, using either the reaction of *cis*-$[Os(NH_3)_4(N_2)_2]^{2+}$ with $[Os(NH_3)_5(N_2)]^{2+}$, or self-condensation of *cis*-$[Os(NH_3)_4(N_2)_2]^{2+}$ (*78, 79, 537*). *cis,cis*-$[(CO)(NH_3)_4Os(N_2)Os(NH_3)_4(CO)]^{4+}$ has also been prepared by the oxidative coupling of amine ligands (*58*). Other coupling reactions have been used to prepare *trans,trans*-$[(CH_3CN)(NH_3)_4Os(N_2)Os(NH_3)_4(NCCH_3)]^{5+}$ from $[Os^{VI}(NH_3)_4(N)]$ under UV/Vis irradiation in acetonitrile (*269*).

$[(4\text{-pic})Cl_3Os(N_2)OsCl(4\text{-pic})_4]^+$ has been prepared from the coupling reactions of *trans*-$[Os(N)Cl_3(4\text{-pic})_2]$ in the presence of 4-pic (*42, 226*).

In addition to the homodinuclear nonaammine complexes, the heterodinuclear complexes *cis*-$[(NH_3)_5Os(N_2)Ru(NH_3)_4Cl]^{5+/4+/3+}$ (*79, 537, 539*) and *cis*-$[(NH_3)_5Ru(N_2)Os(NH_3)_4(isn)]^{6+/5+/4+}$ (*79, 537*) have also been reported.

b. μ-Pyrazine Complexes. The preparations of the μ-pyrazine/ammine complexes are outlined in Scheme 13 (*80*), using the general methods previously described. Details about their spectroscopic and electrochemical properties are given in Sections III,C,1 and IV, B,1, respectively. The heterodinuclear $[A_5RupzOsA_5]^{n+}$ complexes have also been isolated recently.

cis-$[(bpy)_2ClOspzRu(NH_3)_5]^{5+/4+/3+}$ complexes have been prepared and characterized by UV/Vis and near-infrared spectroscopies and electrochemistry. Different electronic isomers are obtained for the ground state of the mixed-valence ion, depending on the solvent (Section III,C,2) (*541*). The Os(VI) dimer, *trans,trans*-$[Cl_4(N)Os(pz)Os(N)Cl_4]^{2-}$, has been prepared and characterized by X-ray crystallography (*226*).

c. μ-Isonicotinamide and μ-Isonicotinamidepoly(proline) Complexes. Complexes of the type $[(NH_3)_5Os(\mu\text{-L})M(NH_3)_5]^{n+}$ (L = isn or $isn(pro)_m$; M = Co or Ru) have been prepared and characterized by a variety of techniques (*429, 430*). They have been used in studying intramolecular electron transfer along protein chains (Section V,C,2).

d. μ-Pyrimidine, μ-4,4′-Bipyridine, and Related Complexes. The decaammine complexes have been prepared by methods analogous to those described for the μ-pyrazine complexes. Although most have not been fully characterized as yet, the μ-4,4′-bipyridine complex has been studied by UV/Vis/NIR, MCD, and IR spectroscopies and electrochemistry (*182, 428, 542–544*).

$[(trpy)(bpy)Os^{II}(4,4'\text{-bpy})Ru^{II}(OH_2)(bpy)_2]^{4+}$ has been prepared and its redox chemistry and spectroelectrochemistry studied. All oxidation states up to $[Os^{III}Ru^{IV}(O)]^{5+}$ are observed, and in some oxidation states proton-induced intramolecular electron transfer is observed (Section V,C,2) (*228*). The preparation, electrochemistry, and photophysics of $[(bpy)_2(CO)Os^{II}(\mu\text{-L})Os^{II}(phen)(dppene)Cl]^{3+}$ (L = 4,4′-bpy or bpa) were reported (*545*).

e. μ-2,3-Bis(2′-pyridyl)pyrazine and μ-2,3-Bis(2′-pyridyl)quinoxaline Complexes. $[(bpy)_2Os(\mu\text{-dpp})M(bpy)_2]^{4+}$ complexes (M = Ru or

[Os(NH3)5(O3SCF3)](CF3SO3)2 + pz
(two fold excess)
25°C 8d in acetone
[Os(NH3)5(O3SCF3)]2+ + pz (excess)
[(NH3)5Os(pz)Os(NH3)5]6+
[Os(NH3)5O3SCF3]2+ in acetone
[Os(NH3)5pz]3+
Ru(NH3)6 2+
E0 0.32 V
E0 -0.09 V
[(NH3)5Os(pz)Os(NH3)5]5+
[Os(NH3)5O3SCF3]2+ in acetone
[Os(NH3)5pz]2+
Zn(Hg)
E0 -0.44 V
[(NH3)5Os(pz)Os(NH3)5]4+

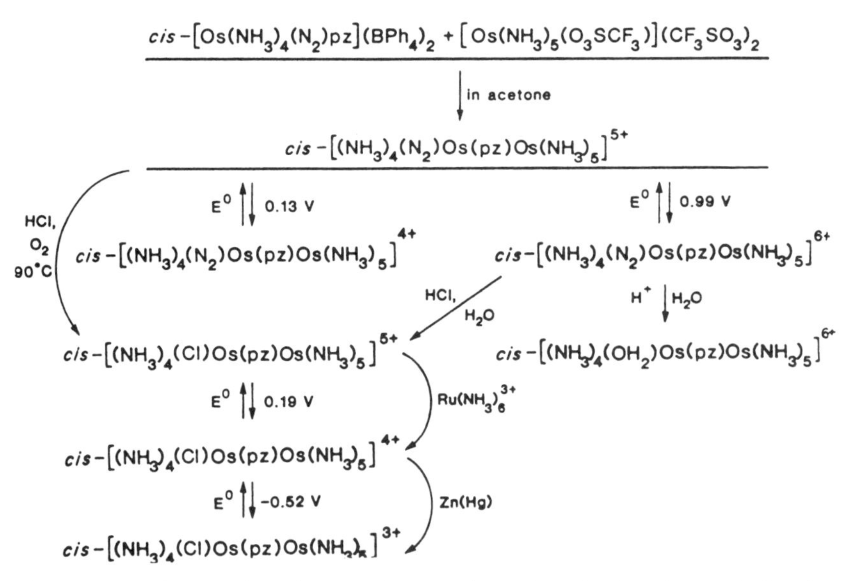

SCHEME 13. Preparation of μ-pyrazine complexes. Reprinted with permission from *Inorganic Chemistry*, Ref. *80*. Copyright 1988, American Chemical Society.

Os) have been prepared and characterized by electronic absorption and emission spectroscopies and electrochemistry (*546*). $[(OC)_3ClRe(L\text{-}L)Os(bpy)_2]^{2+}$ (L-L = dpp or dpq) were prepared by the reaction of $[Os(bpy)_2(L\text{-}L)]^{2+}$ with $[ReCl(CO)_5]$ and characterized by UV/Vis spectroscopy and electrochemistry (*232*). The tetranuclear complex [Os{(μ-

dpp)Ru(bpy)$_2$}$_3$]$^{8+}$ has been prepared and its electrochemistry and UV/Vis and luminescence spectra have been studied (*547*).

f. μ-3,5-Bis(2-pyridyl)-1,2,4-triazolato and μ-3,6-Bis(2-pyridyl)-1,2,4,5-tetraazine Complexes. [(bpy)$_2$M(bpt)M′(bpy)$_2$]$^{n+}$ complexes [M = M′ = Ru; M = M′ = Os; M = Ru, M′ = Os (two isomers)] have been prepared and characterized by UV/Vis/NIR and ^{1}H NMR spectroscopies, electrochemistry, and an X-ray structure of one of the heterodinuclear complexes (*548*). Similar complexes with bptz have also been prepared and characterized by electrochemistry, UV/Vis spectroscopy, and Hückel calculations (*549*).

g. μ-trans-1,2-bis(4′-methyl-2,2′-bipyridyl-4-yl)ethene and Related Complexes. The complexes [(4,4′-Me$_2$bpy)$_2$M(bbpe)M(4,4′-Me$_2$bpy)$_2$]$^{4+}$ [bbpe = *trans*-{4-(4′-Mebpy)}$_2$-CH=CH; M = Ru or Os] have been prepared and characterized by electrochemistry, UV/Vis/NIR, and emission spectroscopies. Although details of the Os$_2$ complex have not been given, the complexes exhibit interesting electrochemical and photophysical properties (*550*). Similar studies have also been reported on [(bpy)$_2$M(μ-L)M′(bpy)$_2$]$^{4+}$ [L = {4-(4′-Mebpy)}$_2$CH$_2$CHOHCH$_2$, M = M′ = Ru; M = Ru, M′ = Os; M = M′ = Os] (*551*).

h. μ-Bis(benzylimidazolato) Complexes. [(bpy)$_2$M(bibzim)M′(bpy)$_2$]$^{2+}$ (M = M′ = Ru; M = Ru, M′ = Os; M = M′ = Os) have been prepared and characterized by CV. They exhibit several reversible one-electron oxidations and reductions that are both metal and ligand centered. The UV/Vis/NIR spectra have also been recorded for the M(II)-M′(II), M(II)-M′(III) and M(III)-M′(III) oxidation states (*552*).

i. μ-Diacetylhydrazinato(2−) Complexes. [(bpy)$_2$M(adc-Me)M(bpy)$_2$]$^{n+}$ (M = Ru or Os) has been prepared and characterized by electrochemistry, UV/Vis spectroscopy, and Hückel calculations (*549*).

j. Porphyrin Dimers. Reactions of the dinuclear porphyrin, H$_4$dpb, with [Os$_3$(CO)$_{12}$] in the presence of py gives [(CO)(py)Os(dpb)-Os(CO)(py)]. On irradiation, it reacts further with py to give [(py)$_2$Os(dpb)Os(py)$_2$]. The heterodinuclear complexes [(CO)(CH$_3$-OH)Ru(dpb)Os(CO)(CH$_3$OH)] and [(CO)(OH$_2$)Os(dpb)Zn] are prepared by the reaction of [Os$_3$(CO)$_{12}$] with [Ru(CO)(CH$_3$OH)(H$_2$dpb)] and [Zn(H$_2$dpb)], respectively. The heterodinuclear Ru/Os complexes are prepared under a set of reactions similar to those of the Os/Os dimers, to produce [(py)$_2$Ru(dpb)Os(py)$_2$] (*141*).

k. μ-Cyanogen Complexes. $[(NH_3)_5OsNCCNOs(NH_3)_5]^{5+}$ is prepared from the oxidative dehydrogenations of the dinuclear 1,2-ethanediamine complex (Section V,E,2,b). It has been studied by UV/Vis/NIR and IR spectroscopies and electrochemistry, and the susceptibilities of the +4, +5, and +6 complexes toward ligand hydrolysis are discussed in Section V,E,3.

l. μ-Dicyanoamide Complexes. $[(NH_3)_5Os(NCNCN)Os(NH_3)_5]^{5+}$ has been prepared by standard methods and characterized by UV/Vis, NIR, and IR spectroscopies and electrochemistry (*542*).

m. μ-Dicyanobenzene and μ-Dicyanobicyclo[2.2.2]octane Complexes. Decaammine complexes have been prepared, but to date their characterization is incomplete (*542, 553*).

n. Bicyclo- and Tricyclo-N Heterocycles. Dinuclear adducts of dabco or tatd with $[OsO_4]$ and $[Os(O)_3(NR)]$ have been known for some time. Recent examples that have been prepared and characterized by ^1H NMR, Raman, IR, and ESCA spectroscopies include $[\{Os(O)_3(NC_8H_{17})\}_2(\mu\text{-L})]$ (*208*).

o. μ-Imido Complexes. One of the products of the reaction between $Bu^tNH(SiMe_3)$ and $[OsO_4]$ is the tetrameric complex *anti*-$[(Bu^tN)_2Os(\mu\text{-}Bu^tN)_2Os(NBu^t)(\mu\text{-}O)_2Os(NBu^t)(\mu\text{-}Bu^tN)_2Os(NBu^t)_2]$. The terminal Os atoms are four coordinate, whereas the central osmiums are five coordinate, as shown from an X-ray structure analysis (*265*).

Reduction of $[Os(NBu^t)_4]$ with PPh_3 or $PMePh_2$ yields a dimeric Os(VI) complex, $[(Bu^tN)_2Os(\mu\text{-}Bu^tN)_2Os(NBu^t)_2]$, whereas reduction with $Me_3O^+BF_4^-$ yields the analogous Os(VII) dimer, $[(Bu^tN)_2Os(\mu\text{-}Bu^tN)_2Os(NBu^t)_2](BF_4)_2$. Both have a tetrahedral arrangement of ligands about the Os centers, as shown by X-ray crystallography. The complexes were further characterized by ^1H NMR, mass, and IR spectroscopies, and electrochemistry. The Os(VII) dimer undergoes a reversible two-electron reduction to Os(VI) dimer and a further reversible two-electron reduction to presumably the Os(V) dimer (*265a*).

p. μ-Nitrido Complexes. The preparation of these complexes is given in Scheme 2 (Section II,B,3,b). $[Os_2(NH_3)_{10}N]^{5+}$ has been characterized by UV/Vis and IR spectroscopies and the electronic structure of this and related complexes has been discussed with the aid of a molecular orbital diagram (*47*). The properties of $[Os_2(N)(NH_3)_8Cl_2]Cl_3\cdot 2H_2O$ and $[Os_2(N)(NH_3)_7Cl_3]Cl_2\cdot H_2O$ have been reported, as well as their aquations to $[Os_2(N)(NH_3)_8(OH_2)_2]^{5+}$ and $[Os_2(N)(NH_3)_7(OH_2)_3]^{5+}$, respectively. The later undergo successive deprotonations to give the mixed aqua/hydroxo and hydroxo complexes. The complexes have been characterized by IR and UV/Vis spectroscopies and electrochemistry (*46*). *trans,trans,trans*-$[\{(4\text{-pic})_2Cl_3Os(\mu\text{-N})\}_2Os(4\text{-pic})_2Cl_2]$ is a minor product of the coupling of $[Os(N)(4\text{-pic})_2Cl_3]$ (Section II,D,3,a) (*42, 226*).

4. *Phosphide-Bridged Complexes.*

The diphosphide-bridged *cis,cis*-$[(H)_2(PMe_2Ph)_2Os(PMePh)_2Os(PMe_2Ph)_2(H)_2]$ complex is one of the products of the photolysis of $[Os(PMe_2Ph)_3(H)_4]$ in benzene (see Sections II,C,5,a and II,D,8,a for other products). This complex has been characterized by X-xay diffraction and ^1H and ^{31}P NMR spectroscopies (*307*). Deprotonation of $[OsCl(PH_3)(NCMe)(CO)(PPh_3)_2]^+$ results in the formation of $[\{Os_2(\mu\text{-}PH_2)Cl(CO)(PPh_3)_2\}_2]$. It has been characterized by X-ray crystallography (*261*).

5. *Oxygen Donor Ligands*

a. Mono(μ-oxo) Complexes. The crystal structure of the complex $[\{trans\text{-}[Os(\eta^4\text{-chba-Et})(OPPh_3)]\}_2O]^{2-}$ has been reported (Section II,C,4,x) (*290*). *cis,cis*-$[(bpy)_2(H_2O)Os^{III}(\mu\text{-O})Os^{IV}(OH)(bpy)_2]^{4+}$ has been characterized by UV/Vis spectroscopy and electrochemistry (*318, 554*).

b. Bis(μ-oxo) Complexes. *trans,trans*-$[(py)_2(O)_2Os(\mu\text{-O})_2Os(O)_2(py)_2]$ has been characterized by ^1H NMR, IR, and Raman spectroscopies (*235*). A variety of bis(μ-oxo) complexes of the general formulas $[\{Os(O)(OCR'R''CH_2NR)(\mu\text{-O})\}_2]$, $[\{Os(O)OC_6H_{10}NR)(\mu\text{-O})\}_2]$, and $[\{Os(O)(OCR'R''CH_2NR)L(\mu\text{-O})\}_2]$ are obtained when either $[Os(O)_3(NR)]$, $[Os(O)_3(NR)L]$, or $[\{Os(O)_3(NR)\}_2(\mu\text{-L})]$ ($R = Bu^t$, Ad, C_5H_{11}, or C_8H_{17}; L = qncd, tatd, or dabco) reacts with an alkene. The complexes have been characterized by IR, Raman, ESCA, and ^1H NMR spectroscopies (*208*).

c. μ-Hydroxo Complexes. Treatment of the cymene complex $[Os_2(\eta^6\text{-cym})Cl_2(\mu\text{-Cl})_2]$ with NaOH results in the formation of $[(\eta^6\text{-cym})Os(\mu\text{-OH})_3Os(\eta^6\text{-cym})]^+$. This reacts with RCO_2H (R = *trans*-PhCH=CH, Ph, Me, or H) or RCHO in acetone (R = *trans*-PhCH=CH, Ph, or Bu^t) to give $[(\eta^6\text{-cym})Os(\mu\text{-OH})_2(\mu\text{-}O_2CR)Os(\eta^6\text{-cym})]^+$ and with RCHO in water to give $[(\eta^6\text{-cym})Os(\mu\text{-H})(\mu\text{-OH})(\mu\text{-}O_2CR)Os(\eta^6\text{-cym})]^+$ (R = H, Me, or Et). The latter complex with R = H has been characterized by X-ray crystallography and all complexes were characterized by ^1H NMR spectroscopy (*555*).

d. μ-RO⁻ Complexes. CO inserts into $[Os(O)_2(OMe)_4]^{2-}$ to form the transdioxo dimer $[(MeCO_2)_2(O)_2Os(\mu\text{-}OMe)_2Os(O)_2(OCOMe)_2]^{2-}$, which has been characterized by IR and NMR (^1H and ^{13}C) spectroscopies and an X-ray structure (*405*). $[(\eta^6\text{-cym})Os(\mu\text{-}OMe)_3Os(\eta^6\text{-cym})]^+$ has been prepared and characterized by IR, mass, and NMR [$^1H(^{187}Os)$ reverse INEPT] spectroscopies (*556*).

e. μ-Dioxane Complexes. *trans,trans*-$[Cl_4(N)Os(dioxane)Os(N)Cl_4]^{2-}$ has been prepared and characterized (*226*).

6. *Sulfur Donor Ligands*

a. μ-Thiolato Complexes. Thermolysis of *trans*-$[Os(S\text{-}dmso)_4Cl_2]$ or *trans*-$[Os(O\text{-}dmso)(S\text{-}dmso)Cl_2]$ at 150°C *in vacuo* results in the formation of the dimer species, $[Os_2(\mu\text{-}MeS)_2(Me_2S)_2Cl_4]$ and $[Os_2(\mu\text{-}MeS)_2(Me_2S)(O\text{-}dmso)Cl_4]\cdot Me_2SO$. These diamagnetic Os(III) dimers have been characterized by IR, electronic absorption, and ^1H NMR spectroscopies, in addition to their conductivities and XPS spectra (*456*).

$[Os(N)(CH_2SiMe_3)_2Cl_2]$ reacts with excess pySH to give the dimeric *cis,cis*-$[Os(N)(CH_2SiMe_3)_2(\mu\text{-}pyS)]_2$ in which each pyS ligand is both a chelate and provides a μ-RS⁻ bridge. In solution, NMR evidence shows the presence of two geometric isomers (syn and anti) and they crystallize in different forms, which are separated manually. The structure of one of the isomers has been determined by X-ray crystallography, and when pure crystals of either of the isomers are redissolved in solution, they equilibrate to a mixture of isomers over several hours (*156*).

fac-$[Co(SCH_2CH_2NH_2)_3]$ reacts with $OsCl_3$ to give the trimer in which the Os(III) center is bound via six thiolato bridging ligands, $[\{fac\text{-}Co(SCH_2CH_2NH_2)_3\}_2Os]^{3+}$. It has been characterized by IR, UV/Vis, and ^1H NMR spectroscopies (*557*).

b. μ-Disulfide and μ-Supersulfide Complexes. $[(cp)_2Os{\equiv}Os(cp)_2]^{2+}$ reacts with elemental sulfur to give the Os(IV) dimer, $[(cp)_2Os(SS)Os(cp)_2]^{2+}$, which has been characterized spectroscopically and by X-ray crystallography. It undergoes a reversible one-electron oxidation to its supersulfide analog (*462*).

7. *Halide-Bridged Complexes*

Oxidation of $[Os(NC_6H_3Me_2\text{-}2,6)_3]$ by $AgBF_4$ results in the Os(VI) dimer, $[(RN)_2Os(\mu\text{-}F)_2Os(NR)_2](BF_4)_2$, but no details of its characterization have been given (*265a*).

$[Os_2X_9]^{n-}$ ($X = Cl^-$ or Br^-, $n = 1$–4) have been prepared and characterized recently by low-temperature electrochemistry and UV/Vis spectroscopy. The far-IR data of $(Bu_4N)_2[Os_2Br_{10}]$, $(Bu_4N)[Os_2Br_9]$, and $(Bu_4N)_3[Os_2Br_9]$ have also been reported. The redox series $[Os_2X_{10}]^{n-}$ ($X = Cl^-$ or Br^-, $n = 1$–5) have also been characterized for the first time using low-temperature cyclic voltammetry (*558*). Detailed Raman, IR, and electronic absorption (UV/Vis/NIR) spectra have also been recorded at ambient and low temperature for the -2 species (*559*).

$[Os_2(\mu\text{-}X)_3(PR_3)_6]^+$ ($X = Cl^-$ or Br^-) are prepared from the reactions of $[Os_2X_8]^{2-}$ with PR_3, and the structure of $[Os_2(\mu\text{-}Cl)_3(PEt_3)_6]^+$ has been determined by X-ray crystallography (*302*). $[Os_2(\mu\text{-}Cl)_3(PMe_2Ph)_6]^+$ and $[Os_2(\mu\text{-}Cl)_2Cl_2(PEt_3)_6]$ are formed from $[OsCl_2(PMe_2Ph)_3]$ in noncoordinating solvents (*145*).

$[(\eta^6\text{-cym})ClOs(\mu\text{-}Cl)_2OsCl(\eta^6\text{-cym})]$ and related complexes are often used as synthetic intermediates in organometallic chemistry, but they are also useful intermediates for the synthesis of a variety of dinuclear Os complexes with inorganic bridges (*555, 556*). On heating, they form $[(\eta^6\text{-cym})Os(\mu\text{-}Cl)_3Os(\eta^6\text{-cym})]^+$, which reacts with propanol to give $[(\eta^6\text{-cym})ClOs(\mu\text{-}Cl)(\mu\text{-}H)OsCl(\eta^6\text{-cym})]$. All complexes have been characterized by IR, mass, and NMR [$^1H(^{187}Os)$ reverse INEPT] spectroscopies (*556*).

8. *Hydride-Bridged Complexes*

a. Hydride-Bridged Complexes. The hydride-bridged complexes, *fac,fac*-$[(H)(PMe_2Ph)_3Os(H)_2Os(PMe_2Ph)_3(H)]$ and *cis,fac*-$[(H)(PMePh_2)_2Os(H)_3Os(PMePh_2)_3]$, are prepared from the photolysis of a saturated solution of the appropriate $[Os(H)_4(PR_3)_3]$ complex, in thf or benzene for the former dimer, and in ethanol for the latter. The dihydride dimer is photosensitive and undergoes a slow photolysis reaction in thf to produce $[(PMe_2Ph)_3Os(H)_3Os(PMe_2Ph)_3]$. All of these dimers have been characterized by X-ray diffraction and ^1H and ^{31}P NMR spectroscopies. All of the dimeric complexes contain Os—Os bonds in addition to the hydride bridges (*307*).

The heterodinuclear complexes $[(cp)_2(H)Zr(\mu\text{-}H)_3Os(PMe_2Ph)_3]$ and $[(1,5\text{-cod})Rh(\mu\text{-}H)_3Os(PMe_2Ph)_3]$ are obtained from treating [Os-

$(H)_3(PMe_2Ph)_3]$ with $[Zr(cp)_2(H)Cl]$ or $[\{Rh(1,5\text{-cod})Cl\}_2]$, respectively (*560*). Protonation of $[Os_2(O_2CMe)_2(CO)_4L_2]$ (L = PMe_2Ph, $PMePh_2$, PPh_3, or py) gives $[Os_2(\mu\text{-H})(\mu\text{-}O_2CMe)_2(CO)_4L_2]^+$, which have been characterized by NMR spectroscopy, and for L = PMe_2Ph, by X-ray crystallography (*561*).

$[(\eta^6\text{-cym})Os(\mu\text{-H})_3Os(\eta^6\text{-cym})]^+$ is prepared by treating $[(\eta^6\text{-cym})Os(\mu\text{-OH})_3Os(\eta^6\text{-cym})]^+$ with 2-propanol, whereas $[(\eta^6\text{-cym})_4Os_4(\mu\text{-H})_4]^{2+}$ is obtained under reflux conditions. Along with $[(\eta^6\text{-cym})ClOs(\mu\text{-Cl})(\mu\text{-H})OsCl(\eta^6\text{-cym})]$, $[(\eta^6\text{-cym})Os(\mu\text{-Cl})(\mu\text{-H})(\mu\text{-}O_2CMe)Os(\eta^6\text{-cym})]^+$, $[(\eta^6\text{-cym})Os(\mu\text{-H})_2(\mu\text{-}O_2CMe)Os(\eta^6\text{-cym})]^+$, and $[(\eta^6\text{-cym})Os(\mu\text{-H})(\mu\text{-}O_2CMe)_2Os(\eta^6\text{-cym})]^+$, all complexes have been characterized by IR, mass, and NMR $[^1H(^{187}Os)$ reverse INEPT] spectroscopies (*556*).

b. Aluminohydride Bridges. Reaction of $[OsCl_2(PPh_3)_3]$ with $LiAlH_4$ in diethyl ether or $[OsCl_2(PMe_3)_4]$ with $LiAlH_4$ in thf yields the aluminohydride-bridged complexes, *fac,fac*-$[(R_3P)_3(H)Os(\mu\text{-H})_2Al(H)(\mu\text{-H})_2Al(H)(\mu\text{-H})_2Os(H)(PR_3)_3]$. The complexes were characterized by IR and ^1H, ^{31}P, and ^{27}Al NMR spectroscopies (*562*).

E. Polymers

1. *Porphyrin Polymers*

The conducting polymers $[Os(oep)(L\text{-}L)]_n$ (L-L = pz, dabco, or 4,4′-bpy) have been prepared by the reaction of $[Os(oep)]_2$ with L-L in a 1:2 molar ratio. They are partially oxidized to form mixed-valence ions (Section IV,D,1) (*563–565*).

2. *μ-Phosphine Oxide Polymers*

trans-$[Os^{VI}(bpy)_2(O)_2]^{2+}$ reacts with dppm or dppene to give polymers of the type *trans*-$[\{Os(bpy)_2(\mu\text{-}OPPh_2RPPh_2O)^{2+}\}_n]$ (*415*).

3. *4-Vinylpyridine and 4-Vinyl-2,2′-bipyridine Polymers*

A considerable amount of research has been performed on polyvinylpyridine and polyvinyl-2,2′-bipyridine polymers of $[Os(bpy)_3]^{2+}$-like complexes. These polymers are prepared by electropolymerization of complexes such as *cis*-$[Os(bpy)_2(vpy)_2]^{2+}$ (*566*), $[Os(vbpy)_nL_m]^{x+}$ (*567, 568*), and $[M(bpy)_2(4\text{-cinn})_2]^{2+}$ (*568*).

4. *Copolymers*

A polymer containing anthracene, $[Ru(bpy)_3]^{2+}$, and $[Os(bpy)_3]^{2+}$ [all covalently linked to a 1:1 copolymer of styrene and *m,p*-(chloromethyl)styrene] has been prepared, and its emission spectrum and intramolecular electron-transfer properties have been studied (*569*).

5. *[OsO_4]-Derivatized Polymers*

Recently, polymers in which [OsO_4] is covalently bound to either a pyridine group, a dabco group in pvp, or a dabco-derivatized cross-linked styrene–divinylbenzene copolymer have been prepared and used as selective catalysts for the oxidation of alkenes (*570, 571*).

III. Electrochemistry of Coordination Complexes

A. General

1. *Books and Reviews*

There have been two books that contain compilations of the electrochemistry of Os (*572, 573*). There have also been reviews that cover the electrochemistry of certain classes of complexes with ligands such as porphyrins (*142*), dithiocarbamates (*463*), and macrocyclic complexes (*39, 93*). The purpose of this section is not to provide a comprehensive review of electrochemical studies over recent years, but rather to give some insight into the factors that affect the redox potentials and their use in obtaining information about π bonding and backbonding. Particular emphasis is placed on the similarities and differences between analogous Os and Ru complexes.

2. *Ionization Potentials*

Correlations have been made between gas-phase ionization potentials of free ions and the redox potentials of isostructural $[MX_6]^{n-}$ complexes of the elements of the same row of the periodic table (*476*). Despite the observation of such correlations, caution must be taken, because they ignore both σ and π ligand field effects. The latter are often more important in influencing the relative oxidizing or reducing strength of complexes.

3. *Ligand Field Effects*

The strength of ligand field interactions increases down a group of the periodic table. Therefore, Os(III/II) redox couples are expected, and have been observed, to be shifted to more negative potentials by CFSE contributions and stronger σ bonding, compared to analogous Ru(III/II) couples. For example, the $[Os(NH_3)_6]^{3+/2+}$ redox couple is 0.83 V more negative than the Ru analog (*69, 574*), although the interpretation is complicated by the different gas-phase ionization potentials of Ru^{2+} and Os^{2+}.

4. *π Bonding and π Backbonding Effects*

The study of redox potentials is a valuable tool in understanding π bonding and backbonding stabilization of oxidation states. Lower oxidation states tend to be more strongly stabilized by π backbonding than are higher oxidation states, moving redox potentials to more positive values with increases in the electron acceptor strength of π acid ligands bound to a metal center. Conversely, π bases tend to stabilize higher oxidation states, resulting in negative shifts in redox potentials as the electron-donating strength increases. It needs to be emphasized, however, that redox properties are a function of both oxidation states. Therefore, any deductions made about the strength of π interactions between a ligand and a metal ion on the basis of redox potentials must consider the effects of σ and π interactions in both oxidation states of the redox couple. The importance of this is illustrated in Section III,B.

5. *Spin-Orbit Coupling*

Spin-orbit coupling stabilizes the ground state of the low-spin d^5 electronic configuration of Os(III) and Ru(III). The ground states of the diamagnetic d^6 M(II) complexes are not influenced by spin-orbit coupling. This results in a lowering of the energy of the M(III) oxidation state with respect to M(II). Because the spin-orbit coupling constant is much larger for Os(III) than Ru(III), this will also tend to make Os(III/II) couples more negative than their Ru(III/II) counterparts. However, the spin-orbit coupling constant is dependent on the degree of delocalization of the valence electrons onto the ligands. This, in turn, is dependent on the strengths of σ and π interactions between the ligand and metal ion; a decrease in the spin-orbit coupling constant results in a positive shift in the redox potential.

6. *Solvent Effects*

Solvent effects on the redox potentials can be very large and arise from both nonspecific electrostatic and specific hydrogen-bonding and π-stacking contributions (*574, 575*). These contributions are particularly important for highly charged species or complexes with ligands that undergo strong hydrogen-bonding interactions with the solvent. Therefore, for meaningful comparisons to be made between redox potentials, either they should be measured under the same sets of conditions, or the redox potentials must be corrected for the solvent contributions (see Section III,B). Because analogous Ru and Os complexes have virtually the same size and hydrogen-bonding abilities (*501*), solvent contributions to the electrochemistry will cancel if the electrochemistry is performed under identical conditions.

B. MONOMERS

The developments in Os pentaammine chemistry described in this review have enabled an extensive library of $E^{\circ\prime}$ values to be obtained, spanning a range of ~2 V (~200 kJ mol^{-1}) (Fig. 7). Table XIV compares the redox potentials of the Os(III/II) complexes with Ru(III/II) analogs versus the appropriate $[M(NH_3)_6]^{3+/2+}$ couples. The data are presented in this manner in order to correct for solvent effects, which can be comparable with those induced by the π effects (*574*). The correction of

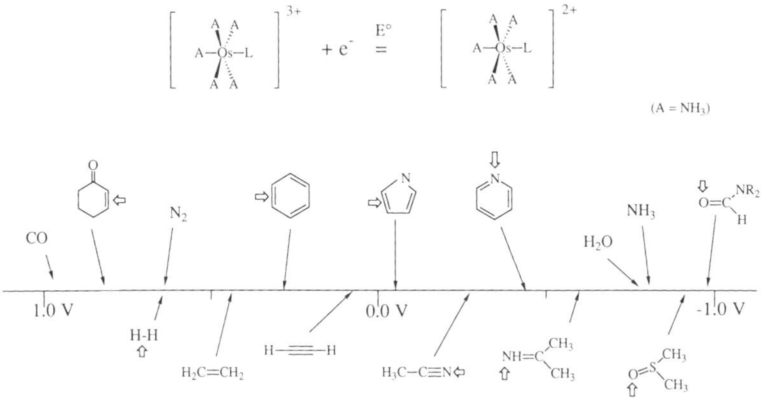

FIG. 7. Redox potentials of $[Os(NH_3)_5L]^{3+/2+}$ couples. In this electrochemical series for pentaammineosmium, the values of $E^{\circ\prime}$ are reported versus NHE as a function of L.

TABLE XIV

COMPARISON OF REDOX POTENTIALS OF $[M(NH_3)_5X]^{n+}$ COMPLEXES VERSUS $[M(NH_3)_6]^{3+/2+}$

Ligand	Ru			Os		
	$E_{\frac{1}{2}}$[a]	Solvent	Ref.	$E_{\frac{1}{2}}$[a]	Solvent	Ref.
SO_2	—	—	—	>1.80	H_2O	*60*
η^2-(*C,C*)-pyH	—	—	—	1.77	dme	*90*
CO	1.35	H_2O	*71*	1.70	H_2O	*83*
η^2-(*C,O*)-Ph_2CO	—	—	—	~1.6	dme	*177*
η^2-(*C,C*)-furan	—	—	—	1.46	an	*179*
η^2(*C,C*)-lut	—	—	—	⩾1.39	dme	*85*
η^2-(*C,C*)-$MeCH_2C{=}C(OMe)Me$	—	—	—	1.37	ac	*168*
η^1-N_2	1.05	H_2O	*576*	1.36	H_2O	*576*
η^1-(*S*)-dmso	0.95	H_2O	*577*			
	0.89	dmso	*68*	~1.35	dmso	*68*
η^2-(*C,C*)-cyclohexene	—	—	—	1.34	an	*432*
η^2-(*C,C*)-$PhCF_3$	—	—	—	1.34	nmp	*75*
η^1-(*N*)-Mepz	0.82	H_2O	*578*	1.33	H_2O	*68*
η^2-(*C,C*)-lutidine	—	—	—	1.30	dme	*90*
η^2-(alkene)-$CH_2{=}CHPh$	0.93	H_2O	*71*	1.27	H_2O	*71*
η^2-(*C,C*)-naphthalene	—	—	—	1.25	H_2O	*172*
η^2-$CH_2{=}CH_2$	0.88	H_2O	*579*	1.25	ac	*120*
η^2-(*C,C*)-thiophene	—	—	—	1.24	an	*179*
η^2-(*C,C*)-$CH_2{=}CHCH{=}CH_2$	0.89	H_2O	*70*	1.22	H_2O	*70*
η^2-(*C,C*)-$MeCH{=}C(OH)Me$[b]	—	—	—	1.21	ac	*168*
η^2-(*C,O*)-Me_2CO	0.203	ac	*182*	~1.2	dme	*177*
η^2-(*C,O*)-CH_3CHO	—	—	—	~1.2	dme	*177*
η^2-(*C,O*)-$Ph(Bu^t)CO$	—	—	—	~1.2	dma	*177*
η^2-(*C,C*)-$Ph(Bu^t)CO$	—	—	—	1.17	nmp	*177*
η^2-$CH_2{=}CH(CH_2)_2CH{=}CH_2$	0.80	H_2O	*70*	1.15	H_2O	*70*
η^2-$CH_2{=}CHCH_3$	0.78	H_2O	*70*	1.13	H_2O	*70*
η^2-(*C,C*)-$PhOCH_3$	—	—	—	1.11	nmp	*75*
η^2-(*C,C*)-C_6H_6	—	—	—	1.10	nmp	*75*
η^2-(*arene*)-$PhC{\equiv}CPh$	—	—	—	~1.10	ac	*171*
η^2-($C{\equiv}C$)-$PhC{\equiv}CPh$	—	—	—	1.05	ac	*171*
η^2-(*arene*)-$CH_2{=}CHPh$[c]	—	—	—	~1.0	H_2O	*71*
η^2-(*C,C*)-$PhNH_2$	—	—	—	1.00	ac	*175*
η^2-$HC{\equiv}CH$	—	—	—	0.95	ac	*171*
η^2-(*C,C*)-$PhNMe_2$	—	—	—	0.90	ac	*175*
η^1-(*S*)-thiophene	0.55	H_2O	*580*	~0.95	an	*120*
η^1-(N)-NCC_6F_5	—	—	—	0.94	dme	*176*
η^1-(N)-$NCC_{14}H_9$[d]	—	—	—	0.75	dme	*176*
η^2-$CH_3C{\equiv}CCH_3$	—	—	—	0.74	ac	*168*

TABLE XIV (*Continued*)

Ligand	Ru			Os		
	$E_{\frac{1}{2}}^{a}$	Solvent	Ref.	$E_{\frac{1}{2}}^{a}$	Solvent	Ref.
η^2-(*C,C*)-pyrrole	—	—	—	0.74	an	*179*
η^1-(*N*)-pyrazine	0.44	H_2O	*581*	0.69	H_2O	*60*
η^1-(*N*)-NCCH=CH$_2$	0.46	H_2O	*582*	0.68	dme	*176*
η^1-(*N*)-NCPh	0.44	H_2O	*582*	0.66	dme	*176*
η^1-pyridazine	—	—	—	0.57	H_2O	*60*
$NCCH_3$	0.324	an	*182*	0.56	dme	*176*
	0.38	H_2O	*582*			
η^1-(*N*)-isn	0.33	H_2O	*582*	0.54	H_2O	*60*
$NCBu^t$	—	—	—	0.52	dme	*176*
η^1-(*N*)-pyrimidine	0.37	H_2O	*583*	0.52	H_2O	*60*
η^1-(*N*)-NMe$_2$Ph	—	—	—	0.46	ac	*175*
η^1-(*N*)-3-pic	—	—	—	0.45	dme	*90*
η^1-(*N*)-pyridine	0.25	H_2O	*582*	0.45	dme	*90*
				0.39	H_2O	*60*
η^1-(*C*)-lutidinium	—	—	—	0.35	dme	*85*
η^1-(*O*)-Me$_2$CO	0.061	ac	*182*	0.32	ac	*177*
η^1-(*N*)-NH=CHMe	—	—	—	0.31	acet	*194*
η^1-(*N*)-NH$_2$Ph	0.16	H_2O	*583*	0.30	ac	*175*
η^1-(*N*)-imidazole	0.06	H_2O	*584*	0.29	dme	*90*
η^1-(*N*)-NH=CMe$_2$	—	—	—	0.23	ac	*194*
η^1-(*N*)-NCSe$^-$	—	—	—	0.17	H_2O	*198*
η^1-(*N*)-NCS$^-$	0.08	H_2O	*581*	0.16	H_2O	*197*
				0.12	H_2O	*198*
dme	—	—	—	~0.1	dme	*90*
dmf	0.061	dmf	*182*	~0.2	dmf	*68*
OH_2	0.02	H_2O	*582*	0.05	H_2O	*69*
NH_3	0.000			0.00		
η^1-(*O*)-dmso	−0.001	dmso	*182*	~0.0	dmso	*68*
	−0.04	H_2O	*577*			
Cl^-	−0.09	H_2O	*582*	−0.07	H_2O	*199*
Br^-	−0.08	H_2O	*585*			
I^-	—	—	—	+0.01	H_2O	*199*

[a] V; potential versus appropriate $[M(NH_3)_6]^{3+/2+}$ couple. The redox potentials of the hexaammine complexes are +0.05 and −0.78 V, respectively, versus NHE in water.

[b] The complexes of the cis and trans ligands have the same redox potentials.

[c] The presence of this couple was not recognized in the original paper, but it is clearly present in Fig. 6 of Ref. *71* as a small response at potentials more negative than the main response that is due to the $[Os(NH_3)_5(\eta^2\text{-}(alkene)\text{-}CH_2{=}CHPh)]^{3+/2+}$ couple. As can be seen from its position, its redox potential is consistent with an η^2-arene structure.

[d] 9-Anthracenecarbonitrile.

the solvent contribution enables meaningful comparisons of the effects of π backbonding and π bonding, both within the Os series, and between the Ru and Os series of complexes. Such comparisons show the extent of stabilization of the M(II) oxidation state, with respect to the M(III) oxidation state, when an ammine ligand is replaced by a π acid or a π base. Because the remaining five ligands are innocent with respect to π interactions (except hyperconjugation), the effect of the π acceptance or donation is amplified to the maximum extent, enabling both subtle differences in ligands and the considerable effect on the electronic distribution within the ligand to be assessed. The latter is very important in understanding the new chemistry that has been outlined in Section V.

Table XIV (*60, 68–71, 75, 83, 85, 90, 120, 168, 171, 172, 175–177, 179, 182, 194, 197–199, 432, 576–585*) shows that, in general, the shift in $[M(NH_3)_5L]^{3+/2+}$ redox potentials from the $[M(NH_3)_6]^{3+/2+}$ couple is larger by 30–50% for Os compared to Ru. This does not, however, give a true indication of the relative extent of π backbonding in the M(II) oxidation states because π backbonding is also important for Os(III), but not for Ru(III) (*67, 586*). Therefore, the use of redox potentials alone underestimates the relative strengths of π bonding in analogous Ru and Os complexes. There are other problems in using redox potentials to quantify π backbonding, even if the relevant contributions of the two oxidation states to shifts in redox potentials could be evaluated. This is because π backbonding in Os(III) decreases the spin-orbit coupling constant, thereby providing a destabilizing effect, which partially offsets the stabilization of Os(III) by stronger metal–ligand bonding. Nonetheless, the electrochemical data establish that π backbonding is much more significant in Os(II) complexes than in Ru(II) analogs. This is rationalized in terms of better spatial and energy overlaps between the metal d orbitals and the ligand π^* orbitals for Os(II), as opposed to Ru(II) (*67*).

By contrast, π bonding will tend to stabilize the M(III) oxidation state with respect to the M(II) oxidation state, thereby leading to negative shifts in redox potentials. This may be important in understanding the much larger shifts in the Os(III/II) couples, compared to Ru(III/II) couples with imidazole, aldehyde, and ketone ligands. These ligands tend to be better π donors than most of the other ligands in Table XIV, and therefore would tend to provide π stabilization of the M(III) oxidation state, which will make the shifts in the M(III/II) oxidation states less positive than would otherwise be the case. Because the energy overlap is greater between π orbitals and Ru(III) d orbitals compared to Os(III), this would result in much less positive shifts for the Ru(III/II) couple.

Given the shifts of over a volt in M(III/II) redox potentials that can be achieved by changing one ligand in the coordination sphere, it is not surprising that a large range of stable oxidation states, from 0 with strong π acceptors [e.g., $[Os(PMe_3)_5]$ (*15*)] to VIII with strong π donors [e.g., $[OsO_4]$], are observed. A simple example of the change in redox potential that can be achieved even with the addition of a moderate π acid is the redox potentials of $[Os(NH_3)_6]^{3+/2+}$ and of $[Os(bpy)_3]^{3+/2+}$, which differ by over 1.5 V. This makes $[Os(NH_3)_6]^{2+}$ a strong reductant, whereas $[Os(bpy)_3]^{3+}$ is a strong oxidant, even though both complexes are surrounded by six N donors. Similarly, the oxidation of the bpy complex to the Os(IV) species (*587*) occurs at a potential ~2 V more positive than that for the hexaammine complex (*83*).

Conversely, the stabilization of Os(IV) by ~ 1 V per deprotonation in $[Os(en)_3]^{n+}$ (*204, 205*) and the stabilization of higher oxidation states by deprotonation of aqua ligands (*325*) illustrate well the large effects brought about by π donation.

C. Dinuclear and Polynuclear Complexes

1. *Comparison of K_{com} Values for Dinuclear Ruthenium and Osmium Complexes*

Table XV (*80, 427, 518, 536, 537, 542, 544, 552, 588–591*) summarizes redox potential data and derived comproportionation constants for Os complexes and their Ru analogs. Although the number of Os complexes that have been prepared, and hence for which data are known, is much less extensive, some clear patterns emerge. The first is that the values of K_{com} for the ammine complexes of Os are approximately the square of the value of the Ru analogs. This dramatic difference is not solely due to stabilization of the mixed-valence ion, but rather to a variety of factors that require consideration. The first of these is the extent of spin-orbit coupling and π backbonding stabilization of the Os(III)–Os(III) dimers, compared to the Ru(III)–Ru(III) dimers. Unlike the Ru(III)–Ru(III) dimers, which are paramagnetic, many of the Os dimers are diamagnetic or exist as an equilibrium between a low-lying triplet excited state and a singlet ground state (Section IV,B,1,c). The latter is a manifestation of the degree of π backbonding and/or direct metal–metal bonding, which enables the metal ions to couple more strongly with respect to the analogous Ru(III) complexes. In the absence of coupling interactions, the triplet states of both the Ru(III) and Os(III) dimers will be stabilized by spin-orbit coupling with respect to the singlet states, but the effect will be significantly greater for the latter. The fact that most of the +6 ions have a diamagnetic ground

TABLE XV

COMPARISON OF REDOX POTENTIALS AND COMPROPORTIONATION CONSTANTS FOR SYMMETRICALLY SUBSTITUTED DINUCLEAR OSMIUM AND RUTHENIUM COMPLEXES AT 20°C

Complex	$E_{\frac{1}{2}}(1)^{a,b}$	$E_{\frac{1}{2}}(2)^{a,c}$	K_{com}	Ref.
$[A_5RupzRuA_5]^{n+}$	0.772	0.376	6.6×10^6	*80*[d]
$[A_5OspzOsA_5]^{n+}$	0.324	−0.438	1.0×10^{13}	*80*[d]
$[A_5RuN_2RuA_5]^{n+}$	0.73	~1.2	$\sim 10^8$	*536*[d]
$[A_5OsN_2OsA_5]^{n+}$	−0.16	1.05	10^{21}	*426*[d]
$[A_5RuNCCNRuA_5]^{n+}$	0.71	≥1.2	$\geq 10^{13}$	*588*[d]
$[A_5OsNCCNOsA_5]^{n+}$	0.37_4	~1.3	$\geq 10^{22}$	*542*[d]
$[A_5Ru(4,4'\text{-bpy})RuA_5]^{n+}$	0.121	0.041	2.3×10^1	*589*[e]
$[A_5Os(4,4'\text{-bpy})OsA_5]^{n+}$	−0.257	−0.415	6.1×10^2	*537, 544*[d]
$[\{(bpy)_2Ru\}_2(bpz)]^{n+}$	2.26	1.76	3.0×10^8	*549*[g]
$[\{(bpy)_2Os\}_2(bpz)]^{n+}$	1.96	1.24	1.6×10^{12}	*549*[g]
$[\{(bpy)_2Ru\}_2(bibzim)]^{n+}$	1.30	1.01	8.2×10^4	*552*[g]
$[\{(bpy)_2Os\}_2(bibzim)]^{n+}$	0.82	0.64	1.1×10^3	*552*[g]
$[\{(bpy)_2Ru\}_2(adc\text{-}Me)]^{n+}$	1.23	0.67	3.1×10^9	*549*[g]
$[\{(bpy)_2Os\}_2(adc\text{-}Me)]^{n+}$	0.79	0.45	5.8×10^5	*549*[g]
$[\{(bpy)_2ClRu\}_2(dppm)]^{n+}$	1.06	0.92	2.3×10^2	*590*[g]
$[\{(bpy)_2ClOs\}_2(dppm)]^{n+}$	0.67	0.54	1.6×10^2	*591*[g]
[(oep)RuRu(oep)]	0.62	−0.11	4×10^{12}	*518*[f]
[(oep)OsOs(oep)]	0.09	−0.53	5×10^{10}	*518*[f]

[a] V versus NHE.
[b] M_2^{6+}/M_2^{5+} couple.
[c] M_2^{5+}/M_2^{4+} couple.
[d] 0.1 *M* HCl.
[e] 1 *M* HCl.
[f] dme; V versus Ag/AgCl.
[g] an; V versus SCE.

state, whereas the analogous Ru complexes have triplet ground states, illustrates the much larger degree of π backbonding interactions in the Os complexes. Therefore, the + 6/+ 5 redox potentials are less positive than they would be in the absence of Os(III) π backbonding. This means that the values of K_{com} underestimate the degree of stabilization of the mixed-valence Os complexes compared with their Ru analogs. For this reason and other factors that influence redox potentials (Section III,A), the use of K_{com} values to quantify the degree of intermetallic interaction must be treated with caution. This is well illustrated by the fact that the values of K_{com} for $[(NH_3)_5Os(\mu\text{-L})Os(NH_3)_5]^{n+}$ (L = N_2 or NCCN) are 10 orders of magnitude larger than those for $[(oep)MM(oep)]^{m+}$. Because the latter have direct metal–metal bonds, the interactions must be greater than with the decaammine complexes (Table XV).

The differences in the abilities of the M(III) and M(II) ions to act as π donors and π acceptors have also been used to rationalize the differences in the values of K_{com} for analogous $[(bpy)_2M(\mu\text{-}L)M(bpy)_2]^{n+}$ complexes (M = Ru or Os). When L is a good π donor, K_{com} is larger for Ru than for Os, but when L is a good π acceptor, the opposite is the case (*549*).

2. *Solvent Effects on the Stabilization of Electronic Isomers of Mixed-Valence Ions*

As a consequence of the solvent dependence of the redox potentials alluded to in Section III,A,6, the ground electronic state of mixed-valence ions with dissimilar redox centers can be controlled by the appropriate choice of solvent. An eloquent example of the design of electronic isomers utilizes the *cis*-$[(bpy)_2ClOspzRu(NH_3)_5]^{4+}$ ion. Os^{III}–Ru^{II} is the ground electronic state in CD_3NO_2 and CD_3CN, but in d_6-dmso, Os^{II}–Ru^{III} is the ground-state electronic configuration (*541*).

D. Polymers

A considerable amount of work has been performed on electrodes coated with $\{cis\text{-}[Os(bpy)_2(vpy)_2]^{2+}\}_n$ and $\{[Os(bpy)_2(vbpy)]^{2+}\}_n$ polymers in order to delineate the controlling factors of electron transfer rates within such polymers (e.g., *566, 592*). Similar work has been performed on nafion polymers in which $[Os(bpy)_3]^{2+}$ has been incorporated (e.g., *593, 594*). Such electrodes have been used for the electrocatalytic oxidation of organic and biological substrates in sensor applications (e.g., *595, 596*). There have been many other studies using such electrodes, but they will not be discussed here.

IV. Spectroscopic and Magnetic Properties of Coordination Complexes

A. Monomers

1. *Osmium(III) Ammine and Amine Complexes*

Magnetic circular dichroism (MCD) spectra of $[Os(NH_3)_6]^{3+}$ (*193*), $[Os(NH_3)_5L]^{3+}$ (*428, 597*), $[Os(NH_3)_5X]^{2+}$ (*597*), and *cis*- and *trans*-$[Os(NH_3)_4Cl_2]^{2+}$ (*543*) have been studied from 4 K to room temperature, and in a range of different media (KCl and KBr disks, PVA films, nafion films, and single crystals). The near-IR region of the spectra, where the intra-t_2 electronic transitions are observed, is of particular interest because of the vibrational fine structure.

Low-temperature polarized single-crystal studies of the near-IR transitions of Os(III) ammine complexes have also been studied. These include $[Os(NH_3)_6](ClO_4)_2Cl \cdot KCl$ (*193*), $[Os(en)_3]_2Cl_6 \cdot KCl \cdot 6H_2O$ (*193*), $[Os(NH_3)_5X]SiF_6$ (*193*), $[Os(NH_3)_5Cl]Cl_2$ (*598*), and $[Os(NH_3)_5Cl](S_2O_6)$ (*598*). In some systems it is possible to find the two origins, which can be assigned to the two transitions in Fig. 8 (*428*), but in many others the spectra are not amenable to simple analyses and the second origin is inferred from EPR spectroscopy. There are also some inconsistencies in the evaluation of the energy differences in the spin-orbit states from optical and EPR spectroscopy. It is conceivable that these inconsistencies are due to the quenching of the low-symmetry splittings by vibronic activity in the first excited (pseudo) Γ_8 state (*193*).

There are also charge-transfer characteristics within the electronic transitions between the spin-orbit states of the pentaammine complexes. This is evidenced by order of magnitude increases in the intensities of these transitions when an ammine or aqua ligand is replaced by a π-base (e.g., Cl^-) or a π-acid (e.g., N heterocycle) (*67*).

Detailed studies of the solvent dependencies of the charge-transfer transitions in the UV/Vis region of $[Os(NH_3)_5(Mepz)]^{3+}$ and a variety of

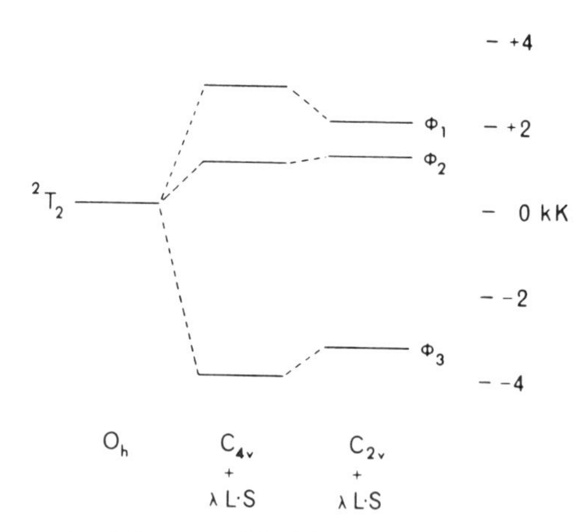

FIG. 8. State energy level diagram of Os(III) (Δ = 1290, β = 160, and λ = 3200 cm^{-1}). Reprinted with permission from the *Journal of Physical Chemistry*, Ref. *428*. Copyright 1984, American Chemical Society.

$[Ru(NH_3)_5(L)]^{n+}$ complexes have been used as an aid to assign the various transitions that occur in this region (*599*).

2. *Tris(2,2'-bipyridine)osmium(II) and Related Complexes*

In order to gain an appreciation of the problems that arise in interpreting the electronic spectroscopy of $[Os(bpy)_3]^{2+}$ and related complexes, it is necessary to understand the controversies in the analogous Ru spectroscopy. More papers on the spectroscopy and photochemistry of $[Ru(bpy)_3]^{2+}$ have been published than for any other compound (*223, 600, 601*). The research in this area, together with complexes with other N-heterocyclic ligands, has been covered in recent reviews (*222–224, 600–604*). This extensive literature has arisen because of the inherently interesting spectroscopy and photophysics of these complexes and their potential uses in solar energy conversion. Though interest in potential practical applications has waned, controversy still rages about the most appropriate descriptions of the electronic structures of the initially formed and lowest energy triplet states of these complexes. This controversy is well summarized in two recent reviews (*600, 601*). The strongest evidence available is that, upon excitation, the initial excited state has a structure in which all of the bpy ligands are equivalent (i.e., the electron is delocalized over all three of the ligands). This excited state then relaxes to one in which the unpaired electron is localized on only one of the bpy ligands (*600, 605–607*).

The literature on the analogous Os spectroscopy and photochemistry is nowhere near as extensive. A major reason for this is the much shorter lifetimes of the triplet excited states, which are a consequence of the much greater degree of spin-orbit coupling in the Os(III)-$bpy^{\overline{\cdot}}$ excited state, in comparison with its Ru analog. The larger spin-orbit coupling in Os(III) increases the mixing of the triplet and singlet states, thereby providing a more efficient pathway for intersystem crossing and, hence, shorter lifetimes of the excited states. At a time when the possibility of solar energy applications was driving the research in this area, this made the study of the Os complexes much less attractive compared to their Ru analogs. Also, the chemistry is more difficult and expensive than that of Ru. Despite this, there is a quite extensive and rapidly growing literature on the spectroscopy and photochemistry of $[Os(bpy)_3]^{2+}$ and related complexes. The literature to 1988 has been covered in recent reviews (*223, 224*). Like the Ru chemistry, controversies regarding the electronic structures of the excited states have raged. However, the most appropriate descriptions of the initially formed excited state of D_3 systems appear to be a delocalized structures.

When the symmetry is reduced by replacement of a bpy ligand by another bidentate, the symmetry restrictions are relaxed and the initially formed excited state is expected to be localized (*600, 601*). The rates of nonradiative decay of a large range of Os(II)–polypyridine have been calculated successfully in terms of a modified "energy gap law" in which low-frequency modes are explicitly considered (*148*).

The charge transfer that occurs on photoexcitation changes the pK_a values of substituents on bpy or related ligands. Because the substituents become more basic as a result of the MLCT nature of the triplet states, excited-state proton transfers can occur. There have been many studies of such reactions of Ru complexes, but few on analogous Os complexes until recently (*233*).

3. *Nitrido and Oxo Complexes*

A feature of the numerous Os(VI) complexes is the presence of highly vibronically coupled charge-transfer spectra in the UV/Vis region, even at room temperature. The emission spectra are also highly structured and the charge-transfer excited states are sufficiently long-lived to undergo both outer-sphere and atom-transfer quenching reactions. A detailed review of this area is beyond the goals of this article, but some representative publications are given (*107, 114, 211, 212, 608*).

B. Dimers

1. *Ammine Complexes*

a. Os_2^{4+} Dimers. Little work has been performed on the decaammine $Os(II)_2$ complexes, because they are very air sensitive and their electronic absorption spectra are not as rich as their oxidized counterparts. Their spectra are dominated by intense MLCT bnads in the visible region. In all cases, the MLCT bands are more intense and are at lower frequencies than in their oxidized counterparts. Complexes that have been studied to date are those with N_2, NCCN, and pz (*80, 536, 543*).

b. Os_2^{5+} Dimers. MCD and EPR spectra have been measured on the mixed-valence $[(NH_3)_5Os(pz)Os(NH_3)_5]^{5+}$ ion in nafion film, and in an ethylene glycol/water glass, respectively. The position of the three absorption bands in the NIR are consistent with both the predictions made using the delocalized electronic coupling model (*609*) and the published absorption spectra in water (Fig. 9) (*80*). The remarkable

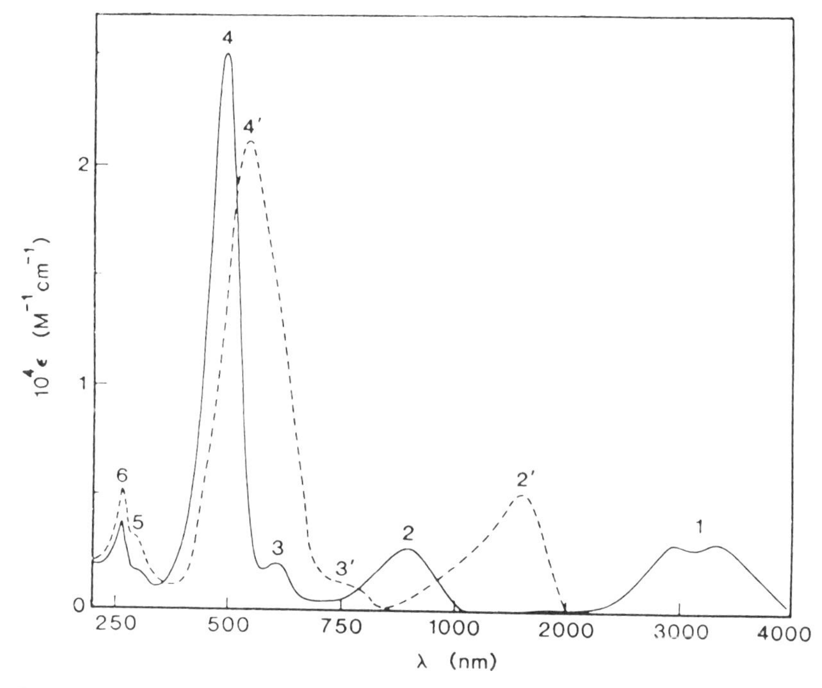

FIG. 9. Ultraviolet, visible (0.1 *M* HCl), near-infrared (0.1 *M* DCl/D_2O), and infrared (KBr) absorption spectra of the electronic transitions of $[(NH_3)_5OspzOs(NH_3)_5]^{5+}$ (solid line) and $[(NH_3)_5RupzRu(NH_3)_5]^{5+}$ (dashed line). The numbers refer to analogous transitions in Ru and Os complexes. The value of ϵ_{max} for transition 1 is approximate because the absorption intensity has been measured in a KBr disk. Reprinted with permission from *Inorganic Chemistry*, Ref. *80*. Copyright 1988, American Chemical Society.

feature of this complex is the intense electronic transitions observed in the IR region of the spectrum. These are an order of magnitude more intense than the N—H stretches that occur at the same frequencies. The normal intra-t_2 bands that occur in the NIR region of mononuclear and dinuclear Os(III) complexes, and trapped-valence Os(III)–Os(II) dimers, e.g., *cis*-$[(NH_3)_4(N_2)Os(pz)Os(NH_3)_5]^{5+}$, are absent in the symmetric mixed-valence ion, which supports a delocalized electronic structure (*80*).

The delocalized electronic coupling model also quantifies the MCD, absorption, and EPR spectra of the N_2 complex $[(NH_3)_5Os(N_2)$-$Os(NH_3)_5]^{5+}$ (*610*), which is delocalized and has a particularly rich electronic absorption spectrum (Fig. 10) (*426, 536, 610*).

$[(NH_3)_5Os(4,4'\text{-bpy})Os(NH_3)_5]^{5+}$ has a weak and broad intervalence

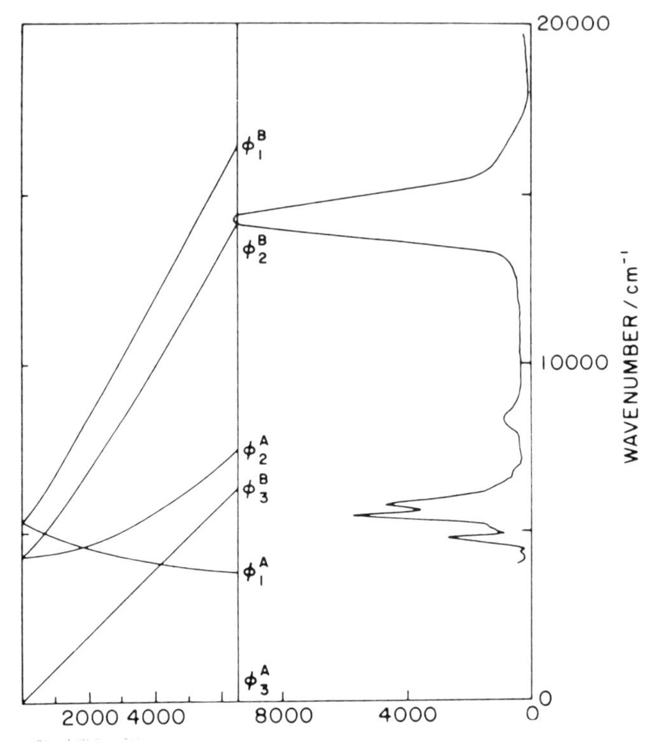

FIG. 10. Calculated energy level scheme for $[(NH_3)_5Os(N_2)Os(NH_3)_5]^{5+}$ as a function of W for a value of $\Delta = 1800\ cm^{-1}$ and $\lambda = 3000\ cm^{-1}$. The observed 8 K absorption spectrum is shown on the right-hand side. Reprinted with permission from the *Journal of the American Chemical Society,* Ref. *610*. Copyright 1985, American Chemical Society.

transition on the edge of the charge-transfer bands. This, together with the presence of normal intra-t_2 Os(III) transitions in the NIR spectrum, indicates that this is a trapped valence Os(III)–Os(II) dimer (*542*).

Resonance Raman spectroscopy is also being used both to aid the assignment of their electronic spectra and to delineate the degree of electronic delocalization (*611*).

c. Os_2^{6+} Dimers. The MCD spectra of $[(NH_3)_5Os(4,4'\text{-bpy})Os(NH_3)_5]^{6+}$ exhibit normal C term behavior down to 4 K, whereas the temperature dependence of the MCD spectra of the pz complex does not because of a temperature-dependent equilibrium between the singlet ground state and the low-lying triplet excited state. The intensity of the C terms actually decreases with decreasing temperature due to the depopulation of the excited triplet state. Such changes in population of these two states are also evident in the temperature dependence of the electronic absorption spectra (Fig. 11) (*428*). By contrast, the μ-N_2 complex is diamagnetic, even at room temperature, and no C terms are observed (*426*). Thus, the extent of coupling between the metal ions increases in the order 4,4′-bpy < pz < N_2. The much weaker coupling in the Ru analogs (4,4′-bpy and pz) is apparent from the fact that they are triplets even down to 1 K, whereas the N_2 complex is too unstable to measure.

2. *Bpy and Related Complexes*

The relatively long lifetimes of the excited states of these complexes have made them particularly attractive in the study of electron and energy-transfer quenching of excited states through organic bridges. In addition, the more positive redox potentials of these ions, compared with their pentaammine counterparts, mean that the mixed-valence ions are not air sensitive, thus facilitating spectroscopic measurements.

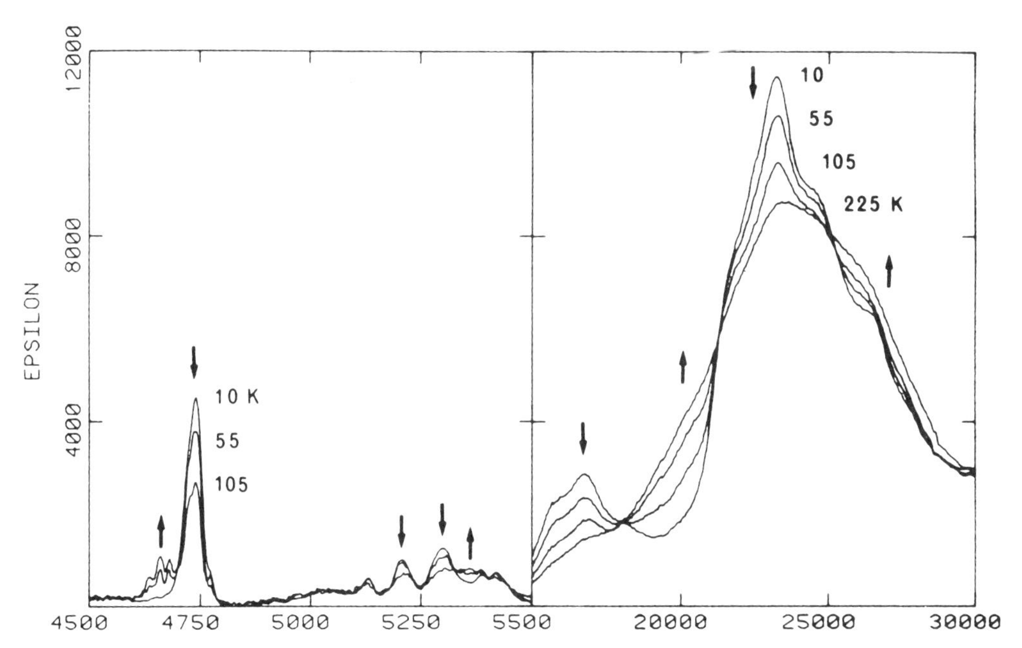

FIG. 11. Absorption spectra of $[(NH_3)_5OspzOs(NH_3)_5]^{6+}$ at various temperatures in PVA. The arrows indicate the direction of intensity change with increasing temperature. Reprinted with permission the from *Journal of Physical Chemistry*, Ref. *428*. Copyright 1984, American Chemical Society.

For these reasons, and the relative ease with which the complexes are prepared, an increasing number of studies have been performed on the photophysical and photochemical properties of such complexes. Though most of the studies in this area have been performed on dinuclear ruthenium complexes, an increasing number of Ru/Os and Os/Os dimers have been studied in recent years (*228, 541–552*) and this subject has been reviewed (*222*).

3. *μ-Nitrido Os(IV) Complexes*

In addition to intense charge-transfer bands that trail into the visible region, these diamagnetic complexes have weaker d–d transitions at lower energies. The presence of these previously unrecognized transitions, and other properties, are rationalized by the MO description shown in Fig. 12, where the d–d transitions are indicated (*47*).

C. Polymers

1. *Porphyrin Polymers*

The intervalence transitions of doped $[M(oep)(L\text{-}L)]_n^{m+}$ polymers (M = Fe, Ru, or Os and L-L = pz, dabco, or 4,4′-bpy) have been studied. The metal–metal coupling and the intensity of the intervalence transition increase in the order dabco < 4,4′-bpy < pz, and the energy of the intervalence band decreases in the same order. Within a series with the same bridging ligand, the strength of metal–metal coupling and the intensity of the intervalence transition increase in the order Fe < Ru < Os, and the energies of intervalence bands decrease in the same order. With $[Os(oep)pz]_n^{m+}$, an intense and relatively sharp electronic transition is observed at ~2000 cm^{-1} in the IR (*563–565*), which is reminiscent of a similar transition in $[(NH_3)_5Os(pz)Os(NH_3)_5]^{5+}$ (Section IV,B,1,b).

2. *Bpy Polymers*

The photophysics and photochemical properties of polymers containing Os–polypyridine or Os/Ru and Os/Zn copolymers are an area of increasing interest. Recent applications include photochemical ligand loss as a basis for imaging and microstructure formation (*612*) and the storage of photochemical redox equivalents in polystyrene polymers, derivatized with bpy (*613*). Long-range energy transfer has also been studied (*569*).

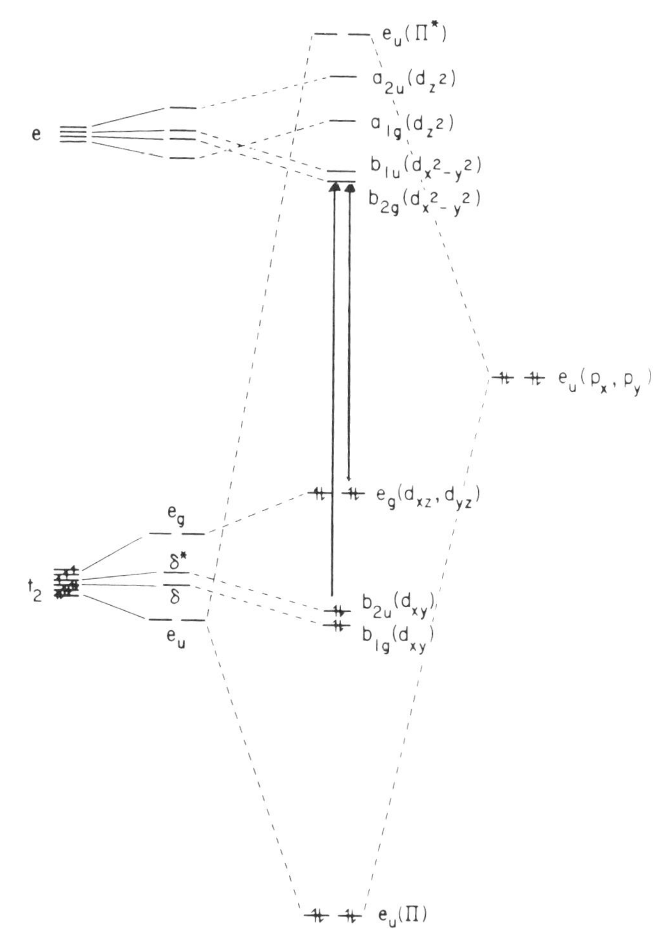

FIG. 12. Qualitative molecular orbital description of bonding within the Os^{IV}–N^{3-}–Os^{IV} moiety. The d–d transitions are indicated by arrows. Reprinted with permission from *Inorganic Chemistry*, Ref. *47*. Copyright 1989, American Chemical Society.

V. Reactivity of Coordination Complexes

A. ADDITION REACTIONS OF OSMIUM(VIII)

The kinetics and thermodynamics of the addition reactions of amine and N-heterocyclic ligands to [OsO_4] have been studied using UV/Vis and NMR spectroscopies. With py, 4-pic, and bpy, the kinetics are too fast to be studied without using stopped-flow techniques, whereas the kinetics of tmen, bpds, phen, and imidazoles in water are monitored by conventional mixing techniques. These are the first studies of such kinetics, and as expected for associative reactions, the activation enthalpies are low (15–60 kJ mol^{-1}) (*335, 336*).

B. Substitution Reactions

Although there have been few studies performed on the substitution kinetics of pentaammineosmium complexes, compared to other pentaammine complexes, the importance of π-bonding and π-backbonding effects on the stabilization of metal–ligand bonds in different oxidation states is amply illustrated in Table XVI (*65, 67, 83, 199, 536, 576, 614–617*). Thus, though N_2 bound to M(II) complexes is inert toward substitution, it is labile when bound to M(III) complexes. Similar observations are made for the $[Os(NH_3)_5(\eta^2\text{-benzene})]^{3+/2+}$ complexes in acetonitrile, for which the specific rates for substitution are 3.5×10^2 and 3.5×10^{-5} sec^{-1}, respectively, for the +3 and +2 ions (*75*). By contrast, M(III) complexes are much more inert than M(II) complexes when the leaving group is a π-donor ligand such as Cl^-.

Another important factor that separates π-acceptor ligands in traditional organometallic complexes, as compared to pentaammine com-

TABLE XVI

Spontaneous Aquation Rates of Pentaammineosmium(III/II) and Analogous Pentaammineruthenium(III/II) Complexes[a]

	k (sec^{-1})			
Ligand	Ru(II)	Ru(III)	Os(II)	Os(III)
N_2	2×10^{-6} [b]	70 [c]	$<10^{-8}$ [d]	2×10^{-2} [c]
Cl^-	6.3 [e]	9.3×10^{-7} [f]	0.2 [g]	$\sim10^{-8}$ [h]
$CF_3SO_3^-$	—	93 [i]	—	8.8×10^{-4} [j]

[a] 25°C.
[b] Ref. *614*.
[c] Refs. *536* and *576*.
[d] Ref. *615*.
[e] Ref. *616*.
[f] Ref. *617*.
[g] Ref. *199*.
[h] Ref. *83*.
[i] Ref. *65*.
[j] Ref. *67*.

plexes, is the relative lability of the ligand. Thus, alkene (Section II,C,2,h) and N_2 (Section II,C,4,j) ligands bound to Os(II) phosphine complexes tend to be rather easily displaced, but are much more inert when bound to $[Os(NH_3)_5]^{2+}$. This is because in the latter, other ligands are not competing for metal π electrons. Such a phenomenon also manifests itself in the greater sensitivity of the electrochemistry of $[Os(NH_3)_5L]^{n+}$ complexes to the nature of L (Section III,B) and the often unusual reactivities of coordinated ligands (Section V,D and V,E).

Though few studies have been performed on the mechanisms of substitution of Os(III) and Os(II) complexes, some deductions can be made about probable mechanisms. In the case of Os(III), with charged leaving groups such as Cl^-, it is expected that the preferred mechanism will be one in which bond breaking substantially precedes bond making (i.e., predominantly dissociatively activated). By contrast, reactions will be more associative with neutral leaving groups, although it is difficult to ascertain where they are expected to occur along the continuum of interchange mechanisms (*618, 619*). In the case of substitution of $[Os(NH_3)_5(\eta^2\text{-benzene})]^{2+}$, the rate constant only varies slightly with the nature of the incoming ligand (L = isn, py, CH_3CN, acetone, CH_3-COPh) in d_6-acetone at 25°C (*75*). This is a strong indication of an I_d mechanism for the substitution reactions, which is to be expected for a sterically crowded d^6 complex.

Aqua exchange in $[Os(\eta^6\text{-}C_6H_6)(OH_2)_3]^{2+}$ has been subjected to detailed mechanistic studies, along with the Ru analog. The volumes of activation are small and positive. This has been taken as an indication of an interchange mechanism that is near the middle of the continuum, but that is more dissociatively than associatively activated (*620*).

C. Electron Transfer Reactions

1. *Intermolecular Electron Transfer*

Electron self-exchange reactions of $[Os(L\text{-}L)_3]^{3+/2+}$ (L-L = phen, 4,7-Me_2phen, 3,5,6,8-Me_4phen, or 3,4,7,8-Me_4phen) were studied by ^1H NMR line broadening. There are two pathways, one involving the +3 ion, and the other involving the PF_6^- ion pair of the +3 ion. Rate constants are $\sim 10^5$ and $\sim 10^6$ M^{-1} sec^{-1}, respectively (*621*).

The reversibility of the $[Os(bpy)_3]^{3+/2+}$ couple makes it useful for the determination of the electron self-exchange rates of other couples by application of the Marcus cross-reaction equation. Recently, this has been applied to the oxidation of SO_3^{2-} to SO_4^{2-} (*622*). The new rate constant for this reaction of 1.63×10^7 M^{-1} sec^{-1} is consistent with the expectations of Marcus theory, but three orders of magnitude greater than that determined previously (*623*). Other self-exchange rates that have been determined using $[Os(bpy)_3]^{3+/2+}$, $[Os(phen)_3]^{3+/2+}$, and related couples include $V(O)(OH)^{2+/+}$ (aq) (*624*) and $[Mn(edta)]^{1-/2-}$ (*625*). The rate of electron transfer from $[Fe(CN)_6]^{2-}$ to $[Os(bpy)_3]^{3+}$, in the presence of reversed micelles, has been used as a probe of coalescence of the reversed micelles (*626*).

2. *Intramolecular Electron Transfer*

As indicated in Section III,A, the Os(III/II) redox potentials are generally much more negative than analogous Ru(III/II) redox potentials. This has important implications in the study of intramolecular electron transfer through organic chains and, in particular, through polypeptides. This is an area that has been under intensive study in Ru chemistry because of the desire to understand such electron transfer, so important in many biological processes (*427, 429*). These processes have been studied by monitoring the rate of intramolecular electron transfer from Ru(II) or Os(II) to Co(III), or Os(II) to Ru(III). Os(II) is a much stronger reductant than Ru(II), and Os(II) π backbonding is also much stronger. Because all other factors, such as inner- and outer-sphere reorganizational energy terms, are approximately equal, the rates of intramolecular electron transfer are much faster in the Os(II) complexes, compared with their Ru(II) analogs (Table XVII). This is an important recent development because it extends the range over which intramolecular electron transfer is observable and also illustrates the importance of conformational changes of the polypeptide backbone in

TABLE XVII

COMPARISONS OF THE RATES OF INTRAMOLECULAR ELECTRON TRANSFER AT 25°C THROUGH ORGANIC MOLECULES IN CO(III)/RU(II), CO(III/OS(II), AND RU(III)/OS(II) DIMERS [a]

Complex		k(sec^{-1})	
M(III)LM′(II)	Co(III)/Ru(II)	Co(III/Os(II)	Ru(III)/Os(II)
isn	1.2×10^{-2}	2.4×10^{5}	$>5 \times 10^{9}$
proisn	1.04×10^{-3}	2.9×10^{2}	3.1×10^{6}
$(pro)_2$isn	6.4×10^{-6}	0.6	3.7×10^{4}
$(pro)_3$ isn	5.6×10^{-5}	0.08	3.2×10^{2}
$(pro)_4$isn	1.4×10^{-4}	0.09	0.5

[a] Refs. *429* and *430*.

affecting the rate of intramolecular electron transfer. Thus, for the Ru complexes, the addition of a third or more prolines to the polypeptide chain results in a increase in the rate of intramolecular electron transfer. This has been attributed to a cis–trans isomerization of the proline groups, which enables the Co(III) center to get nearer to the Ru(II) center than is the case for the complexes with only two proline groups (*427*). The kinetic behavior observed for the Os analogs differs in that the rates of intramolecular electron transfer are much faster than the rate of cis–trans isomerization of the proline chains. This results in the rate of intramolecular electron transfer decreasing as the chain length increases.

More recently the same reactions have been monitored for Ru(III)–Os(II) analogs. These complexes are generated by the reduction of the Ru(III)–Os(III) dimers by e^- (aq) or CO_2^- (aq) radicals in pulse radiolysis experiments (*429*). Because of the much lower inner-sphere reorganizational energy terms for the $[Ru(NH_3)_5L]^{3+/2+}$ couples compared with Co(III/II) analogs, the rates of intramolecular electron transfer in the Ru(III)–Os(II) dimers are much larger than those of Co(III)–Os(II) dimers (*429*).

Upon deprotonation, $[(trpy)(bpy)Os^{III}(4,4'\text{-}bpy)Ru^{II}(OH_2)(bpy)_2]^{5+}$ and $[(trpy)(bpy)Os^{III}(4,4'\text{-}bpy)Ru^{III}(OH)(bpy)_2]^{5+}$ undergo rapid intramolecular electron transfer to form $[(trpy)(bpy)Os^{II}(4,4'\text{-}bpy)\text{-}Ru^{III}(OH)(bpy)_2]^{4+}$ and $[(trpy)(bpy)Os^{II}(4,4'\text{-}bpy)Ru^{IV}(O)(bpy)_2]^{4+}$, respectively, but no details have been given about the rates (*228*). Fast intramolecular electron transfer from the excited states of $[Os(trpy)_2]^{2+}$ complexes with donor or acceptor substituents have also been observed (*229*).

D. Linkage Isomerizations

As indicated earlier, a dramatic difference is observed in the chemistry of Os(II) and Os(III), the former gaining stabilization from π acids and the latter from σ and π donors and to a lesser extent π acids. This difference is most amplified when the ligand set is predominantly saturated, e.g., pentaammineosmium. In this configuration, the electron-rich Os(II) must depend entirely on the sixth ligand for stabilization through π backbonding.

The sharp contrast in the chemical nature of Os(II) and Os(III) often results in a linkage isomerization of an ambidentate ligand accompanying such a redox change. To date, such isomerizations have been reported solely for the pentaammine moiety. Typically, through electrochemical measurements, a cyclic process is observed: the oxidation of a species Os(II)–AB to Os(III)–AB is proceeded by an intramolecular isomerization to Os(III)–BA. Subsequently, the latter species is reduced, at some new potential to Os(II)–BA, which then reisomerizes to Os(II)–AB, thus completing the cycle. Some ligands have several different binding sites, and in these cases three or more isomers can be observed. The thermodynamically stable Os(III)–BA complex normally binds the ligand via the best σ donor and, in this respect, forms typical coordination complexes. By contrast, the thermodynamically favored Os(II)–AB complex binds the ligand normally in an η^2 fashion typical of organometallic compounds. Of particular significance in these systems are the thermodynamic parameters obtained that relate the relative affinities of the metal for the bond sites A and B.

1. $\eta^1 \leftrightarrow \eta^2$ *Linkage Isomerizations for Aldehydes and Ketones*

First reported in 1986 (*181*), the complex $[Os(NH_3)_5(acetone)]^{2+}$ and related aldehyde and ketone complexes (*177*) were the first examples of linkage isomerizations on Os(III/II). In acetone solution, a detailed electrochemical and chemical investigation revealed that the substitutionally inert complex, $[Os(NH_3)_5(\eta^2\text{-acetone})]^{2+}$, is in facile equilibrium with the η^1 form, the former being favored by 21 kJ mol^{-1}. Upon oxidation, the Os—C bond is ruptured, but the Os—O bond remains intact, even in good donor solvents such as dma. Reduction of the η^1–acetone–Os(III) species occurs at a potential of ~750 mV negative of that of the η^2 form in acetone. Subsequent $\eta^1 \rightarrow \eta^2$ isomerization of the ketone occurs with a specific rate of 6×10^3 sec^{-1} at 20 ± 2°C.

Of the variety of aldehydes and ketones investigated (*177*), all were found to be η^2 coordinated to $[Os(NH_3)_5]^{2+}$ except for the bulky pinacolone. However, it is unclear whether this is a thermodynamic or kinetic limitation.

At the time of its original report (*627*), the labile complex $[Ru(NH_3)_5(acetone)]^{2+}$ was considered to contain an O-bound ketone. However, recent studies (*182–184*) revealed both η^2 and η^1 bonding modes in solution. In this case the former mode of binding is favored by 7 kJ mol^{-1}. Astonishingly, even in the case of Ru(III), a significant population of π-bound acetone is present. However, this type of η^2 binding is clearly more stable relative to the η^1 mode in Os(II) complexes compared to the Ru(II) analogs in a variety of ketone ligands (*177, 181–184*).

2. η^2-*C═S to* η^2-*C═Se Linkage Isomerization Reactions of CSSe, and* η^2-*(C═S) to* η^1-*(C) Reactions of S═CSMe*$^-$

Given the facile linkage isomerization reactions for a large number of η^2-bound ligands coordinated to $[Os(NH_3)_5]^{2+}$, it is astonishing that there is no low-energy pathway for the η^2-C,S to η^2-C,Se linkage isomerization of [$Os(\eta^2\text{-CSSe})(CO)(CNR)(PPh_3)$]. Both linkage isomers are kinetically stable and the η^2-C,S isomer is converted to the η^2-C,Se linkage isomer only by a set of addition and elimination reactions (*152, 628*). The η^2-(S,C) to η^1-(C) linkage isomerizations of coordinated $SCSMe^-$ are also induced by addition reactions (*152*) (Section II,C,2,e).

3. *Linkage Isomerizations with Arenes and Heterocycles*

A series of $[Os(NH_3)_5(\eta^2\text{-arene})]^{2+}$ complexes has been investigated in order to determine the effect of substituent on the stability of the Os–arene bond (*75*). As part of this study, information was gained regarding the ability of the substituent to direct the metal to a specific π-bond site within the ring. Surprisingly, both electron-donating and electron-withdrawing groups were found to direct the metal to the 2,3-η^2 position relative to this substituent. Bulky groups, on the other hand, tended to direct the metal to the 3,4-η^2 position relative to the bulky group. In cases in which steric effects can be discounted, the free energies of 2,3-$\eta^2 \rightarrow$ 3,4-η^2 isomerizations range from 0 to 10 kJ mol^{-1}. All of the substituted benzene complexes exhibit migration around the ring, with specific rates ranging from 1 sec^{-1} (L = anisole) to 10^4 sec^{-1} (L = benzene), at room temperature. For monosubstituted benzene ligands, electron-donating substituents were most effective at retarding this rate (*75*). As observed for the η^2-arene complexes, η^2-heterocyclic complexes often show migration around the ring at ambient temperature. The only cases in which such isomerizations have not been observed are when the ligand is η^2-furan or η^2-thiophene (*90*).

Oxidation of these materials often resulted in a linkage isomerization of the organic ligand from the ring to a basic site of the substituent. Examples include arene-to-amine (anilines) (*175*), arene-to-ketone (phenones) (*177*), arene-to-ether (anisole) (*120*), and arene-to-alkyne (diphenylacetylene) (*497*) isomerizations. In the case of the 2,2-dimethylpropiophenone (*177*), a sequence of one-electron oxidations and reductions resulted in the Os complex isomerizing from an arene to an η^1-ketone to an η^2-ketone, then back to the ring to form an Os—η^2-C,C bond. Although most of these reactions were not investigated in any detail, in two cases, aniline and *N,N*-dimethylaniline, a complete kinetic and thermodynamic analysis has been undertaken (Scheme 14) (*175*).

For aniline, an equilibrium with K_{eq} close to unity exists between the N-bound and 2,3-C,C-η^2-bound isomers of Os(II). The one-electron oxidation of the latter isomer ($E^{\circ\prime}$ = 0.16 V versus NHE in acetone) is followed by an arene-to-nitrogen linkage isomerization, proceeding with a specific rate constant of 180 sec^{-1} at 20 ± 2°C. Initially, on reduction of $[Os(NH_3)_5(N\text{-aniline})]^{3+}$ (−0.54 V), no change in coordination geometry occurs, but with time, rearrangement to the ring isomer occurs with a specific rate constant of 4.6×10^{-5} sec^{-1} at 20 ± 2°C. The data indicate free energies of N → π isomerization for Os(III) and Os(II) of 69 and 16.7 kJ mol^{-1}, respectively, at 20 ± 2°C.

Similar experiments with the *N,N*-dimethylaniline analog reveal that the addition of the two methyl groups destabilizes the N-bound form by 33 kJ mol^{-1} on Os(II) and >50 kJ mol^{-1} for Os(III). The increased sensitivity to steric effects for Os(III) is thought to be a result of the increased demand for electron density by this higher oxidation state. The differences in isomerization energy for both oxidation states are reflected almost entirely in the rate constants of isomerization from nitrogen-bound to arene-bound complexes.

Heterocyclic complexes of pentaammineosmium(II) have also been reported for pyridinium (*85*), 2,6-lutidine (*85*), 2,6-lutidinium (*85*), pyrrole (*179*), furan (*179*), and thiophene (*179*), in which the organic ligand is dihapto coordinated via a C═C bond (*90*). These ligands are thought to rearrange upon oxidation to coordinate through the heteroatom.

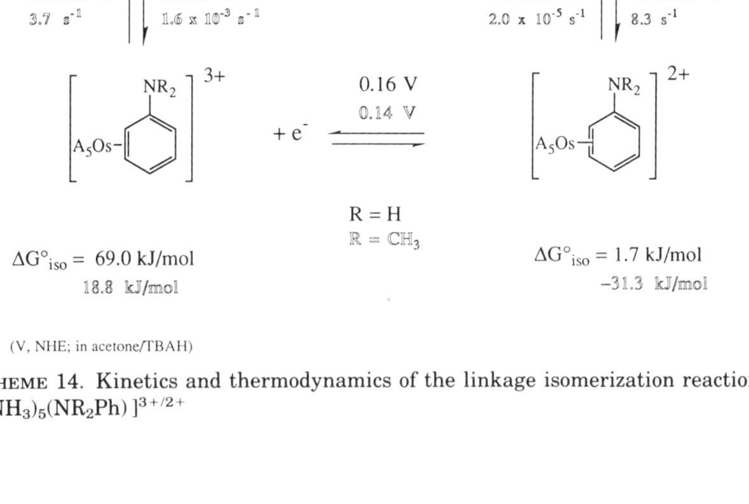

$[A_5Os\text{-}NR_2(C_6H_5)]^{3+}$ + e^- ⇌ $[A_5Os\text{-}NR_2(C_6H_5)]^{2+}$ (−0.54 V; −0.38 V)

1.8 x 10^2 s^{-1} / 3.7 s^{-1}; 8 x 10^{-11} s^{-1} / 1.6 x 10^{-3} s^{-1}

4.6 x 10^{-5} s^{-1} / 2.0 x 10^{-5} s^{-1}; 2.3 x 10^{-5} s^{-1} / 8.3 s^{-1}

$[A_5Os(\eta^2\text{-}C_6H_5NR_2)]^{3+}$ + e^- ⇌ $[A_5Os(\eta^2\text{-}C_6H_5NR_2)]^{2+}$ (0.16 V; 0.14 V)

R = H

R = CH_3

$\Delta G°_{iso}$ = 69.0 kJ/mol; 18.8 kJ/mol

$\Delta G°_{iso}$ = 1.7 kJ/mol; −31.3 kJ/mol

(V, NHE; in acetone/TBAH)

SCHEME 14. Kinetics and thermodynamics of the linkage isomerization reactions of $[Os(NH_3)_5(NR_2Ph)]^{3+/2+}$

Only for lutidine has this been confirmed, wherein the isomerizations occur with specific rate constants of $\geqslant 10^3$ sec^{-1} ($\eta^2 \rightarrow \eta^1$) and 36 ± 10 sec^{-1} ($\eta^1 \rightarrow \eta^2$) for the Os(III) and Os(II) complexes, respectively. It is worth noting that the resulting complex, $[Os(NH_3)_5(N\text{-}2,6\text{-lutidine})]^{3+}$, cannot be made by conventional, substitution-based, synthetic methods for Os(III), and is likely to be the only reported example of this molecule bonding to a transition metal through nitrogen.

When pentaammineosmium(II) complexes of pyridinium salts or 2,6-lutidine are allowed to stand, an activation of the C4—H bond is observed, yielding σ-bound pyridinium complexes (*85*). Such C—H activation is unusual for aromatic heterocycles with six-membered rings and has not been observed for Os(II) complexes of arenes. Its nearest parallel is the C—H bond activation that accompanies the linkage isomerization of N-bound imidazole to the C-bound isomers of Ru(II) ammine complexes (*441*). It is remarkable that the latter type of isomerization has not as yet been observed in analogous Os chemistry, and the former has not been observed in Ru chemistry. Thus, $[Os^{II}(NH_3)_5(N\text{-im})]^{2+}$ has been prepared, but it does not undergo isomerization to $[Os^{II}(NH_3)_5(C\text{-im-H})]^{+}$ over 2 days (*90*) under conditions that facilitate the isomerization of the Ru(II) analog (*441*). This is not due to any inherent instability of the C-bound imidazole complexes, because $[Os^{II}(NH_3)_5(C\text{-diMeim-H})]^{2+}$ is readily prepared (*90*). This again highlights some fundamental differences in the reactivity of Os and Ru complexes.

4. *S ↔ O Linkage Isomerizations with Dimethyl Sulfoxide*

As earlier reported for Ru (*577*), $[Os(NH_3)_5(dmso)]^{2+}$ preferentially binds the ligand via the S atom (*67, 120, 200*), an action that allows for considerable π back-donation. Upon oxidation of this species (0.36 V, NHE) a linkage isomerization ensues with a specific rate of $\leqslant 0.1$ sec^{-1}, in which the sulfoxide ligand shifts from sulfur to oxygen coordination. The reduction potential of $[Os(NH_3)_5(O\text{-dmso})]^{3+}$ (−0.90 V, NHE) is dramatically shifted due both to stabilization of Os(III) and to destabilization of Os(II), as compared to sulfur coordination. The cycle is completed by an O → S isomerization on Os(II) at a specific rate of >100 sec^{-1}.

5. *S-Sulfinato to O-Sulfinato Linkage Isomerizations*

$[OsCl\{(S)\text{-}SO_2C_6H_4Me\text{-}4\}(CO)(PPh_3)_2]$ undergoes CO addition with concomitant linkage isomerization to form $[OsCl\{(O)\text{-}OS(O)C_6H_4Me\text{-}4\}(CO)_2(PPh_3)_2]$ (*434*).

E. Reactions of Ligands on Osmium

1. *Atom-Transfer Reactions*

Reports of atom-transfer reactions for complexes of Os in high oxidation states chiefly appear in three areas: hydride transfer, oxotransfer, and nitrogen transfer.

a. Hydrogen Atom Transfer. In general, H^- and H_2 complexes are uncommon for polypyridine and polyamine systems lacking Os—C or Os—P bonds, although a few examples are known. The chemistry of hydride and dihydrogen complexes containing Os—C and Os—P bonds are covered in a number of reviews (*4, 8–10*) and will not be discussed further here. Recently, $[OsA_4(H)_2]^{2+}$ complexes were observed to undergo reductive coupling in the presence of π acids or halides (X^-) to form $[OsA_4(H_2)(\pi \text{ acid})]^{2+}$ and $[OsA_4(H_2)X]^{2+}$, respectively (*89*). Such activity has been observed for both *cis*- and *trans*-dihydride configurations. Similar reactions probably occur in the elimination of H_2 from $[Os(L\text{-}L)(PPh_3)_2(CO)(H)_2]^{2+}$ (L-L = bpy, 4,4′-Me_2bpy, or 5,5′-Me_2bpy) on treatment with CF_3SO_3H to form $[Os(L\text{-}L)(PPh_3)_2(CO)(OSO_2CF_3)]^+$ (*19*). $[Os(NH_3)_5(H_2)]^{2+}$ has been shown to transfer dihydrogen to acetone upon the one-electron oxidation to Os(III) (*201*). $[Os(NH_3)_5(H_2)]^{3+}$ is stable only in strongly acidic media, and is thought to readily deprotonate to form the Os(III) hydride, $[Os(NH_3)_5H]^{2+}$, the species suspected of being the active reducing agent for the ketone.

b. Oxygen Atom Transfer. These are extremely numerous in Os(VIII) chemistry, in which oxo-transfer reactions of $[OsO_4]$, $[Os(O)_n(NR)_{4-n}]$, and their adducts with N-donor ligands are widely used in organic chemistry, interactions with DNA, and biological fixing and staining (Section II,C,6,b). The cis dihydroxylation reactions with alkenes, acetylenes, arenes, and unsaturated heterocycles have been known for many years in organic chemistry. Literally hundreds of different osmate esters have been prepared by these atom-transfer reactions, although a much smaller number have been characterized prior to hydrolysis of the *cis*-diol. Despite the long history of the applications of these reactions to organic chemistry, it is only recently that the mechanisms have been elucidated, particularly for those reactions that are asymmetric, promoted by N heterocycles or light, and/or are catalytic. It is not possible to detail all of these advances here, but they are well documented in recent reviews and papers (*35, 36, 327–332, 334–336*). Recent studies on the mechanisms of oxo-transfer reactions to other organic and inorganic substrates are summarized in Table IX.

Such reactions are not only interesting in their own right but are germane to the synthesis of many Os complexes.

The reverse reactions, in which Os(VI) complexes undergo an oxo atom transfer with organic substrates, are also very important. They are the key steps in the regeneration of the catalyst in the Os(VIII)-catalyzed cis-dihydroxylation reactions of organic N oxides (*27, 35, 329, 390, 391, 394*). Although the trigonal-planar $[Os^{VI}(NAr)_3]$ (Ar = 2,3-$Pr^i{}_2C_6H_3$) is air-stable, it undergoes an atom-transfer reaction with Me_3NO to give $[Os^{VIII}(O)(NAr)_3]$ (*22*).

Oxotransfer is well documented for complexes of Ru(IV) → Ru(VI) (*629, 630*). In addition to oxotransfer to organic substrates, Ru complexes have been shown to be electrocatalysts for the oxidation of water (*631*). Though not as strong oxidizing agents as their Ru analogs, Os(VI) complexes have also been reported to undergo oxo-transfer reactions. *cis*- and *trans*-$[Os(bpy)_2(O)_2]^{2+}$ react with two equivalents of PPh_3 to give either the *cis*- or *trans*-$[Os(bpy)_2(OPPh_3)_2]^{2+}$, respectively (*415*). Similar activity is shown by the porphyrin complex, $[Os(oep)(O)_2]$ (*218*), the complex containing the tetradentate amide/pyridine ligand, *trans*-$[Os(bpbH_2)(O)_2]$ (*283*), and the macrocyclic complex, *trans*-$[Os(14\text{-tmc})(O)_2]$ (*39*). Indeed, oxo-transfer reactions of dioxoOs(VI) complexes with PPh_3 are standard preparative methods for the syntheses of Os(II), Os(III), and Os(IV) complexes. Oxo complexes of Os(VI) in saturated ligand environments are less reactive than with bpy or porphyrinato ligands, but have been shown to be powerful photo-oxidants (*107, 114*). Similarly, the oxo-bridged dimers of Os are not sufficiently oxidizing to oxidize water to dioxygen (*554*). *trans*-$[Os(bpbH_2)(O)_2]$ also reacts with cyclohexene to form cyclohexenol (*283*).

c. Nitrogen Atom Transfer. Although not as numerous as oxo-transfer reactions, the nitrogen atom-transfer reactions of $[Os^{VIII}(O)_n(NR)_{4-n}]$ with unsaturated substrates such as alkenes have received recent attention in organic synthesis. These reactions result in Os(VI) imido complexes that ultimately yield diamines and aminoalcohols (*22, 35, 208*). Other atom-transfer reactions of Os(VIII) include the reaction of $[Os(O)(NAr)_3]$ with PR_3 to form $[Os^{IV}(NAr)_2(PR_3)_2]$, OPR_3, and $ArNPR_3$ (*22*), and $[Os(O)_3(NBu^t)]$ with mesMgBr to give $[Os^{VI}(O)_2(mes)_2]$ and $mesNH_2$ (*30*).

In the late 1970s, Buhr and Taube demonstrated a nitrido coupling reaction in which treatment of $[Os(NH_3)_5(CO)]^{2+}$ by Ce(IV) generated $[\{Os(NH_3)_4(CO)\}_2(\mu\text{-}N_2)]^{4+}$ in high yield (*58*). Work in this area has recently been extended to complexes of aromatic nitrogen ligands:

when *trans*-[$Os(py)_2Cl_3N$] is heated in py for several hours, N_2 is eliminated in good yield concurrent with the formation of [$Os(py)_3Cl_3$] (*42, 226*). Experiments with substituted pyridines indicate that coupling is triggered by substitution of the heterocycle into a coordination position trans to the nitrido group. Similar coupling reactions occur with [$Os^{VI}(NH_3)_4N$] under UV/Vis irradiation in acetonitrile to form the mixed-valence *trans,trans*-[$Os_2(\mu\text{-}N_2)(NH_3)_8(NCCH_3)_2]^{5+}$ complex (*269*). Such reactions are not only of fundamental interest, but they are also the reverse reactions involved in nitrogen fixation. A knowledge of the factors that affect the coupling reaction may lead to the rational design of nitrogen-fixing reactions.

[$Os^{VI}(N)Cl_3(pic)_2$] and [$Os^{VI}(N)Cl_2(trpy)]^+$ undergo atom-transfer reactions with CN^- to form [$Os^{IV}(pic)_2(NCN)Cl_3]^-$ and [Os^{IV}-(tryp)(NCN)Cl_2], and with N_3^- to form [$Os^{II}(pic)_2(N_2)Cl_3]^-$ and [Os^{II}-(trpy)(N_2)Cl_2], respectively (*632*). [$Os^V(trpy)(bpy)(N)]^{2+}$ is believed to undergo an atom transfer with H_2O to form [$Os^{II}(trpy)(bpy)(NH_2O)$] in the oxidative conversion of [$Os^{II}(trpy)(bpy)(NH_3)]^{2+}$ to [Os^V(trpy)(bpy)(NO)$]^{3+}$ (*43*).

Although not strictly atom-transfer reactions, because there are no changes in formal oxidation states, the nucleophilic substitution reactions of the nitrido complexes are parallel reactions. The nitrido ligands of [$Os(N)R_4]^-$ are alkylated readily to give [$Os(NR')R_4$], and the kinetics of the reaction between [$Os(N)(CH_2SiMe_3)_4]^-$ and MeI have been studied (*155*). This illustrates that even in the Os(VI) oxidation state, N^{3-} retains some nucleophilic characteristics. Whether it acts as a nucleophile or electrophile depends critically on the nature of the other ligands and the substrate with which it reacts. Even though this ligand can act as nucleophile when bound to Os(VI), it is not as good a nucleophile as RS^-, because [$Os(N)(CH_2SiMe_3)_2(SCH_2CH_2S)]^-$ is preferentially methylated at the sulfur to form [$Os(N)(CH_2SiMe_3)_2(SCH_2CH_2$-SMe)] (*156*).

d. Halogen Atom Transfer. Rapid halide atom transfer occurs in the system [$Os^{II}(cp)_2$]/[$Os^{IV}(cp)_2I]^+$, although much slower than the analogous Ru chemistry. There are two independent reaction pathways. One is first order in both reactions and is the only pathway available in polar solvents, and the other is first order with respect to the Os(II) complex and zero order with respect to the Os(IV) complex. This second pathway is only observed at low concentrations of Os and in weakly polar solvents. The rate constants of the overall second-order reactions vary by over an order of magnitude in going from dmso (3.5 M^{-1} sec^{-1}) to acetone (78.3 M^{-1} sec^{-1}) at 20°C, although the activation enthalpies and entropies show a much bigger variation. The cross-reaction $[Ru(cp)_2Cl]^+/[Os(cp)_2]$ has also been studied (*633, 634*).

2. *Ligand-Centered Reactions on Osmium(IV)*

a. Allyl Complexes. As is common for most η^3-allyl complexes of higher oxidation states (*180*), $[Os^{IV}(NH_3)_5(\eta^3\text{-allyl})]^{3+}$ readily undergoes nucleophilic attack to give substituted $[Os^{II}(NH_3)_5(\eta^2\text{-olefin})]^{2+}$ complexes (*180*). Suitable nucleophiles include PPh_3, py, and MeO^-. The cyclic allyl complex, $[Os(NH_3)_5(C_6H_9)]^{3+}$, reacts with base to generate an Os(II) 1,3-cyclohexadiene species. The tendency of allyl complexes to deprotonate is markedly enhanced when the product is aromatic, hence, even in acidic media, $[Os(NH_3)_5(3,6\text{-dimethoxycyclohexene})]^{2+}$ eliminates two equivalents of methanol to generate $[Os(NH_3)_5(\eta^2\text{-benzene})]^{2+}$. This sequence is shown in Scheme 15, starting from the hydrogenation of *p*-dimethoxybenzene (*180*).

b. Oxidative Dehydrogenation. The Os(IV) oxidation state is also important in oxidative dehydrogenation reactions. $[Os^{IV}(en)(en\text{-}H)_2]^{2+}$ (*204, 205*) is an intermediate in the oxidative dehydrogenation reaction of $[Os(en)_3]^{3+}$. The reaction proceeds by base-catalyzed disproportionation of the Os(III) compound to Os(II) and deprotonated Os(IV) complexes. Such disproportionation is facilitated by the stabilization of Os(IV) by π bonding between the deprotonated amines and the electron-deficient Os(IV) center (*204, 205*). The subsequent intra-

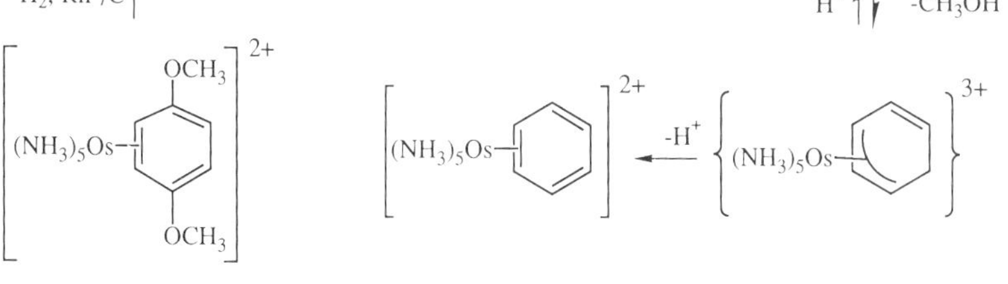

SCHEME 15. Reactions of $[Os(NH_3)_5(\eta^3\text{-allyl})]^{3+}$ complexes.

molecular two-electron transfer to produce $[Os(en)_2(enim)]^{2+}$ is driven by the π backbonding stabilizing of Os(II). The reaction continues in air via a series of oxidations, disproportionations, and intramolecular redox reactions to produce $[Os(en)(diim)_2]^{2+}$ as the air-stable product (*204, 205*). This chemistry is summarized in Schemes 11 and 16. Parallel chemistry is observed for $[Ru(en)_3]^{3+}$, but the Ru(IV) intermediates are too unstable to be characterized and the aerial oxidation stops at $[Ru(en)_2(diim)]^{2+}$ (*635*). The extent of oxidation depends on the ability of O_2 to oxidize $[M^{II}(en)_2(diim)]^{2+}$ to the M(III) counterpart. The Ru(II) complex has a more positive redox potential than does its Os(II) counterpart and hence is not easily oxidized to Ru(III) in order that the disproportionation and subsequent oxidative dehydrogenations can occur to give $[Ru(en)(diim)_2]^{2+}$.

Similar reactions occur in the oxidative dehydrogenation reactions of $[Os(bpy)_2(ampy)]^{2+}$ to $[Os(bpy)_2(impy)]^{2+}$ and related ligand oxidations (*256*) that parallel the well-studied Ru chemistry (*636*). Such reactions are also used in the synthesis of $[(NH_3)_5Os(NCCN)Os(NH_3)_5]^{5+}$ from $[(NH_3)_5Os(NH_2CH_2CH_2NH_2)Os(NH_3)_5]^{6+}$ (*542*), and the oxidations both of amines to nitriles and of alcohols to aldehydes and ketones in $[Os(NH_3)_5L]^{n+}$ complexes (*637*). Os(II) carbonyl complexes are also prepared from the oxidative dehydrogenation of methanol, to form $[Os(PPr^i_3)_3(CO)(H)(Cl)]$ (*18*). $MCl_3 \cdot xH_2O$ reacts with the tetradentate ligand, picstien, in air to give *trans*-$[M(\text{picstien-dii})Cl_2]$ (M = Ru or Os) in which the diamine ligand has undergone an oxidative dehydrogenation reaction to form a M(II) diimine complex (*638*).

c. Ligand Deprotonation. Base-catalyzed disproportionations of $[Os(NH_3)_5(\text{N-heterocycle})]^{3+}$ complexes present a problem in the direct synthesis of these complexes from the reaction $[Os(NH_3)_5(OSO_2CF_3)]^{2+}$ with excess ligand. Upon deprotonation and disproportionation of the Os(III) complex, it is presumed that amido ligand further deprotonates to form imido or nitrido Os(IV) species. These doubly or triply deprotonated ammine ligands both stabilize the higher oxidation states and labilize cis ammine ligands to substitution, resulting in multiple cis substitution of the N heterocycles (*68*).

3. *Ligand-Centered Reactions on Osmium(III)*

For Os(III), the origin of most ligand-centered reactions is due to the Lewis acidic nature of the metal ion. This effect raises the acidity of $[Os(NH_3)_6]^{3+}$ significantly ($pK_a \approx 16$) (*87*). Hence, trace amounts of this material in an acetone solution of $[Os(NH_3)_6]^{2+}$ act to catalyze the

$$[Os(en)_2(en-H)]^{3+} + OH^- \rightleftharpoons [Os(en)(en-H)_2]^{2+} + H_2O$$

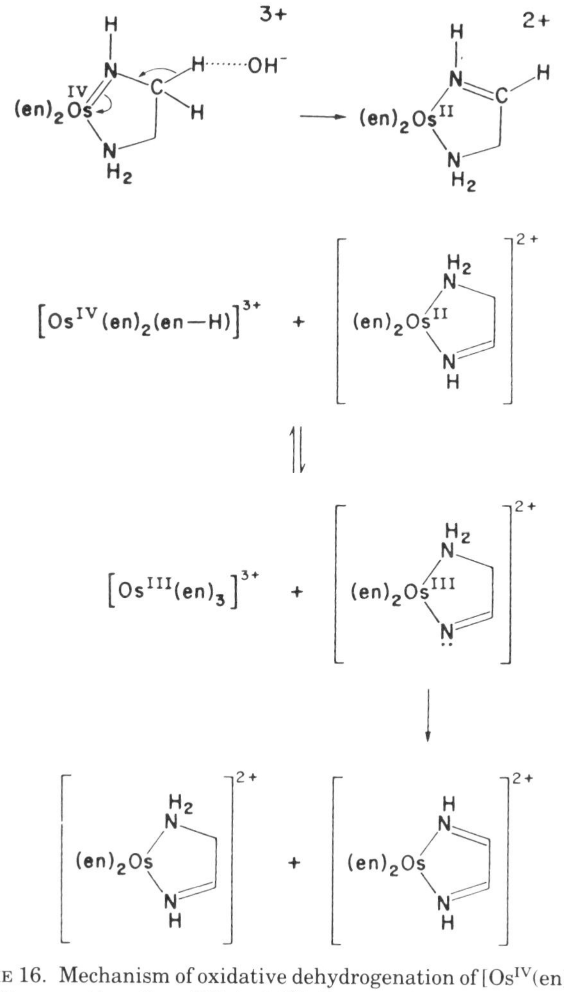

SCHEME 16. Mechanism of oxidative dehydrogenation of $[Os^{IV}(en)_2(en\text{-}H)]^{3+}$ produced from the disproportionation of $[Os(en)_3]^{3+}$. Reprinted with permission from the *Journal of the American Chemical Society*, Ref. *204*. Copyright 1982, American Chemical Society.

formation of the imine species $[Os(NH_3)_5\{NHC(CH_3)_2\}]^{2+}$, wherein the imine acts to stabilize the electron-rich Os(II) center (*194*).

The acidity of the Os(III) ammine complexes is enhanced considerably by replacement of an ammine ligand by N-heterocyclic, imine, or other π-acid ligands. This is due to the transferral of electron density from the π-donor ligand, NH_2^-, to the π^* orbital of the π-acid ligand, via the Os(III) center. This aids considerably in catalyzing reactions that proceed via base-catalyzed disproportionation reactions (Sections V,E,1,c and V,E,2.

Although Os(III) is a Lewis acid, it undergoes stronger π backbonding than does Ru(III) (*196, 426, 586*). For example, acid hydrolysis of $[Ru(NH_3)_5(NCMe)]^{3+}$ gives $[Ru(NH_3)_5(NHCOMe)]^{2+}$ (or its protonated analog at low pH values) with a specific rate constant of 1.2×10^{-5} sec^{-1} at 25°C. This rate is independent of pH and ionic strength, which is indicative of OH_2 being the nucleophile (*639*). The analogous chemistry occurs for Os(III), but at a much slower rate ($k \ll 1 \times 10^{-5}$ sec^{-1} at 25°C) (*67, 119*). A similar significant stabilization is apparent in the acid hydrolysis of the cyanogen-bridged complexes. Thus, hydrolysis of the cyanogen ligand in $[(NH_3)_5Ru(NCCN)Ru(NH_3)_5]^{5+}$ gives $[(NH_3)_5Ru(NHCOCN)Ru(NH_3)_5]^{4+}$ with a specific rate constant of 7.0×10^{-3} sec^{-1} at 25°C (*588, 640*). By contrast, the same reaction at the mixed-valence Os complex occurs with a specific rate constant of $\sim 2 \times 10^{-4}$ sec^{-1} at 25°C (*542*). The acid hydrolyses of the cyanogen ligand bound to both of the +6 ions is too fast to be measured using conventional cyclic voltammetry (up to 50 V sec^{-1} at 25°C). The Os_2^{4+} and Ru_2^{4+} ions are both very resistant to nucleophilic attack at the cyanogen ligand, due to the very substantial degree of transferral of π electron density by π backbonding. In summary, bonding of nitriles to Os(III) activates them to nucleophilic attack, but this activation appears to be moderated by a greater degree of π backbonding for Os(III), compared to Ru(III). Of course the results can also be interpreted in terms of stronger π bonding of the nitriles to Ru(III) than Os(III). This would also activate the nitriles to a greater extent when bound to the former, but the balance of evidence strongly points to π backbonding deactivation by Os(III) as the major source of the differences in reactivities. Confirmation of this awaits crystal structures of isomorphous complexes. If π bonding activation is the most important factor, the Ru-N(nitrile) bond will be shortened with respect to the anologous Os-N bond, whereas the converse will be true if π backbonding deactivation is most important.

A further demonstration of the Lewis acidity of Os(III) is the aldol condensation reaction of $[Os(NH_3)_5(\eta^1\text{-acetone})]^{3+}$, to form the diacetone alcohol complex (*67, 90*). The catalysis of this reaction can occur in one of two ways. Either Os(III) catalyzes the deprotonation of a methyl group of the bound acetone ligand to produce a nucleophile for attack at a second acetone ligand, or the Os(III) center polarizes the C=O bond of the acetone ligand, which facilitates the attack of a $CH_3COCH_2^-$ nucleophile. The kinetics and mechanisms were not studied in detail and isotopic labeling experiments would be required to distinguish between these two possibilities or one in which both forms of activation are occurring. Similar reactions occur with the Co(III) (*90*) and Ru(III) (*182–184*) analogs.

There is also other evidence for the activation toward nucleophilic attack of ligands coordinated to Os(III). One example is the formation of $[Os^{III}(NH_3)_5(N\text{-Etpz})]^{4+}$ as a by-product in the reaction of $[Os(NH_3)_5(OSO_2CF_3)]^{2+}$ with pz in triethylphosphate. The by-product is thought to occur via the nucleophilic attack of $[Os(NH_3)_5(pz)]^{3+/2+}$ on the $[Os(NH_3)_5\{OP(OEt)_3\}]^{3+}$ intermediate (*68, 80*). If the $[Os(NH_3)_5(pz)]^{3+}$ complex is the nucleophile, then it is a further indication of the importance of π backbonding in the Os(III) complex. In fact, it has been shown that the pyrazine ligand bound to Os(III) is surprisingly basic for a diazine coordinated to a M(III) center. Dissolution of yellow $[Os(NH_3)_5(pz)]^{3+}$ in 0.1 *M* HCl results in an immediate color change to the dark brown $[Os^{III}(NH_3)_5(pzH)]^{4+}$ ion (*66, 67*).

4. Ligand-Centered Reactions on Osmium(II)

a. Protonations. The vast majority of ligand-centered reactivity on Os(II) ammine complexes is attributed to the strong tendency of this metal ion to undergo π backbonding with π-acid ligands. Perhaps the most fundamental demonstration of this is the enhancement in proton affinity reported for various organic bases upon coordination to the metal. A striking example is the pyrazine complex $[Os(NH_3)_4Cl(pz)]^+$ (*77*), which is greater than seven orders of magnitude *more* basic than for the free ligand. Such increases in basicity are also observed for the $[M(NH_3)_5(pz)]^{2+}$ complexes (*60, 67*). The fact that the increase in basicity is 6.8 pK_b units when M = Os, but only 1.9 pK_b units when M = Ru, quantifies the much greater π basicity of Os(II), compared to Ru(II) (*60*). In contrast, $[Os(NH_3)_5(3,4\text{-}\eta^2\text{-lutidine})]^{2+}$, in which the heterocycle is dihapto bound (*85*), shows virtually no change in proton affinity upon ligation. However, dihapto coordination of aromatic molecules can also result in a thermodynamic protonation site that differs from the free ligand, e.g., free pyrrole, which protonates preferentially at the 2-position ($pK_a \approx -4$), protonates at the 4-position in $[Os(NH_3)_5(2,3\text{-}\eta^2\text{-pyrrole})]^{2+}$, but at considerably higher pH values (>4) (*430*). Similarly, $[Os(NH_3)_5(\eta^2\text{-aniline})]^{2+}$ and related complexes show enhanced basicity at the para-position of the ring, an observation which suggests that the ligand possesses a considerable dienamine character

(*431*). Finally, upon treatment with acid, both diene and allyl ether complexes of $[Os(NH_3)_5]^{2+}$ yield Os(IV) η^3-allyl species, the latter reaction proceeding by the elimination of alcohol (*180*).

b. Alkylations. Reactions analogous to the protonation of π-acid ligand bound to Os ammine complexes have been observed with other electrophiles. The pyrazine complexes, $[Os(NH_3)_5(pz)]^{2+}$ and $[Os(NH_3)_4(N_2)(pz)]^{2+}$, and other diazine complexes are N-methylated by the reaction of CH_3I in dmso (*195*). $[Os(NH_3)_5(\eta^2\text{-pyrrole})]^{2+}$ complexes are methylated at the 4-position or the N-position by CH_3OTf in dme (*85*), and 2,3,-η^2-coordinated phenol is methylated at the oxygen (*178*). When this phenol complex or the aniline derivative is treated with maleic anhydride, a Michael reaction is observed exclusively at the para position, and in high yield (*178, 431*). For the aniline complex, oxidative work-up followed by esterification in MeOH results in dimethyl (4-aminophenyl)succinate ester, free of ortho-substituted contaminants (Scheme 17) (*431*).

cis-$[Os(bpy)_2(CN)_2]$ is methylated with MeI or benzylated with $PhCH_2Br$ in MeCN to give the isonitrile complexes, *cis*-$[Os(bpy)_2(CNR)_2]^{2+}$ (R = Me or $PhCH_2$) (*97*).

c. Arene Activation. One of the most significant outcomes of the formation of stable $[Os(NH_3)_5(\eta^2\text{-arene})]^{2+}$ complexes is the manifestation of dienelike reactivity in the organic ligand. $[Os(NH_3)_5(\eta^2\text{-benzene})]^{2+}$ is readily hydrogenated under 1 atm of hydrogen at 20°C to

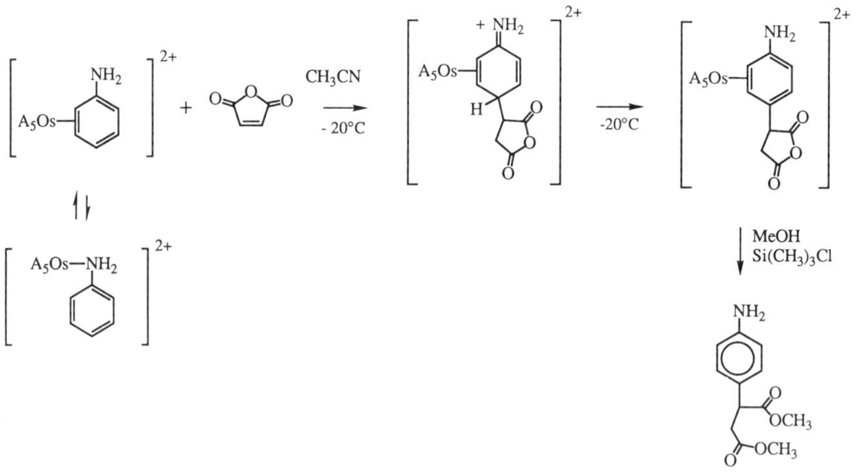

2+
NH2
A5Os
+
CH3CN
- 20°C
+ NH2
A5Os
H
-20°C
MeOH
Si(CH3)3Cl
A5Os—NH2
OCH3
OCH3

SCHEME 17. Reaction of $[Os(NH_3)_5(\eta^2\text{-aniline})]^{3+}$ to give dimethyl (4-aminophenyl)succinate ester.

the cyclohexene analog in the presence of a Pd^0 catalyst (*432*). When the reaction is carried out under D_2 in CD_3OD, all of the deuterium is incorporated at a common ring face. Thus, the metal serves to activate the arene, protect the olefin product from further hydrogenation, and direct the stereochemistry of the reaction. Such a hydrogenation can be carried out on substituted benzene complexes as well (*169*). For the case of the 2,3-η^2-anisole species, hydrogenation in dry methanol yields the 3-methoxycyclohexene isomer in high yield. When this reaction is repeated in aqueous methanol with a trace of acid, the major product is a 2-cyclohexene-1-one-complex, i.e., the hydrolysis product of a 2-methoxy-1,3-diene intermediate (Scheme 18). The dihapto coordination of arenes also results in their activation toward additional metalation, even by weaker π-bases such as $[Ru(NH_3)_5]^{2+}$ (*172*).

One of the most exciting findings concerning η^2-arene activation is the enhancement of nucleophilic character at the ortho and para positions of phenols, phenyl ethers, and anilines (*178, 431*). As mentioned above in the context of alkylations, Michael additions readily occur at the para position of phenol and aniline. The action of the methylnitrillium ion on anisole or phenol in CH_3CN leads to the formation of 4-methoxy- and 4-hydroxyacetophenone imines in excellent yields (*178*). When a methyl group occupies the para position, imine formation takes place exclusively ortho to the electron-donating substituent. The redistribution in electron density upon coordination of these arenes is significant enough that the complexes $[Os(NH_3)_5(o\text{-cresol})]^{2+}$ and

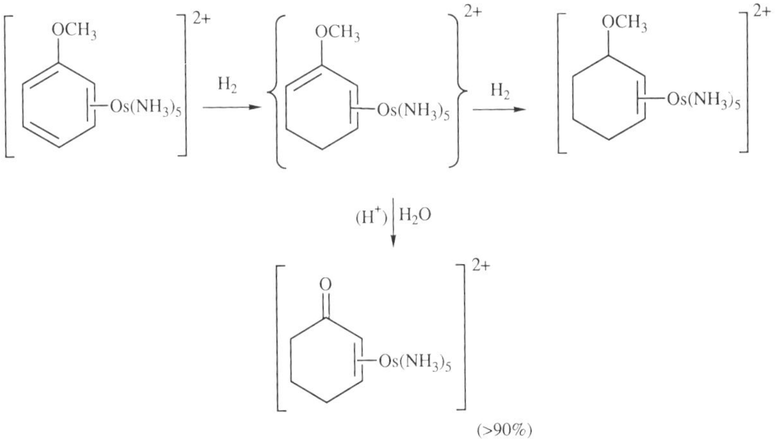

(>90%)

SCHEME 18. Preparation of a 2-cyclohexene-1-one complex from a 2,3-η^2-anisole complex.

$[Os(NH_3)_5(m\text{-cresol})]^{2+}$ exist primarily as the 2,5-cyclohexadien-1-one tautomers under equilibrium conditions in methanol (*178*).

d. Activation of N-Heterocycles. The reduction of $[Os^{III}(NH_3)_5(N\text{-heterocycle})]^{3+}$ complexes by Zn(Hg) in acidic solution initially produces the Os(II) analogs. However, the charge-transfer bands of these complexes slowly disappear upon further reduction and this is attributed to hydrogenation of the ligands (*60*). Given the recent work on hydrogenation of the arenes discussed previously (*169, 172, 432*) and the formation of η^2-bound N-methylated complexes (*85*), a likely mechanism is as follows. Following reduction to Os(II), the complex is protonated to form the diazenium complex, which rearranges to form an equilibrium concentration of the η^2 linkage isomer. This isomer would be expected to be susceptible to hydrogenation in chemistry that parallels that observed for the η^2-arenes (*169*).

e. Addition Reactions. $[Os(NH_3)_5(\eta^2\text{-pyrrole})]^{2+}$ is 2,3-$\eta^2 \leftrightarrow$ 3,4-η^2 fluxional at room temperature and shows markedly different reactivity toward cycloaddition to maleic anhydride than is observed for the free ligand (*179*). Whereas pyrrole itself is quite resistant to this dienophile and undergoes a Michael addition only at high temperatures, the complex readily forms a mixture of exo and endo cycloaddition products at room temperature. The reaction is thought to proceed through a dipolar cycloaddition in which the active intermediate is an azomethine ylide stabilized by metal coordination at the 3,4-positions. By contrast, no reaction is observed when $[Os(NH_3)_5(\text{pyrrole})]^{2+}$ is replaced by the furan analog, even though cycloaddition to maleic anyhydride readily occurs with the free ligand.

Alkyne complexes of Os(II) are observed to undergo addition of water and methanol across the alkyne bond, resulting in stable vinyl alcohol and vinyl ether complexes, respectively (*168*). When an aqueous solution of $[Os(NH_3)_5(CH_3C{\equiv}CCH_3)]^{2+}$ is allowed to stand, the initial product observed is the *cis*-2-hydroxy-2-butene complex. Over a period of several days, this species equilibrates with its trans stereoisomer, the latter being slightly favored in aqueous solution ($K_{eq} = 1.5$).

The Os(II) alkene complexes $[OsCl(NO)(R_2C{=}CR_2)(PPh_3)_2]$ are quite reactive toward addition of $OSNSO_2C_6H_4Me\text{-}4$ (Section II,C,2,f) (*163, 164*).

Elliott and Shepherd (*70*) have investigated the effect of Os(II) coordination on dienes and report that the metal acts as a protecting group for electrophilic addition of bromine. Thus, the action of bromine on $[Os(NH_3)_5\{\eta^2\text{-}1,2\text{-}(1,3\text{-butadiene})\}]^{2+}$ is thought to result in the 3,4-dibromo-1-butene analog.

f. Condensation of Acetone. The reduction of *cis*-[Os(NH_3)$_4$(O-SO_2CF_3)$_2$]$^+$ in acetone results in the substitution-inert [Os-(NH_3)$_4$(daa)]$^+$ complex (daa = diacetone alcohol). The fact that the complex is substitution inert shows that one site is occupied by a η^2-ketone group. The other site is expected to be occupied by the deprotonated alcohol, although the coordination mode has not been positively identified. It appears that this reaction occurs via an intramolecular condensation of *cis*-acetone ligands, but the details of the mechanism are uncertain and the condensation may have occurred prior to reduction of the Os(III) complex *cf.* Section V,E,3 (*81, 89*).

g. Dehydration Reactions. The propensity of Os(II) to bind strong π acceptor ligands drives a number of dehydration reactions. These include the dehydration of formate to carbon monoxide (*119*), hydrogen sulfite to sulfur dioxide (*197*), and oximes to nitriles (*195*). The former, in particular, is a rapid reaction and must also involve a linkage isomerization reaction because the formate ligand is initially O bound, whereas the product has the C-bound carbonyl ligand. A likely intermediate is the η^2-*C,O*-formic acid complex, by analogy with the preferred linkage isomers of ketones and aldehydes. As shown in the following section, such an intermediate is expected to activate the ligand to dehydration by analogy with the elinination reaction of the η^2-*C,O*-acetaldehyde complex. A second possible intermediate is the C-bound formate isomer.

Upon reoxidation of the Os(II) complexes, hydration of the ligands bound to Os(III) is much slower than the dehydration of the ligands bound to Os(II). Thus, reversible redox couples are found for the oxidation of the Os(II) dehydration products. Though the SO_2 and nitrile ligand will hydrate in the Os(III) oxidation state, the latter forms amides and is quite slow. In the case of the CO complex, the Os(III) complex is too labile for any appreciable CO hydration to occur within the lifetime of the complex.

Similar dehydration reactions are observed at Ru(III), but they are somewhat slower (*195, 641–643*), which is consistent with the smaller degree of stabilization of the products by π backbonding and, by implication, the transition states/intermediates in the reactions.

h. Other Elimination Reactions. Though Os chemistry parallels Ru chemistry in the dehydration/elimination reactions, Os(II) exhibits other elimination reactions that have no known parallels in pentaammineruthenium(II) chemistry. A particularly surprising reaction is the rapid extrusion of CO from [Os(NH_3)$_5$(dmf)]$^{2+}$ to form [Os-(NH_3)$_5$(CO)]$^{2+}$ and NH(CH_3)$_2$ (*120*). Other similar reactions that occur are the elimination of methane from coordinated acetaldehyde to form $[Os(NH_3)_5(CO)]^{2+}$, and the elimination of H_2 from the methylimine complex, $[Os(NH_3)_5(NH{=}CHCH_3)]^{2+}$, to form $[Os(NH_3)_5(NCCH_3)]^{2+}$ (*120*). Similar reactions are used in the preparation of $[Os(H)(X)(CO)(PR_3)_n]$ complexes (Section II,C,2,b), but these require much greater forcing conditions, illustrating the greater degree of π stabilization of the Os(II) π-acid complexes with the pentaammine moiety as opposed to complexes with other π acids, such as phosphines. The pentaammine chemistry also shows the much greater driving force of π-backbonding stabilization of Os(II) compared to RU(II), although facile elimination reactions of $[Ru(NH_3)_5(HON{=}C(R)CH_3)]^{2+}$ to produce $[Ru(NH_3)_5(N{\equiv}CCH_3)]^{2+}$ and ROH have been reported (*644*).

Abbreviations and Trivial Nomenclature

A	ammine or amine
$(aa)_2en$	*N,N*-bis(acetylacetone) ethylenediimine = 3,3′-(1,2-ethanediylnitrilo) bis(1-methyl-1-butanonato)
abn	aminobenzonitrile
ac	acetone
acac	acetylacetonato(1−) = 2,4-pentanedionato(1−)
Acdhqd	acetyldihydroquinidine
acet	acetaldehyde
AcOH	acetic acid
adc-Me	1,2-diacetylhydrazido(2−)
ampy	2-(aminomethyl)pyridine
an	acetonitrile
$[9]aneS_3$	1,4,7-trithiacyclononane
$[14]aneS_4$	1,4,8,11-tetrathiacyclotetradecane
$[14]aneN_4$	1,4,8,11-tetraazacyclotetradecane
$[15]aneN_4$	1,4,8,12-tetraazacyclopentadecane
$[16]aneN_4$	1,5,9,13-tetraazacyclohexadecane
$[18]aneS_6$	1,4,7,10,13,16-hexathiacyclooctadecane
anisole	methoxybenzene
$(ba)_2en$	*N,N′*-bis(benzoylacetone) ethylenediimine = 3,3′-(1,2-ethanediylnitrilo)-bis(1-phenyl-1-butanonato)

bbpe	*trans*-1,2-bis(4′-methyl-2,2′-bipyridyl-4-yl)ethene
bibzim	2,2′-bis(benzimidazolate)(2−)
bim	2,2′-biimidazolato(2−)
bimH	2,2′-biimidazolato(1−)
bpa	1,2-bis(4-pyridyl)ethane
$bpbH_2$	*N,N*′-bis(2′-pyridinecarboxamide)-1,2-benzene
bpb	*N,N*′-bis[2′-pyridinecarboxamido(1−)]-1,2-benzene
bpds	4,7-diphenýl-1,10-phenanthroline disulfonate
bpt	3,5-bis(pyridin-2-yl)-1,2,4-triazolate(1−)
bptz	3,6-bis(2-pyridyl)-1,2,4,5-tetraazine
bpy	2,2′-bipyridine
4,4′-bpy	4,4′-bipyridine
bsd	2,1,3-benzoselenadiazole
bta	benzotriazolato(1−)
btd	2,1,3-benzothiadiazole
Bu^n	*n*-butyl = 1-butyl
Bu^t	*tert*-butyl = 2-(2-methylpropyl)
$(Bu^i)_2en$	*N,N*′-bis(isobutyrylacetone) ethylenediimine = 3,3′-(1,2-enthanediylnitrilo)bis(1-(2-butyl)-1-butanonato)
4-Bu^tpy	4-*tert*-butylpyridine = 4-[2-(2-methylpropyl)]pyridine
3-Bu^t-saltmen	*N,N*′-(1,1,2,2-tetramethylethylene)bis-(3-*tert*-butylsalicylideneaminato)(2−)
chd	1,2-cyclohexanediol
chp	6-chloro-2-hydroxypyridinato(1−)
4-cinn	*N*-(4-pyridyl)cinnamamide
ClBzdhq	(4-chlorobenzoyl)dihydroquinine
ClBzdhqd	(4-chlorobenzoyl)dihydroquinidine
cod	1,5-cyclooctadiene
cp	cyclopentadienyl
o-cresol	2-methylphenol
p-cresol	4-methylphenol
$crMe_3$	*meso*-1,2,6,10,11-pentamethyl-2,6,10-triaza[11](2,6)-pyridinophane

CT	charge transfer
Ctmen	2,3-dimethylbutane-2,3-diamine
Ctmen-H	2,3-dimethylbutane-2,3-diaminato(1−)-*N*
Ctmen-2H	2,3-dimethylbutane-2,3-diaminato(2−)-*N,N′*
cym	cymene = 1-methyl-4-(1′-methylethyl)benzene
CV	cyclic voltammetry
daa	diacetone alcohol = 4-hydroxy-4-methylpenta-2-one
dabco	1,4-diazabicyclo[2.2.2]octane
dabcoMeps/dvp	dabco-methylated polystyrene/divinylbenzene copolymer
das	*cis*-1,2-bis(dimethylarsino)benzene
dbcat	3,5-di-*tert*-butylcatechol
dca	dicyanoamide(1−)
1,2-dcb	1,2-dicyanobenzene
1,3-dcb	1,3-dicyanobenzene
1,4-dcb	1,4-dicyanobenzene
dcbpy	2,2′-bipyridine-4,4′-dicarboxylate(2−)
dcpe	1,2-bis(dicyclohexylphosphino)ethane
ddq	dichlorodicyanobenzoquinone
depe	1,2-bis(diethylphosphino)ethane
dhc	*cis*-1,2-dihydrocatecholate(2−) = *cis*-5,6-dihydroxy-1,3-cyclohexadienate(2−)
dhdhp	*cis*-9,10-dihydro-9,10-dihydroxyphenanthrenato(2−)
diaa	di(4-anisyl)amine = bis(4-methoxyphenyl)amine
diim	ethanediimine
diMeim	*N,N′*-dimethylimidazolinium
diMeim-H	*N,N′*-dimethylimidazolinium ion deprotonated at the 2-position
dma	*N,N*-dimethylacetamide
dmcdhqd	dimethylcarbamoyl dihydroquinidine
dme	1,2-dimethoxyethane
dmf	*N,N*-dimethylformamide
dmpe	1,2-bis(dimethylphosphino)ethane
dmps	2,3-dimercaptopropanesulfonate(3−)
dmso	dimethyl sulfoxide

dpae	1,2-bis(diphenylarsino)ethane
dpb	1,8-bis[5-(2,8,13,17-tetraethyl-3,7,12, 18-tetramethyl)porphyrinato(2-)] biphenylene
dpp	2,3-bis(2′-pyridyl)pyrazine
dppe	1,2-bis(diphenylphosphino)ethane
dppene	*cis*-1,2-bis(diphenylphosphino)ethene
dppm	bis(diphenylphosphino)methane
dppmO	(diphenylphosphinomethyl)diphenylphosphine oxide
dppmS	(diphenylphosphinomethyl)diphenylphosphine sulfide
dpq	2,3-bis(2′-pyridyl)quinoxaline
ehba	2-ethyl-2-hydroxybutanoato(2−)
en	1,2-ethanediamine
en-H	1,2-ethanediaminato(1−)
enim	2-aminoethan-1-imine
EPR	electron paramagnetic resonance
Etpz	*N*-ethylpyrazinium
Fc	ferrocene
Fc^+	ferricenium
fhp	6-fluoro-2-hydroxypyridinato(1−)
gn	glutaronitrile
glyc	glycolate(2−)
H_2chba	3,5-dichloro-2-hydroxybenzamide
H_4chba-dcb	1,2-bis(2-hydroxy-3,5-dichlorobenzamido)-4,5-dichlorobenzene
H_4chba-Et	1,2-bis(2-hydroxy-3,5-dichlorobenzamido)ethane
H_4chba-ethylene	*cis*-1,2-bis(2-hydroxy-3,5-dichlorobenzamido)ethylene
H_4chba-*t*-1,2-diEtO-Et	1,2-bis(2-hydroxy-3,5-dichlorobenzamido)-*trans*-1,2-diethoxyethane
H_4chba-*t*-1,2-diHO-Et	1,2-bis(2-hydroxy-3,5-dichlorobenzamido)-*trans*-1,2-dihydroxyethane
H_4chba-*t*-1-OH-2-MeO-Et	1,2-bis(2-hydroxy-3,5-dichlorobenzamido)-*trans*-1-hydroxy-2-methoxyethane
hhch	*cis,cis,cis,cis,trans,trans*-1,2,3,4,5,6-hexahydroxycyclohexanate(6−)
hmt	hexamethylenetetraamine
hp	2-hydroxypyridinate(1−)

Hpy	pyridinium
HpyS	pyridinium-2-thiolate
H_2dpd	1-[5-(2,8,13,17-tetraethyl-3,7,12,18-tetramethyl)porphyrinato(2−)]-8-[5-(2,8,13,17-tetraethyl-3,7,12,18-tetramethyl)porphyrin]biphenylene
H_4dpd	1,8-bis[5-(2,8,13,17-tetraethyl-3,7,12,18-tetramethyl)porphyrin]biphenylene
H_2fo-chba	*N*-formyl-3,5-dichloro-2-hydroxybenzamide
H_4hba-b	1,2-bis(2-hydroxybenzamido)benzene
im	imidazole
impy	2-(carboxyimine)pyridine
isn	*iso*-nicotinamide = 4-(carboamide) pyridine
4-lutdm	4-(2,6-lutidinium)
malt	maltolato(1−) = 3-hydroxy-2-methyl-1-oxacyclohexa-2,5-dien-4-onato(1−)
Mb	myoglobin
MCD	magnetic circular dichroism
4,4′-Me_2bpy	4,4′-dimethyl-2,2′-bipyridine
5,5′-Me_2bpy	5,5′-dimethyl-2,2′-bipyridine
Me_4bpy	4,4′,5,5′-tetramethyl-2,2′-bipyridine
MeOdhqd	methoxydihydroquinidine
4,7-Me_2phen	4,7-dimethyl-1,10-phenanthroline
3,5,6,8-Me_4phen	3,5,6,8-tetramethyl-1,10-phenanthroline
3,4,7,8-Me_4phen	3,4,7,8-tetramethyl-1,10-phenanthroline
Mepy	*N*-methylpyridinium(1+)
Me_2PymS	4,6-dimethylpyrimidine-2-thiolato(1−)
Me_2PymSH	4,6-dimethylpyrimidine-2-thiol
Mepz	*N*-methylpyrazinium
mes	mesityl = 2,4,6-trimethylphenyl
mix	*meso*-porphyrinato(2−) IX-dicarboxylic acid
mix-dme	*meso*-porphyrinato(2−) IX-dimethyl ester
MO	molecular orbital
m.p.	melting point

mv^{2+}	methyl viologen = *N,N'*-dimethyl-4,4'-bipyridinium(2+)
NBu^t	*tert*-butylimide = 2-methylpropan-2-imide
nd	2,3-naphthalenediolato(2−)
$(NEt)_2bpy$	4,4'-bis(diethylamino)-2,2'-bipyridine
neopentyl	2,2-dimethylpropyl
NIR	near-infrared
nmp	*N*-methylpyrrolidine
NQR	nuclear quadrapole resonance
Ntmen	*N,N,N',N'*-tetramethyl-1,2-ethanediamine
oep	octaethylporphyrinato(2−)
oep^+	octaethylporphyrinato(1−) radical
oep^-	octaethylporphyrinato(3−) radical
OTf	triflate = trifluoromethanesulfonate
Pc	phthalocyanato(2−)
pda	1,2-phenylenediaminato(2−)
Ph	phenyl
phen	1,10-phenanthroline
phenba	*N,N'*-1,2-phenylenebis[2-acetyl-1-amino-1-buten-3-onato(1−)]
4-pic	4-picoline = 4-methylpyridine
picstien	3,4-diphenyl-1,6-bis(2'-pyridyl)-2,5-diazahexane
picstiendii	3,4-diphenyl-1,6-bis(2'-pyridyl)-2,5-diazahexan-1,5-diene
Pr^i	*iso*-propyl = 2-propyl
Pro	proline
ptz	phenothiazine
pvp	polyvinyl pyridine
py	pyridine
pyca	pyridine-2-carboxylato(1−)
pycaH	pyridine-2-carboxylic acid
pyO	pyridine *N*-oxide
pyr	pyrimidine
pyS	pyridine-2-thiolato(1−)
pySH	pyridine-2-thiol
pySSpy	2,2'-bis(pyridyl) disulfide

pz	pyrazine
pzH	pyrazinium(1+)
qncd	quinuclidine
salen	*N,N'*-ethylenebis(salicylideneaminato)(2−)
5-SO_3^--bpy	2,2'-bipyridine-5-sulfonate(1−)
tatd	1,3,5,7-tetraazatricyclo[3.3.1.1^{3,7}]decane
tcne	tetracyanoethene
teta	*C-meso*-5,5,7,12,12,14-hexamethyl-1,4,8,11-tetraazacyclotetradecane
tetraphos	3,6-diphenyl-1,8-bis(diphenylphosphino)-3,6-diphosphaoctane
thch	*cis,trans,trans*-3,4,5,6-tetrahydroxycyclohex-1-enato(4−)
thf	tetrahydrofuran
thio	thiourea
ththa	*cis,trans,trans*-1,2,3,4-tetrahydro-1,2,3,4-tetrahydroxyanthracenate(4−)
14-tmc	1,4,8,11-tetramethyl-1,4,8,11-tetraazacyclotetradecane
15-tmc	1,4,8,12-tetramethyl-1,4,8,12-tetraazacyclopentadecane
16-tmc	1,5,9,13-tetramethyl-1,5,9,13-tetraazacyclohexadecane
tmp	*meso*-tetramesitylporphyrinato(2−)
o-tolyl	2-tolyl = 2-methylphenyl
tpp	*meso*-tetraphenylporphyrinato(2−)
triflate	trifluoromethanesulfonate
triflato	trifluoromethanesulfonato
trop	tropolonato(−)
trpy	2,2':6',2''-terpyridine
tterpy	4'-phenyl-2,2':6',2''-terpyridine
vbpy	4-vinyl-4'-methyl-2,2'-bipyridine
vpy	4-vinylpyridine
XPS	X-ray photoelectron spectroscopy
p-Xtpp	*meso*-tetra(4-Xphenyl)porphyrinato(2−)
X-tterpy	4'-(4'''-Xphenyl)-2,2':6',2''-terpyridine
2-xylyl	2,6-dimethylphenyl

ACKNOWLEDGMENTS

The authors would like to thank Zai-Wei Li, Henry Taube, Andreas Ludi, Kim Finnie, Elmars Krausz, and David Ware for supplying the results of certain studies prior to publication. PAL is grateful for funding from the Australian Research Council for some aspects of the work described and for a CSIRO Postdoctoral Fellowship, which enabled the development of the Os triflato chemistry at Stanford University.

REFERENCES

1. Tennant, S., *Philos. Trans. R. Soc. London* **94,** 411 (1804).
2. Griffith, W. P., *in* "Comprehensive Coordination Chemistry" (G. Wilkinson, R. D. Gillard, and J. A. McCleverty, eds.), Vol. 4, Chapter 46, pp. 519–633. Pergamon, Oxford, 1987.
3. Bretherick, L., "Hazards in the Chemical Laboratory," 3rd ed., p. 422. The Royal Society of Chemistry, London, 1981.
4. Adams, R. D., and Selegue, J. P., *in* "Comprehensive Organometallic Chemistry" (G. Wilkinson, F. G. A. Stone, and E. W. Abel, eds.), Vol. 4, Chapter 33 pp. 967–1064. Pergamon, Oxford, 1982.
5. "Gmelin's Handbuch der Anorganischen Chemie," 8th ed., Vol. 66. Verlag Chemie, Berlin, 1939.
6. "Gmelin's Handbuch der Anorganischen Chemie, Supplement Volume, 'Osmium'," Vol. 1, No. 66. Springer-Verlag, Berlin, 1980.
7. Robinson, S. D., *Annu. Rep. Prog. Chem., Sect. A: Inorg. Chem.* **83,** 310 (1987); **84,** 227 (1988); **85,** 219 (1989).
8. Shapley, P. A., *J. Organomet. Chem.* **351,** 145 (1988).
9. Shapley, P. A., *J. Organomet. Chem.* **318,** 409 (1987).
10. Keister, J. B., *J. Organomet. Chem.* **318,** 297 (1987).
11. Le Bozec, H., Touchard, D., and Dixneuf, P. H., *Adv. Organomet. Chem.* **29,** 163 (1989).
12. Roper, W. R., *NATO ASI Ser., Ser. C* **269,** 27 (1989).
13. Kallmann, S., *Talanta* **34,** 677 (1987).
14. Singh, S., Mathur, S. P., Thakur, R. S., and Katyal, M., *Acta Cienc. Indica [Ser.] Chem.* **13,** 40 (1987).
15. Ermer, S. P., Shinomoto, R. S., Deming, M. A., and Flood, T. C., *Organometallics* **8,** 1377 (1989).
16. Martins, S., Jr., and Vugman, N. V., *Chem. Phys Lett.* **141,** 548 (1987).
17. Elliott, G. P., McAuley, N. M., and Roper, W. R., *Inorg. Synth.* **26,** 184 (1989).
18. Esteruelas, M. A., and Werner, H., *J. Organomet. Chem.* **303,** 221 (1986).
19. Sullivan, B. P., Lumpkin, R. S., and Meyer, T. J., *Inorg. Chem.* **26,** 1247 (1987).
20. Cotton, F. A., and Wilkinson, G., "Advanced Inorganic Chemistry," 5th ed., Chapter 17, pp. 632–633. Wiley, New York.
21. Tooze, R. P., Stavropoulos, P., Motevalli, M., Hursthouse, M. B., and Wilkinson, G., *J. Chem. Soc., Chem. Commun.* p. 1139 (1985).
22. Anhaus, J. T., Kee, T. P., Schofield, M. H., and Schrock, R. R., *J. Am. Chem. Soc.* **112,** 1642 (1990).
23. Arnold, J., Wilkinson, G., Hussain, B., and Hursthouse, M. B., *J. Chem. Soc., Chem. Commun.* p. 1349 (1988).

23a. Dengel, A. C., and Griffith, W. P., *Inorg. Chem.* **30,** 869 (1991).
24. Danopoulos, A. A., Wong, A. C. C., Wilkinson, G., Hursthouse, M. B., and Hussain, B., *J. Chem. Soc., Dalton Trans.* p. 315 (1990).
25. Che, C. M., Lam, M. H. W., Wang, R. J., and Mak, T. C. W., *J. Chem. Soc., Chem. Commun.* p. 820 (1990).
26. Marshman, R. W., Bigham, W. S., Wilson, S. R., and Shapley, P. A., *Organometallics* **9,** 1341 (1990).
27. Sivik, M. R., Gallucci, J. D., and Paquette, L. A., *J. Org. Chem.* **55,** 391 (1990).
28. Shapley, P. A. B., Own, Z. Y., and Huffman, J. C., *Organometallics* **5,** 1269 (1986).
29. Pearlstein, R. M., Blackburn, B. K., Davis, W. A., and Sharpless, K. B., *Angew. Chem.* **102,** 710 (1990); *Angew Chem., Int. Ed. Engl.* **29,** 639 (1990).
30. McGilligan, B. S., Arnold, J., Wilkinson, G., Hussain-Bates, B., and Hursthouse, M. B., *J. Chem. Soc., Dalton Trans.* p. 2465 (1990).
31. Longley, C. J., Savage, P. D., Wilkinson, G., Hussain, B., and Hursthouse, M. B., *Polyhedron* **7,** 1079 (1988).
32. Stravopoulos, P. G., Edwards, P. G., Behling, T., Wilkinson, G., Motevalli, M., and Hursthouse, M. B., *J. Chem. Soc., Dalton Trans.* p. 169 (1987).
33. Edwards, C. F., Griffith, W. P., and Williams, D. J., *J. Chem. Soc., Chem. Commun.* p. 1523 (1990).
34. Katti, K. V., Roesky, H. W., and Rietzel, M., *Z. Anorg. Allg. Chem.* **553,** 123 (1987).
35. Griffith, W. P., *Transition Met. Chem. (London)* **15,** 251 (1990).
36. Griffith, W. P., and White, A. D., *Proc. Indian Natl. Sci. Acad., Part A* **52,** 804 (1986).
37. Dobson, J. C., Takeuchi, K. J., Pipes, D. W., Geselowitz, D. A., and Meyer, T. J., *Inorg. Chem.* **25,** 2357 (1986).
38. Mosseri, S., Neta, P., Hambright, P., Sabry, D. Y., and Harriman, A., *J. Chem. Soc., Dalton Trans.* p. 2705 (1988).
39. Che, C., and Poon, C., *Pure Appl. Chem.* **60,** 1201 (1988).
40. Che, C. M., Chung, W. C., and Lai, T. F., *Inorg. Chem.* **27,** 2801 (1988).
41. Pipes, D. W., Bakir, M., Vitols, S. E., Hodgson, D. J., and Meyer, T. J., *J. Am. Chem. Soc.* **112,** 5507 (1990).
42. Ware, D. C., Ph.D. Thesis, Stanford University, Stanford, California (1986).
43. Murphy, W. R., Jr., Takeuchi, K., Barley, M. H., and Meyer, T. J., *Inorg. Chem.* **25,** 1041 (1986).
44. Barner, C. J., Collins, T. J., Mapes, B. E., and Santarsiero, B. D., *Inorg. Chem.* **25,** 4322 (1986).
45. Cleare, M. J., and Griffith, W. P., *J. Chem. Soc. A,* p. 1117 (1970).
46. Kim, S. H., Moyer, B. A., Azan, S., Brown, G. M., Olins, A. L., and Allison, D. P., *Inorg. Chem.* **28,** 4648 (1989).
47. Lay, P. A., and Taube, H., *Inorg. Chem.* **28,** 3561 (1989).
48. Dwyer, F. P., and Hogarth, J. H., *J. Proc. R. Soc. N.S.W.* **84,** 117 (1951).
49. Dwyer, F. P., and Hogarth, J. H., *J. Proc. R. Soc. N.S.W.* **85,** 113 (1952).
50. Watt, G. W., and Vaska, L., *J. Inorg. Nucl. Chem.* **5,** 304 (1958).
51. Watt, G. W., and Vaska, L., *J. Inorg. Nucl. Chem.* **5,** 308 (1958).
52. Watt, G. W., and Vaska, L., *J. Inorg. Nucl. Chem.* **6,** 246 (1958).
53. Watt, G. W., and Vaska, L., *J. Inorg. Nucl. Chem.* **77,** 66 (1958).
54. Allen, A. D., and Stevens, J. R., *J. Chem. Soc., Chem. Commun.* p. 1147 (1967).
55. Allen, A. D., and Stevens, J. R., *Can. J. Chem.* **50,** 3093 (1972).
56. Bottomley, F., and Tong, S. B., *Inorg. Synth.* **16,** 9 (1976).
57. Elson, C. M., Gulens, J., and Page, J. A., *Can. J. Chem.* **49,** 297 (1971).
58. Buhr, J. D., and Taube, H., *Inorg. Chem.* **18,** 2208 (1979).

59. Lay, P. A., Magnuson, R. H., and Taube, H., *Inorg. Synth.* **24,** 269 (1986).
60. Sen, J., and Taube, H., *Acta Chem. Scand., Ser. A* **A33,** 125 (1979).
61. Allen, A. D., and Stevens, J. R., *Can. J. Chem.* **51,** 92 (1973).
62. Taube, H., *Pure Appl. Chem.* **51,** 901 (1979).
63. Taube, H., *Comments Inorg. Chem.* **1,** 127 (1981).
64. Dixon, N. E., Jackson, W. G., Lancaster, M. J., Lawrance, G. A., and Sargeson, A. M., *Inorg. Chem.* **20,** 470 (1981).
65. Dixon, N. E., Lawrance, G. A., Lay, P. A., and Sargeson, A. M., *Inorg. Chem.* **23,** 2940 (1984).
66. Lay, P. A., Magnuson, R. H., Sen. J., and Taube, H., *J. Am. Chem. Soc.* **104,** 7658 (1982).
67. Lay, P. A., Magnuson, R. H., and Taube, H., *Inorg. Chem.* **28,** 3001 (1989).
68. Lay, P. A., Magnuson, R. H., and Taube, H., *Inorg. Chem.* **27,** 2848 (1988).
69. Gulens, J., and Page, J. A., *J. Electroanal. Chem., Interfacial Electrochem.* **55,** 239 (1974).
70. Elliott, M. G., and Shepherd, R. E., *Inorg. Chem.* **27,** 3332 (1988).
71. Elliott, M. G., Zhang, S., and Shepherd, R. E., *Inorg. Chem.* **28,** 3036 (1989).
72. Zhang, S., and Shepherd, R. E., *Inorg. Chim. Acta* **163,** 237 (1989).
73. Finn, M. G., and Harman, W. D., unpublished results.
74. Harman, W. D., and Taube, H., *Inorg. Chem.* **26,** 2917 (1987).
75. Harman, W. D., Sekine, M., and Taube, H., *J. Am. Chem. Soc.* **110,** 5725 (1988).
76. Scheidegger, H., Armor, J. N., and Taube, H., *J. Am. Chem. Soc.* **90,** 3263 (1968).
77. Magnuson, R. H., and Taube, H., *J. Am. Chem. Soc.* **97,** 5129 (1975).
78. Magnuson, R. H., and Taube, H., *J. Am. Chem. Soc.* **94,** 7213 (1972).
79. Magnuson, R. H., Ph.D. Thesis, Stanford University, Stanford, California (1974).
80. Lay, P. A., Magnuson, R. H., and Taube, H., *Inorg. Chem.* **27,** 2364 (1988).
81. Li, Z.-W., Lay, P. A., Taube, H., and Harman, W. D., *Inorg. Chem.,* in press.
82. Svetlov, A. A., and Sinitsyn, N. M., *Zh. Neorg. Khim.* **31,** 2902 (1986); *Russ. J. Inorg. Chem. (Engl. Transl.)* **31,** 1667. (1986).
83. Buhr, J. D., Winkler, J. R., and Taube, H., *Inorg. Chem.* **19,** 2416 (1980).
84. Lay, P. A., unpublished results.
85. Cordone, R., and Taube, H., *J. Am. Chem. Soc.* **109,** 8101 (1987).
86. Hall, J. P., and Griffith, W. P., *Inorg. Chim. Acta* **48,** 65 (1981).
87. Buhr, J. D., and Taube, H., *Inorg. Chem.* **19,** 2425 (1980).
88. Harman, W. D., Gebhard, M., and Taube, H., *Inorg. Chem.* **29,** 567 (1990).
89. Li, Z.-W., and Taube, H., work in progress.
90. Cordone, R., Ph.D. Thesis, Stanford University, Stanford, California (1988).
91. Buchler, J. W., *in* "The Porphyrins" (D. Dolphin, ed.), Vol. 1, p. 463. Academic Press, New York, 1978.
92. Buchler, J. W., Kokisch, E., and Smith, P. D., *Struct. Bonding (Berlin)* **34,** 79 (1978).
93. Che, C. M., and Poon, C. K., *Pure Appl. Chem.* **60,** 495 (1988).
94. Lay, P. A., Sargeson, A. M., and Taube, H., *Inorg. Synth.* **24,** 291 (1986).
95. Ware, D. C., Lay, P. A., and Taube, H., *Inorg. Synth.* **24,** 299 (1986).
96. Takeuchi, K. J., Thompson, M. S., Pipes, D. W., and Meyer, T. J., *Inorg. Chem.* **23,** 1845 (1984).
97. Kober, E. M., Caspar, J. V., Sullivan, B. P., and Meyer, T. J., *Inorg. Chem.* **27,** 4587 (1988).
98. Popov, A. M., and Egorova, M. B., *Zh. Obschch. Khim.* **58,** 1673 (1988); *Chem. Abstr.* **109,** 242938y (1988).
99. Dwyer, F. P., and Hogarth, J. W., *Inorg. Synth.* **5,** 206 (1957).

100. Bould, J., Greenwood, N. N., and Kennedy, J. D., *J. Organomet. Chem.* **249,** 11 (1983).
101. Bould, J., Crook, J. E., Greenwood, N. N., and Kennedy, J. D., *J. Chem. Soc., Chem. Commun.* p. 951 (1983).
102. Beckett, M. A., Greenwood, N. N., Kennedy, J. D., and Thornton-Pett, M., *J. Chem. Soc., Dalton Trans.* p. 795 (1986).
103. Elrington, M., Greenwood, N. N., Kennedy, J. D., and Thornton-Pett, M., *J. Chem. Soc., Dalton Trans.* p. 2277 (1986).
104. Brown, M., Greenwood, N. N., and Kennedy, J. D., *J. Organomet. Chem.* **309,** C67 (1986).
105. Hosmane, N. S., and Sirmokadam, N. N., *Organometallics* **3,** 1119 (1984).
106. Alcock, N., Jasztal, M. J., and Wallbridge, M. G. H., *J. Chem. Soc., Dalton Trans.* p. 2793 (1987).
107. Yam, V. W. W., and Che, C. M., *New J. Chem.* **13,** 707 (1989).
108. Weigard, W., Nagel, U., and Beck, W., *Z. Naturforsch., B: Chem. Sci.* **43,** 328 (1988).
109. Matsumura-Inoue, T., Ikemoto, I., and Umezawa, Y., *J. Electroanal. Chem. Interfacial Electrochem.* **209,** 135 (1986).
110. Afanas'ev, M. L., Kubarev, Yu. G., and Zeer, E. P., *Sovrem. Metody YaMR EPR Khim. Tverd. Tela [Mater. Vses. Koord. Soveshch.], 4th, 1985* pp. 63–65. Akad. Nauk SSR, Inst. Khim. Fiz., Chernogolovka, USSR., *Chem. Abstr.* **104,** 99844d (1986).
110a. Xu, X. L., and Hulliger, F., *J. Solid State Chem.* **80,** 120 (1989).
110b. Gentil, L. A., Nauaza, A., Olabe, J. A., and Rigoiti, G. E., *Inorg. Chim. Acta* **179,** 89 (1991).
111. Oyama, N., Ohsaka, T., Yamamoto, N., Matsui, J., and Hatozaki, O., *J. Electroanal. Chem. Interfacial Electrochem.* **265,** 297 (1989).
111a. Vogler, A., and Kunkely, H., *Inorg. Chim. Acta* **150,** 3 (1988).
112. Opekar F., and Beran, P., *Electrochim. Acta* **22,** 249 (1977).
113. Sartori, C., Preetz, W., *Z. Anorg. Allg. Chem.* **572,** 151 (1989).
114. Yam, V. W. W., and Che, C., *Coord. Chem. Rev.* **97,** 93 (1990).
115. Sartori, C., and Preetz, W., *Z. Naturforsch., A: Phys. Sci.* **43,** 239 (1988).
116. Thomas, N. C., *Coord. Chem. Rev.* **93,** 225 (1989).
117. Bruns, M., and Preetz, W., *Z. Naturforsch., B: Anorg. Chem., Org. Chem.* **41B,** 25 (1986).
118. Lay, P. A., and Taube, H., unpublished results (1982).
119. Lay, P. A., and Taube, H., *Inorg. Chem.,* to be submitted.
120. Harman, W. D., Ph.D. Thesis, Stanford University, Stanford, California (1987).
121. Moers, F. G., and Langhout, J. P., *Recl. Trav. Chim., Pays. Bas* **91,** 591 (1972).
122. Vaska, L., *J. Am. Chem. Soc.* **86,** 1943 (1964).
123. Ahmad, N., Robinson, S. D., and Uttley, M. F., *J. Chem. Soc., Dalton Trans.* p. 843 (1972).
124. Meyer, U., and Werner, H., *Chem. Ber.* **123,** 697 (1990); Werner, H., Meyer, U., Peters, K., and von Schnering, H. G., *ibid.* **122,** 2097 (1989).
125. Esteruelas, M. A., Sola, E., Oro, L. A., Werner, H., and Meyer, U., *J. Mol. Catal.* **45,** 1 (1988).
126. Harding, P. A., Robinson, S. D., and Henrick, K., *J. Chem. Soc., Dalton Trans.* p. 415 (1988).
127. Bohle, D. S., Clark, G. R., Rickard, C. E. F., and Roper, W. R., *Chem. Aust.* **54,** 293 (1987).
128. Werner, H., Esteruelas, M. A., Meyer, U., and Wrackmeyer, B., *Chem. Ber.* **120,** 11 (1987).

129. Alteparmakian, V., and Robinson, S. D., *Inorg. Chim. Acta* **116,** L37 (1986).
130. Bohle, D. S., Rickard, C. E. F., and Roper, W. R., *J. Chem. Soc., Chem. Commun.* p. 1594 (1985).
131. Bohle, D. S., and Roper, W. R., *Organometallics* **5,** 1607 (1986).
132. Bohle, D. S., Jones, T. C., Rickard, C. E. F., and Roper, W. R., *Organometallics* **5,** 1612 (1986).
133. Conway, C., Kemmitt, R. D. W., Platt, A. W. G., Russell, D. R., and Sherry, L. J. S., *J. Organomet. Chem.* **292,** 419 (1985).
134. Sanchez-Delgado, R. A., Thewalt, U., Valencia, N., Andriollo, A., Marquez-Silva, R. L., Puga, J., Schoellhorn, H., Klein, H. P., and Fontal, B., *Inorg. Chem.* **25,** 1097 (1986).
135. Sanchez-Delgado, R. A., Valencia, N., Marquez-Silva, R., Andriollo, A., and Medina, M., *Inorg. Chem.* **25,** 1106 (1986).
136. Mura, P., Olby, B. G., and Robinson, S. D., *J. Chem. Soc., Dalton Trans.* p. 2101 (1985).
137. Sanchez-Delgado, R. A., Andriollo, A., Gonzalez, E., Valencia, N., Leon, V., and Espidel, J., *J. Chem. Soc., Dalton Trans.* p. 1859 (1985).
138. Lundquist, E. G., Huffman, J. C., and Caulton, K. G., *J. Am. Chem. Soc.* **108,** 8309 (1986).
139. Buchler, J. W., and Rohbock, K., *J. Organomet. Chem.* **65,** 223 (1974).
140. Che, C.-M., Poon, C.-K., Chung, W.-C., and Gray, H. B., *Inorg. Chem.* **24,** 1277 (1985).
141. Collman, J. P., and Garner, J. M., *J. Am. Chem. Soc.* **112,** 166 (1990).
142. Kadish, K. M., *Prog. Inorg. Chem.* **34,** 435 (1986).
143. Che, C. M., and Chung, W. C., *J. Chem. Soc., Chem. Commun.* p. 386 (1986).
144. Che, C. M., Chiang, H. J., Margalit, R., and Gray, H. B., *Catal. Lett.* **1,** 51 (1988).
145. Coombe, V. T., Heath, G. A., Stephenson, T. A., Whitelock, J. D., and Yellowlees, L. J., *J. Chem. Soc., Dalton Trans.* p. 947 (1985).
146. Collins, T. J., Coots, R. J., Furutani, T. T., Keech, J. T., Peake, G. T., and Santarsiero, B. D., *J. Am. Chem. Soc.* **108,** 5333 (1986).
147. Megehee, E. G., and Meyer, T. J., *Inorg. Chem.* **28,** 4084 (1989).
148. Kober, E. M., Caspar, J. V., Lumpkin, R., and Meyer, T. J., *J. Phys. Chem.* **90,** 3722 (1986).
149. Bruce, M. R. M., Megehee, E., Sullivan, B. P., Thorp, H., O'Toole, T. R., Downard, A., and Meyer, T. J., *Organometallics* **7,** 238 (1988).
150. Herberhold, M., and Hill, A. F., *J. Organomet. Chem.* **377,** 151 (1989).
151. Clark, G. R., Marsden, K., Rickard, C. E. F., Roper, W. R., and Wright, L. J., *J. Organomet Chem.* **338,** 393 (1988).
152. Herberhold, M., Hill, A. F., McAuley, N., and Roper, W. R., *J. Organomet. Chem.* **310,** 95 (1986).
153. Herberhold, M., and Hill, A. F., *J. Organomet. Chem.* **315,** 105 (1986).
154. Zhang, N., Mann, C. M., and Shapley, P. A., *J. Am. Chem. Soc.* **110,** 6591 (1988).
155. Shapley, P. A., and Own, Z. Y., *J. Organomet. Chem.* **335,** 269 (1987).
156. Zhang, N., Wilson, S. R., and Shapley, P. A., *Organometallics* **7,** 1126 (1988).
157. Zhang, N., and Shapley, P. A., *Inorg. Chem.* **27,** 976 (1988).
158. Own, Z. Y. B., Ph.D. Thesis, University of Illinois, Chicago (1986).
159. Marshman, R. W., and Shapley, P. A., *J. Am. Chem. Soc.* **112,** 8369 (1990).
160. Arnold, J., Wilkinson, G., Hussain, B., and Hursthouse, M. B., *Organometallics* **8,** 1362 (1989).
161. Stavropolous, P., Savage, P. D., Tooze, R. P., Wilkinson, G., Hussain, B., Motevalli, M., and Hursthouse, M. B., *J. Chem. Soc., Dalton Trans.* p. 557 (1987).

162. Arnold, J., Wilkinson, G., Hussain, B., and Hursthouse, M. B., *J. Chem. Soc., Dalton Trans.* p. 2149 (1989).
163. Herberhold, M., Hill, A. F., Clark, G. R., Rickard, C. E. F., Roper, W. R., and Wright, A. H., *Organometallics* **8,** 2483 (1989).
164. Herberhold, M., and Hill, A. F., *J. Organomet. Chem.* **395,** 315 (1990).
165. Harper, T. G. P., Shinomoto, R. S., Deming, M. A., and Flood, T. C., *J. Am. Chem. Soc.* **110,** 7915 (1988).
166. Clark, G. R., Greene, T. R., and Roper, W. R., *Aust. J. Chem.* **39,** 1315 (1986).
167. Cordone, R., Harman, W. D., and Taube, H., *J. Am. Chem. Soc.* **111,** 2896 (1989).
168. Harman, W. D., Dobson, J. C., and Taube, H., *J. Am. Chem. Soc.* **111,** 3061 (1989).
169. Harman, W. D., Schaefer, W. P., and Taube, H., *J. Am. Chem. Soc.* **112,** 2682 (1990).
170. Collman, J. P., Brothers, P. J., McElwee-White, L., and Rose, E., *J. Am. Chem. Soc.* **107,** 6110 (1985).
171. Harman, W. D., Wishart, J. F., and Taube, H., *Inorg. Chem.* **28,** 2411 (1989).
172. Harman, W. D., and Taube, H., *J. Am. Chem. Soc.* **110,** 7555 (1988).
173. Harman, W. D., and Taube, H., *J. Am. Chem. Soc.* **109,** 1883 (1987).
174. Hasegawa, T., Sekine, M., Schaefer, W. P., and Taube, H., *Inorg. Chem.* **30,** 449 (1991).
174a. Taube, H., and Sekine, M., unpublished results.
175. Harman, W. D., and Taube, H., *J. Am. Chem. Soc.* **110,** 5403 (1988).
176. Sekine, M., Harman, W. D., and Taube, H., *Inorg. Chem.* **27,** 3604 (1988).
177. Harman, W. D., Sekine, M., and Taube, H., *J. Am. Chem. Soc.* **110,** 2439 (1988).
178. Harman, W. D., Hipple, W. G., and Kopach, M. E., in preparation.
179. Cordone, R., Harman, W. D., and Taube, H., *J. Am. Chem. Soc.* **111,** 5969 (1989).
180. Harman, W. D., Hasegawa, T., and Taube, H., *Inorg. Chem.* **30,** 453 (1991).
181. Harman, W. D., Fairlie, D. P., and Taube, H., *J. Am. Chem. Soc.* **108,** 8223 (1986).
181a. Esteuruelas, M. A., Valero, C., Oro, L. A., Meyer, U., and Werner, H., *Inorg. Chem.* **30,** 1159 (1991).
182. Powell, D. W., Ph.D. Thesis, University of Sydney (1990).
183. Lay, P. A., and Powell, D. W., *in* "Proceedings of the Seventh Australian Electrochemistry Conference (Electrochemistry: Current and Potential Applications)" (T. Tran, and M. Skyllas-Kazacos, eds.), pp. 237–240. Royal Australian Chemical Institute, Electrochemical Division, Sydney, 1988.
184. Powell, D. W., and Lay, P. A., *Inorg. Chem.*, submitted for publication.
185. Gieren, A., Ruiz-Perez, C., Huebner, T., Herberhold, M., and Hill, A. F., *J. Chem. Soc., Dalton Trans.* p. 1693 (1988).
186. Hanack, M., and Vermehren, P., *Inorg. Chem.* **29,** 134 (1990).
187. Amendola, P., Antoniutti, S., Albertin, G., and Bordignon, E., *Inorg. Chem.* **29,** 318 (1990).
188. Albertin, G., Antoniutti, S., and Bordignon, E., *J. Chem. Soc., Dalton Trans.* p. 2353 (1989).
189. Barratt, D. S., Glidewell, C., and Cole-Hamilton, D. J., *J. Chem. Soc., Dalton Trans.* p. 1079 (1988).
190. Clark, G. R., Rickard, C. E. F., Roper, W. R., Salter, D. M., and Wright, L. J., *Pure Appl. Chem.* **62,** 1039 (1990).
191. Holmes-Smith, R. D., Stobart, S. R., Vefghi, R., Zaworotko, M. J., Jochem, K., and Cameron, T. S., *J. Chem. Soc., Dalton Trans.* p. 969 (1987).
192. Hu, X., *Gaodeng Xuexiao Huaxue Xuebao* **11,** 1 (1990); *Chem. Abstr.* **113,** 90295d (1990).
192a. Esteruelas, M. A., Oro, L. A., and Valero, C., *Organometallics* **10,** 462 (1991).

193. Finnie, K., and Krausz, E. R., unpublished results (1989).
194. Harman, W. D., and Taube, H., *Inorg. Chem.* **27,** 3261 (1988).
195. Lay, P. A., and Taube, H., unpublished results.
196. Johnson, A., and Taube, H., *J. Indian Chem. Soc.* **66,** 503 (1989).
197. Buhr, J. D., Ph.D. Thesis, Stanford University, Stanford, California (1978).
198. Preetz, W., and Buetje, K., *Z. Anorg. Allg. Chem.* **557,** 112 (1988).
199. Gulens, J., and Page, J. A., *J. Electroanal. Chem. Interfacial Electrochem.* **67,** 215 (1976).
200. Lay, P. A., Gulyas, P., Harman, W. D., and Taube, H., in preparation.
201. Harman, W. D., and Taube, H., *J. Am. Chem. Soc.* **112,** 2261 (1990).
202. Dwyer, F. P., and Hogarth, J. W., *J. Am. Chem. Soc.* **75,** 1008 (1953).
203. Dwyer, F. P., and Hogarth, J. W., *J. Am. Chem. Soc.* **77,** 6152 (1955).
204. Lay, P. A., Sargeson, A. M., Skelton, B. W., and White, A. H., *J. Am. Chem. Soc.* **104,** 6161 (1984).
205. Lay, P. A., McLaughlin, G. M., and Sargeson, A. M., *Aust. J. Chem.* **40,** 1267 (1987); Lay, P. A., Ph.D. Thesis, Australian National University, Canberra (1981).
206. Patel, A., Ludi, A., Bürgi, H.-B., Raselli, A., and Bigler, P., *Inorg. Chem.* (submitted for publication).
207. Malin, J. M., Schlemper, E. O., and Murmann, R. K., *Inorg. Chem.* **16,** 615 (1977).
208. Griffith, W. P., McManus, N. T., and White, A. D., *J. Chem. Soc., Dalton Trans.* p. 1035 (1986).
209. Harbron, S. K., and Levason, W., *J. Chem. Soc., Dalton Trans.* p. 633 (1987).
210. Che, C.-M., and Cheng, W. K., *J. Am. Chem. Soc.* **108,** 4644 (1986).
211. Wong, K.-Y., and Anson, F. C., *J. Electronanal. Chem. Interfacial Electrochem.* **237,** 69 (1987).
211a. Yam, V. W.-W., and Che, C.-M., *J. Chem. Soc., Dalton Trans.* 3741 (1990).
212. Che, C. M., and Cheng, W.-K., *J. Chem. Soc., Chem. Commun.* p. 1519 (1986).
213. Che, C. M., Cheng, W.-K., Lai, T.-F., Poon, C.-K., and Mak, T. C. W., *Inorg. Chem.* **26,** 1678 (1987).
214. Che, C. M., Leung, W. H., and Chung, W. C., *Inorg. Chem.* **29,** 1841 (1990).
215. Nasri, H., and Scheidt, W. R., *Acta Crystallogr., Sect. C: Cryst. Struct. Commun.* **C46,** 1096 (1990).
216. Buchler, J. W., and Smith, P. D., *Chem. Ber.* **109,** 1465 (1976).
217. Che, C. M., Lai, T. F., Tong, W. F., and Marsh, R. E., unpublished results.
218. Che, C. M., Lai, T. F., Chung, W. C., Schaefer, W. P., and Gray, H. B., *Inorg. Chem.* **26,** 3907 (1987).
219. Groves, J. T., and Quinn, R., U.S. Pat. 4,822,899 A18 (1989); *Chem. Abstr.* **112,** 7351e (1990).
220. Collman, J. P., McDevitt, J. T., Yee, G. T., Leidner, C. R., McCullough, L. G., Little, W. A., and Torrance, J. B., *Proc. Natl. Acad. Sci. U.S.A.* **83,** 4581 (1986).
221. Bolt, N. J., Goodwill, K. E., and Bocian, D. F., *Inorg. Chem.* **27,** 1188 (1988).
222. Meyer, T. J., *Pure Appl. Chem.* **58,** 1193 (1986); **62,** 1003 (1990).
223. Yersin, H., Braun, D., Hensler, G., and Gallhuber, E., *NATO ASI Ser., Ser. C* **288,** 195 (1989).
224. Constable, E. C., *Adv. Inorg. Chem.* **34,** 1 (1989).
225. Constable, E. C., *Adv. Inorg. Chem. Radiochem.* **30,** 69 (1986).
226. Ware, D. W., and Taube, H., *Inorg. Chem.* (in press).
227. Garcia, M. P., López, A. M., Esteruelas, M. A., Lahoz, F. J., and Oro, L. A., *J. Chem. Soc., Dalton Trans.* p. 3465 (1990).
228. Neyhart, G. A., and Meyer, T. J., *Inorg. Chem.* **25,** 4807 (1986); Loeb, L. B., Neyhart,

G. A., Worl, L. A., Danielson, E., Sullivan, B. P., and Meyer, T. J., *J. Phys. Chem.* **93,** 717 (1989).
229. Collin, J. P., Guillerez, S., and Sauvage, J. P., *J. Chem. Soc., Chem. Commun.* p. 776 (1989).
230. Collin, J.-P., Guillerez, S., and Sauvage, J.-P., *Inorg. Chem.* **29,** 5009 (1990).
231. Della Ciana, L., Dressick, W. J., Sandrini, D., Maestri, M., and Ciano, M., *Inorg. Chem.* **29,** 2792 (1990).
232. Kalyanasundaram, K., and Nazeeruddin, M. K., *J. Chem. Soc., Dalton Trans.* p. 1657 (1990).
233. Kalyanasundaram, K., and Nazeeruddin, M. K., *Inorg. Chim. Acta* **171,** 213 (1990).
234. Cipriano, R. A., Levason, W., Mould, R. A. S., Pletcher, D., and Webster, M., *J. Chem. Soc., Dalton Trans.* p. 2609 (1990); Blake, A. J., Heath, G. A., Smith, G., Yellowlees, L. J., and Sharp, D. W. A., *Acta Crystallogr., Sect. C: Cryst. Struct. Commun.* **C44,** 1836 (1988).
235. El-Hendawy, P. A. M., Griffith, W. P., Tahu, F. I., and Moussa, M. N., *J. Chem. Soc., Dalton Trans.* p. 901 (1989).
236. Herrmann, W. A., Thiel, W. R., and Kuchler, J. G., *Chem. Ber.* **123,** 1945 (1990).
237. Figge, R., Patt-Siebel, V., Conradi, E., Mueller, U., and Dehnicke, K., *Z. Anorg. Allg. Chem.* **558,** 107 (1988).
238. Kanishcheva, A. S., Mikhailov, Yu. N., Sinitsyn, M. N., Svetlov, A. A., Kokunov, Yu. V., and Buslaev, Yu. A., *Dokl. Akad. Nauk SSSR* **308,** 381 (1989); Sinitsyn, M. N., Svetlov, A. A., Kokunov, Yu. V., Fal'kengof, A. T., Larin, G. M., Minin, V. V., and Buslaev, Yu. A., *ibid.* **293,** 1144 (1987).
239. Bobkova, E. Yu., Svetlov, A. A., Rogalevich, N. L., Novitskii, G. G., and Borkovskii, N. B., *Zh. Neorg. Khim.* **35,** 981 (1990); *Russ. J. Inorg. Chem.* (*Engl. Transl.*) **35,** 549 (1990); Salomov, A. S., Mikhailov, Yu. N., Kanishcheva, A. S., Svetlov, A. A., Sinitsyn, N. M., Porai-Koshits, M. A., and Parpiev, N. A., *Zh. Neorg. Khim.* **33,** 2608 (1988); *Russ. J. Inorg. Chem.* (*Engl. Transl.*) **33,** 1496 (1988).
240. Pandey, K. K., Sharma, R. B., and Pandit, P. K., *Inorg. Chim. Acta* **169,** 207 (1990).
241. Pandey, K. K., Ahuja, S. R., and Goyal, M., *Indian J. Chem., Sect. A* **24A,** 1059 (1985).
242. Bottomley, F., Hahn, E., Pickardt, J., Schumann, H., Mukaida, M., and Kakihana, H., *J. Chem. Soc., Dalton Trans.* p. 2427 (1985).
243. Bobkova, E. Yu., Svetlov, A. A., Rogalevich, N. L., Novitskii, G. G., Borkovskii, N. B., and Sinitsyn, M. N., *Zh. Neorg. Khim.* **35,** 979 (1990); *Russ. J. Inorg. Chem.* (*Engl. Transl.*) **35,** 546 (1990).
244. Svetlov, A. A., Sinitsyn, N. M., and Kravchenko, V. V., *Zh. Neorg. Khim.* **35,** 336 (1990); *Russ. J. Inorg. Chem.* (*Engl. Transl.*) **35,** 189 (1990).
245. Sinitsyn, M. N., Svetlov, A. A., Kanishcheva, A. S., Mikhailov, Yu. N., Sadikov, G. G., Kokunov, Yu. V., and Buslaev, Yu. A., *Zh. Neorg. Khim.* **34,** 2795 (1989); *Russ. J. Inorg. Chem.* (*Engl. Transl*) **34,** 1599 (1989).
246. Pandey, K. K., Nehete, D. T., Tewari, S. K., Rewari, S., and Bharduraj, R., *Polyhedron* **7,** 709 (1988).
247. Nikol'skii, A. B., Popov, A. M., Simonenko, N. G., Khoninzhii, V. V., and Egorova, M. B., *Zh. Obshch. Khim.* **58,** 930 (1988); *J. Gen. Chem. USSR* (*Engl. Transl.*) **58,** 825 (1988).
248. Herberhold, M., and Hill, A. F., *J. Organomet. Chem.* **363,** 371 (1989).
248a. Mingos, D. M. P., Sherman, D. J., and Bott, S. G., *Transition Met. Chem.* (*London*) **12,** 471 (1987).
249. Stershic, M. T., Keefer, L. K., Sullivan, B. P., and Meyer, T. J., *J. Am. Chem. Soc.* **110,** 6884 (1988).

250. Vorob'eva, N. E., Pavlov, P. T., and Zhivopistsev, V. P., *Izv. Vyssh. Uchebn. Zaved., Khim. Khim. Tekhnol.* **32**, 33 (1989); *Chem. Abstr.* **112**, 234900b (1990).
251. Gowda, H. S., Ahmed, S. M., and Raj, J. B., *Indian J. Chem., Sect. A* **29A**, 95 (1990).
252. Vorob'eva, N. E., Zhivopistsev, V. P., and Pavlov, P. T., *Zh. Anal. Khim.* **44**, 467 (1989); *Chem. Abstr.* **112**, 29922u (1990).
253. Vorob'eva, N. E., Zhivopistsev, V. P., Zhdanova, T. N., Pavlov, P. T., Veretennikova, O. V., and Berdinskii, I. S., *Otkrytiy, Izobret., Prom. Obraztsy, Tovarnye Znaki* p. 159 (1987); *Chem. Abstr.* **108**, 160580r (1988).
254. Cruz-Garritz, D., Gelover, S., Torrens, H., Leal, J., and Richards, R. L., *J. Chem. Soc., Dalton Trans.* p. 2393 (1988).
255. Armstrong, J. E., and Walton, R. A., *Inorg. Chem.* **22**, 1545 (1983).
256. Lay, P. A., and Sargeson, A. M., unpublished results.
257. Fenske, D., Baum, G., Swidersky, H. W., and Dehnicke, K., *Z. Naturforsch., B: Chem. Sci.* **45**, 1210 (1990).
258. Sacksteder, L. A., Demas, J. N., and DeGraff, B. A., *Inorg. Chem.* **28**, 1787 (1989).
259. Perkins, T. A., Pourreau, D. B., Netzel, T. L., and Schanze, K. S., *J. Phys. Chem.* **93**, 4511 (1989).
260. Popov, A. M., Egorova, M. B., Khorunzhii, V. V., and Drobachenko, A. V., *Zh. Neorg. Khim.* **33**, 2319 (1988); *Russ. J. Inorg. Chem. (Engl. Transl.)* **33**, 1324 (1988).
261. Bohle, D. S., Clark, G. R., Rickard, C. E. F., Roper, W. R., and Taylor, M. J., *J. Organomet. Chem.* **348**, 385 (1988).
262. Kravchenko, E. A., Burstev, M. Yu., Morgunov, V. G., Sinitsyn, M. N., Kokunov, Yu. V., and Buslaev, Y. A., *Koord. Khim.* **13**, 1520 (1987).
263. Kravchenko, E. A., Burstev, M. Yu., Sinitsyn, M. N., Svetlov, A. A., Kokunov, Yu. V., and Buslaev, Yu. A., *Dokl. Akad. Nauk SSSR* **294**, 130 (1987).
264. Droege, M. W., Harman, W. D., and Taube, H., *Inorg. Chem.* **26**, 1309 (1987).
265. Danopoulos, A. A., and Wilkinson, G., *Polyhedron* **9**, 1009 (1990).
265a. Danopoulos, A. A., Wilkinson, G., Hussain-Bates, B., and Hursthouse, M. B., *J. Chem. Soc., Dalton Trans.* 269 (1991).
266. Che, C.-M., Lam, M. H.-W., and Mak, T. C. W., *J. Chem. Soc., Chem. Commun.* p. 1529 (1989).
267. Nath, N., and Singh, L. P., *Rev. Roum. Chim.* **31**, 489 (1986).
268. Nefedov, V. I., Trishkina, E. M., Sinitsyn, M. N., Svetlov, A. A., Kokunov, Yu. V., and Buslaev, Yu. A., *Dokl. Akad. Nauk SSSR* **291**, 614 (1986).
269. Che, C. M., Lam, H. W., Tong, W. F., Lai, T. F., and Lau, T. C., *J. Chem. Soc., Chem. Commun.* p. 1883 (1989).
270. Che, C. M., Lau, T. C., Lam, H. W., and Poon, C. K., *J. Chem. Soc., Chem. Commun.* p. 114 (1989).
271. Ta, N. C., *Synth. React. Inorg. Met.-Org. Chem.* **16**, 1357 (1986).
272. Sartori, C., and Preetz, W., *Z. Anorg. Allg. Chem.* **565**, 23 (1988).
273. Cockman, R. W., and Peacock, R. D., *J. Flourine Chem.* **30**, 469 (1986).
274. Buetje, K., and Preetz, W., *Z. Naturforsch., B: Chem. Sci.* **43**, 371 (1988).
275. Buetje, K., and Preetz, W., *Z. Naturforsch., B: Chem. Sci.* **43**, 382 (1988).
276. Bhattacharyya, R., and Saha, A. M., *Transition Met. Chem. (London)* **12**, 85 (1987).
277. Bütje, K., and Preetz, W., *Z. Naturforsch., B: Chem. Sci.* **43**, 574 (1988).
278. Kukushkin, V. Yu., Egorova, M. B., and Popov, A. M., *Koord. Khim.* **13**, 1507 (1987).
279. Sinitsyn, N. M., Kokunova, V. N., and Svetlov, A. A., *Zh. Neorg. Khim.* **33**, 2340 (1988); *Russ. J. Inorg. Chem. (Engl. Transl.)* **33**, 1336 (1988).
280. Che, C.-M., Cheng, W.-K., and Mak, T. C. W., *Inorg. Chem.* **27**, 250 (1988).
281. Che, C.-M., Cheng, W.-K., and Mak, T. C. W., *Inorg. Chem.* **25**, 703 (1986).
281a. Lynch, W. E., Lintvedt, R. L., and Shui, X. O., *Inorg. Chem.* **30**, 1014 (1991).

282. Lahiri, G. K., Bhattacharya, S., Ghosh, B. K., and Chakravorty, A., *Inorg. Chem.* **26,** 4324 (1987).
283. Che, C.-M., Cheng, W.-K., and Mak, T. C. W., *J. Chem. Soc., Chem. Commun.* p. 200 (1986).
284. Anson, F. C., Collins, T. J., Gipson, S. L., Keech, J. T., and Krafft, T. E., *Inorg. Chem.* **26,** 1157 (1987).
285. Collins, T. J., and Keech, J. T., *J. Am. Chem. Soc.* **110,** 1162 (1988).
286. Collins, T. J., Lai, T., and Peake, G. T., *Inorg. Chem.* **26,** 1674 (1987).
287. Anson, F. C., Collins, T. J., Coots, R. J., Gipson, S. L., Krafft, T. E., Santarsiero, B. D., and Spies, G. H., *Inorg. Chem.* **26,** 1161 (1987).
288. Anson, F. C., Collins, T. J., Gipson, S. L., Keech, J. T., Krafft, T. E., and Peake, G. T., *J. Am. Chem. Soc.* **108,** 6593 (1986).
289. Anson, F. C., Christie, J. A., Collins, T. J., Coots, R. J., Furutani, T. T., Gipson, S. L., Keech, J. T., Krafft, T. E., Santarsiero, B. D., and Spies, G. H., *J. Am. Chem. Soc.* **106,** 4460 (1984).
290. Christie, J. A., Collins, T. J., Krafft, T. E., Santarsiero, B. D., and Spies, G. H., *J. Chem. Soc., Chem. Commun.* p. 198 (1984).
291. Muller, J. G., and Takeuchi, K. J., *Inorg. Chem.* **26,** 3634 (1987).
292. Olby, B. G., Robinson, S. D., Hursthouse, M. B., and Short, R. C., *Polyhedron* **7,** 1781 (1988).
293. Colson, S. F., and Robinson, S. D., *Polyhedron* **9,** 1737 (1990); *Inorg. Chim. Acta* **149,** 13 (1988).
294. Mahapatra, A. K., Ghosh, B. K., Goswami, S., and Chakravorty, A., *J. Indian Chem. Soc.* **63,** 101 (1986).
295. Mukhopadyay, A., and Ray, S., *Acta Crystallogr., Sect. C: Cryst. Struct. Commun.* **C43,** 14 (1987).
296. Ghosh, B. K., Mukherjee, R., and Chakravorty, A., *Inorg. Chem.* **26,** 1946 (1987).
297. Desrosiers, P. J., Shinomoto, R. S., Deming, M. A., and Flood, T. C., *Organometallics* **8,** 2861 (1989).
298. Gotzig, J., Werner, R., and Werner, H., *J. Organomet. Chem.* **290,** 99 (1985).
299. Cappellani, E. P., Maltby, P. A., Morris, R. H., Schweitzer, C. T., and Steele, M. R., *Inorg. Chem.* **28,** 4437 (1989).
300. Mezzetti, A., Del Zotto, A., and Rigo, P., *J. Chem. Soc., Dalton Trans.* p. 2515 (1990).
301. Bautista, M., Earl, K. A., Maltby, P. A., and Morris, R. H., *J. Am. Chem. Soc.* **110,** 4056 (1988).
302. Fanwick, P. E., Fraser, I. F., Tetrick, S. M., and Walton, R. A., *Inorg. Chem.* **26,** 3786 (1987).
302a. Cipriano, R. A., Levason, W., Mould, R. A. S., Pletcher, D., and Webster, M., *J. Chem. Soc., Dalton Trans.* p. 339 (1990).
303. Cotton, F. A., Diebold, M. P., and Matusz, M., *Polyhedron* **6,** 1131 (1987).
304. Robinson, P. D., Hinckley, C. C., and Ikuo, A., *Acta Crystallogr., Sect. C: Cryst. Struct. Commun.* **C44,** 1491 (1988).
305. Bressan, M., and Morvillo, A. *Stud. Org. Chem.* **33,** 277 (1988).
306. Desrosiers, P. J., Shinomoto, R. S., and Flood, T. C., *J. Am. Chem. Soc.* **108,** 7964 (1986).
307. Bruno, J., Huffman, J. C., Green, M. A., Zubkowski, J. D., Hatfield, W. E., and Caulton, K.G., *Organometallics* **9,** 2556 (1990).
308. Siedle, A. R., Newmark, R. A., and Pignolet, L. H., *Inorg. Chem.* **25,** 3412 (1986).
309. Howard, J. A. K., Johnson, O., Koetzle, T. F., and Spencer, J. L., *Inorg. Chem.* **26,** 2930 (1987).

310. Shinomoto, R. S., Desrosiers, P. J., Harper, T. G. P., and Flood, T. C., *J. Am. Chem. Soc.* **112,** 704 (1990).

310a. Desrosiers, P. J., Shinomoto, R. S., and Flood, T.C., *J. Am. Chem. Soc.* **108,** 1346 (1986).

311. Hinckley, C. C., Matusz, M., and Robinson, P. D. *Acta Crystallogr., Sect. C: Cryst. Struct. Commun.* **C44,** 371 (1988).

312. Otruba, J. P., Neyhart, G. A., Dressick, W. J., Marshall, J. L., Sullivan, B. P., Watkins, P. A., and Meyer, T. J., *J. Photochem.* **35,** 133 (1986).

313. Hinckley, C. C., Matusz, M., and Robinson, P. D. *Acta Crystallogr., Sect. C: Cryst. Struct. Commun.* **C44,** 1829 (1988).

314. Lay, P. A., Ludi, A., and Taube, H., work in progress.

315. Griffith, W. P., *J. Chem. Soc. A* p. 211 (1969).

316. Scagliarini, G., and Masetti-Zannini, A., *Gazz. Chim. Ital.* **53,** 504 (1923).

317. Svetlov, A. A., Sinitsyn, M. M., and Kravchenko, V. V., *Zh. Neorg. Khim.* **34,** 953 (1989); *Russ J. Inorg. Chem. (Engl. Transl.)* **34,** 535 (1989).

318. Vining, W. J., and Meyer, T. J., *Inorg. Chem.* **25,** 2023 (1986).

319. Cotton, F. A., Dunbar, K. R., and Matusz, M., *Inorg. Chem* **25,** 1589 (1986).

320. Bardin, M. B., Goncharenko, V. P., and Ketrush, P. M., *Zh. Anal. Khim.* **42,** 2013 (1987).

321. Lay, P. A., and Sasse, W. H. F., *Inorg. Chem.* **24,** 4707 (1985).

322. Cotton, F. A., Dunbar, K. R., and Matusz, M., *Inorg. Chem.* **25,** 1585 (1986).

323. Moskvin, L. N., and Shmatko, A. G., *Zh. Neorg. Khim.* **33,** 1229 (1988); *Russ. J. Inorg. Chem. (Engl. Transl.)* **33,** 695 (1988).

324. Singh, A. K., Tewari, A., and Sisodia, A. K., *Natl. Acad. Sci. Lett. (India)* **8,** 209 (1985).

325. Pipes, D. W., and Meyer, T. J., *Inorg. Chem.* **25,** 4042 (1986).

326. Filina, G. G., Shcheglov, O. F., and Chebotarev, O. V., *Zh. Vses. Khim. O-va.* **31**(5), 119 (1986); *Mendcleev Chem. J. (Engl. Transl.)* **31**(4), 67 (1986).

327. Nakajima, M., Tomioka, K., and Koga, K., *Yuki Gosei Kagaku Kyokaishi* **47,** 878 (1989); *Chem Abstr.* **112,** 138295n (1990).

328. Lohray, B. B., Kalantar, T. H., Kim, B. M., Park, C. Y., Shibata, T., Wai, J. S. M., and Sharpless, K. B., *Tetrahedrom Lett.* **30,** 2041 (1989).

329. Griffith, W. P., *Platinum Met. Rev.* **33,** 181 (1989).

330. Nakajima, M., Tomioka, K, and Koga, K., *Kagaku (Kyoto)* **42,** 422 (1987); *Chem. Abstr.* **108,** 55141a (1988).

331. Kobs, S. F., Ph.D. Thesis, Ohio State University, Columbus (1986).

332. Singh, H. S., *in* "Organic Synthesis by Oxidation with Metal Compounds" (W. J. Mijs and C. R. H. I. de Jonge, eds.), p. 633. Plenum Press, New York, 1986.

333. Ruegger, U. P., and Tassera, J., *Swiss Chem.* **8,** 43 (1986).

334. Svendsen, J. S., Marko, I., Jacobsen, E. N., Rao, C. P., Bott, S., and Sharpless, K. B., *J. Org. Chem.* **54,** 2263 (1989).

335. Kobs, S. F., and Behrman, E. J., *Inorg. Chim. Acta* **138,** 113 (1987).

336. Kobs, S. F., and Behrman, E. J., *Inorg. Chim. Acta* **128,** 21 (1987).

337. Peschke, T., Wollweber, L., Gabert, A., Augsten, K., and Stracke, R., *Histochemistry* **93,** 443 (1990).

338. Vandeputte, D. F., Jacob, W. A., and Van Grieken, R. E., *J. Histochem. Cytochem.* **38,** 331 (1990).

339. Anderson, L. J., Boyles, J. K., and Hussain, M. M., *J. Lipid Res.* **30,** 1819 (1989).

340. Nandi, K. N., Beal, J. A., and Knight, D. S., *J. Neurosci. Methods* **25,** 159 (1988).

341. Lascano, E. F., and Berria, M. I., *J. Histochem. Cytochem.* **36,** 697 (1988).

342. Hearn, S. A., *J. Histochem. Cytochem.* **35,** 795 (1987).
343. Caceci, T., and Frankum, K. E., *J. Microsc.* (*Oxford*) **147,** 109 (1987).
344. Van Dort, J. B., Zeelen, J. P., and De Bruijn, W. C., *Histochemistry* **87,** 71 (1987).
345. Krause, C. R., Ichida, J. M., and Dochinger, L. S., *Scanning Electron Microsc.* p. 975 (1986).
346. Tetzlaff, W., and Hofbauer, A., *Histochemistry* **85,** 295 (1986).
347. Derenzini, M., Farabegoli, F., and Marinozzi, V., *J. Histochem. Cytochem* **34,** 1161 (1986).
348. Hosoda, S., and Kojima, K., *Polym. Commun.* **30,** 83 (1989).
349. Kakugo, M., Sadatoshi, H., Yokoyama, M., and Kojima, K., *Macromolecules* **22,** 547 (1989).
350. Friedrich, J., Pohl, M., Elsner, T., and Altrichter, B., *Acta Polym.* **39,** 594 (1988).
351. Palecek, E., Boublikova, P., and Nejedley, K., *Biophys. Chem.* **34,** 63 (1989).
352. Boublikova, P., *Stud. Biophys.* **131,** 185 (1989).
353. Campbell, R. D., and Cotton, R. G. H., Eur. Pat. Appl. EP 329311 A2 23 (1989); *Chem. Abstr.* **112,** 213550x (1990).
354. Hara, T., and Tsukagoshi, K., *Bull. Chem. Soc. Jpn.* **60,** 2031 (1987).
355. Bakhvalova, I. P., Volkova, G. V., and Ivanov, V. M., *Zh. Anal. Khim.* **45,** 496 (1990); *Chem. Abstr.* **113,** 34049q (1990).
356. Csanyi, L. J., Galbacs, Z. M., Nagy, L., and Horváth, I., *Acta Chim. Hung.* **123,** 123 (1986).
357. Csanyi, L. J., Galbacs, Z. M., Nagy, L., and Marek, N., *Transition Met. Chem.* (*Weinheim, Ger.*) **11,** 319 (1986).
358. Singh, V., *J. Inst. Chem.* (*India*) **59,** 248 (1987).
358a. Sattar, S., and Kustin, K. *Inorg. Chem.* **30,** 1668 (1991).
359. Sarala, G., Rao, P. J., Sethuram, B., and Rao, T. N., *Transition Met. Chem.* (*London*) **13,** 113 (1988).
360. Khomutova, E. G., and Khvorostukhina, N. A., *Zh. Anal. Khim.* **41,** 1647 (1986).
361. Jørgensen, K. A., and Hoffmann, R., *J. Am. Chem. Soc.* **108,** 1867 (1986).
362. Shul'man, R. S., Marochkina, L. Ya., Torgov, V. G., Drozdova, M. K., Glukhikh, L. K., and Yatsenko, V. T., *Zh. Neorg. Khim.* **33,** 166 (1988); *Russ. J. Inorg. Chem.* (*Engl. Transl.*) **33,** 91 (1988).
363. Veerasomaiah, P., Reddy, K., Sethuram, B., and Rao, T. N., *J. Indian Chem. Soc.* **66,** 755 (1989).
364. Somaiah, P. V., Reddy, K. B., Sethuram, B., and Rao, T. N., *Indian J. Chem., Sect. A* **27A,** 876 (1988).
365. Singh, A. K., Saxena, S., Saxena, M., Gupta, R., and Mishra, R. K., *Indian J. Chem., Sect. A* **27A** 438 (1988).
366. Singh, B., Singh, A. K., Singh, M. B., and Singh, A. P., *Tetrahedron* **42,** 715 (1986).
367. Singh, B., Singh, A. K., and Singh, A. P., *Natl. Acad. Sci. Lett.* (*India*) **8,** 111 (1985).
368. Singh, A. K., Sisodia, A. K., Saxena, S., and Saxena, M., *Indian J. Chem., Sect. A* **26A,** 600 (1987).
369. Singh, M. B., Singh, A. P., Singh, A. K., and Singh, B., *Natl. Acad. Sci. Lett.* (*India*) **9,** 175 (1986).
370. Singh, A. K., Parmar, A., and Sisodia, A. K., *Natl. Acad. Sci. Lett.* (*India*) **9,** 273 (1986).
371. Misra, R. K., Saxena, S., and Singh, A. K., *Viznana. Parishad Anusandhan. Patrika* **31,** 83 (1989); *Chem. Abstr.* **212,** 118038c (1990).
372. Rao, P. J. P., Sethuram, B., and Rao, T. N., *J. Indian Chem. Soc.* **67,** 101 (1990).
373. Sarala, G., Rao, P. J. P., Sethuram, B., and Rao, T. N., *Indian J. Chem., Sect. A* **26A,** 475 (1987).

374. Swarnalakshmi, N., Uma, V., Sethuram, B., and Rao, T. N., *Indian J. Chem., Sect. A* **26A,** 592 (1987).
375. Naidu, H. M. K., Yamuna, B., and Mahadevappa, D. S., *Indian J. Chem., Sect. A* **26A,** 114 (1987).
376. Hingorani, H., Goyal, H. C., Chandra, G., and Srivastava, S. N., *Rev. Roum. Chim.* **30,** 725 (1985).
377. Mythily, C. K., Rangappa, K. S., and Mahadevappa, D. S., *Indian J. Chem., Sect. A* **29A,** 676 (1990).
377a. Hussain, J., Mishra, S. K., and Sharma, P. D., *J. Chem. Soc., Dalton Trans.* p. 89 (1991).
378. Mangalam, G., and Meenakshisundaram, S., *J. Indian Chem. Soc.* **66,** 368 (1989).
379. Rao, K. L. V., Rao, M. P., Sethuram, B., and Rao, T. N., *Transition Met. Chem. (London)* **14,** 165 (1989).
380. Reddy, K. B., Sethuram, B., and Rao, T. N., *Z. Phys. Chem. (Leipzig)* **270,** 732 (1989).
381. Volkova, T. Ya., Rozovskii, G. I., and Sakharov, A. A., *Liet. TSR Mokslu Akad. Darb., Ser. B* p. 3 (1985); *Chem. Abstr.* **104,** 206679c (1986).
382. Khetawat, G. K., and Menghani, G.D., *J. Indian Chem. Soc.* **64,** 766 (1987).
383. Shukla, R. K., and Anandam, N., *Acta Cienc. Indica [Ser.] Chem.* **11,** 242 (1985).
384. Mishra, P., and Khandual, N. C., *Indian J. Chem., Sect. A* **25A,** 902 (1986).
385. Mishra, P., Mohapatra, R.C., and Khandual, N. C., *J. Indian Chem. Soc.* **63,** 291 (1986).
386. Radhakrishnamurti, P. S., and Tripathy, K. S., *Indian J. Chem., Sect. A* **25A,** 762 (1986).
387. Saroja, P., Rao, M. P., and Kandlikar, S., *Transition Met. Chem. (London)* **15,** 351 (1990).
388. Radhakrishnamurti, P. S., Panda, R. K., and Panigrahi, J. C., *Indian J. Chem., Sect. A* **26A,** 124 (1987).
389. Hamilton, J. G., Mackey, O. N., Rooney, J. J., and Gilheany, D. G., *J. Chem. Soc., Chem. Commun.* p. 1600 (1990).
390. Erdik, E., and Matteson, D. S., *J. Org. Chem.* **54,** 2742 (1989).
391. Wai, J. S. M., Marko, I., Svendsen, J. S., Finn, M. G., Jacobsen, E. N., and Sharpless, K. B., *J. Am. Chem. Soc.* **111,** 1123 (1989).
392. Wallis, J. M., and Kochi, J. K., *J. Am. Chem. Soc.* **110,** 8207 (1988).
393. Corey, E. J., Jardine, P.D., Virgil, S., Yuen, P. W., and Connell, R. D., *J. Am. Chem. Soc.* **111,** 9243 (1989).
394. Jacobsen, E. N., Marko, I., France, M. B., Svendsen, J. S., and Sharpless, K. B., *J. Am. Chem. Soc.* **111,** 737 (1989).
395. Dijkstra, G. D. H., Kellogg, R. M., Wynberg, H., Svendsen, J. S., Marko, I., and Sharpless, K. B., *J. Am. Chem. Soc.* **111,** 8069 (1989).
396. Gupta, S., Ali, V., and Upadhyay, S. K., *Int. J. Chem. Kinet.* **21,** 315 (1989).
397. Gupta, S., Ali, V., and Upadhyay, S. K. *Transition Met. Chem.* (London) **13,** 573 (1988).
398. Dengel, A. C., El-Hendawy, A. M., Griffith, W. P., and White, A. D., *Transition Met. Chem. (London)* **14,** 230 (1989).
399. Shaplygin, I. S., and Lazarev, V. B., *Zh. Neorg. Khim.* **31,** 3181 (1986); *Russ. J. Inorg. Chem. (Engl. Transl.)* **31,** 1827 (1986).
400. Hagen, K., Hobson, R. J., Holwill, C. J., and Rice, D. A., *Inorg. Chem.* **25,** 3659 (1986).
401. Robinson, P. D., Hinckley, C. C., and Kibala, P. A., *Acta Crystallogr., Sect. C: Cryst. Struct. Commun.* **C44,** 1365 (1988).

402. Hinckley, C. C., Kibala, P. A., and Robinson, P. D., *Acta Crystallogr., Sect. C: Cryst. Struct. Commun.* **C43,** 842 (1987).
403. Harbron, S. K., Jewiss, H. C., Levason, W., and Webster, M., *Acta Crystallogr., Sect. C: Cryst. Struct. Commun.* **C43,** 37 (1987).
404. Oishi, T., and Hirama, M., *J. Org. Chem.* **54,** 5834 (1989).
405. Arnold, J., Wilkinson, G., Hussain, B., and Hursthouse, M. B., *Polyhedron* **8,** 597 (1989).
406. Hinckley, C. C., Ali, I. A., and Robinson, P. D., *Acta Crystallogr., Sect. C: Cryst. Struct. Commun.* **C46,** 697 (1990).
407. El-Hendawy, A. M., *Polyhedron* **9,** 2309 (1990).
408. Hinckley, C. C., and Kibala, P. A., *Polyhedron* **5,** 1119 (1986).
409. Bhattacharya, S., Boone, S. R., Fox, G. A., and Pierpont, C. G., *J. Am. Chem. Soc.* **112,** 1088 (1990).
410. Griffith, W. P., Pumphrey, C. A., and Skapski, A. C., *Polyhedron* **6,** 891 (1987).
411. Haga, M., Isobe, K., Boone, S. R., and Pierpont, C. G., *Inorg. Chem.* **29,** 3795 (1990).
412. El-Hendawy, A. M., Griffith, W. P., and Pumphrey, C. A., *J. Chem. Soc., Dalton Trans.* p. 1817 (1988).
413. Greaves, S. J., and Griffith, W. P., *Polyhedron* **7,** 1973 (1988).
414. El-Hendawy, A. M., Griffith, W. P., O'Mahoney, C. A., and Williams, D. J., *Polyhedron* **8,** 519 (1989).
414a. El-Hendawy, A. M. *Inorg. Chim. Acta* **179,** 223 (1991).
415. Dobson, J. C., and Meyer, T. J., *Inorg. Chem.* **28,** 2013 (1989).
416. Colson, S. F., and Robinson, S. D., *Polyhedron* **8,** 2179 (1989).
417. Queiros, M. A. M., and Robinson, S. D., *Inorg. Chem.* **17,** 310 (1978).
418. Deeming, A. J., Randle, N. P., Hursthouse, M. B., and Short, R. L., *J. Chem. Soc., Dalton Trans.* p. 2473 (1987).
419. Andriollo, A., Esteruelas, M., A., Meyer, U., Oro, L. A., Sanchez-Delgado, R. A., Sola, E., Valero, C., and Werner, H., *J. Am. Chem. Soc.* **111,** 7431 (1989).
420. Esteruelas, M. A., Sola, E., Oro, L. A., Meyer, U., and Werner, H., *Angew. Chem.* **100,** 1621 (1988); *Angew. Chem., Int. Ed. Engl.* **27,** 1563 (1988).
421. Hinckley, C. C., Ikuo, A., and Robinson, P. D., *Acta Crystallogr., Sect. C: Cryst. Struct. Commun.* **C44,** 1827 (1988).
422. Hinckley, C. C., Matusz, M., Kibala, P. A., and Robinson, P. D., *Acta Crystallogr., Sect. C: Cryst. Struct. Commun.* **C43,** 1880 (1987).
423. Kubas, G. J., *Acc. Chem. Res.* **21,** 120 (1988).
424. Dobson, A., Moore, D. S., Robinson, S. D., Hursthouse, M. B., and New, L., *Polyhedron* **4,** 1119 (1985).
425. Packard, A. B., O'Brien, G. M., and Treves, S., *Nucl. Med. Biol.* **13,** 519 (1986).
426. Lay, P. A., Magnuson, R. H., Taube, H., Ferguson, J., and Krausz, E. R., *J. Am. Chem. Soc.* **107,** 2551 (1985).
427. Isied, S. S., *Prog. Inorg. Chem.* **32,** 443 (1984).
428. Dubicki, L., Ferguson, J., Krausz, E. R., Lay, P. A., Maeder, M., and Taube, H., *J. Phys. Chem.* **88,** 3940 (1984).
429. Vassilian, A., Wishart, J. F., van Hemelryck, B., Schwartz, H., and Isied, S. S., *J. Am. Chem. Soc.* **112,** 7278 (1990).
430. Harman, W. D., and Myers, W. H., *J. Am. Chem. Soc.* (in press).
431. Kopach, M. E., Gonzalez, J., and Harman, W. D. Submitted.
432. Harman, W. D., and Taube, H., *J. Am. Chem. Soc.***110,** 7906 (1988).
433. Harding, P. A., Preece, M., Robinson, S. D., and Henrick, K., *Inorg. Chim Acta* **118,** L31 (1986).
434. Herberhold, M., and Hill, A. F., *J. Organomet. Chem.* **353,** 243 (1988).

435. Graziosi, G., Preti, C., Tosi, G., and Zannini, P., *Aust. J. Chem.* **38,** 1675 (1985).
436. Svetlov, A. A., Sinitsyn, N. M., and Kravchenko, V. V., *Zh. Neorg. Khim.* **33,** 1220 (1988); *Russ. J. Inorg. Chem. (Engl. Transl.)* **33,** 690 (1988).
437. Ali, R., Higgins, S. J., and Levason, W., *Inorg. Chim. Acta* **84,** 65 (1984).
438. Hope, E. G., Levason, W., Murray, S. G., and Marshall, G. L., *J. Chem. Soc., Dalton Trans.* p. 2185 (1985).
439. Hope, E. G., Levason, W., Webster, M., and Murray, S. G., *J. Chem. Soc., Dalton Trans.* p. 1003 (1986).
440. Blake, A. J., and Schröder, M., *Adv. Inorg. Chem.* **35,** 1 (1990).
441. Tweedle, M. F., and Taube, H., *Inorg. Chem.* **21,** 3361 (1982).
442. Catala, R. M., Cruz-Garritz, D., Hills, A., Hughes, D. L., Richards, R. L., Sosa, P., and Torrens, H., *J. Chem. Soc., Chem. Commun.* p. 261 (1987).
443. Cruz-Garritz, D., Sosa, P., Torrens, H., Hills, A., Hughes, D. L., and Richards, R. L., *J. Chem. Soc., Dalton Trans.* p. 419 (1989).
444. Cruz-Garritz, D., Torrens, H., Leal, J., and Richards, R. L., *Transition Met. Chem. (Weinheim Ger.)* **8,** 127 (1983).
445. Utegulov, R. N., Kamysbaev, D. Kh., Ospanov, Kh. K., and Kozlovskii, E. V., *Koord. Khim.* **14,** 1529 (1988).
446. Herberhold, M., and Hill, A. F., *J. Chem. Soc., Dalton Trans.* p. 2027 (1988).
447. Pekhn'o, V. I., Fokina, Z. A., and Volkov, S. V., *Zh. Neorg. Khim.* **33,** 1214 (1988); *Russ. J. Inorg. Chem. (Engl. Transl.)* **33,** 687 (1988).
448. Angelici, R. J., *Coord. Chem. Rev.* **105,** 61 (1990).
449. Bol'shakov, K. A., Sinitsyn, N. M., Pichkov, V. N., Gribanovskaya, M. G., and Rudnitskaya, O. V., *Zh. Neorg. Khim.* **31,** 720 (1986); *Russ. J. Inorg. Chem. (Engl. Transl.)* **31,** 410 (1986).
450. Rudnitskaya, O. V., Lin'ko, I. V., Ivanova, T. M., Pichkov, V. N., Gribanovskaya, M. G., and Linko, R. V. *Zh. Neorg. Khim.* **31,** 3173 (1986); *Russ. J. Inorg. Chem. (Engl. Transl.)* **31,** 1822 (1986).
451. Gribanovskaya, M. G., Rudnitskaya, O. V., Pichkov, V. N., Lipnitskii, I. V., and Nobitskii, G. G., *Koord. Khim.* **15,** 991 (1989).
452. Morelli, B., *Analyst (London)* **111,** 1289 (1986).
453. Morelli, B., *Anal. Lett.* **18,** 2453 (1985).
454. Calderazzo, F., Pampaloni, G., Vitali, D., Collamati, I., Dessy, G., and Fares, V. *J. Chem. Soc., Dalton Trans.* 1965 (1980).
455. Antonov, P. G., Kukushkin, Yu. N., Konnov, V. I., and Kostikov, Yu. P., *Koord. Khim.* **6,** 1585 (1980).
456. Antonov, P. G., and Amantova, I. A., *Zh. Obshch. Khim* **58,** 2523 (1988); *J. Gen. Chem. USSR (Engl. Transl.)* **58,** 2245 (1988).
457. Robinson, P. D., Hinckley, C. C., and Ikuo, A., *Acta Crystallogr., Sect. C: Cryst. Struct. Commun.* **C45,** 1079 (1989).
458. Cabeza, J. A., Adams, H., and Smith, A. J., *Inorg. Chim. Acta* **114,** L17 (1986).
459. Cabeza, J. A., and Maitlis, P. M., *J. Chem. Soc., Dalton Trans.* p. 573 (1985).
460. Kinsenyi, J. M., Cabeza, J. A., Smith, A. J., Adams, H., Sunley, G. J., Salt, N. J. S., and Maitlis, P. M., *J. Chem. Soc., Chem. Commun.* p 770 (1985).
461. Kotov, V. Yu., Nikol'skii, A. B., and Popov, A. M., *Vestn. Leningr. Univ., Fiz., Khim.* p. 39 (1985); *Chem. Abstr.* **104,** 80859r (1986).
462. Droege, M. W., and Taube, H., *Inorg. Chem.* **26,** 3316 (1987).
463. Bond, A. M., and Martin, R. L., *Coord. Chem. Rev.* **54,** 23 (1984).
464. Cole-Hamilton, D. J., and Stephenson, T. A., *J. Chem. Soc., Dalton Trans.* p. 2396 (1976).
465. Craciunescu, D. G., Parrondo-Iglesias, E., Alonso, M. P., Molina, C., Doadrio-Lopez,

A., Gomez, A., Mosquerra, R. M., Ghirvu, C., and Gaston de Iriarte, E., *An. R. Acad. Farm.* **54,** 16 (1988).

465a. Pramanik, A., Bag, N., Ray, D., Lahiri, G. K., and Chakravorty, A., *Inorg. Chem.* **30,** 410 (1991).

465b. Pramanik, A., Bag, N., Ray, D. Lahiri, G. K., and Chakravorty, A., *J. Chem. Soc., Chem. Commun.* p. 139 (1991).

466. Mukherjee, R. M., and Shastri, B. B. S., *Indian J. Chem., Sect. A* **29A,** 809 (1990).

467. McQueen, E. W. D., Blake, A. J., Stephenson, T. A., Schröder, M., and Yellowlees, L. J., *J. Chem. Soc., Chem. Commun.* p. 1533 (1988).

468. Sharma, C. L., and Narvi, S. S., *J. Indian Chem. Soc.* **62,** 397 (1985).

469. Lomakina, L. N., Ivanov, V. M., and Belyaeva, V. K., *Koord. Khim.* **14,** 1098 (1988).

470. Chakrabarty, A. K., *Indian J. Chem., Sect. A.* **27A,** 366 (1988).

471. Olby, B. G., and Robinson, S. D., *Inorg. Chim. Acta* **165,** 153 (1989).

472. Singh, A. K., Roy, B., and Singh, R. P., *J. Indian Chem. Soc.* **62,** 316 (1985).

473. Andzhaparaidze, D. I., and Akimov, V. K., *Izv. Akad. Nauk Gruz. SSR, Ser. Khim.* **14,** 179 (1988); *Chem. Abstr.* **111,** 4950t (1989).

474. Akhond, M., Safavi, A., and Massoumi, A., *Anal. Lett.* **20,** 29 (1987).

475. Morelli, B., *Analyst (London)* **112,** 1395 (1987).

476. Heath, G. A., Moock, K. A., Sharp, D. W. A., and Yellowlees, L. J., *J. Chem. Soc., Chem. Commun.* p. 1503 (1985).

477. Sun, I. W., Ward, E. H., and Hussey, C. L., *J. Electrochem. Soc.* **135,** 3035 (1988).

478. Rybakov, V. B., Aslanov, L. A., Volkov, S. V., and Pekhn'o, V. I., *Koord. Khim.* **15,** 700 (1989).

479. Thiele, G., Wochner, H., and Wagner, H., *Z. Anorg. Allg. Chem.* **530,** 178 (1985).

480. Alyoubi, A. O., Greenslade, D. J., Forster, M. J., Preetz, W., *J. Chem. Soc., Dalton Trans.* p. 381 (1990).

481. Preetz, W., and Groth, Th., *Z. Naturforsch., B: Anorg. Chem., Org. Chem.* **41B,** 885 (1986).

482. Bakhvalova, I. P., and Volkova, G. V., *Zh. Neorg. Khim.* **33,** 665 (1988); *Russ. J. Inorg. Chem. (Engl. Transl.)* **33,** 372 (1988).

483. Schoenen, N., and Schmidtke, H. H., *Mol. Phys.* **57,** 983 (1986).

484. Robinson, P. D., Hinckley, C. C., Matusz, M., and Kibala, P. A., *Acta Crystallogr., Sect. C: Cryst. Struct. Commun.* **C44,** 619 (1988).

485. Takazawa, H., Ohba, S., Saito, Y., and Sano, M., *Acta Crystallogr. Sect. B: Struct. Sci.* **B46,** 166 (1990).

486. Singh, M. A., and Armstrong, R. L., *J. Magn. Reson.* **84,** 448 (1989).

487. Singh, M. A., and Armstrong, R. L., *J. Magn. Reson.* **78,** 538 (1988).

488. Singh, M. A., and Armstrong, R. L., *J. Phys. C: Solid State Phys.* **19,** L221 (1986).

489. Krupski, M., and Armstrong, R. L., *Can. J. Phys.* **67,** 566 (1989).

490. Bihaye, C., Butler, T. A., and Knapp, F. F., Jr., *J. Radioanal. Nucl. Chem.* **102,** 399 (1986).

491. Peng, Y. B., Vekeman, G., Maenhout-Van der Vorst, W., and Cardon, F., *J. Imaging Sci.* **31,** 55 (1987).

492. Preetz, W., and Irmer, K., *Z. Naturforsch., B: Chem. Sci.* **45,** 283 (1990).

493. Schmidtke, H. H., Grzonka, C., and Schoenherr, T., *Spectrochim. Acta, Part A* **45A,** 129 (1989).

494. Hope, E. G., Levason, W., and Ogden, J. S., *J. Chem. Soc., Dalton Trans.* p. 61 (1988).

495. Chakravarty, A. S., *Indian J. Phys., Sect. A* **63A,** 115 (1989).

496. Kjems, J. K., Vaknin, D., Davidov, D., Selig, H., and Yeshurun, Y., *Synth. Met.* **23,** 113 (1988).

497. Vaknin, D., Davidov, D., Zevin, V., and Selig, H., *Phys. Rev. B: Condens. Matter* [3] **35,** 6423 (1987).
498. Ohana, I., Vaknin, D., Selig, H., Yacoby, Y., and Davidov, D., *Phys. Rev. B: Condens. Matter* [3] **35,** 4522 (1987); Kjems, J. K., Yeshurun, Y., Vaknin, D., Davidov, D., and Selig, H., *ibid.* **36,** 6981 (1987).
499. Gnanasekaran, S., Ranganayaki, S., Gnanasekaran, P., and Krishnan, S. S., *Asian J. Chem.* **1,** 173 (1989).
500. Hope, E. G., Levason, W., and Ogden, J. S., *J. Chem. Soc., Dalton Trans.* p. 997 (1988).
501. Hambley, T. W., and Lay, P. A., *Inorg. Chem.* **25,** 4553 (1986).
502. Fokina, Z. A., Bryukhova, E. V., Pekhn'o, V. I., and Kiznetsov, S. I., *Izv. Akad. Nauk SSSR, Ser. Khim.* p. 1409 (1986).
503. Holloway, J. H., and Rook, J., *J. Chem. Soc., Dalton Trans.* p. 2285 (1987).
504. Robinson, P. D., Hinckley, C. C., Matusz, M., and Kibala, P. A., *Polyhedron* **6,** 1695 (1987).
505. Kritikos, M., Noreus, D., Bogdanovic, B., and Wilczok, U., *J. Less-Common Met.* **161,** 337 (1990).
506. Frost, P. W., Howard, J. A. K., and Spencer, J. L., *J. Chem. Soc., Chem. Commun.* p. 1362 (1984).
507. Collman, J. P., Wagenknecht, P. S., Hembre, R. T., and Lewis, N. S., *J. Am. Chem. Soc.* **112,** 1294 (1990).
508. Bautista, M., Earl, K. A., Morris, R. H., and Sella, A., *J. Am. Chem. Soc.* **109,** 3780 (1987).
509. Bautista, M. T., Earl, K. A., Maltby, P. A., Morris, R. H., Schweitzer, C. T., and Sella, A., *J. Am. Chem. Soc.* **110,** 7031 (1988).
510. Earl, K. A., Morris, R. H., and Sawyer, J. F., *Acta Crystallogr., Sect. C: Cryst. Struct. Commun.* **C45,** 1137 (1989).
511. Bautista, M. T., Earl, K. A., and Morris, R. H., *Inorg. Chem.* **27,** 1124 (1988).
512. Antoniutti, S., Albertin, G., Amendola, P., and Bordignon, E., *J. Chem. Soc., Chem. Commun.* p. 279 (1989).
513. Bruno, J. W., Huffman, J. C., and Caulton, K. G., *J. Am. Chem. Soc.* **106,** 1663 (1984).
514. Hamilton, D. G., and Crabtree, R. H., *J. Am. Chem. Soc.* **110,** 4126 (1988).
515. Johnson, T. J., Huffman, J. C., Caulton, K. G., Jackson, S. A., and Eisenstein, O. *Organometallics* **8,** 2073 (1989).
516. Desrosiers, P. J., Cai, L., and Halpern, J., *J. Am. Chem. Soc.* **111,** 8513 (1989).
517. Losada, J., Alvarez, S., Novoa, J. J., Mota, F., Hoffmann, R., and Silvestre, J. *J. Am. Chem. Soc.* **112,** 8998 (1990).
518. Collman, J. P., Prodolliet, J. W., and Leidner, C. R., *J. Am. Chem. Soc.* **108,** 2916 (1986).
519. Tait, C. D., Garner, J. M., Collman, J. P., Sattelberger, A. P., and Woodruff, W. H., *J. Am. Chem. Soc.* **111,** 9072 (1989).
520. Cotton, F. A., and Vidyasagar, K., *Inorg. Chem.* **29,** 3197 (1990).
521. Agaskar, P. A., Cotton, F. A., Dunbar, K. R., Falvello, L. R., Tetrick, S. M., and Walton, R. A., *J. Am. Chem. Soc.* **108,** 4850 (1986).
522. Fanwick, P. E., Tetrick, S. M., and Walton, R. A., *Inorg. Chem.* **25,** 4546 (1986).
523. Fanwick, P. E., King, M. K., Tetrick, S. M., and Walton, R. A., *J. Am. Chem. Soc.* **107,** 5009 (1985).
524. Cotton, F. A., and Vidyasagar, K., *Inorg. Chim. Acta* **166,** 109 (1989).
525. Johnson, T. W., Tetrick, S. M., and Walton, R. A., *Inorg. Chim. Acta* **167,** 133 (1990).

526. Cotton, F. A., and Matusz, M., *Inorg. Chim. Acta* **143,** 45 (1988).
527. Tooze, R. P., Wilkinson, G., Motevalli, M., and Hursthouse, M. B., *J. Chem. Soc., Dalton Trans.* p. 2711 (1986).
528. Cotton, F. A., and Matusz, M., *Polyhedron* **6,** 1439 (1987).
529. Cotton, F. A., Dunbar, K. R., and Matusz, M., *Polyhedron* **5,** 903 (1986).
530. Clark, R. J., and Hempleman, A. J., *J. Chem. Soc., Dalton Trans.* p. 2601 (1988).
531. Clark, R. J. H., Hempleman, A. J., and Tocher, D. A., *J. Am. Chem. Soc.* **110,** 5968 (1988).
532. Hilts, R. W., Sherlock, S. J., Cowie, M., Singleton, E., and Steyn, M. M. de V., *Inorg. Chem.* **29,** 3161 (1990).
533. Cotton, F. A., and Dunbar, K. R., *J. Am. Chem. Soc.* **109,** 2199 (1987).
534. Lu, K. L., Lin, Y. C., Cheng, M. C., and Wang, Y., *Acta Crystallogr., Sect. C: Cryst. Struct. Commun.* **C44,** 979 (1988).
535. Vogler, A., Osman, A. H., and Kunkely, H., *Inorg. Chem.* **26,** 2337 (1987).
536. Richardson, D. E., Sen. J. P., Buhr, J. D., and Taube, H., *Inorg. Chem.* **21,** 3136 (1982).
537. Lay, P. A., Magnuson, R. H., and Taube, H., *Inorg. Chem.* (to be submitted).
538. Senoff, C. V., *Chem. Can.* **21,** 31 (1969).
539. Elson, C. M., Gulens, J., and Page, J. A., *Can. J. Chem.* **49,** 207 (1971).
540. Emilsson, T., and Srinivasan, V. S., *Inorg. Chem.* **17,** 491 (1978).
541. Hupp, J. T., Neyhart, G. A., and Meyer, T. J., *J. Am. Chem. Soc.* **108,** 5349 (1986).
542. Lay, P. A., and Taube, H., in preparation.
543. Ferguson, J., Krausz, E. R., Lay, P. A., and Maeder, M., unpublished results (1984).
544. Powell, D. W., and Lay, P. A., to be submitted.
545. Schanze, K. S., Neyhart, G. A., and Meyer, T. J., *J. Phys. Chem.* **90,** 2182 (1986).
546. Kalyanasundaram, K., and Nazeeruddin, M. K., *Chem. Phys. Lett.* **158,** 45 (1989).
547. Campagna, S., Denti, G., Sabatino, L., Serroni, S., Ciano, M., and Balzani, V., *J. Chem. Soc., Chem. Commun.* p. 1500 (1989).
548. Hage, R., Haasnoot, J. G., Nieuwenhuis, H. A., Reedijk, J., De Ridder, D. J. A., and Vos, J. G., *J. Am. Chem. Soc.* **112,** 9245 (1990).
548a. Barigelletti, F., De Cola, L., Balzani, V., Hage, R., Haasnoot, J. G., Reedijk, J., and Vos, J. G., *Inorg. Chem.* **30,** 641 (1991).
549. Kaim, W., and Kasack, V., *Inorg. Chem.* **29,** 4696 (1990).
550. Boyde, S., Strouse, G. F., Jones, W. E., Jr., and Meyer, T. J., *J. Am. Chem. Soc.* **112,** 7395 (1990).
551. Furue, M., Kinoshita, S., and Kushida, T., *Chem. Lett.* p. 2355 (1987).
552. Haga, M., Matsumura-Inoue, T., and Yamabe, S., *Inorg. Chem.* **26,** 4148 (1987).
552a. Haga, M., and Bond, A. M., *Inorg. Chem.* **30,** 475 (1991).
553. Hammershøi, A., Lay, P. A., and Taube, H., to be published.
554. Gilbert, J. A., Geselowitz, D., and Meyer, T. J., *J. Am. Chem. Soc.* **108,** 1493 (1986).
555. Cabeza, J. A., Smith, A. J., Adams, H., and Maitlis, P. M., *J. Chem. Soc., Dalton Trans.* p. 1155 (1986).
556. Cabeza, J. A., Mann, B. E., Maitlis, P. M., and Brevard, C., *J. Chem. Soc., Dalton Trans.* p. 629 (1988).
557. Johson, D. W., and Brewer, T. R., *Inorg. Chim. Acta* **154,** 221 (1988).
558. Heath, G. A., and Humphrey, D. G., *J. Chem. Soc., Chem. Commun.* p. 672 (1990).
559. Bruns, M., and Preetz, W., *Z. Anorg. Allg. Chem.* **537,** 88 (1986).
560. Lundquist, E. G., and Caulton, K. G., *Inorg. Synth.* **27,** 26 (1990).
561. Deeming, A. J., Randle, N. P., Bates, P. A., and Hursthouse, M. B., *J. Chem. Soc., Dalton Trans.* p. 2753 (1988).

562. Barron, A. R., and Wilkinson, G., *J. Chem. Soc., Dalton Trans.* p. 287 (1986).
563. Collman, J. P., McCullough, L., and Zisk, M. B., *Gov. Rep. Announce Index (U.S.)* **89,** Abstr. No 905918 (1989).
564. Collman, J. P., McDevitt, J. T., Leidner, C. R., Yee, G. T., Torrance, J. B., and Little, W. A., *J. Am. Chem. Soc.* **109,** 4606 (1987).
565. Collman, J. P., McDevitt, J. T., Yee, G. T., Zisk, M. B., Torrance, J. B., and Little, W. A., *Synth. Met.* **15,** 129 (1986).
566. Jernigan, J. C., and Murray, R. W., *J. Am. Chem. Soc.* **109,** 1738 (1987).
567. Abruña, H. D., and Hurrell, H. C., *Inorg. Chem.* **29,** 736 (1990).
568. Surridge, N. A., Keene, F. R., White, B. A., Facci, J. S., Silver, M., and Murray, R. W., *Inorg. Chem.* **29,** 4950 (1990).
569. Strouse, G. F., Worl, L. A., Younathan, J. N., and Meyer, T. J., *J. Am. Chem. Soc.* **111,** 9101 (1989).
570. Cainelli, G., Contento, M., Manescalchi, F., and Plessi, L., *Synthesis* p. 47 (1989).
571. Cainelli, G., Contento, M., Manescalchi, F., and Plessi, L., *Synthesis* p. 45 (1989).
572. Meites, L., Zuman, P., Rupp, E. B., and Narayanan, A., eds., "CRC Handbook Series in Inorganic Electrochemistry," Vol. 5. CRC Press, Boca Raton, Florida, 1985.
573. Colom, F., *in* "Standard Potential of Ruthenium and Osmium" (A. J. Bard, R. Parsons, and J. Jordan, eds.), pp. 413–427. Dekker, New York, 1985.
574. Lay, P. A., McAlpine, N. S., Hupp, J. T., Weaver, M. J., and Sargeson, A. M., *Inorg. Chem.* **29,** 4322 (1990).
575. Lay, P. A., *J. Phys. Chem.* **90,** 878 (1986).
576. Elson, C. M., Gulens, J., Itzkovitch, I. J., and Page, J. A., *J. Chem. Soc., Chem. Commun.* p. 875 (1970).
577. Yeh, A., Scott, N., and Taube, H., *Inorg. Chem.* **21,** 2542 (1982).
578. Creutz, C., and Taube, H., *J. Am. Chem. Soc.* **95,** 1086 (1973).
579. Lehmann, H., Schenk, K. J., Chapuis, G., and Ludi, A., *J. Am. Chem. Soc.* **101,** 6197 (1979).
580. Kuehn, C. G., and Taube, H., *J. Am. Chem. Soc.* **98,** 689 (1976).
581. Lim, H. S., Barclay, D. J., and Anson, F. C., *Inorg. Chem.* **11,** 1460 (1972).
582. Matsubara, T., and Ford, P. C., *Inorg. Chem.* **15,** 1107 (1976).
583. Richardson, D. E., unpublished results.
584. Sundberg, R. J., Bryan, R. F., Taylor, I. F., Jr., and Taube, H., *J. Am. Chem. Soc.* **96,** 381 (1974).
585. Marchant, J. A., Matsubara, T., and Ford, P. C., *Inorg. Chem.* **16,** 2160 (1977).
586. Bino, A., Lay, P. A., Taube, H., and Wishart, J. F., *Inorg. Chem.* **24,** 3969 (1985).
587. Garcia, E., Kwak, J., and Bard, A. J., *Inorg. Chem.* **27,** 4377 (1988).
588. Tom, G., and Taube, H., *J. Am. Chem. Soc.* **95,** 5310 (1975).
589. Richardson, D. E., and Taube, H., *Inorg. Chem.* **20,** 1278 (1981).
590. Sullivan, B. P., and Meyer, T. J., *Inorg. Chem.* **19,** 752 (1980).
591. Kober, E. M., Goldsby, K. A., Narayana, D. N. S., and Meyer, T. J., *J. Am. Chem. Soc.* **105,** 4303 (1983).
592. Jernigan, J. C., and Murray, R. W., *J. Am. Chem. Soc.* **112,** 1034 (1990).
593. Naegeli, R., Redepenning, J., and Anson, F. C., *J. Phys. Chem.* **90,** 6227 (1986).
594. He, P., and Chen, X., *J. Electroanal. Chem. Interfacial Electrochem.* **256,** 353 (1988).
595. Chen, X., Xia, B., and He, P., *J. Electroanal. Chem. Interfacial Electrochem.* **281,** 185 (1990).
596. Gregg, B. A., and Heller, A., *Anal. Chem.* **62,** 258 (1990).
597. Dubicki, L., Ferguson, J., Finnie, K., Krausz, E. R., Lay, P. A., and Maeder, M., unpublished results.

598. Ferguson, J., Krausz, E. R., and Lay, P. A., unpublished results.
599. Creutz, C., and Chou, M. H., *Inorg. Chem.* **26,** 2995 (1987).
600. Krausz, E., *Comments Inorg. Chem.* **7,** 139 (1988).
601. Krausz, E., and Ferguson, J., *Prog. Inorg. Chem.* **37,** 293 (1989).
602. Juris, A., Barigelletti, F., Campagna, S., Balzani, V., Belzer, P., and von Zelewsky, A., *Coord. Chem. Rev.* **84,** 85 (1988).
603. Krausz, R. A., *Struct. Bonding (Berlin)* **67,** 1 (1987).
604. Kalyanasundaram, K., *Coord. Chem. Rev.* **46,** 159 (1982).
605. Krausz, E., *J. Lumin.* **42,** 283 (1988).
606. Krausz, E., Moran, G., and Riesen, H., *Chem. Phys. Lett.* **165,** 401 (1990).
607. Krausz, E., *Chem. Phys. Lett.* **165,** 407 (1990).
608. Hopkins, M. D., Miskowski, V. M., and Gray, H. B., *J. Am. Chem. Soc.* **108,** 6908 (1986).
609. Dubicki, L., Ferguson, J., and Krausz, E. R., *J. Am. Chem. Soc.* **107,** 179 (1985).
610. Dubicki, L., Ferguson, J., Krausz, E. R., Lay, P. A., Maeder, M., Magnuson, R. H., and Taube, H., *J. Am. Chem. Soc.* **107,** 2167 (1985).
611. Armstrong, R. S., Horsfield, W., and Lay, P. A., work in progress.
612. Gould, S., O'Toole, T. R., and Meyer, T. J., *J. Am. Chem. Soc.* **112,** 9490 (1990).
613. Worl, L. A., Strouse, G. F., Younathan, J. N., Baxter, S. M., and Meyer, T. J., *J. Am. Chem. Soc.* **112,** 7571 (1990).
614. Armor, J. N., and Taube, H., *J. Am. Chem. Soc.* **92,** 6170 (1970).
615. Folkesson, B., *Acta Chem. Scand.* **26,** 4157 (1972).
616. Coleman, G. N., Gesler, J. W., Shirley, F. A., and Kuempel, J. R., *Inorg. Chem.* **12,** 1036 (1973).
617. Broomhead, J. A., Basolo, F., and Pearson, R. G., *Inorg. Chem.* **3,** 826 (1964).
618. Lay, P. A., *Comments Inorg. Chem.* **11,** 235 (1991).
619. Lay, P. A., *Coord. Chem. Rev.* In press.
620. Stebler-Röthlisberger, M., Hummel, W., Pittet, P.-A., Bürgi, H.-B., Ludi, A., and Merbach, A. E., *Inorg. Chem.* **27,** 1358 (1988).
621. Triegaardt, D. M., and Wahl, A. C., *J. Phys. Chem.* **90,** 1957 (1986).
622. Sarala, R., and Stanbury, D. M., *Inorg. Chem.* **29,** 3456 (1990).
623. Creutz, C., Sutin, N., and Brunschwig, B. S., *J. Am. Chem. Soc.* **101,** 1297 (1979).
624. Macartney, D. H., *Inorg. Chem.* **25,** 2222 (1986).
625. Macartney, D. H., and Thompson, D. W., *Inorg. Chem.* **28,** 2195 (1989).
626. Bommarius, A. S., Holzwarth, J. F., Wang, D. I. C., and Hatton, T. A., *J. Phys. Chem.* **94,** 7232 (1990).
627. Baumann, J. A., and Meyer, T. J., *Inorg. Chem.* **19,** 345 (1980).
628. Brothers, P. J., Headford, C. E. L., and Roper, W. R., *J. Organomet. Chem.* **105,** C29 (1980).
629. Moyer, B. A., Sipe, B. K., and Meyer, T. J., *Inorg. Chem.* **20,** 1475 (1981).
630. Groves, J. T., and Ahn, K. H., *Inorg. Chem.* **26,** 3831 (1987).
631. Gilbert, J. A., Eggleston, D. S., Murphy, W. R., Geselowitz, D. A., Gersten, S. W., Hodgson, D. J., and Meyer, T. J., *J. Am. Chem. Soc.* **107,** 3855 (1985).
632. Ware, D. C., and Taube, H., unpublished results.
633. Kirchner, K., Han, L.-F., Dodgen, H. W., Wherland, S., and Hunt, J. P., *Inorg. Chem.* **29,** 4556 (1990).
634. Smith, T. P., Iverson, D. J., Droege, M. W., Kwan, K. S., and Taube, H., *Inorg. Chem.* **26,** 2882 (1987).
635. Lane, B. C., Lester, J. E., and Basolo, F., *J. Chem. Soc., Chem. Commun.* p. 1618 (1971).

636. Ridd, M. J., and Keene, F. R., *J. Am. Chem. Soc.* **103,** 5733 (1981).
637. Lay, P. A., and Taube, H., unpublished results.
638. Fenton, R. R., Honours Thesis, Macquarie University, Sydney, Australia (1985).
639. Anderes, B., and Lavallee, D. K., *Inorg. Chem.* **22,** 3724 (1983).
640. Tom, G., Ph.D. Thesis, Stanford University, Stanford, California, (1975).
641. Guengerich, G. P., and Schug, K., *Inorg. Chem.* **22,** 1401 (1983).
642. Isied, S., and Taube, H., *Inorg. Chem.* **13,** 1545 (1974).
643. Franco, D. W., and Taube, H., *Inorg. Chem.* **17,** 571 (1978).
644. Guengerich, C. P., and Schug, K., *Inorg. Chem.* **22,** 181 (1983).

OXIDATION OF COORDINATED DIIMINE LIGANDS IN BASIC SOLUTIONS OF TRIS(DIIMINE)IRON(III), -RUTHENIUM(III), AND -OSMIUM(III)

O. MØNSTED and G. NORD

Department of Inorganic Chemistry, University of Copenhagen, H.C. Ørsted Institute, DK-2100 Copenhagen, Denmark

I. Introduction
II. Rate Law and Stoichiometry for the Reduction of Dilute Solutions of Tris(diimine)iron(III), -ruthenium(III), and -osmium(III) in Base
III. Stoichiometry and Identification of Oxidized Reaction Products in Concentrated Solutions
  A. Iron(III) Complexes
  B. Ruthenium(III) Complexes
  C. Osmium(III) Complexes
IV. Intimate Mechanism of Formation of the First Intermediate
  A. Empirical Correlations of Rate and Equilibrium Data
  B. Survey of Literature Mechanisms
  C. Metal–Ligand Bond Rupture as a Prerequisite for Ligand Oxidation
V. Conclusion
  References

I. Introduction

The possibility of the practical application of the catalytic photodecomposition of water based on the reactivity of the excited states of tris(2,2′-bipyridine) complexes of ruthenium(III) and ruthenium(II) has attracted considerable interest, but it is now clear that the efficiency of this process is limited not only by the lack of efficient catalysts, particularly for the dioxygen-evolving path, but also by both thermal and photochemical ligand oxidation (*1, 2*) and ligand substitution reactions (*3*) of the 2,2′-bipyridine complexes. The stoichiometrically analogous tris(2,2′-bipyridine) and tris(1,10-phenanthroline) complexes of both iron(III) and osmium(III) have also been found to undergo thermal redox and substitution reactions, particularly in aqueous base (*4–6*), and the data from the literature for these reactions are critically examined in this review article. Figure 1 shows ligand structures and ligand numbering schemes, and the ligand name abbreviations are given in Table I.

When tris(diimine)iron(III), -ruthenium(III), or -osmium(III) salts are dissolved in water, reduction to the intensely colored tris(diimine)metal(II) complexes soon becomes apparent. In acid solution this process is slow, but in alkaline solution it appears to be significantly accelerated. It is known that the kinetics of this reduction process is a complicated plethora of parallel and consecutive reactions (*2, 4–6*), and that metal(II) complexes of differently oxidized diimine ligands are formed in addition to the dominant tris(diimine)metal(II) product. These different metal(II) complexes frequently have similar visible absorption spectra, and the accurate stoichiometry of the reactions is therefore not easily established. Mixed monomeric and polymeric complexes containing diimine, hydroxo, and oxo ligands are also formed by diimine ligand aquation and polymerization reactions, and although the accurate compositions of these hydrolyzed solutions are not known, it has been found that one or more species in such solutions are responsible for the observed production of dioxygen in such systems. Figure 2 illustrates this for the tris(2,2′-bipyridine)iron(III) complex in basic solution. The figure also shows that this dioxygen-producing reaction is not catalytic, and that dioxygen is not produced in the absence of hydrolysis products. Destruction of the partly hydrolyzed (diimine)-iron(III)complex, which is responsible for the dioxygen production, by reduction to a stoichiometrically similar but much more labile (diimine)iron(II) complex would be in accord with these observations.

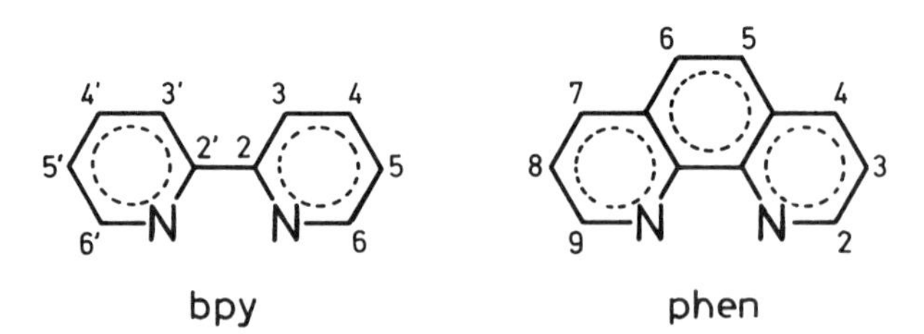

FIG. 1. 2,2′-Bipyridine and 1,10-phenanthroline.

TABLE I

LIGAND NAME ABBREVIATIONS

Abbreviation	Ligand
diim	Diimine Ligand in general
bpy	2,2′-Bipyridine
4,4′-me_2bpy	4,4′-Dimethyl-2,2′-bipyridine
phen	1,10-Phenanthroline
5-mephen	5-Methyl-1,10-phenanthroline
5,6-me_2phen	5,6-Dimethyl-1,10-phenanthroline
4,7-me_2phen	4,7-Dimethyl-1,10-phenanthroline
3,4,7,8-me_4phen	3,4,7,8-Tetramethyl-1,10-phenanthroline
py	Pyridine
trpy	2,2′:6′, 2″-Terpyridine

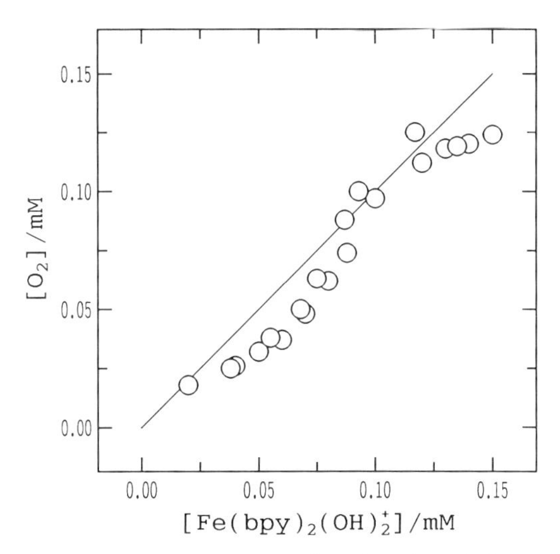

FIG. 2. Concentration of dioxygen produced as function of the concentration of Fe-$(bpy)_2(OH)_2{}^+$. The experiments were performed by rapidly adding base to an acid solution of $Fe(bpy)_3{}^{3+}$ that had been allowed partly to hydrolyze. C[Fe(III)] = 1.00 m*M* and C[OH^-] = 0.100 *M* in the reacting solution (*5*).

II. Rate Law and Stoichiometry for the Reduction of Dilute Solutions of Tris(diimine)iron(III), -ruthenium(III), and -osmium(III) in Base

Reduction of these tris(diimine)metal(III) complexes in basic solution is easily monitored by conventional spectrophotometry in the visible range. The molar absorption coefficients of the reduced species are very large, and at most wavelengths are significantly larger than for the tris(diimine)metal(III) initial reactants. The traditional experimental approach within this field of chemistry has therefore been stopped-flow visible spectrophotometry, pseudo-first-order conditions with base in large excess over complex, and complex concentrations in a range between 1 and 0.01 mM. Under such conditions there is usually a range of concentrations wherein the experimental rate law corresponds to

$$-d[\mathrm{M(III)}]/dt = d[\mathrm{M(II)}]/dt = k'[\mathrm{M(III)}][\mathrm{OH^-}] \quad (1)$$

This rate law requires a rate-determining formation of an intermediate that has either added hydroxide or lost a proton (cf. Section IV,B), and as the overall stoichiometry to a good approximation conforms to Eq. (2)

$$n\mathrm{M(III)} \rightarrow (n-1)\mathrm{M(diim)_3}^{2+} + \mathrm{M(diim)}_{3-x}(\text{oxidized diim})_x{}^{2+} \quad (2)$$

the initially formed intermediate is required to be oxidized in a number of rapid successive steps. In sufficiently dilute solution, n has a value around 6, but this appears to be a lower limit, and in solutions of a higher metal ion concentration, n has been found to increase significantly (cf. Section III), in particular for the most oxidizing systems. Table II contains rate constants, k', and activation parameters as well as the reagent concentration ranges for which the rate law conforms to Eq. (1) and the stoichiometry conforms to Eq. (2) with n about 6.

The oxidized ligand(s) have not been well characterized in most systems, but from the limited evidence available (*2, 7*), introduction of hydroxy groups in the aromatic ring systems according to the generalized stoichiometry

$$-\mathrm{H} + 2\mathrm{OH^-} \rightarrow -\mathrm{OH} + \mathrm{H_2O} + 2e^- \quad (3)$$

is not an unreasonable supposition. In view of the fact that n is about 6, introduction of one hydroxy group in each of the three diimine ligands is a distinct possibility (cf. Section III).

TABLE II

KINETIC AND THERMODYNAMIC PARAMETERS FOR THE REACTION OF TRIS(DIIMINE) METAL(III) COMPLEXES IN BASIC SOLUTION[a]

Complex	k' (25°C) (M^{-1} sec^{-1})	ΔH^* (kJ mol^{-1})	ΔS^* (J K^{-1}mol^{-1})	$E°$ [Eq. (4)] (V)	Ref.
$Fe(bpy)_3^{3+}$	15.8(0.1), 12, 8[b]	62.4(0.8)	−17(3)	0.98	*4, 6*
$Ru(bpy)_3^{3+}$	148[b]	64(4)	+12(13)	1.25	*1, 2*
$Os(bpy)_3^{3+}$	4.7(0.6)	65.4(1.6)	−13(6)	0.84	*4, 6*
$Fe(4,4'\text{-}me_2bpy)_3^{3+}$	1.65(0.21)	54(6)	−60(20)	—	*6*
$Os(4,4'\text{-}me_2bpy)_3^{3+}$	0.32(0.04)[c]	64(1.6)	−46(5)	—	*6*
$Fe(phen)_3^{3+}$	420(25)	46.5(0.8)	−42.(3)	1.01	*4, 6*
$Os(phen)_3^{3+}$	156(16)	50(5)	−35(16)	0.84	*4, 6*
$Fe(5\text{-}mephen)_3^{3+}$	191(33)	—	—	1.02	*6*
$Os(5\text{-}mephen)_3^{3+}$	93(10)	36(4)	−80(16)	—	*6*
$Fe(5,6\text{-}me_2phen)_3^{3+}$	117(17)	38.0(0.6)	−78(2)	0.97	*6*
$Fe(4,7\text{-}me_2phen)_3^{3+}$	35(1.2)	43.0(0.6)	−72(2)	—	*6*
$Fe(3,4,7,8\text{-}me_4phen)_3^{3+}$	17(1.2)	63(8)	−12(2)	0.81	*6*

[a] See Eqs. (1) and (2). Standard deviations are given in parentheses. Conditions for kinetic experiments: C[Fe(III)], 0.1–1mM; C[OH^-], 0.005–0.10 M; C[Os(III)], 0.01–0.1 mM; C[OH^-], 0.005–0.075 M; C[Ru(III)], 0.03–0.17 mM; C[OH^-], 0.01 M.
[b] Ionic strength adjusted to 1 M with Na_2SO_4, others with NaCl.
[c] H/D isotope effect: $k'(H_2O)/k'(D_2O) = 2.25$.

III. Stoichiometry and Identification of Oxidized Reaction Products in Concentrated Solutions

Detailed information on mechanistic aspects of the ligand oxidation reactions is limited by the fact that well-defined tractable kinetics is only found for systems so very dilute in the metal ion reactants that stoichiometric studies including isolation of reaction products have not yet been practicable. Some selected systems have, however, been studied in some detail, but at significantly higher metal ion concentrations than used for the kinetic studies. It is relevant to recall, however, that under such conditions the rate usually does not follow Eq. (1) and the stoichiometry does not conform to Eq. (2) with a value of n about 6.

A. IRON(III) COMPLEXES

Iron(III) complexes are generally more labile than analogous complexes of both ruthenium(III) and osmium(III), and this also holds for the diimine complexes discussed here. In acidic and neutral solution, there is thus an acid-independent diimine ligand aquation reaction of tris(diimine)iron(III) complexes to give diaquabis(diimine)iron(III) or deprotonation products thereof. This puts a lower limit on the hydroxide concentration if complications due to ligand dissociation reactions competing with the metal ion reduction reactions are to be avoided. In sufficiently basic solution, however, this is not a problem, as demonstrated by the data in Table III.

When the basic solutions are allowed to stand, the product iron(II) complexes dissociate and eventually iron(III)hydroxide precipitates. From such solutions, free diimine ligand and a free oxidized ligand product, which has not been further characterized, were isolated. In solutions in which dioxygen was formed (cf. Fig. 2), free diimine-*N*-oxide was also detected (*4*).

B. Ruthenium(III) Complexes

Tris(diimine)ruthenium(III) complexes are significantly more oxidizing than the analogous complexes of both iron(III) and osmium(III). This correlates well with the observation that rates of reduction in base are also faster for the tris(diimine)ruthenium(III) complexes. The tris(1,10-phenanthroline)ruthenium(III) reduction is significantly faster than the tris(2,2′-bipyridine)ruthenium(III) reduction, and this may be the reason why it is only the latter reaction that has been investigated in detail (*1, 2*). This system is particularly complex, and the rate law given by Eq. (1) holds only for very small concentrations of ruthenium complex. In contrast to the iron(III) systems, simple kinetics

TABLE III

Comparison of Kinetic Data for Ligand Dissociation and Metal Ion Reduction Reactions[a]

Reaction	Medium	Half-life (Seconds)
$Fe(bpy)_3^{3+} \rightarrow Fe(bpy)_2(OH_2)_2^{3+}$	0.030 *M* HCl	8.6×10^3
$Fe(phen)_3^{3+} \rightarrow Fe(phen)_2(OH_2)_2^{3+}$	0.030 *M* HCl	15.0×10^3
$Fe(bpy)_3^{2+} \rightarrow Fe(bpy)_2(OH)_2$	0.100 *M* NaOH	6.9×10^2
$Fe(phen)_3^{2+} \rightarrow Fe(phen)_2(OH)_2$	0.100 *M* NaOH	7.6×10^2
$Fe(bpy)_3^{3+} \rightarrow$ Fe(II) [cf. Eq.(2)]	0.100 *M* NaOH	0.44
$Fe(phen)_3^{3+} \rightarrow$ Fe(II) [cf. Eq.(2)]	0.100 *M* NaOH	0.017

[a] Reactions take place in aqueous acidic and basic solutions at 25°C. Data from Pedersen (*5*).

has also an upper limit in the hydroxide concentration, as there is an extra term second order in hydroxide, which even in 0.01*M* hydroxide accounts for about 1% of the observed rate of reduction (*2*).

Increase in the ruthenium concentration increases the stoichiometric factor, *n* in Eq. (2), from about 6 up to about 20, and in these more concentrated solutions rates of ruthenium(III) reduction are no longer first order in ruthenium(III). Under these conditions reaction products depend on the hydroxide concentration and include hydroxy-aromatic ligands [cf. Eq. (3)], carbonate, and trace amounts of dioxygen. Ruthenium complexes of ligands in which one pyridine ring had been completely oxidized were also characterized (*2*). This accounts for the carbonate, and the minor dioxygen yields could originate from complexes oxidized to ruthenium(IV) (*8*). Unlike the iron(III) system, neither free 2,2′-bipyridine nor the *N*-oxide was detected.

C. Osmium(III) Complexes

The osmium(III) complexes are very robust and the free ligand has never been observed as a reaction product. Under the conditions wherein Eq. (1) holds, so does Eq. (2) with the value of *n* about 6. Increase in the osmium complex concentrations results in the formation of relatively long-lived intermediates with osmium in a higher oxidation state, and this seems to be a characteristic difference from the iron and ruthenium systems. The formation of higher oxidation states of osmium were studied in some detail using a 0.5 m*M* solution of tris(2,2′-bipyridine)osmium(III) in 0.1 *M* sodium hydroxide (*5*). Kinetic studies were combined with a multitude of physical measurements, which were interpreted with three intermediates. The initial intermediate, formed according to Eq. (1), was further oxidized to give an osmium(IV) product, which in turn decomposed to give (2,2′-bipyridine)dihydroxodioxoosmium(VI). This latter complex has subsequently been characterized further (*9*).

IV. Intimate Mechanism of Formation of the First Intermediate

The present review is chiefly concerned with the mechanism of the formation of the first intermediate and not with the very complicated kinetics of its further fast oxidation. In this section an attempt will be made first to analyze relevant data for the initial slowest step for all the three metal centers with the ligands shown in Table II, and second, to discuss mechanisms that have been suggested in the literature.

A. Empirical Correlations of Rate and Equilibrium Data

As illustrated in Fig. 3, there is a linear correlation between the free energy of activation for the conversions as in Eq. (1) and the corresponding standard reduction potentials

$$M(diim)_3{}^{3+} + e^- \rightarrow M(diim)_3{}^{2+}; \qquad M = Fe, Ru, Os \tag{4}$$

Rates constants are seen to be consistently higher for the tris(1,10-phenanthroline) complexes of iron(III) and osmium(III) than for the analogous tris(2,2′-bipyridine) complexes, in accord with the fact that the reaction of tris(1,10-phenanthroline)ruthenium(III) is known to be qualitatively faster than that of tris(2,2′-bipyridine)ruthenium(III) (*10*). Figure 4 illustrates that the variation in rate with diimine ligand is the same for iron(III) as for osmium(III) complexes, and that there is a good linear free-energy relationship between the two sets of data with a slope of 1. These correlations all point toward significant mechanistic similarities between the reactions independent not only of the ligand system but also of the metal center. Figure 5 shows correlation of the rate constant for reduction with the acid dissociation constants of the di- and monoprotonated diimine ligands. The left part of the figure illustrates that, within the limits of available experimental data, the

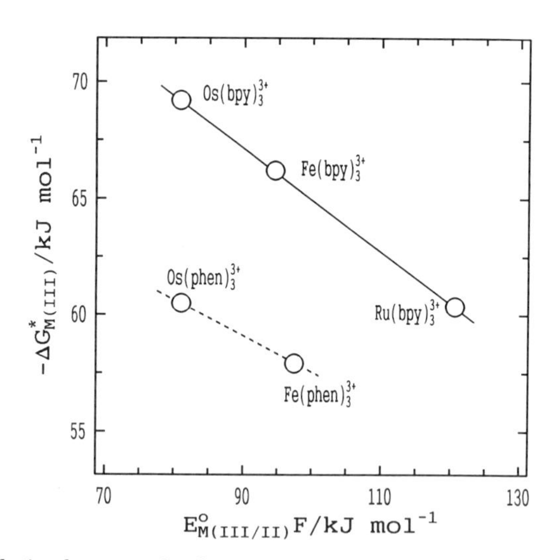

Fig. 3. Correlation between the free energies of activation for the reactions given by Eq. (2) and the standard reduction potentials for the tris(diimine)metal(III) complexes [cf. Eq. (4)]. The data are from Table II.

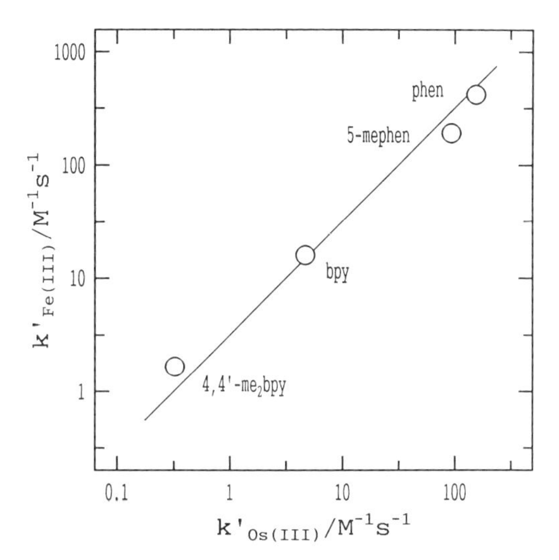

FIG. 4. Linear free energy correlation between the rates of reduction of iron(III) and osmium(III) complexes [cf. Eq. (2)]. The line is drawn with a slope of 1.00. The data are from Table II.

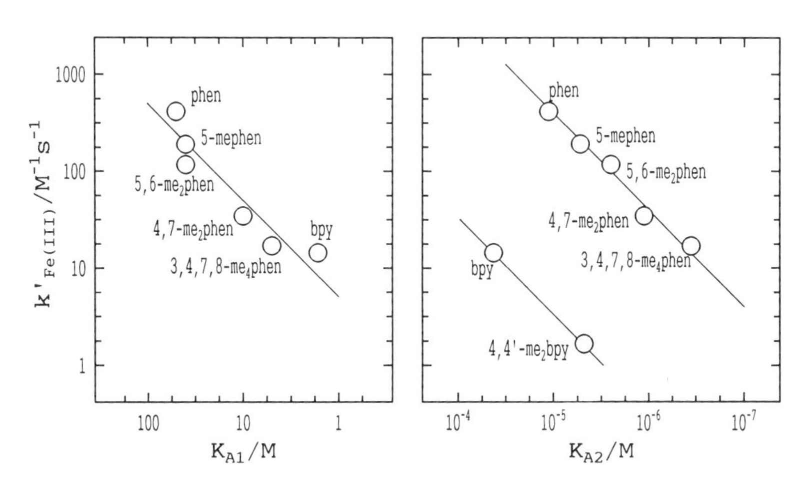

FIG. 5. Free energy correlations between the rates of reduction of tris(diimine)iron(III) complexes [cf. Eq. (2)] and the acid dissociation constants of the di- and monoprotonated free ligands. The lines are drawn with slopes of 1.00. The data are from McBryde (*11*) and from Table II.

difference between the 2,2′-bipyridine and the 1,10-phenanthroline complexes apparent in Fig. 3 disappears when the reduction rates are correlated with the first acid dissociation constant of the diprotonated diimine ligand. In the right part of the figure, however, the differences between the two ligand types are again apparent, and it is tempting to correlate this difference with the stereochemical rigidity of the 1,10-phenanthroline ligand in contrast to the rotational freedom around the C(2) — C(2′) bond in the 2,2′-bipyridine ligand system, which particularly for the acid dissociation constant of the monoprotonated 2,2′-bipyridine gives an energy contribution not present for the 1,10-phenanthroline system. This is illustrated in Fig. 6.

B. Survey of Literature Mechanisms

The suggested literature mechanisms for the first step of the diimine ligand oxidation in basic solution include (1) simple electron transfer, (2) dissociation of a ligand proton, (3) adduct formation at a ligand carbon atom, and (4) dissociation of a metal–nitrogen bond. These four mechanistic possibilities are shown in Scheme 1 and are discussed below.

Mechanism 1 is a simple rate-determining outer-sphere electron transfer between hydroxide and the tris(diimine)metal(III) reactant. This mechanism is, however, not very plausible for the reactants in Table II, because, first, the enthalpy of activation is significantly less than the enthalpy of reaction (*6*). Second, the standard reduction potential for Eq. (5)

$$OH + e^- \rightarrow OH^-; \qquad E^\circ = 1.89\text{V} \tag{5}$$

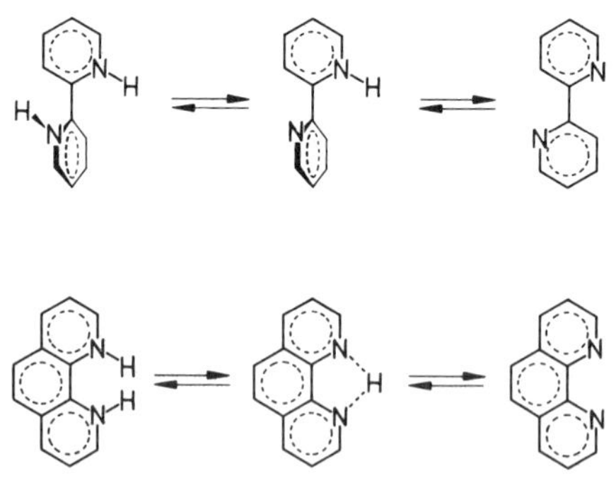

FIG. 6. Illustration of the difference in the conformational behavior of 2,2′-bipyridine and 1,10-phenanthroline on protonation.

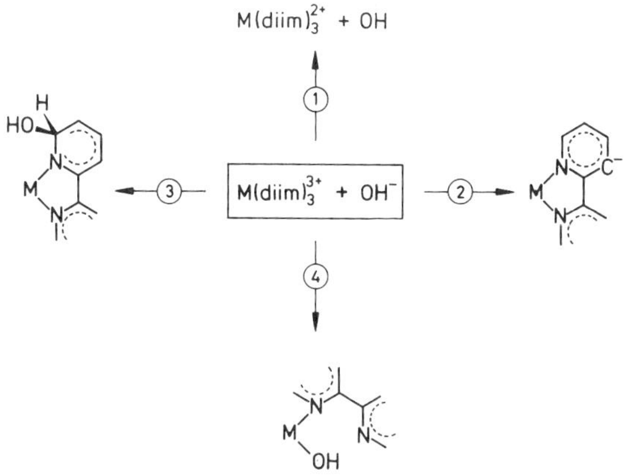

SCHEME 1.

in combination with the standard reduction potentials for the tris(diimine)metal(III) reductions [cf. Eq. (4)] and Table II gives reaction rates for the reverse process, rates that are all much greater than the value usually ascribed to a diffusion-controlled process. Third, the general increase in rate for the tris(1,10-phenanthroline) complexes over that of the tris(2,2′-bipyridine) complexes is significantly larger than that generally found for outer-sphere electron transfer reactions of such reactants. For these latter reactions, the small increase in rate for complexes of 1,10-phenanthroline over 2,2′-bipyridine presumably derives from the small difference in standard reduction potentials. Two such examples are oxidation of hexaaquairon(II) for which $k[Fe(phen)_3^{3+}] = 3.0 \times 10^5\ M^{-1}\ sec^{-1}$ and $k[Fe(bpy)_3^{3+}] = 2.2 \times 10^5\ M^{-1}\ sec^{-1}$ (*6*), and oxidation of iodide for which $k[Os(phen)_3^{3+}] = 26\ M^{-1}\ sec^{-1}$ and $k[Os(bpy)_3^{3+}] = 17 M^{-1}\ sec^{-1}$ (*5*), respectively. Thus the rate constant ratios for these two pairs of tris(diimine) reactants are significantly smaller than the ratios of reduction rate constants for the same tris(diimine) reactants in Table II, which are both more than 10 times larger.

Mechanism 2 requires deprotonation of a ligand, and dissociation of the hydrogen bound to the C(3) atom was suggested to be the slow step for the reaction of tris(2,2′-bipyridine)ruthenium(III) in base (*12*). This would require a mechanism for the 1,10-phenanthroline complexes different from that of the 2,2′-bipyridine complexes, but, from the data in Table II as illustrated in Figs. 3–5, this seems unlikely. The requirement of a different mechanism is based upon the significant differences in rates of D/H exchange as measured by ^1H NMR for the tris(diimine) complexes of ruthenium(II) and osmium(II) (*13,14*), under the assumption that these rates reflect the relative acidities of the ligand hydrogens. For both tris(2,2′-bipyridine)ruthenium(II) and -osmium(II), the hydrogen atoms at the C(3) and C(3′) atoms exchange much faster than do the hydrogen atoms at the other carbon atoms, when measured in basic mixtures of D_2O and nonaqueous solvents. The much slower rates of exchange at the other positions of the 2,2′-bipyridine ligands are comparable with all the rates seen for coordinated 1,10-phenanthroline, the largest difference being exchange of the 1,10-phenanthroline ligand C(2) and C(9) hydrogen atoms, which is about five times as fast as exchange at the C(6) and C(6′) atoms of coordinated 2,2′-bipyridine (*13, 14*). In aqueous solution, the rate of exchange at all positions in the tris(diimine)metal(II) complexes is very slow and has not been measured. In the tris(diimine)metal(III) systems it is likely that the acidity of the ligand hydrogen atoms is increased, but it would seem unlikely that the relative rates will be very different from those found for the metal(II) systems.

Mechanism 3 was originally suggested to explain the kinetics and stoichiometry of tris(diimine)iron(III) and -osmium(III) reactions in base (*6*) and has recently been discussed in some further detail for the tris(2,2′-bipyridine)ruthenium(II) system (*2*). It has the reversible nucleophilic addition of hydroxide to a carbon atom of the coordinated diimine ligand as the first step. This is then supposed to be followed by an intramolecular ligand-to-metal electron transfer and rapid further oxidation of the generated radical ligand. This mechanism was also considered for the overall two-electron reduction in aqueous base of the ruthenium(IV) complexes: $Ru(trpy)(phen)O^{2+}$, $Ru(trpy)(bpy)O^{2+}$, and $Ru(bpy)_2(py)O^{2+}$ (*7*). The first of these was studied in some detail and a number of similarities with the presently considered reactions are apparent. First, the rate law was analogous to that for the reduction of tris(2,2′-bipyridine)ruthenium(III) in base; second, the activation parameters were similar to those shown for the systems in Table I; and third, a ligand was oxidized. This mechanism requires the hydroxide adduct to be formed reversibly, and because there is at present no convincing evidence for such adducts in the present or other similar systems and also the lack of structural evidence for these "covalent hydrates" has earlier been discussed in some detail (*12, 15*), mechanism 3 seems at present highly speculative.

Mechanism 4 requires substitution at the metal center with concomitant dissociation of a metal nitrogen bond as the first step. It is generally accepted that complete dissociation of a bidentate chelate ligand is a two-stage process, and that breaking of the second bond in stereochemically rigid systems is frequently the slow step. The steric arguments that were earlier much used against the tendency of 1,10-phenanthroline to behave as a monodentate ligand were, however, nullified when the nitrate salt of the cation $[Pt(phen)_2(CN)]^+$ was found to have one of the platinum nitrogen bonds about 0.7 Å longer than the remaining three bonds from platinum to the 1,10-phenanthroline ligands (*16*). Also, singly bonded diimine intermediates have been invoked to explain the photosubstitution of not only tris(2,2′-bipyridine)ruthenium(II) but also of the tris(1,10-phenanthroline) complex (*3*). Two-step dissociation is in general agreement with data for the kinetics of complex formation of 1,10-phenanthroline (*17*), and this mechanism is discussed further in the next section.

C. Metal–Ligand Bond Rupture as a Prerequisite for Ligand Oxidation

The mechanism depicted in Scheme 2 involves two main steps. Rupture of the first metal–nitrogen bond accompanied by coordination of a water ligand at the metal center is followed by reversible deprotonation and intramolecular reduction of the metal center. Under the experimental conditions wherein the concentration of base is much larger than the concentration of tris(diimine) complex, and, applying the steady-state approximation to the concentration of the intermediate species with the monodentate diimine ligand, Eq. (6) can be derived as

$$k' = n(k_{d1}/k_{c1})k_r(K_A/K_W) \qquad (6)$$

which holds for the condition $k_{c1} >> k_r K_A(OH^-)/K_W$, K_W being the ionic product of water. Comparison of Eq. (6) with the empirical correlations in Fig. 3–5 makes it likely that differences in ligands are pri-

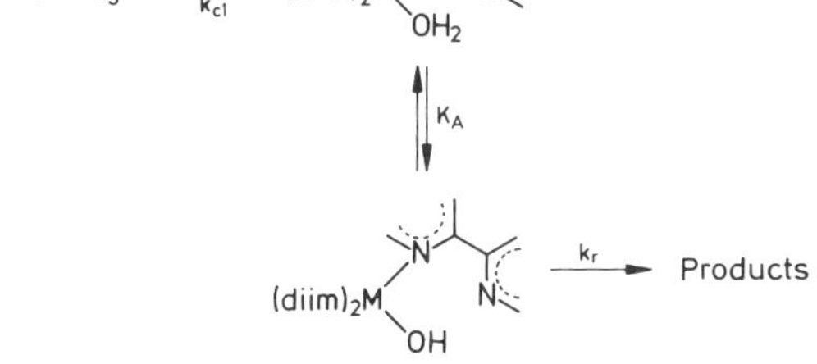

$M(diim)_3^{3+}$ ⇌ (k_{d1}, k_{c1}) $(diim)_2M(OH_2)$ ⇌ (K_A) $(diim)_2M(OH)$ → (k_r) Products

SCHEME 2.

marily exhibited by differences in the k_{d1}/k_{c1} ratio, and that changes in the metal center are reflected by ligand-independent contributions to the parameters.

The consistent variation of k' with diimine ligand at different metal centers apparent in Fig. 4 and the correlation between k' and the equilibrium constant for dissociation and uptake of the second proton of the diprotonated ligand support the mechanism shown in Scheme 2. A further discussion along such lines is, however, impossible at present, because very little detailed information is available about complex formation of these diimine ligands at the metal centers under discussion. The iron(III) complexes are the most labile, but even for these systems there are no kinetic or equilibrium data for the uptake of the third ligand. This is because formation of the dimeric complex $(diim)_2\text{-}Fe(OH)_2Fe(diim)_2^{4+}$ effectively competes with formation of Fe-$(diim)_3^{3+}$. This is also the reason why the tris(diimine) complexes of iron(III), ruthenium(III), osmium(III), and also chromium(III) are synthesized by oxidation of the stoichiometrically analogous divalent metal complexes. The dissociation rates given in Table III chiefly reflect a fast equilibrium between the monodentate and the bidentate diimine ligand followed by a rate-determining breaking of the second metal–nitrogen bond. Consequently, there seems at present no way of reasonably estimating the k_{d1}/k_{c1} ratio for the different ligands.

The acid dissociation constant of a coordinated water molecule is dominated by the overall charge of the metal complex, as demonstrated by the data in Tables IV and V (*18–21*). It is, however, doubtful whether the acidities of the complexes in these tables are representative of the acidities of the complexes with monodentate diimine ligands. This may be seen from a consideration of data for the hydrolysis of tris(2,2′-bipyridine)chromium(III), a complex that in both charge and size resembles tris(2,2′-bipyridine)iron(III). In the acidity range from 1 down to about 10^{-11} M this chromium(III) complex hydrolyzes in a manner that, if considered as arising from formation of a monodentate diimine complex, would lead to a K_A value of about $10^{-7.5}$ M (*22*). This is smaller than the values for tripositive species in Tables IV and V and could reflect significant hydrogen bonding stabilization of the acid form in the monodentate species.

The rate constant for the redox step, k_r, is unlikely to reflect a simple electron transfer from the monodentate diimine ligand to the metal center because replacement of coordinated water with coordinated hydroxide would be expected to decrease the oxidizing power of the metal(III) center. This is well documented by the standard reduction potentials of the aqua and hydroxo complexes in Table IV, and it would seem

TABLE IV

STANDARD REDUCTION POTENTIALS AND ACID DISSOCIATION CONSTANTS FOR SOME AQUAPENTAKIS(IMINE)RUTHENIUM AND -OSMIUM COMPLEXES AT 25°C[a]

M(III)–OH_2 complex	$E°$(M–OH_2) (V)	Ionic medium (M)	$-\log[K_A$(M(III)] (M)	$-\log[K_A$(M(II)] (M)	$E°$(M-OH) (V)[b]
(trpy)(bpy)Os$(OH_2)^{3+}$	0.65	1.0	2.0	7.8	0.32
(trpy)(bpy)Os$(OH_2)^{3+}$	0.60	0.10	—	—	—
(trpy)(bpy)Ru$(OH_2)^{3+}$	1.03	0.10	1.7	9.7	0.56
(trpy)(phen)Ru$(OH_2)^{3+}$	1.05	0.10	1.7	10.0	0.56
(trpy)(phen)Ru$(OH_2)^{3+}$	—	1.0	—	10.2	—
$(bpy)_2$(py)Ru$(OH_2)^{3+}$	1.02	1.0	0.9	10.8	0.42
$(bpy)_2$(py)Ru$(OH_2)^{3+}$	1.04	0.10	0.9	10.2	0.49

[a] See Eq. (4). Data from Roecker *et al.* (*7*), Gilbert *et el.* (*9*), and Pipes and Meyer (*19*).
[b] Standard reduction potentials for M(III)OH^{2+} + $e^- \rightarrow$ M(II)OH^- calculated from the standard reduction potentials for the aquapentakis(imine) species and the acid dissociation constants.

unlikely that the water ligand in the metal(II) intermediates is more acid than in the metal(III) intermediates, as would be required for the hydroxometal(III) complex to be more oxidizing than the analogous aqua complex. Instead, the formal hydrogen atom transfer mechanism depicted in Scheme 3 may operate, in which the reduced oxidizing power of the metal center is compensated for by the presence of a base to accept a proton from a ligand carbon atom. The suggested intermediate is analogous to that detected as a bidentate ligand in bis(2,2′-bipyri-

TABLE V

ACID DISSOCIATION CONSTANTS FOR SOME DIAQUABIS(DIIMINE) METAL(III) COMPLEXES[a]

Complex	$-\log(K_{A1})$ (M)	$-\log(K_{A2})$ (M)	Ref.
cis-$(bpy)_2Os(OH_2)_2^{3+}$	1.9	5.4	*19*
trans-$(bpy)_2Os(OH_2)_2^{3+}$	0.5	4.4	*19*
cis-$(phen)_2Cr(OH_2)_2^{3+}$	3.4	6.0	*18*
cis-$(bpy)_2Cr(OH_2)_2^{3+}$	3.5	6.1	*18*
cis-$(phen)_2Co(OH_2)_2^{3+}$	4.5	6.8	*20*
cis-$(bpy)_2Co(OH_2)_2^{3+}$	4.7	7.1	*21*

[a] Reaction takes place in a 1 M ionic medium at 25°C.

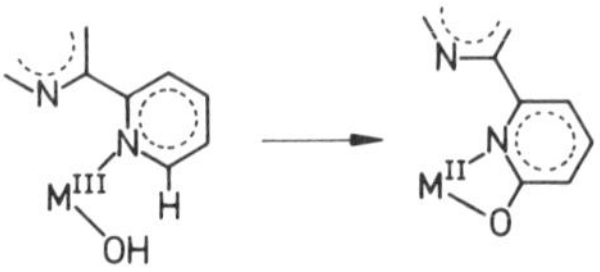

SCHEME 3.

dine)(2-hydroxypyridine)ruthenium(II) (*7*). Also, the D/H isotope effect (cf. Table II) would fit such a mechanism.

V. Conclusion

The rates of oxidation of diimine ligands in basic solutions of tris(diimine)iron(III), -ruthenium(III), and -osmium(III) are in accord with a common mechanism for the first step. A detailed consideration of this is limited at present not only by the complicated and only partly unraveled stoichiometries, but also by the paucity of detailed kinetic and equilibrium data.

A consideration of alternative mechanisms for the common slowest step in the ligand oxidation has led to the conclusion that a metal–ligand dissociation reaction may well operate. It would seem worthwhile to test this by using the newly developed cage complexes (*23, 24*) in which the coordination sphere is effectively unchanged but in which metal–ligand bond breaking would be very difficult. It would seem possible that such caged tris(diimine)ruthenium complexes could be useful for the catalytic photodecomposition of water, but the consequences of the fact that the standard reduction potentials are about 0.3 V higher than for the tris(2,2′-bipyridine)ruthenium(III/II) system (*25*) are not easily foreseeable. Thus, although this higher potential means that production of hydroxyl radicals by electron transfer from hydroxide ions to the metal(III) center is now a distinct possibility (cf. mechanism 1 in Scheme 1, and Section IV,B), and ideally these radicals could form peroxide and finally dioxygen in an uncatalyzed system, they could also cause ligand oxidation. This latter possibility would be in accord with the results of pulse radiolysis studies with tris(1,10-phenanthroline)iron(III) and tris(2,2′-bipyridine)iron(III) and -ruthenium(III), which are known to form "hydroxyl radical adducts." These adducts, on dimerization, disproportionation, or further oxidation, are converted stoichiometrically into hydroxy-substituted aromatic compounds [cf. Eq. (3)]. One obvious way of decreasing the oxidizing power of these new caged diimine ligand complexes would be to replace the ruthenium(III) center by iron(III) or osmium(III), but this would mean that production of dioxygen would still require a catalyst. Alternatively, one of the imine coordination positions could be replaced with a ligand that lowers the standard reduction potential of the ruthenium(III) species. The combination of this aspect with the possibility of direct substrate binding to the metal center, preferably in a multicentered system, remains, however, an intriguing intellectual and synthetic challenge.

References

1. Creutz, C., and Sutin, N., *Proc. Natl. Acad. Sci. U.S.A.* **72,** 2858 (1975).
2. Ghosh, P. K., Brunchwig, B. S., Chou, M., Creutz, C., and Sutin, N., *J. Am. Chem. Soc.* **106,** 4772 (1984).
3. Durham, B., Caspar, J. V., Nagle, J. K., and Meyer, T. J., *J. Am. Chem. Soc.* **104,** *4803 (1982).*
4. Nord, G., Pedersen, B., and Bjergbakke, E., *J. Am. Chem. Soc.* **105,** 1913 (1983).
5. Pedersen, B., Ph.D. Thesis, Copenhagen University (1982).
6. Nord, G., and Wernberg, O., *J. Chem. Soc., Dalton Trans.* p. 866 (1972); p. 845 (1975).
7. Roecker, L., Kutner, W., Gilbert, J. A., Simmons, M., Murray, R. W., and Meyer, T. J., *Inorg. Chem.* **24,** 3784 (1985).
8. Gilbert, J. A., Eggleston, D. S., Murphy, W. R., Geselowitz, D. A., Gersten, S. W., Hodgson, D. J., and Meyer, T. J., *J. Am. Chem. Soc.* **107,** 3855 (1985).
9. Gilbert, J. A., Geselowitz, D., and Meyer, T. J., *J. Am. Chem. Soc.* **108,** 1493 (1986).
10. Miller, J. D., Ph. D. Thesis, Cambridge University (1965).
11. McBryde, W. A. E., *I.U.P.A.C. Chem. Data Ser.* **17** (1978).
12. Serpone, N., Ponterini, G., Jamieson, M. A., Bolletta, F., and Maestri, M., *Coord. Chem. Rev.* **50,** 209 (1983).
13. Nord, G., and Wernberg, O., unpublished work (1984).
14. Wernberg, O., *J. Chem. Soc., Dalton Trans.* p. 1993 (1986).
15. Nord, G., *Comments Inorg. Chem.* **4,** 193 (1985).
16. Wernberg, O., and Hazell, A., *J. Chem. Soc., Dalton Trans.* p. 973 (1980).
17. Margerum, D. W., *J. Am. Chem. Soc.* **79,** 2728 (1956).
18. Inskeep, R. G., and Bjerrum, J., *Acta Chem. Scand.* **15,** 62 (1961).
19. Pipes, D. W., and Meyer, T. J., *Inorg. Chem.* **25,** 4042 (1986).
20. Ablov, A. V., and Palade, D. M., *Russ. J. Inorg. Chem. (Engl. Transl.)* **6,** 567 (1961).
21. Palade, D. M., *Russ. J. Inorg. Chem. (Engl. Transl.)* **14,** 231 (1969).
22. Maestri, M., Bolletta, F., Serpone, N., Maggi, L., and Balzani, V., *Inorg. Chem.* **15,** 2048 (1976).
23. Vögtle, F., Ebmeyer, F., Müller, W. M., Stutte, P., Grammenudi, S., Seel, C., and Walton, A., *Abstr. Int. Symp. Macrocyclic Chem., I.U.P.A.C., 13th,* Hamburg p. 1.12 (1988).
24. De Cola, L., Barigelletti, F., Balzani, V., Belser, R., von Zelewsky, A., Vögtle, F., Ebmeyer, F., and Grammenudi, S., *J. Am. Chem. Soc.* **110,** 7210 (1988).
25. Barigelletti, F., De Cola, L., Balgani, V. Belser, P., von Zelewsky, A., Vögtle, F., Ebmeyer, F., and Grammenudi, S., *J. Am. Chem. Soc.* **111,** 4662 (1989).

INDEX

A

ab initio molecular calculations
  complex lithium amides, 117–123
  uncomplexed lithium amides, 105–107
  iminolithiums, 79–82
Acetone, condensation, osmium(II), 351
Addition reactions
  osmium(II), 350
  osmium(VIII), 331–332
S-Adenosyl-L-homocysteine, formation, 204–206
$[AgCl_3]^{2-}$ anions, packing, 8–9
$[Ag_2I_6]^{4-}$ ion, 20–21
$[Ag_3I_6]^{3-}$ ion, 21
$[Ag_4I_8]^{4-}$ ion, 21–22
$[Ag_2X_4]^{2-}$ species, dinuclear, 16
Aldehyde complexes, osmium, 284–285
  linkage isomerizations, 336–337
η^2-(*C*,*O*)-Aldehydes, osmium, 244–245
Alkene complexes, osmium, 242–243
Alkylations, osmium(II), 348
Alkyl complexes, osmium, 239–242
Alkyllithium compounds, crystal structures, 83–85
Alkylsulfonato complexes, osmium, 286–287
Alkyne ligands, osmium, 243
Alkynyllithium complexes, crystal structures, 85–88
Allyl complexes, osmium, 244
  ligand-centered reactions, 343
Aluminohydride bridges, osmium, 314
Amide complexes, osmium, 284
1-Amidino-2-thiourea complexes, osmium, 290
Amine complexes, osmium, 247
2-Aminobenzenethiol complexes, osmium, 295
Ammine complexes, osmium, 247–248
Antitumor drugs
  cisplatin as, history, 175–179
  platinum compounds
    future studies, 206–208
    resistance to, 192–193
    second-generation, 178
Aqua complexes, osmium, 276–278
Aquapentakis(imine)ruthenium and -osmium complexes, reduction, potentials and dissociation constants, 394–395
Arene activation, osmium(II), 348–350
Arene complexes, osmium, 305
  linkage isomerization, 337–339
η^6-Arene ligands, osmium, 244
Arsenito complexes, osmium, 288
Arsine complexes, osmium, 274–275
Aryl complexes, osmium, 239–242
Aryllithium compounds, crystal structures, 83–85
Arylsulfonato complexes, osmium, 286–287
Atom-transfer reactions, osmium, 340–343
Azido complexes, osmium, 263

B

Benzeneseleninato complexes, osmium, 287
Benzoselenadiazole complexes, osmium, 290–291
Benzotriazole complexes, osmium, 270–271
Bicyclo-*N* heterocycles, osmium, 310
2,2′-Bipyridine, 382
  conformational behavior, 390
μ-4,4′-Bipyridine complexes, osmium, 307

μ-Bis(benzylimidazolato) complexes, osmium, 309
μ-*trans*-1,2-Bis(4′-methyl-2,2″-bipyridyl-4-yl)ethene complexes, osmium, 309
Bis(μ-oxo) complexes, osmium, 311
μ-2,3-Bis(2′-pyridyl)pyrazine complexes, osmium, 307–309
μ-2,3-Bis(2′-pyridyl)quinoxaline complexes, osmium, 307–309
μ-3,5-Bis(2-pyridyl)-1,2,4-triazolato complexes, osmium, 309
Bonding
  N—Li compounds, 52–58
  organonitrogen–lithium compounds, 132
Borohydride complexes, osmium, 299–300
Bpy complexes, osmium, spectroscopic and magnetic properties, 329–330
Bpy polymers, osmium, spectroscopic and magnetic properties, 330
Bromo complexes, osmium, 297–299
Bromocuprates(I), 39
μ-1,3-Butadiene complexes, osmium, 305
$(Bu^t_2C{=}NLi \cdot HMPA)_2$
  MNDO energy difference, 80
  molecular structure, 68–70
  NMR spectra, 77–78
$[Bu^t(Ph)C{=}NLi]_6$, 59
$(Bu^{t2}{}_2(Ph)C{=}NLi)_6$, 59
$[Bu^t(Ph)C{=}N]_6Li_4Na_2$, molecular structure, 67–68

C

Carbene complexes, osmium, 242
Carbonato complexes, osmium, 285
Carbon dioxide complexes, osmium, 239
Carbon disulfide complexes, osmium, 239
Carbonyl complexes, osmium, 236–238
Carboxylato complexes, osmium, 285–286, 303
Catechol complexes, osmium, 283–284
Cation–halogenide ligand packing, 35–36
Cation size, metal(I) coordination number effects, 33–38
C-bound aldehyde and ketone complexes, osmium, 246
Charge, metal(I) coordination number effects, 38–39
Chelating amide ligands, osmium, 266–270
Chloro complexes, osmium, 297–299
Chlorothionitrene complex, osmium, 262
η^2-CH_2PR_2 complexes, osmium, 274
Chromato complexes, osmium, 287
Cisplatin
  as antitumor drug, history, 175–179
  aqueous solution chemistry, 179–180
  DNA binding positions, 183–184
  d(pGpG) and d(CpGpG) adducts, 187
  reaction with
    glutathione, 203
    plasma protein, 202
  structure, 177
Cl—Ag—Cl linkage, 32
Cl—Cu—Cl linkage, 6, 17, 19
^{13}C NMR, complex lithium amides, 129
μ-CO_2 complexes, osmium, 304
Complexed lithium amides, 108–131
  attachment of Lewis base molecules, 113–114
  calculational studies, 117–123
  dimeric, structural parameters, 112
  isolation of ladder structures, 120
  monomeric, structural parameters, 115–116
  solid-state structures, 108–117
  in solution, 123–131
    ^{13}C NMR spectra, 129
    double-labeling NMR, 130
    limited-ladder structures, 124–125
    ^7Li NMR spectra, 124–126
    principal structures, 128–129
  structural types, 109
Coordinated diimine ligands, oxidation, 381–397
  dioxygen production, 382–383
  first intermediate formation mechanism, 387–396
    empirical correlations of rate and equilibrium data, 388–390
    ligand protonation, 391–392
    literature survey, 390–393
    metal center substitution, 392–393
    metal–ligand bond rupture, 393–396
    rate-determining outer-sphere electron transfer, 390–391
    reversible nucleophilic addition, 392
  iron(III) complexes, oxidized reaction products, 385–386

kinetic data for dissociation and reduction reactions, 386
name abbreviations, 383
osmium(III) complexes, oxidized reaction products, 387
ruthenium(III) complexes, oxidized reaction products, 386–387
Copolymers, osmium, 315
Copper(I) coordination number, as function of concentration of halogenide ligand, 33–35
Cubane-type clusters, 156
with Mo_3MS_4 cores, 163–164
with Mo_4O_4/Mo_4OS_3 cores, 158
with Mo_4S_4 cores, 158–161
structural parameters, 158
$[Cu_6Br_9]^{3-}$ anion, 24–25
$[Cu_2Cl_3]^-$, 17, 19
$[Cu_7Cl_{10}]^{3-}$ chain polymer, 17, 19
$[Cu_2I_5]^{3-}$ anion, dinuclear, 20
$[Cu_6I_{11}]^{5-}$ cluster, hexanuclear, 24
$[Cu_8I_{13}]^{5-}$ cluster, 25
$[Cu_{19}I_{27}]^{8-}$ anion, 25
$[Cu_{36}I_{56}]^{20-}$ cluster, 25
$[Cu_4X_6]^{2-}$ clusters, 16–18
$[Cu_5X_7]^{2-}$ cluster, pentanuclear, 22, 24
Cyanato complexes, osmium, 263–264
Cyano complexes, osmium, 235–236, 305
μ-Cyanogen complexes, osmium, 310
μ-η^5,η^1-Cyclopentadienyl(2−) complex, osmium, 304
Cysteine complexes, osmium, 295

D

Damage recognition protein, 207
Decaammineosmium(III) and (II) complexes, synthesis, 226–227
Dehydration reactions, osmium(II), 351
μ-Diacetylhydrazinato(2−) complexes, osmium, 309
Diaquabis(diimine)metal(III) complexes, acid dissociation constants, 394–395
Diarsine ligands, osmium, 275–276
Diaryliminolithium rings, 67
1,3-Diaryltriazene complexes, osmium, 271
Diazene complexes, osmium, 260
Dichlorobis(triphenylphosphine)argentate(I) complexes, osmium, 299
μ-Dicyanoamide complexes, osmium, 310
μ-Dicyanobenzene complexes, osmium, 310
μ-Dicyanobicyclo[2.2.2]octane complexes, osmium, 310
Dihydrogen complexes, osmium, 300–301
Di(hydroxyethyl)dithiocarbamate, nephrotoxicity inhibitor, 197
Diketonate complexes, osmium, 285
Dimers, osmium complexes, spectroscopic and magnetic properties, 326–330
Dimethyl sulfoxide, osmium, S $\leftrightarrow$ O linkage isomerizations, 339
O-Dimethyl sulfoxide complexes, osmium, 284
S-Dimethyl sulfoxide complexes, osmium, 291
Dinitrogen complexes, osmium, 259, 305–307
Dinuclear anions, halogenocuprate(I) and halogenoargentate(I) ions
four-coordinated metal centers, 20–21
three-coordinated metal centers, 8–9, 12–16
Dinuclear complexes, osmium, electrochemistry, 321–323
μ-Dioxane complexes, osmium, 312
Dioxygen complexes, osmium, 285
μ-2-(Diphenylphosphine)pyridine complexes, osmium, 304
Discrete monomeric anions, halogenocuprate and halogenoargentate(I) ions, 2–6
μ-Disulfide complexes, osmium, 312
Disulfur oxide complexes, osmium, 292
Dithiocarbamate complexes, 293–294
Dithiophosphinate complexes, osmium, 294
DNA
distortion by cisplatin, 186
platinum binding, 183–185
DRP, 207

E

Electron transfer, osmium complexes
intermolecular, 332–333
intramolecular, 333–334
Elimination reactions, osmium(II), 351–352

η^2-bound arene ligands, osmium, 243–244
η^2-bound linkages, osmium, linkage isomerization, 337
Ether complexes, osmium, 285

F

Fluoro complexes, osmium, 296–299
Four-coordinated metal centers, 19
Fumarase, inhibition by cisplatin, 195

G

Glutathione
function in cisplatin resistance, 192–193
intracellular, depletion, 191
nephrotoxicity inhibitor, 198–199
[Pt(dien)Cl]Cl reactions, 199–200
reaction with cisplatin, 203
structure, 190

H

Halide-bridged complexes, osmium, 313
Halogen atom transfer, osmium, 342–343
Halogenoargentate(I) ions, 1–2
connectivity relationships, 4–5
dinuclear anions
four-coordinated metal centers, 20–21
three-coordinated metal centers, 8–9, 12–16
discrete clusters with higher nucleicity, 25
discrete monomeric anions, 2–6
five-coordinated metal centers, 32
hexanuclear anions, four-coordinated metal centers, 24–25
ion bonding, 3, 6
metal(I) coordination number, 32–40
cation size effects, 33–38
charge and shape effects, 38–39
mononuclear anions
four-coordinated metal centers, 19
three-coordinated metal centers, 7–11
pentanuclear anions, four-coordinated metal centers, 22, 24
polymeric anions, four-coordinated metal centers, 26–32
polynuclear anions, three-coordinated metal centers, 17, 19
tetranuclear anions
four-coordinated metal centers, 21–23
three-coordinated metal centers, 16–18
trinuclear anions
four-coordinated metal centers, 21
three-coordinated metal centers, 16
X—M—X linkages, 6
Halogenocuprate(I) ions, 1–2
connectivity relationships, 4–5
dinuclear anions
four-coordinated metal centers, 20–21
three-coordinated metal centers, 8–9, 12–16
discrete clusters with higher nucleicity, 25
discrete monomeric anions, 2–6
five-coordinated metal centers, 32
hexanuclear anions, four-coordinated metal centers, 24–25
ion bonding, 3, 6
metal(I) coordination number, 32–40
cation size effects, 33–38
charge and shape effects, 38–39
mononuclear anions
four-coordinated metal centers, 19
three-coordinated metal centers, 7–11
pentanuclear anions, four-coordinated metal centers, 22, 24
polymeric anions, four-coordinated metal centers, 26–32
polynuclear anions, three-coordinated metal centers, 17, 19
tetranuclear anions
four-coordinated metal centers, 21–23
three-coordinated metal centers, 16–18
trinuclear anions
four-coordinated metal centers, 21
three-coordinated metal centers, 6
X—M—X linkages, 6

$[H_2C(CH_2)_5NLi]_6$
molecular structure, 97–98
in solution, 108
$\{[H_2C(CH_2)_3NLi]_3 \cdot PMDETA\}_2$, 109–111
$\{[H_2C(CH_2)_3NLi]_2 \cdot TMEDA\}_2$, 109–111
$(H_2C{=}NLi)_2$, structures, 79–81
Heterocycles, osmium, 231–233, 254–255
activation, 350
linkage isomerization, 337–339
Heterocyclic ligands, osmium, 244
Hexanuclear anions, halogenocuprate(I) and halogenoargentate(I) ions, 24–25
$(c\text{-Hexyl}_2NLi \cdot HMPA)_n$, ^7Li NMR spectrum, 126–127
$(H_2NLi)_4$
bond parameters, 103–105
relative energies and association energies, 102–103
stacking, 103–104
$H_2{=}NLi)_4$, MNDO calculations, 81
$(H_2NLi)_6$, forms, 105–107
$(H_2NLi \cdot H_2O)_4$, structures, 121–122
$(H_2NLi \cdot H_2O)_n$
energy profiles, 118–120
structures, 118–119
$(H_2NLi)_4 \cdot 2H_2O$ complexes, optimized geometry and relative energy, 121
$(H_2NLi)_2$ rings
association energy, 99–101
bond parameters, 100–101
forms, 99–100
relative energy, 99–100
$(H_2NLi)_3$ rings
association energy, 99–101
bond parameters, 100–101
forms, 99–100
relative energy, 99–100
$(H_2NLi)_4$ ring, forms, 101–102, 105
$(H_2NLi)_n$, energy profiles, 118–120
^1H NMR spectra, 77–78
$[(H_2O)_9Mo_3S_4CuCuS_4Mo_3(H_2O)_9]^{8+}$, structure, 167, 168
$[(H_2O)_9Mo_3S_4HgS_4Mo_3(H_2O)_9]^{8+}$, structure, 168, 169
$[(H_2O)_9Mo_3S_4MoS_4Mo_3(H_2O)_9]^{8+}$, structure, 160, 162
Hydrazine complexes, osmium, 258
Hydride-bridged complexes, osmium, 313–314
Hydride complexes, osmium, 299–300
Hydrogenarsenito complexes, osmium, 288
Hydrogen atom transfer, osmium, 340
Hydrogen selenido complexes, osmium, 290
Hydrolysis, cisplatin, 179
Hydroxo complexes, osmium, 276–278, 311–312
Hydroxylamine complexes, osmium, 258, 303–304

I

Imido complexes, osmium, 262, 310
Imine complexes, osmium, 259–260
Iminolithium complexes, solid-state structures, 67–75
dimers, 68–70
$(Ph_2C{=}NLi \cdot pyridine)_4$, 71–72
trimers, 68–70
Iminooxosulfane ligands, osmium, 292
Incomplete cubane-type clusters, 145–154
with Mo_3O_4 cores, 145–146
structural parameters, 146–154
with $Mo_3O_3S/Mo_3O_2S_2/Mo_3OS_3$, 146–147
with Mo_3S_4 cores, 147, 154
with $W_3O_4/W_3O_3S/W_3O_2S_2/W_3OS_3$ cores, 162
with W_3S_4 cores, 163
Intermolecular electron transfer, osmium, 332–333
Intramolecular electron transfer, osmium, 334–335
Iodo complexes, osmium, 297–299
Ionicity, uncomplexed lithium imides, 60
Ionization potentials, osmium complexes, 315
Iron(III) complexes
oxidized reaction products, 385–386
reduction rates, 388–389
μ-Isonicotinamide complexes, osmium, 307
μ-Isonicotinamidepoly(proline) complexes, osmium, 307
Isonitrile complexes, osmium, 245
Isopropylcarbamoyl, as bridging ligand, osmium, 305
Isoselenocyanato complexes, osmium, 293
Isothiocyanato complexes, osmium, 293

K

Ketone complexes, osmium, 284–285
  linkage isomerizations, 336–337
η^2-(C,O)-Ketones, osmium, 244–245

L

Ladder structure, 134
  $(H_2NLi)_4$, 102, 105
  organolithiums, 53–55
Lewis acidity, osmium(III), 346–347
Ligand deprotonation, osmium, 344
Ligand field interactions, osmium complexes, 316
Li—H—C linkages, 56–57, 75, 84
Li—H_{3C} interactions, 85
$(LiH)_6$ hexamers, *ab initio* calculations, 105
$[Li(HMPA)_4]^+ \cdot [(Ph_2C{=}N)_6Li_5 \cdot HMPA]^-$, structure, 73–74
Li—Li bond, 56, 58, 132
^6Li–^{15}N double labeling, 130
Linkage isomerizations, osmium, 335–339
Li—N—Li linkages, 100–101
^7Li NMR, 75–79
  complex lithium amides, 124–125
  $[(PhCH_2)_2NLi]_3$, 107–108
Lithium amides, 48–50, 92–131, 134
  complexed, *see* Complexed lithium amides
  structural types, 93
  uncomplexed, 94–108
    calculational studies, 99–107
    dimers, 96–97
    key parameters, 95
    ring laddering, 97–99
    solid-state structures, 94–99
    in solution, 107–108
Lithium arylamides, 129
Lithium halide, crystal structures, 90–92
Lithium halide–organolithium complexes, crystal structures, 90–92
Lithium imides, 49
  hexameric, structure, 81–82
  molecular orbital calculations, 79–82
  solution structures, 75–79
  tetrameric, structure, 81
  uncomplexed, 58–67
    diaryliminolithium rings, 67
    geometries, 61–63
    imino ligand orientation, 61–63
    ionicity, 60
    mean angles, 59–60
    N—Li bond, 63–65
    preparation, 59
    trimeric rings, 63, 65–66
Lithium phosphide rings, 111–112

M

Macrocyclic complexes, osmium, 230–231, 250–251
Malonate complexes, osmium, 286
$[Me_2N)_2C{=}NLi]_6$, 59
$[Me_2N)_2C{=}N]_4LiNa_3 \cdot 3HMPA$, structure, 74–75
$[Me_2N(Ph)C{=}NLi]_9$, 76–77
Metal(I) coordination number, halogenocuprate(I) and halogenoargentate(I) ions, 32–40
  cation size effects, 33–38
  charge and shape effects, 38–39
Metal–ligand bond rupture, coordinated diimine ligands, 393–396
Metallothionein, function in cisplatin resistance, 192–193
Methyllithium, 56–57
$[M_2I_6]^{4-}$ ion, dinuclear, 20
Minimum neglect of differential overlap calculations
  complex lithium amides, 117–123
  iminolithiums, 79–82
  N—Li compounds, 56
  uncomplexed lithium amides, 99–105
MO calculations, $(LiH)_4$ tetramers, 105
Mo_3MS_4 cores
  cluster preparation, 169
  cubane-type mixed-metal clusters, 163—164
Monomeric anions, discrete, halogenocuprate and halogenoargentate(I) ions, 2–6
Monomers, osmium complexes
  electrochemistry, 317–321
  spectroscopic and magnetic properties, 323–326

Mononuclear anions, halogenocuprate(I) and halogenoargentate(I) ions
four-coordinated metal centers, 19
three-coordinated metal centers, 7–11
Mono(μ-oxo) complexes, osmium, 311
Mo_3O_4 cores
cluster preparation, 165
incomplete cubane-type clusters, 145–146
Mo_4O_4 cores, cubane-type clusters, 158
$[(Mo_3O_4)_2(edta)_3]^{4-}$, structure, 145–146, 150
$[Mo_3O_3(Hnta)_3]^{2-}$, structure, 146–147, 151
$Mo_3O_{4-n}S_n$ cores, cluster preparation, 167
$[Mo_3O_4(ox)_3(H_2O)_3]^{2-}$, structure, 150
Mo_3OS_3 cores, incomplete cubane-type clusters, 146–147, 153
Mo_4OS_3, cubane-type clusters, 158
Mo_3O_3S cores, incomplete cubane-type clusters, 146–147, 151–152
$Mo_3O_2S_2$ cores, incomplete cubane-type clusters, 146–147, 152
$[Mo_3O_3S(cys)_3]^{2-}$, structure, 147, 151
$[Mo_3OS_3(dtp)_4(im)]$, structure, 147, 153
$[Mo_4OS_3(H_2O)_{12}]^{5+}$, structure, 156
$[Mo_3OS_3(ida)_3]^{2-}$, structure, 147, 152
$[Mo_3O_2S_2(NCS)_9]^{5-}$, structure, 147, 152
$[Mo_3S_4([9]aneN_3)_3^{4+}$, structure, 154
Mo_3S_4 cores, incomplete cubane-type clusters, 154–156
Mo_4S_4 cores
cluster preparation, 167
cubane-type clusters, 157–162
$[Mo_3S_4Cp_3]^+$, structure, 153
$[Mo_3S_4(dtp)_4(H_2O)]$, structure, 156
$[Mo_4S_4(edta)_2]^{3-}$, structure, 161–162
$[Mo_3S_4(H_2O)_9]^{4+}$, structure, 155
$[Mo_4S_4(H_2O)_{12}]^{5+}$, structure, 158, 161
$[Mo_4S_4(NCS)_{12}]^{6-}$, structures, 157, 159, 161
$[Mo_4S_4(NH_3)_{12}]^{4+}$, structure, 157, 161
$[Mo_3S_4(SCH_2CH_2S)_3]^{2-}$, structure, 155
MX_2^-
chain, 38
single chains, distances and angles, 27
stoichiometry, 26
MX_3^{2-}
chain, 38
monomer, 7
mononuclear, bond distances and angles, 10–11
stoichiometry, 26, 28
$M_2X_3^-$, stoichiometry, 26, 28–30
$[M_2X_4]^{2-}$ species, dinuclear, planar and folded, 9, 12
$[M_2X_4]^{2a}$ species, connectivity relationships, 13–15
$[M_2X_5]^{3-}$ species, dinuclear, planar and folded, 8–9, 12
$M_3X_4^-$, stoichiometry, 29, 31
$M_4X_5^-$, stoichiometry, 29, 31
$[M_4X_8]^{4-}$ clusters
containing three- and four-coordinated metal centers, distances, 23
tetranuclear, 21–22

N

Naddtc
exchange reactions with platinated sulfur compounds, 196–197
nephrotoxicity inhibitor, 194–195
side effects, 197
Nephrotoxicity, cisplatin, 194–199
Nitrato complexes, osmium, 287
Nitrido-bridged complexes, osmium, 225
Nitrido complexes, osmium, 225, 263, 310–311
spectroscopic and magnetic properties, 326, 330–331
Nitrile complexes, osmium, 260–262
Nitro complexes, osmium, 265
Nitrogen atom transfer, osmium, 341–342
Nitrosoamine complexes, osmium, 254
Nitrosoarene complexes, osmium, 254
Nitrosonium complexes, osmium, 254, 256–258
Nitrosyl complexes, osmium, 254, 256–258
N—Li bond, 63–65
length
complex lithium amides, 111
pattern, ring ladder, 97–99

$(NLi)_2$ rings, 97, 99
$\eta^{1-}(N)^-$ oxime complexes, osmium, 260
$\eta^{1-}(N,0)^-$ oxime complexes, osmium, 260
Nucleic acids, platinum compound binding, 187–189
Nucleobases, platinum binding, 181–183

O

Oligonucleotides, platinum binding, 185–187
Organofluorine complexes, osmium, 296–299
Organolithium compounds
  ring stacking, 82–92
    systems capable of, 82–83
  uncomplexed, structure, 53–54
  X-ray crystal structure, 48
Organonitrogen–lithium compounds, 47–135
  bonding, 132
  C—Li bond, 53
  complexed, structures, 134–135
  definitions and nomenclature, 131–132
  laddered structure, 53–55
  lithium amides, 48–50
  organic reaction mechanisms, 132–133
  stacked structure, 53–55
  structural types, 50–52
  structure and bonding, 52–58
  uncomplexed, structures, 133–134
  uses, 48–52
μ-Orthometalated triphenylphosphine complexes, osmium, 304
$[Os\{CH_2CH_2S(NSO_2C_6H_4Me-4)O\}Cl(NO)(PPh_3)_2]$, structure, 241–242
Os_2^{4+} dimers, spectroscopic and magnetic properties, 326
Os_2^{5+} dimers, spectroscopic and magnetic properties, 326–328
Os_2^{6+} dimers, spectroscopic and magnetic properties, 328–329
Osmaboranes, 232–235
Osmacarboranes, 232–235
Osmium, 219–352
  atom-transfer reactions, 340–343
  carbon ligands, 304–305
  dimers with Os—Os Bonds, 302–304
  halide-bridged complexes, 313
  halogen atom transfer, 342–343
  history, 219–220
  hydride-bridged complexes, 313–314
  hydrogen atom transfer, 340
  intermolecular electron transfer, 332–333
  intramolecular electron transfer, 334–335
  ligand deprotonation, 344
  linkage isomerizations, 335–339
  nitrogen atom transfer, 341–342
  nitrogen donor ligands, 305–311
  oxidation states, 221–223
  oxidative dehydrogenation, 343–345
  oxygen atom transfer, 340–341
  oxygen donor ligands, 311–312
  phosphide-bridged complexes, 311
  polymers, 314–315
  reviews, 220
  S- to O-sulfinato linkage isomerizations, 339
  substitution reactions, 332–333
  sulfur donor ligands, 312
  synthetic methods, 223–232
    ammine complexes, 226–230
    complexes with bpy, trpy, phen, and related ligands, 231–232
    N-macrocyclic complexes, 230–231
    nitrido and nitrido-bridged complexes, 225
    osmium(III) complexes, 232
    oxo ligand complexes, 224
Osmium(0), 221
Osmium(I), 221
Osmium(II), 221–222
  addition reactions, 350
  alkylations, 348
  arene activation, 348–350
  dehydration reactions, 351
  elimination reactions, 351–352
  ligand-centered reactions, 347–352
  protonation, 347–348
Osmium(III), 222
  Lewis acidity, 346–347
  ligand-centered reactions, 344–347
Osmium(IV), 222
  ligand-centered reactions, 343–344
Osmium(V), 222
Osmium(VI), 222–223
Osmium(VII), 223

Osmium(VIII), 223
addition reactions, 331–332
Osmium ammine complexes, synthesis, 226–230
Osmium complexes, 315
bpy and related complexes, spectroscopic and magnetic properties, 329–330
electrochemistry, 315–323
dinuclear and polynuclear complexes, 321–323
ionization potentials, 315
ligand field effects, 316
monomers, 317–321
oxidized reaction products, 387
π bonding and backbonding effects, 316, 320
polymers, 323
redox potentials, 317–321
reduction rates, 388–389
solvent effects, 317
spin-orbit coupling, 316
stabilization of electronic isomers of mixed–valence ions, 323
spectroscopic and magnetic properties
dimers, 326–329
monomers, 323–326
synthesis
alkene complexes, 242–243
alkyl and aryl complexes, 239–242
alkyl- and arylsulfonato complexes, 286–287
alkyne ligands, 243
allyl complexes, 244
amide complexes, 284
amine complexes, 247, 249–250
2-aminobenzenethiol complexes, 295
ammine complexes, 247–248
aqua and hydroxo complexes, 276–278
arsenito and hydrogenarsenito complexes, 288
arsine complexes, 274–275
azido complexes, 263
benzeneseleninato complexes, 287
benzothiadiazole complexes, 290–291
benzotriazole, 1,3-diaryltriene, 2-(phenylazo)pyridine, and (phenylazo)acetaldoxime complexes, 270–271
borohydride complexes, 299–300
carbene complexes, 242
carbonato complexes, 285
carbon dioxide complexes, 239
carbon disulfide complexes, 239
carbonyl complexes, 236–238
carboxylato complexes, 285–286
catechol and quinone complexes, 283–284
C-bound aldehyde and ketone complexes, 246
chelating amide ligands, 266–270
chloro, bromo, and iodo complexes, 297–299
chlorothionitrene complex, 262
complexes with thiocarbonyl selenocarbonyl, and tellurocarbonyl ligands, 238–239
cyanato complexes, 263–264
cyano complexes, 235–236
cysteine complexes, 295
diarsine ligands, 275–276
diazene complexes, 260
dichlorobis(triphenylphosphine)-argentate(I) complexes, 299
diethylphosphate and phosphine oxide complexes, 284
dihydrogen complexes, 300–301
diketonate complexes, 285
O-dimethyl sulfoxide complexes, 284
S-dimethyl sulfoxide complexes, 291
dinitrogen complexes, 259
dioxygen complexes, 285
disulfur oxide and sulfur dioxide complexes, 292
dithiocarbamate and xanthate complexes, 293–294
dithiophosphinate complexes, 294
η^6-arene ligands, 244
η^2-bound arene and heterocyclic ligands, 243–244
η^2-CH_2PR_2 complexes, 274
η^2-(*C*,*O*)-ketones and-aldehydes, 244–245
$\eta^{1-}(N)^-$ and $\eta^{1-}(N,O)^-$ oxime complexes, 260
η^1-(O)-aldehyde and -ketone complexes, 284–285
ether complexes, 285

fluoro complexes, 296–299
N-heterocyclic complexes, 254–255
hydride complexes, 299–300
hydrogen selenido complexes, 290
hydroxylamine and hydrazine complexes, 254, 258
imido complexes, 262
imine complexes, 259–260
iminooxosulfane ligands, 292
isonitride compounds, 245
isothiocyanato and isoselenocyanato complexes, 293
macrocyclic complexes, 250–151
nitrato complexes, 287
nitrido complexes, 263
nitrile complexes, 260–262
nitro and thionitro complexes, 265
nitrosarene and nitrosoamine complexes, 254
nitrosyl and nitrosonium complexes, 254, 256–258
organofluorine complexes, 296–299
osmaboranes and osmacarboranes, 232–234
oxalate and malonate complexes, 286
oxo complexes, 278–282
phosphine complexes, 271–274
phosphite complexes, 272, 274
porphyrin and phthalocyanine complexes, 251–253
pyridine oxide complexes, 284
2-pyridinethiolato, 2-pyrimidinethiolato, and thiopyrine complexes, 295–296
pyrimidinethione and thiobarbituric acid complexes, 296
quinolol complexes, 270
ROH, RO^-, and R_3SiO^- complexes, 279–281, 283
Schiff base complexes, 265–266
selenocyanato complexes, 265
silyl complexes, 246–247
stibine complexes, 276
sulfato, chromato, and perrhenato complexes, 287
sulfide and polysulfide complexes, 290
O-sulfinato complex, 287
sulfito complexes, 293
sulfur, selenium, and tellurium tetrahalide complexes, 299
thiocyanato complexes, 264
thioether and selenoether complexes, 288
thioether macrocycle complexes, 288
thiolato complexes, 288–289
thionitrosoamine complexes, 291–292
thionitrosyl complexes, 259
thiophene complexes, 290
S-thiosulfato complexes, 292–293
thiourea and 1-amidino-2-thiourea complexes, 290
2-(tolylthio)picolinamide, 295
Osmium(III) amine and ammine complexes, 323–325
$[OsO_4]$-derivatized polymers, osmium, 315
Oxalate complexes, osmium, 286
Oxidation, *see also* Coordinated diimine ligands
ligand, metal–ligand bond rupture, 393–396
Oxidative dehydrogenation, osmium, 343–345
Oxo complexes, osmium, 278–282
spectroscopic and magnetic properties, 326
Oxo ligands, osmium complexes, 224
Oxycarbanion–alkali metal compounds, structures, 88–90
Oxygen atom transfer, osmium, 340–341

P

Pentaammineosmium complexes, 248
Pentaammineosmium(II) complexes, linkage isomerization, 338–339
Pentaammineosmium(III/II) complexes
aquation rates, 332
synthesis, 226–227
Pentanuclear anions, halogenocuprate(I) and halogenoargentate(I) ions, 22, 24
Perrhenato complexes, osmium, 287
$[(PhCH_2)_2NLi]_3$
molecular structure, 94, 96
in solution, 107–108
$[(PhCH_2)_2NLi \cdot OEt_2]_2$
molecular structure, 112–113
in solution, 126

[(PhCH$_2$)$_2$NNa]$_n$, 97
(Ph$_2$C=NLi·pyridine)$_4$, 79
  structure, 71–72
1,10-Phenanthroline, 382, 390
(Phenylazo)acetaldoxime complexes, osmium, 270
2-(Phenylazo)pyridine complexes, osmium, 271
[Ph(H)C=C=NLi·TMEDA]$_2$, structure, 70
[Ph(Me$_2$N)C=NLi]$_6$, 59–65
  orientation of imino ligand, 61–63
Ph(naphthyl)NLi·PMDETA, molecular structure, 116
Ph(naphthyl)NLi·TMEDA, molecular structure, 116–117
[PhN(Me)Li·TMEDA]$_2$, molecular structure, 113–115
Phosphide-bridged complexes, osmium, 311
Phosphine complexes, osmium, 271–274
Phosphine/halide complexes, osmium, 273
Phosphine/hydride complexes, osmium, 274
Phosphine oxide complexes, osmium, 284, 314
Phosphite complexes, osmium, 272, 274
Phthalocyanine complexes, osmium, 251–253
π backbonding, osmium complexes, 316, 320
π bonding, osmium complexes, 316, 320
Platinum
  DNA binding, 183–185
  nucleobase binding, 181–183
  oligonucleotide binding, 185–187
Platinum amine compounds, 175–208
  inactivation, 190–192
  reaction products, sulfur-containing biomolecules, 202–206
Platinum antitumor compounds
  future studies, 206–208
  resistance to, 192–193
  Pt(II) compounds, reactivation, 201
  Pt(IV) compounds, 177, 201
  second-generation, 178
Platinum compounds, nucleic acid binding, 187–189
Platinum–DNA interactions
  antitumor activity, 181
  reaction products, 184–185
Platinum–protein interactions, models, 196
Platinum–sulfur interactions, 189–206, 208
  inactivation of platinum amine compounds, 190–192
  nephrotoxicity and rescue agents, 194–199
  Pt(II) compound reactivation, 201
  Pt(IV) compound reduction, 201
  rate-determining step, 199–201
Polymeric anions, four-coordinated metal centers, halogenocuprate(I) and halogenoargentate(I) ions, 26–32
  miscellaneous, 31–32
  with stoichiometry
    MX_2^-, 26–28
    MX_3^{2-}, 26, 28
    $M_2X_3^-$, 26, 28–30
    $M_3X_4^-$, 29, 31
    $M_4X_5^-$, 29, 31
Polymers, osmium, 314–315
  spectroscopic and magnetic properties, 330
  electrochemistry, 323
Polynuclear anions, three-coordinated metal centers, halogenocuprate(I) and halogenoargentate(I) ions, 17, 19
Polynuclear complexes, osmium, electrochemistry, 321–323
Polysulfide complexes, osmium, 290
μ-Porphyrinato complexes, osmium, 303
Porphyrin complexes, osmium, 251–253
Porphyrin dimers, osmium, 309
Porphyrin polymers, osmium, 314
  spectroscopic and magnetic properties, 330
Protonation, osmium(II), 347–348
[Pt(diam)(R′R″SO)Cl](NO$_3$), 187–188
[Pt(dien)Cl]Cl, reactions with glutathione, 199–200
[*trans*-Pt(Met-*N*,*S*)$_2$], structure, 204–205
[*cis*-Pt(NH$_3$)$_2$(Cys-*N*,*S*)], structure, 204
[*cis*-Pt(NH$_3$)$_2$(Cys-*S*)$_2$], structure, 204
[Pt$_2$(NH$_3$)$_4$(GS)$_2$], structure, 203
[*cis*-Pt(NH$_3$)$_2$(GS)Cl], structure, 203
[*cis*-Pt(NH$_3$)$_2$(4-mepy)Cl]Cl, 188–189

[*trans*-Pt(NH_3)(Met-*S*)(Met-*N*,*S*)],
structure, 204–205
[*cis*-PT(NH_3)$_2$(N-het)Cl]Cl, 187–188
^{195}Pt NMR spectroscopy, 180
cis-Pt, *see*Cisplatin
μ-Pyrazine complexes, osmium, 307
Pyridine oxide complexes, osmium, 284
2-Pyridinethiolato complexes, osmium,
295–296
μ-Pyrimidine complexes, osmium, 307
2-Pyrimidinethiolato complexes, osmium,
295–296
Pyrimidinethione complexes, osmium,
296
Pyrrolidine, lithiation, 109–110

Q

Quadrupole splitting constants, 129
Quinolol complexes, osmium, 270
Quinone complexes, osmium, 283–284

R

Redox potentials
comparison of dinuclear ruthenium and
osmium complexes, 321–323
osmium complexes, 317–321
Ring laddering, uncomplexed lithium
amides, 97–99
Ring stacking, *see* Organolithium
compounds, ring stacking
μ-RO^- complexes, osmium, 312
(RR′C=NLi)$_2$ dimers, 68–70
(RR′C=NLi)$_3$ trimers, 68–70
R_3SiO^- complexes, osmium, 279–281, 283
Ruthenium, dinuclear complexes, redox
potentials, 321–323
Ruthenium(III) complexes, oxidized
reaction products, 386–387

S

Schiff base complexes, osmium, 265–266
Selenium tetrahalide complexes, osmium,
299
Selenocarbonyl complexes, osmium,
238–239
Selenocyanato complexes, osmium, 265
Selenoether complexes, osmium, 288
Silver(I) coordination number, as
function of concentration of
halogenide ligand, 36–38
Silyl complexes, osmium, 246–247
Sodium thiosulfate, 194, 198
exchange reactions with platinated
sulfur compounds, 196–197
Solvent effects, redox potentials, osmium
complexes, 317
Spectroscopic and magnetic properties,
323
Spin-orbit coupling, osmium complexes,
316
Stacked structure, organolithiums, 53–55
Stacking, 134, *see also* Organolithium
compounds, ring stacking
Stibine complexes, osmium, 276
Substitution reactions, osmium, 332–333
Sulfato complexes, osmium, 287
Sulfide complexes, osmium, 290
O-Sulfinato complex, osmium, 287
Sulfito complexes, osmium, 293
Sulfur-containing biomolecules, reaction
products, platinum amine
compounds, 202–206
Sulfur dioxide complexes, osmium, 292
Sulfur tetrahalide complexes, osmium,
299
μ-Supersulfide complexes, osmium, 312

T

Tellurium tetrahalide complexes,
osmium, 299
Tellurocarbonyl complexes, osmium,
238–239
Tetraammineosmium complexes,
synthesis, 228–230
Tetrahedral oxo/aryl complexes, osmium,
240–241
Tetranuclear anions, halogenocuprate(I)
and halogenoargentate(I) ions
four-coordinated metal centers, 21–23
three-coordinated metal centers, 16–18

Thiobarbituric acid complexes, osmium, 296
Thiocarbonyl complexes, osmium, 238–239
Thiocyanato complexes, osmium, 264
Thioether complexes, osmium, 288
Thioether macrocycle complexes, osmium, 288
Thiolato complexes, osmium, 288–289, 312
Thionitro complexes, osmium, 265
Thionitrosoamine complexes, osmium, 291–292
Thionitrosyl complexes, osmium, 259
Thiophene complexes, osmium, 290
Thiopyrine complexes, osmium, 295–296
S-Thiosulfato complexes, osmium, 292–293
Thiourea, exchange reactions with platinated sulfur compounds, 196–197
Thiourea complexes, osmium, 290
Three-coordinated metal centers, 7
2-(Tolylthio)picolinamide complexes, osmium, 295
Triammineosmium complexes, synthesis, 229–231
Tricyclo-*N* heterocycles, osmium, 310
Triethylphosphate oxide complexes, osmium, 284
Triidoargentate(I) anion, 8
Triiodocuprate(I) ion, 7
Trimeric rings, uncomplexed lithium imides, 63, 65–66
Trinuclear anions, halogenocuprate(I) and halogenoargentate(I) ions
  four-coordinated metal centers, 21
  three-coordinated metal centers, 16
Tris(2,2′-bipyridine)iron(III) complex, dioxygen production, 382–383
Tris(2,2′-bipyridine)osmium(II) complexes, spectroscopic and magnetic properties, 325–326
Tris(diimine)iron(III) complexes, reduction, 384–385, 388–389
Tris(diimine)metal(III) complexes, reduction potentials, 388
Tris(diimine)osmium(III), reduction, 384–385
Tris(diimine)ruthenium(III), reduction, 384–385

V

4-Vinyl-2,2′-bipyridine polymers, osmium, 314
4-Vinylpyridine polymers, osmium, 314

W

$[W_3O_4([9]aneN_3)_3]^{4+}$, structure, 162
W_3O_4 cores, incomplete cubane-type clusters, 163
$W_3O_{4-n}S_n$ cores, cluster preparation, 168–169
W_3OS_3 cores, incomplete cubane-type clusters, 162
$W_3O_2S_2$ cores, incomplete cubane-type clusters, 162
W_3O_3S cores, incomplete cubane-type clusters, 162
$[W_3O_3S(NCS)_9]^{5-}$, structure, 164
WR-2721, 194, 198
W_3S_4 cores, incomplete cubane-type clusters, 163–164
$[W_3S_4(H_2O)_9]^{4+}$, structure, 165, 166
$[W_3S_4(NCS)_9]^{5-}$, structure, 164, 165

X

Xanthate complexes, osmium, 293–294
X—M—X linkages, halogenocuprate(I) and halogenoargentate(I) ions, 6